Lecture Notes in Earth Sciences

Edited by Somdev Bhattacharji, Gerald M. Friedman, Horst J. Neugebauer and Adolf Seilacher

33

B. Allard H. Borén A. Grimvall (Eds.)

Humic Substances in the Aquatic and Terrestrial Environment

Proceedings of an International Symposium
Linköping, Sweden, August 21–23, 1989

Springer-Verlag
Berlin Heidelberg New York London
Paris Tokyo Hong Kong Barcelona

Editors
Prof. Bert Allard
Dr. Hans Borén
Prof. Anders Grimvall
Department of Water and Environmental Studies
Linköping University
S-581 83 Linköping, Sweden

ISBN 3-540-53702-3 Springer-Verlag Berlin Heidelberg New York
ISBN 0-387-53702-3 Springer-Verlag New York Berlin Heidelberg

Printed in Germany

Printing and binding: Druckhaus Beltz, Hemsbach/Bergstr.
2132/3140-543210 – Printed on acid-free paper

Preface

In 1986 a Nordic conference on humic substances was arranged in Oslo, Norway. The conference was very successful, and it was suggested that there should be continued regular meetings for Scandinavian scientists working within this area. Linköping University accepted the responsibility of hosting the second meeting in this series.

At an early stage it was decided that the meeting should also be open to non-Nordic participants. Thus, the International Symposium on Humic Substances in the Aquatic and Terrestrial Environment, August 21-23, 1989, in Linköping, Sweden, attracted about 120 participants from 19 nations. A total of 71 contributions were presented, orally or as posters. Papers from both the oral and poster sessions have been collected in this proceedings volume, in which the chapters have the same titles as the oral sessions of the symposium. Each chapter (with the exception of Chap. 2) starts with a paper based on the plenary lecture of the session.

The program and final selection of papers were made by the Scientific Committee (Prof Bert Allard, Linköping University; Dr Göran Bengtsson, University of Lund, Sweden; Dr Hans Borén, Linköping University; Dr James Ephraim, Linköping University; Dr Egil Gjessing, National Institute of Public Health, Norway; Prof Anders Grimvall, Linköping University; Prof Bjarne Holmbom, University of Åbo Academy, Finland; Dr Ronald Malcolm, US Geological Survey, USA; and Prof Robert Petersen, University of Lund, Sweden). The assistance of Ms Irina Arsenie, Ms Gunilla Asplund, Ms Maria Nordén and Ms Catharina Pettersson, and also of the members of the Scientific Committee, in editing and preparing the conference proceedings is gratefully acknowledged, as well as the excellent help with word processing and lay-out by Ms Susanne Eriksson, Ms Lisbeth Thornbury and Ms Pia Sandholm.

Financial contribution was obtained from the Swedish National Environmental Protection Board.

Bert Allard, Hans Borén and Anders Grimvall

Table of Contents

Introduction

Session 1: Isolation, Fractionation, and Characterization

Session 2: Biological and Chemical Transformation and Degradation

Session 3: Complex Formation and Interactions with Solids

Session 4: Biological Activity

Session 5: Halogenation of Humic Substances

Appendix

Introduction

The Different Roles of Humic Substances in the Environment

Bert Allard, Hans Borén and Anders Grimvall
Department of Water and Environmental Studies, Linköping University,
S-581 83 Linköping, Sweden

Humic substances comprise a class of biogenic, coloured, organic substances that are ubiquitous in soil, sediment and water. Originally, the occurrence and nature of humic substances were regarded as issues of primarily academic interest. This situation is now rapidly changing, and studies of humics have gained recognition as important contributions to environmental science. In particular it has been shown that humic substances, in several different ways can interact with biologically active substances, thereby modifying their environmental impact (see Table 1).

Whereas the history of soil humus studies goes back to the 19th century, the awareness of aquatic humus is more recent. The brownish colour that, in many surface waters, shows the presence of substantial amounts of humic substances, was long considered to be a harmless phenomenon that did not call for detailed investigations. Humic waters had few known toxic effects, and the refractory character of humic substances indicated the they played a peripheral role in most biochemical processes. In fact, it was not until the mid 70's that aquatic humus was brought into focus in environmental science. The event trigging this was the discovery of the interaction between humic substances and chlorine used for disinfection of drinking water. Toxic substances, such as chloroform, were detected in all chlorinated waters, and humic substances were identified as the main precursors.

The role of humics in the mobilization and subsequent transport of trace elements in the environment was recognized for the first time in the early 80's. This role was considered to be of particular importance in connection with geologic storage of high-level radioactive waste. In water with "normal" concentration levels of humic compounds, the speciation of e.g. the trivalent actinides, would be entirely dominated by the complexation with these agents.

Table 1 Some different roles of humic substances in the environment.

Complexing agent or adsorbent for man-made pollutants	Complexing agent or adsorbent for naturally occurring substances	Precursor of toxic substances in technical processes	Precursor of toxic substances in natural processes

The topics of this conference (Session 1 - Isolation, fractionation and characterization; Session 2 - Biological and chemical transformation and degradation; Session 3 - Complex formation and interactions with solids; Session 4 - Biological activity; and session 5 - Halogenation of humic substances) were selected to represent areas of current environmental interest.

As described by Malcolm (Chapter 1), developement of isolation procedures, as well as fractionation methods and characterization schemes, are still topics of considerable interest. Isolation procedures are being refined and information about elemental composition, functional groups and molecular weight of natural organics is accumulating. The tendency in the obtained results is clear: even if isolates of humic substances are purified by the best techniques available, the resulting mixtures of organics are very complex. In addition, it is obvious that humics of different origin can differ significantly with respect to elemental composition, molecular weight, etc. This implies that one cannot expect to find a simple structural characterization of humic substances. Another implication is that further studies of the humics are likely to reveal new aspects of the chemistry of this heterogeneous group of substances.

Humic substances are affected by a great number of biological and chemical transformation mechanisms, including microbial processes and photochemical degradations reactions. The papers in Chapter 2 demonstrate the greatly diversified character of this subject.

In recent years, the effect of naturally occurring matter on the mobility and distribution of man-made pollutants has attracted particular attention. Several studies have shown that the transport of toxic, hydrophobic organic substances in soil/water systems can be facilitated by the presence of humic substances. An even larger number of studies have dealt with complexation of metals and sorption of metals to geological materials in the presence or absence of humic substances. Despite the accumulating amount of empirical data concerning the complexing properties of humics, the detailed description of these phenomena is still a matter of controversy. As discussed by Ephraim (Chapter 3), the heterogeneity and the polyelectrolyte character of the humic substances was not properly considered until a few years ago, and many problems remain unsolved.

The review by Petersen (Chapter 4) demonstrates that humic substances may have both beneficial and detrimental effects on the environment. Furthermore, biological effects of humic substances are often a result of interactions with other substances. The complex relationship between the toxicity of aluminium and the presence of humic substances can illustrate this fact.

It has already been mentioned that humic substances can act as precursors of toxic substances during drinking water disinfection. Further details of this type of interaction are given in the review by Holmbom (Chapter 5). Halogenation by-products can also be formed naturally, and humic substances may play an important role in the formation of substances normally considered to be xenobiotic.

Considering that numerous pollution problems and biological processes that can be affected or modified due to the presence of humic substances, it is of great interest to note that there are clear indications of upward trends in the concentration of humic substances in e.g. Swedish rivers. This provides another example of the crucial role of humic substances in the understanding of the present state and current trends in water and soil quality.

The articles published in this book demonstrate how humic substances now attract the attention of scientists from a large number of different disciplines. Several of the studied problems are of such a character that they can only be solved through multidisciplinary work. At the same time, numerous basic studies of humic substances and other natural organics are still being performed and are prerequisites for a better understanding of the role of these compounds in the environment.

Session 1:
Isolation, Fractionation, and Characterization

Factors to be Considered in the Isolation and Characterization of Aquatic Humic Substances

Ronald L. Malcolm
U.S. Geological Survey, Denver Federal Center
M.S. 408, Denver, Colorado 80225

Abstract

A detailed procedure using XAD-8 resin is presented for the isolation of dissolved fulvic acids and humic acids from water. The procedure entails pressure filtration to remove suspended sediment, sorption of humic substances onto XAD-8 resin at pH 2, desorption of humic substances in base, fulvic/humic separation at pH 1, desalting on XAD-8 resin, hydrogen saturation on cation exchange resin, and freeze-drying. Careful attention must be given to thorough resin cleaning and many procedural details in order to obtain relatively ash-free humic isolates. The equipment required for the procedure is expensive and the method is time consuming, but no other isolation method is known to produce quantitative and unaltered humic isolates from water. The procedure can be used to isolate small quantities (less than 100 mg) of humic substances from water, or it can be scaled to produce large quantities (100 g or more) of humic substances from water. Humic substances may be characterized by several methods. The more useful traditional characterization methods include elemental analysis, ash content, functional group analysis by titration and infrared spectroscopy, and molecular weight analysis. The new characterization methods of ^{1}H-NMR, ^{13}C-NMR, pyrolysis/mass spectroscopy, amino acid analysis, saccharide analysis, and carbon isotopic analysis (^{14}C and δ ^{13}C content) are usually more definitive than traditional characterizations.

Introduction

The isolation of humic substances from water has progressed rapidly during the last two decades. Previous to 1970, few studies were conducted on aquatic humic substances because isolation and concentration procedures were poorly developed. Humic substances could be sorbed from water onto charcoal and anion exchange resins [1], but recovery of humic substances from these sorbents was low. To obtain isolates containing high concentrations of aquatic humic substances, some investigators used freeze-drying of selected, colored, whole river waters, such as the Sopchoppy and Suwannee River waters [2], which contained high concentrations of humic substances and low concentrations of inorganic suspended sediments.

The use of XAD resins to concentrate humic substances from water has been a major breakthrough in organic hydrology. These resins were first used by a research group at Iowa State University in the early 1970's [3-5] for isolation of trace organic contaminants from groundwater. Riley and Taylor [6] and Mantoura and Riley [7] were the first to use XAD-2 resin to isolate humic substances from water. During the middle 1970's the organic geochemistry researchers of the U.S. Geological Survey in Denver, Colorado, evaluated several XAD resins, studied in depth their sorptive behavior, and developed several procedures for isolating organic constituents, including humic substances, from water [8-13]. Other resins such as Duolite A-7 [14] and DEAE cellulose [15] have recently been advocated for isolation of humic substances from water.

XAD-2 and XAD-8 resins have been the two most popular resins used to isolate humic substances from water. After extensive comparative studies, Aiken, *et al.* [12] concluded that XAD-8 was a better resin than XAD-2 for the isolation of humic substances from water because the smaller pores of the XAD-2 resin excluded or were fouled by the relatively large humic solutes and that complete recovery of the sorbed humic substances could not be achieved. For these and other reasons, XAD-8 has been the resin of choice by most freshwater organic hydrologists. XAD-2 resin has continued to be the resin of choice by most marine hydrologists and some freshwater organic chemists and hydrologists.

Although the use of both XAD-8 and XAD-2 resins has proliferated during the last decade, it is the author's opinion that inadequate attention has been given by investigators to limitations in the usage of the resins. The purposes of this paper are 1) to reiterate the general criteria and specific chromatographic principles which must be considered by the user of XAD and other resins, 2) to present a detailed procedure for isolating humic substances from water, 3) to provide detailed information essential to those researchers planning a humic substances isolation effort, 4) to make the reader aware of the cost and scale of humic substances isolation procedures, and 5) to emphasize the necessity for following detailed and time-consuming procedures which are required when using XAD resins if one is to achieve exacting research findings and characterizations of aquatic humic substances.

Materials and Methods

Resins

XAD-2 and XAD-4 resins are styrene-divinylbenzene copolymers. The resins are macroporous of low polarity with a 20 to 50 mesh size. XAD-2 resin has less crosslinkage, larger pore size, and less surface area than XAD-4 resin. Both resins can be obtained from Rohm and Haas Company, 500 Richmond Street, Philadelphia, PA 19137.[1]

XAD-8 resin is an acrylic ester polymer of intermediate polarity. It is macroporous with a 20 to 50 mesh size. It has a larger hydrated pore size and less surface area than XAD-2 resin. The resin can also be obtained from Rohm and Haas.

Duolite A-7 is a weak base, macroporous, anionic-exchange resin which is a phenol-formaldehyde condensation product. The resin can be obtained from Diamond Shamrock Chemical Company, Nopco Chemical Division, 1901 Spring Street, Redwood City, CA 94063.[1]

AG SOW-X8, a non-macroporous cation exchange resin, was obtained from Bio-Rad Laboratories.[1] The resin is 20 to 50 mesh size and is in the hydrogen form.

Stainless-Steel Canisters

If the filtration and water processing is not done in a streamside mobile trailer, 40-l stainless-steel cans are convenient to transport water to the laboratory site. Several dozen of these cans are desirable for this use.

Glassware

Soxhlet Extraction Apparatus of sufficient size to hold 2 l or larger extraction thimbles. Such very large extractors will have a solvent reservoir capacity of 12 to 15 l and stand 1.5 to 2 m in height. Three of four units are desirable.

Glenco glass columns, 3500 series with Teflon fittings and tubing. Column sizes for XAD resin adsorption chromatography are 12x125 cm and 8x90 cm. Column sizes for cation exchange chromatography are 3x60 cm for the cation-exchange resin precolumn and 6x100 cm for the final cation-exchange columns.

Glass jugs of 19 l and 45 l capacity made of Pyrex or Kimex hard glass. Twenty-five units of each size are desirable.

Filtration Apparatus

Millipore stainless-steel plate filter apparatus that holds 142 mm or 293 mm membrane filters. Twenty-four or more units of 142 mm size, or six units or more of 293 mm size, or equivalent. One of the 293 mm units has an equivalent filtration surface area of four of the 142 mm units.

Membrane filter of 142 mm or 293 mm in diameter, 0.45 µm porosity, and free of organic detergent wetting agents. Gelman vinyl-metricel membrane filters #64835 (142 mm in size) and #64838 (293 mm in size) or equivalent are required.

Stainless-steel filtration reservoir tanks of 20 l to 60 l capacity that are capable of being pressurized to 200 psi are required to provide a water supply for every two to four stainless-steel plate firers. A very large stainless-steel tank of capacity (in excess of 100 l) may be desirable. The reservoir tank is connected to the individual

[1]Use of any trade names is for descriptive purposes only and does not constitute endorsement by the U.S. Geological Survey

plate filters by flexible stainless-steel tubing with quick-disconnect fittings on each end.

Nitrogen or helium gas for pressure filtration should be of high quality. Gas pressure for filtration may be accomplished by several small gas bottles or a few large gas bottles with a manifold to each pressure reservoir.

Instrumentation

- A peristaltic pump (Cole-Parmer Masterflex) is needed to deliver filtered water and elution reagents to the resin columns.

- A Homelite AP 220 positive displacement pump with stainless-steel heads can be used to deliver the water from streams to 40 l stainless-steel cans for water transport.

- A portable pH meter and a conductivity meter with flow-through cells are desirable.

- A DOC Carbon Analyzer such as a Beckman 915 or equivalent.

- A large capacity freeze-drier that can freeze-dry 8 to 12 l of liquid each day.

Two types of centrifuges are desirable. One should be a large-capacity, low-speed unit such as an International Model K or equivalent. A second unit should be of moderate capacity and high speed such as can be achieved with a swing-out bucket head on a Sorval RC-2B centrifuge.

Reagents

Reagent grade pellets of NaOH or higher purity. Reagent grade HCl or higher purity. Reagent water with a low organic carbon concentration and specific conductance less than 0.5 S. If reagent water contains a DOC concentration of 0.5 mg C/l or higher, the water should be purified by passage through a column of XAD-8 resin before use.

Occurrence and Abundance of Dissolved Aquatic Humic Substances

Dissolved humic substances are only part of the dissolved organic carbon (DOC) in water and are never equivalent to the total DOC. Because natural waters contain adiversity of dissolved and particulate inorganic and organic constituents, and many of the dissolved organic constituents are not humic substances, water must be filtered and the dissolved humic substances selectively removed. Dissolved humic substances include only dissolved fulvic and humic acids; there is no dissolved humin fraction. Dissolved fulvic and humic acids in uncolored freshwater streams commonly comprise approximately 40 percent and 4.5 percent, respectively, of the DOC content of 3 to 6 mg C/l [16,17]. The ratio of fulvic acids to humic acids is commonly 9:1. In organically colored waters common to Canada, Scandinavia, and Northern Russia, humic substances as a percentage of the DOC increase with increasing DOC concentrations, and often account for 60 to 80 percent of the DOC. In these highly colored waters, the humic acids become a much higher percentage of the DOC and the fulvic acids to humic acids ratio commonly decreases to 4:1 or lower.

Analyses of approximately 100 surface water samples of the United States yield an average of general constituents according to a sixfold DOC fractionation procedure [17] as shown in Fig. 1. Approximately 99+ percent of the DOC are natural constituents and less than 1 percent are contaminants.

The average DOC concentration in potable groundwaters is 1 mg C/l [18]. The average DOC distribution of approximately 25 groundwater samples collected from aquifers in the United States for the same six organic fractions is shown in Fig. 2. Essentially 100 percent of the groundwater DOC are natural organic constituents, although in infrequent instances of groundwater contamination, the percentage reported as contaminants is commonly high and extremely variable. The DOC concentration in groundwaters is much lower than in surface waters, and the organic constituents as part of the total DOC are also very different. For example, humic substances are commonly less than 20 percent of the groundwater DOC; whereas, low-molecular-weight organic acids may comprise 50 percent of the DOC.

The DOC of deep uncontaminated sewater has generally been reported to be near 1 mg C/l [19,20]. The concentration of humic substances in seawater, which is primarily as fulvic acid, is 10 to 20 percent of the DOC. The DOC of deep seawaters is lower (1 mg C/l or lower) than seawater in the euphotic zone or near coastal regions (1.5 mg C/l or higher). The precise DOC value for seawaters is now in question because it has been found to be method dependent [21] with an underestimation of the low-molecular-weight hydrophilic acid fraction.

Choice of Resin for the Concentration of Humic Substances From Water

Several isolation methods, including freeze-concentration of whole water samples [22], freeze-drying of whole water samples [2], strong anion-exchange resins [1], charcoal [23], and flocculation with heavy metal oxides and hydroxides [24-27], have been used in attempts to isolate humic substances from water. These methods were of limited success because only partial isolation and/or recovery of humic substances was achieved, the humic isolates were mixed with many non-humic organic solutes and inorganic constituents (clay minerals and inorganic salts), and because various contaminants were introduced into humic isolates. After the initial usage of XAD resins during the 1970's, it was recognized that XAD resins held great promise in overcoming the limitations of previously used methods for isolation of humic

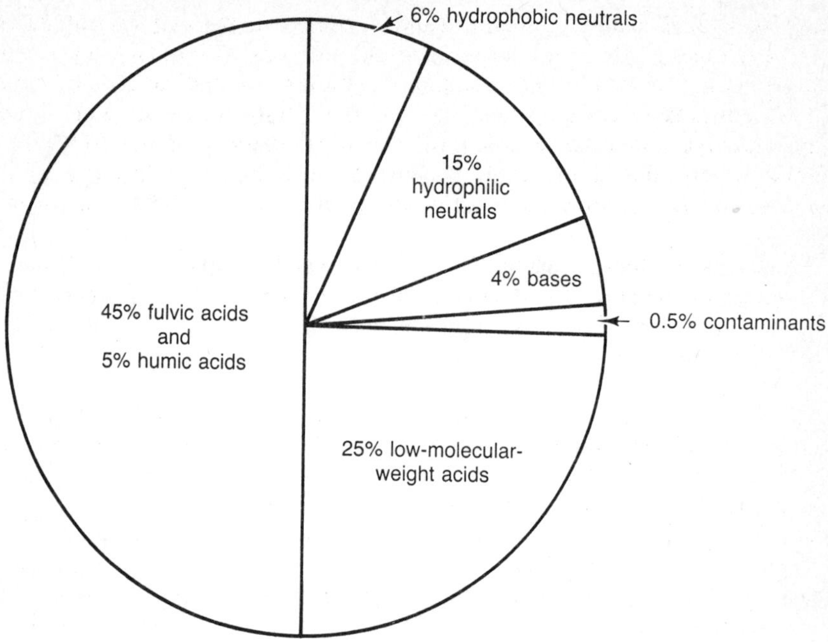

Figure 1 Distribution of surface water DOC in rivers of the United States.

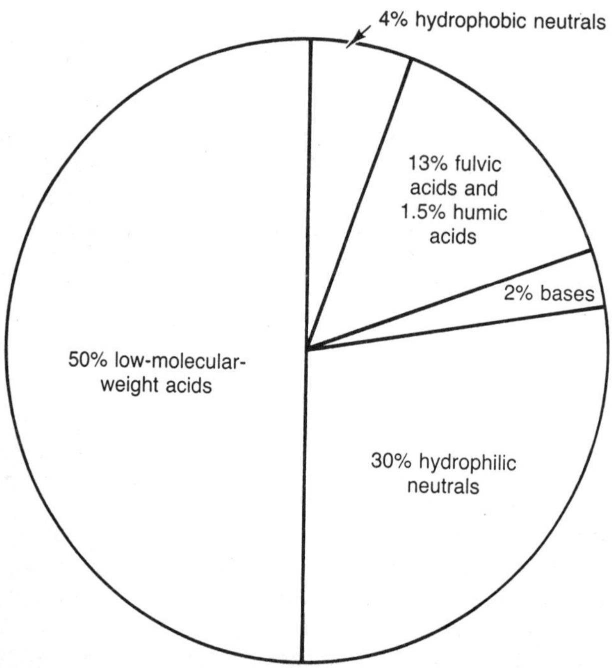

Figure 2 Distribution of groundwater DOC in aquifers of the United States.

substances from water. However, before XAD resins could be advocated, many basic and applied experiments on their sorptive properties had to be conducted.

All the XAD resins are macroporous and uncharged, but they vary in composition, surface area, pore size, cross-linkage, density, and polarity as shown in Table 1. XAD-l, XAD-2, and XAD-4 are aromatic polymer matrices of styrene-divinylbenzene, whereas XAD-7 and XAD-8 are aliphatic polymer matrices of methylacrylic acid (acrylic ester). Because of the high surface area of XAD-4 resin, it was postulated to be the best resin for isolating humic substances from water. Its high surface area was expected to result in a high sorptive capacity.

Table 1 Composition, surface area, pore size, cross-linkage, density and polarity of XAD resin [28].

XAD resin	Surface Area m^2/g	Pore Size Å	Cross-linkage	Density	Polarity
			Styrene-divinylbenzene polymer		
1	100	200	low	1.06	none
2	330	90	moderate	1.08	none
4	750	50	high	1.09	none
			Acrylic ester polymer		
7	450	80	moderate	1.25	slight
8	250	250	low	1.26	slight

Several systematic experiments [8-14] were conducted to elucidate the properties of the resins, their sorptive capacities, the rates of solute sorption, the mechanisms of solute sorption, the effects of solute size on sorption, and many other factors of resin usage. Several of the experimental results were not predicted.

Some of the more important results and conclusions are listed as follows:

1. Excessive resin bleed precluded the use of XAD-7 for concentration of humic substances from water.

2. The testing of XAD-1 was discontinued because its manufacture was discontinued by Rohm and Haas.

3. In experiments with solutions of small organic solutes, the ranking of resin sorptive capacities was XAD-4 > XAD-2 > XAD-8.

4. In experiments with natural waters and solutions of humic substances, the ranking of resin sorptive capacities was XAD-8 > XAD-2 > XAD-4.

5. Due to the large size of humic solutes dissolved in natural waters, the small pores of XAD-4 and XAD-2 excluded humic substances from a large part of the interior macroporous surface of the resin beads. Many macropores were plugged with humic substances, preventing sorption of smaller organic solutes.

6. The recovery of humic solutes from XAD-2 and XAD-4 resins was incomplete, ranging from 75 percent to 85 percent. This poor recovery was postulated to be due to the formation of π-π bonds between the aromatic matrix of these resins and aromatic moieties within the humic substances which lead to an inadvertant fractionation of the humic substances from water.

7. The sorptive rate for humic solutes was most rapid on XAD-8; it exhibited the highest sorptive capacity for humic solutes, and the XAD-8 resin yielded essentially 100 percent recovery of sorbed humic substances with no specific sorption.

The pore size of the hydrated XAD-8 resin is in excess of 250 Å, thus enabling the complete access of its surface area to sorption of natural organic solutes. The theoretically higher surface areas are not used nor is the higher potential sorptive capacities of XAD-2 and XAD-4 achieved in isolation of organic solutes from natural waters because the small pores of the resin are plugged by high-molecular-weight naturally occurring humic solutes that limit the access of other natural organic solutes to a large part of the inner macroporous surface sorptive area [12]. For these and other reasons, XAD-8 is the resin of choice for the isolation of humic substances from natural waters.

The author has performed similar experiments to evaluate XAD-2 and XAD-8 resins for the isolation of dissolved seawater humic substances (primarily fulvic acids). The unpublished results were not as definitive as with freshwater humic substances. The XAD-2 and XAD-8 resins appeared to be almost equally effective in isolation of dissolved seawater humic substances, but there were some compositional differences between humic substances isolated on XAD-2 resin and XAD-8 resin.

Overview of Resin Isolation Procedure

The resin procedure to obtain fulvic acid and humic acid isolates from water can be referred to in two parts. The first part (isolation and concentration of humic substances from water) entails filtration, acidification, resin sorption, resin elution, and resin column regeneration. After a water-sampling site is chosen according to research objectives, a representative sample of the water source is collected into suitable containers for transport to the laboratory, or the water can be pumped continuously (at the sampling site) into a mobile laboratory designed to contain the equipment necessary for water processing. The water sample is pressure filtered through a 0.45 μm membrane filter to remove particulate material. The filtered water is collected in 45 l glass jugs and acidified to pH 2 in order to protonate the humic substances to an uncharged state. The acidified water is pumped onto a preparative glass column containing several liters of cleaned XAD-8 resin. The uncharged humic substances and the neutral organic solutes are sorbed onto the resin column until the column is near saturation and breakthrough of the humic substances occurs in the column effluent. The capacity of the column to concentrate humic substances or other given solutes from water is referred to as k', the column distribution coefficient. A k' of 50 will result in approximately 95+ percent of the humic substances in water being retained on the XAD-8 resin column. For most uncolored stream waters, 306 l of filtered water can be processed through the column before saturation of a 9 l resin column occurs. After saturation, the resin column is flushed with 0.1 M NaOH, the humic substances are ionized, and then eluted from the column in three bed volumes of basic solution. The eluted humic substances are acidified to a pH between 4 and 5, and stored for additional processing. The resin column is acidified to pH 2 with HCl. The process of adsorbing and eluting the humic substances from a given quantity of water (usually 306 l) at a desired k', and the reacidification of the resin column, which is made ready to accept another amount of acidified filtered water, is referred to as "a run" or "one run" in sample isolation and concentration.

The second part of the procedure (humic substances separation and purification) entails the separation of fulvic acids from humic acids, desalting, and hydrogen saturation of humic substances. Humic substances that have been eluted from a number of runs are combined and the pH lowered to 1 with HCl. The humic acids will precipitate; the fulvic acids will remain in solution. Humic acids are separated from fulvic acids by high-speed centrifugation. The fulvic acids are readsorbed onto the XAD-8 resin column and washed with deionized water to remove the inorganic acid. The fulvic acids are back eluted from the resin column with 0.1 M NaOH, the eluate is hydrogen saturated by passage through a column of hydrogen-saturated exchange resin, then freeze-dried. The humic acid is dissolved in dilute NaOH, diluted to a concentration less than 250 mg C/l, acidified to pH 2.0 with HCl, desalted on the XAD-8 resin column the same as the fulvic acids, hydrogen saturated by resin exchange, and then freeze-dried.

Procedural Details and Considerations

The resin isolation procedure for aquatic humic substances is very lengthy and time consuming. There are many details and helpful hints, which if followed, make the procedure more successful, more trouble free, and easier to accomplish. Some of these details will be incorporated in this section along with some of the reasons for various operations involved in the procedure.

Cleaning the Resins

The proper and complete cleaning of the resins for organic solute isolation is as important as resin selection and usage for the success of the proposed research objective. Inadequately and improperly cleaned resin can negate all the time and effort in organic isolation and fractionation due to extensive artifacts and resin contamination in the humic isolates.

An extensive amount of time, effort, and expense is required for adequate resin cleaning. There are no shortcuts; several weeks or months are required for cleaning resins before they can be used for solute isolation.

For efficient time and productive effort, 5 to 20 l of clean resin is required for humic substances isolation from each site. The cleaning of 20 l of resin is not a small task. It requires days of planning, the purchase of special equipment, and a period of months for actual resin cleaning in several solvents.

The resins, as received from the supplier, are commercial grade with an abundance of contaminants and unpolymerized monomers. The resins are also of nominal mesh size and have an abundance of smaller sizes (fines). The first step in the cleaning process is to wash the resin in a large container with 0.1 M NaOH solution. The NaOH solution should be renewed daily for the first 10 days. After the NaOH solution is decanted each day, the resin should be washed several times with deionized water with decantation of the fine resin particles. The removal of fine particles is essential for maintenance of a high flow rate through large resin columns. Otherwise the columns will become clogged with fine resin particles and adequate flow rates can not be maintained. After the 10-day cleaning in NaOH, and the subsequent rinsing with deionized water, rinse the resin three times with methanol. The resin is now ready to be transferred to a Soxhlet cleaning apparatus.

Using three or four Soxhlet units, the solvent cleaning of the resin can be done in an assembly-line process. The extraction thimbles containing resin can be moved in sequence from one solvent to another after each 5-day cleaning period. The recommended cleaning sequence is methanol, acetonitrile, and diethyl ether. Five days in each of these solvents is considered to be one cleaning cycle. The resins can be cleaned in additional solvents, if desired, and should be cleaned in any solvent that will be used to elute sorbed solutes from the resin in an isolation procedure.

For new or unused resin, it is imperative that the resin be cleaned a minimum of two and preferably three cleaning cycles before use because the new resin has a large quantity of impurities. Finally, after cleaning the resin, it should be cleaned once more for two days in a Soxhlet containing methanol and then stored in methanol before use.

Used resin which has been used in organic solute isolation should be saved and cleaned for a minimum of one and preferably two cleaning cycle(s) if time permits before the next usage. It is usually much more desirable to clean and reuse the resin than to start with new or unused resin because the resin becomes cleaner with repeated cleaning cycles and the cleaning with NaOH is not necessary.

Prior to resin usage, the resin should be slurry packed into a glass chromatographic column fitted with Teflon endcap, valves, and tubing. The column should be filled to approximately 90 percent capacity. The resin column must be rinsed with approximately 75 to 100 column volumes of distilled (reagent) water to free the resin of methanol. The final effluent washings of the resin column should be tested for DOC and should contain no higher DOC concentration than is normal for distilled water (<0.5 mg C/l). The column must be rinsed in excess of nasal detection of methanol. It is imperative that the resin is washed free of methanol because the sorptive properties of the resin are decreased for most solutes in water-methanol mixtures.

After the methanol has been rinsed from the resin, the resin should be base rinsed with one column volume of 0.1 M NaOH and then acid rinsed with one column volume of 0.1 M HCl in order to clean the resin of any hydrophobic solute contaminants in the large volume of distilled water. This base-acid rinse should be repeated immediately before column use. This base-acid rinse should be repeated before resin usage during any period which the resin has been standing unused in acid solution for 1 day or longer in order to clean the resin of any solubilized or hydrolyzed resin components.

The XAD resins should not be stored or left standing in basic solution for extended periods because the resin will slowly hydrolyze and contribute resin bleed to organic isolates desorbed from the resin. Between daily usage of the resin, it should be stored in acid solution. The time required for elution of the resins in basic solution should be minimal to prevent resin bleed into the desorbed organic solutes.

Volume of Water Sample

For practical purposes, a minimum of 100 to 200 mg of humic substances is required for adequate characterization. The amount needed in addition to characterization requirements for given proposed research objectives vary greatly, but 150 to 200 mg are usual minimal quantities. Therefore, a total minimal quantity of 300 to 400 mg is required. For surface water with a DOC of 4.5 mg/l, assuming 40 percent of the DOC as fulvic acids, and assuming an anticipated recovery of 80 percent, approximately 275 l of water must be processed to acquire 400 mg of fulvic acids. To acquire 400 mg of humic acids from the same water sample would require approximately 9 times the amount of water to be processed (approximately 2500 l).

Isolation of 300 mg of humic acids from groundwaters requires the processing of a very large volume of water. For example, groundwater with a typical DOC of 0.5 mg/l, containing 3 percent of the DOC as humic acids, and anticipating an 80 percent recovery of humic acids, would require the processing of 25,000 l of water.

The isolation of 300 mg of seawater humic acids seems completely impractical based on the experience of the author. After processing over 32,000 l of seawater, less than 50 mg of ash-free humic acids were isolated. In order to isolate 300 mg of seawater humic acids, approximately 200,000 l of seawater must be sampled and processed. It is a much more reasonable effort to isolate 300 mg of fulvic acids from seawater. Assuming a DOC of 1 mg/l, 10 percent of the DOC as fulvic acids, and 80 percent recovery, approximately 4,000 l of seawater must be processed.

Sampling-Site Selection and Water Sampling

The water-sampling site is often dictated by specific research objectives. If there is some liberty in site selection, it should be free from local influences of contamination and far enough downstream from the confluence of another stream that complete mixing within the stream has occurred. Stream sampling should be the composite of several vertical samples across the stream transect rather than point sampling or sampling from the streambarlk.

When sampling wells, one should empty the casing several times to remove stagnant water before sample collection. To prevent sample contamination from oil-lubricated pumps, one should sample only wells with artesian flow or wells containing water-lubricated pumps.

The sampling of seawater is not an easy task because special equipment is often needed for sampling from ocean-going vessels. Care should be taken to adequately document the sampling depth and location because samples taken from nearshore and shallow depths in the upper euthrophic zone may be greatly influenced by algal and bacterial growth. Because seawater is corrosive to stainless-steel vessels, saline samples should be collected and stored in glass, glass-lined, or Teflon-lined containers.

Filtration of Water Samples

Separation of the dissolved and particulate phases in water is essential for removal of particulate inorganic sediments, particulate organic detritus, and various microorganisms and aquatic invertebrates. It is also well recognized that particulate and dissolved organic substances differ in composition and reactivity. The most universally accepted definition for separation of dissolved and particulate constituents in water is filtration through a 0.45-μm membrane filter. This definition is arbitrary as are other techniques such as centrifugation and density separation. The author's recommendation for separation of dissolved humic substances from whole water samples is filtration through a "partially clogged" non-contaminating 0.45 μm membrane filter. One such membrane filter is the Gelman Metricel without a detergent wetting agent. In the past, this type of membrane filter was only supplied containing an organic detergent as a wetting agent for the filter. The detergent contaminated the filtered water sample. Preleaching the filter to remove the detergent was difficult and time consuming. This type of membrane filter will selectively sorb organic solutes of very low water solubility, but it has a very low capacity for humic or fulvic acids. The ideal membrane filter is not yet available; therefore, knowledgeable choices must be made for water filtration according to the specific objectives of the proposed research. Silver membrane filters were frequently used for filtration in the past before the non-detergent Gelman Metricel membrane filters were available. Silver membrane filters are desirable because they have carefully controlled pore sizes and do not sorb organic substances from water. These membranes are expensive and have been found to result in slight contamination of humic substances with Ag, Hg, and Fe from the membrane filter. Such trace metal contamination may limit humic isolates for trace metal studies.

The major purpose of filtration is to obtain the dissolved organic solutes free of inorganic and organic particulate constituents in the water. Most natural water samples appear to contain a continuum in size of both organic and inorganic constituents from true solution to large particulates. It is often difficult to determine if humic substances are in true solution, micelles, or small aggregates. Clay and silt size particles of crystalline and amorphous minerals are always suspended in water. Their presence in humic isolates results in high ash contents of humic substances and causes numerous interferences with organic analysis; therefore, the best index of filtration effectiveness is low inorganic ash contents. Because many clay-size minerals such as smectites and amorphous oxides are in the size range of 0.2 μm to 0.02 μm, and will readily pass through a 0.45 μm filter, a "clogged" 0.45 μm membrane filter is advocated for water filtration. The "clogged" 0.45 μm membrane filter will effectively remove essentially all the fine colloidal clay particles; therefore, the effective pore size of the "clogged" 0.45 μm membrane filter is in the size range of 0.01 μm or less. In practical use, the "clogged" condition is attained by passing sufficient amounts of water sample through the 0.45 μm filter at a pressure of 5 to 10 psi to reduce the flow rate to a dropwise rate. The period of time usually required to clog the filter is 10 to 15 minutes. The water that passed through the membrane filter during the clogging process is either discarded or refiltered through the clogged membrane filter.

Humic isolates that are obtained after membrane filtration and desalting by cation exchange are routinely low in ash content. Fulvic acids are commonly less than 1 percent ash and humic acids are commonly less than 2.5 percent ash. Filtration almost always obviates the necessity for treatment of humic substances with mixed HCl/HF solutions to dissolve inorganic ash constituents. The HCl/HF treatment is undesirable and should be avoided because of fractionation and large losses of humic substances during the treatment.

Resin Sorption and Desorption of Humic Substances

Filtered water in 45 l glass jugs is acidified to pH 2.0 with concentrated HCl. The acidified water sample is delivered to the top of the 9 l resin column by a peristaltic pump (Cole-Palmer Masterflex) with a number 18 roller head at a rate of 1 l/minute. All the connecting lines and valves are Teflon except for the 20-cm portion of tubing in the roller head; that portion is Tygon. According to Equation (1) [29],

$$V_E = V_O\ (1 + k') \qquad (1)$$

a k' of 50 can be achieved with the passage of 306 l of water. V_E is the breakthrough elution volume of a solute such as fulvic acid, V_O is the void volume of the resin column, and k' is the column distribution coefficient or column capacity factor.

$$K' = \frac{\text{(mass of solute sorbed on XAD-8 resin column)}}{\text{(mass of solute dissolved in column pore water)}} \qquad (2)$$

After 306 l of acidified water is passed through the column, the pump is stopped and the inlet and outlet tubings on the glass column are reversed for back elution of the resin column. Care must be taken to fill the inlet line by pumping with 0.1 M NaOH before connecting the tubing to the bottom connection of the column. Open the valve at the bottom of the column simultaneously at the beginning of pumping the 0.1 M NaOH. The initial pumping rate must be near the maximum pumping capacity of the pump head so that the resin column will be lifted up to the top of the glass column, thus removing the void space at the top of the glass column. After the resin column has moved upward to the top of the glass column, reduce the pumping rate to 400 to 500 ml/min or to that rate which will support the resin column in the top of the glass column. The 0.1 M NaOH elution solution should move slowly and uniformly up the resin column. If the air in the base inlet line were not initially removed, the introduction of air bubbles into the bottom of the column usually would cause channeling of the base flow and occasional overturning of the resin column during elution.

The elution technique enables the collection of a highly concentrated center part or "center cut" of the humic substances elution in a 1 l volume. The humic substances in this part will be the most concentrated and may not require further concentration if the concentration is greater than 500 mg C/l. This part of the eluate should be acidified immediately to a pH between 3 to 4 and stored on ice.

An idealized back-elution curve for humic substances eluted with 0.1 M NaOH from a 9 l column of XAD-8 resin is shown in Fig. 3. The sequential segments of the elution curve are designated as A through E. Color may be observed in the eluate just before the 6 l void volume of the column is attained, even though the pH of the eluate is below pH 7, due to slight channeling of base in the resin column. This 6-l portion of the elution curve is Part A. The color, concentration of eluted humic substances, and the pH of the eluate will gradually increase until the eluate becomes strongly basic (> pH 12) in the second portion of the elution curve from 6 l to 7.5 l. This 1.5 l portion is Part B of the elution curve. Part C of the elution curve is the 1 l "center-cut" previously mentioned. The color and concentration of humic substances gradually decreases in the fourth portion (Part D) of the elution curve from 8.5 l to 14 l. At the 14 l point in the elution curve, the pumping of base is stopped and the pumping of 0.1 M HCl is begun by rapidly changing the inlet pump tube from base to acid containers. In the next portion of the elution curve (Part E) the concentration of humic substances becomes very low and approaches zero after the resin column becomes strongly acidic.

The eluate from Parts A through E of the elution curve are treated in three different ways: stored for separation of fulvic and humic acids, stored for reconcentration, or added to the next column run. The 1.5 l Part B of the eluate prior to the "center-cut" (Part C) and the 5.5 l Part D after the "center-cut" should be combined in a separate glass container, the pH lowered with HCl to a pH between 3

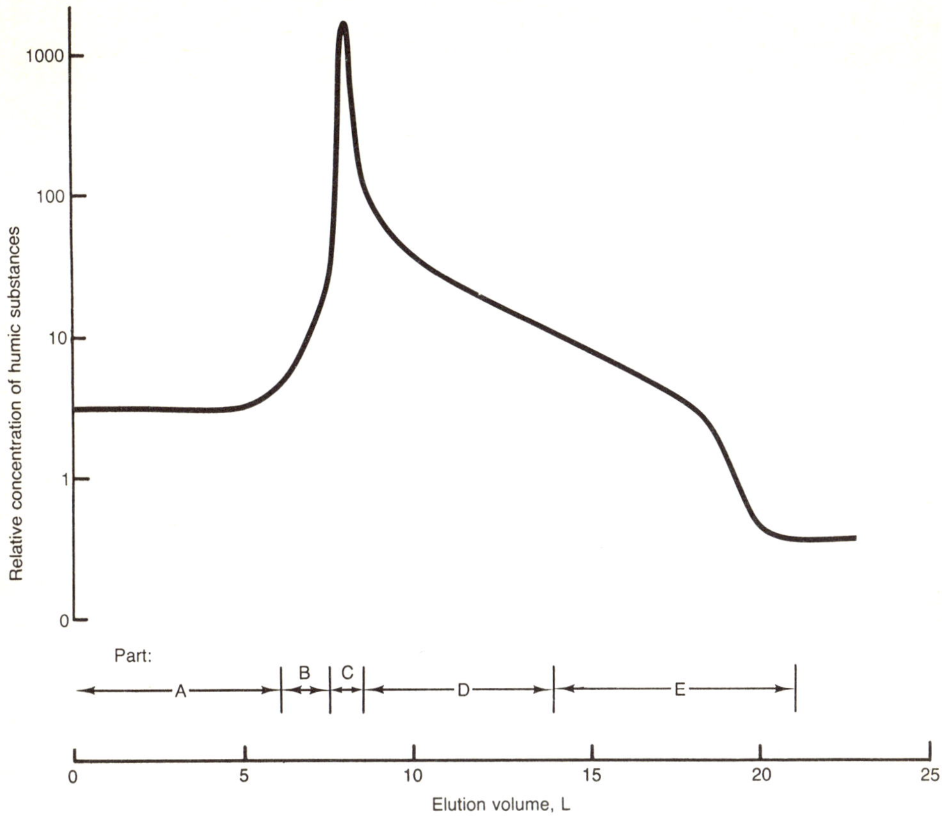

Figure 3 Idealized back-elution curve of humic substances with 0.1 M NaOH from a 9 l column of XAD-8 resin.

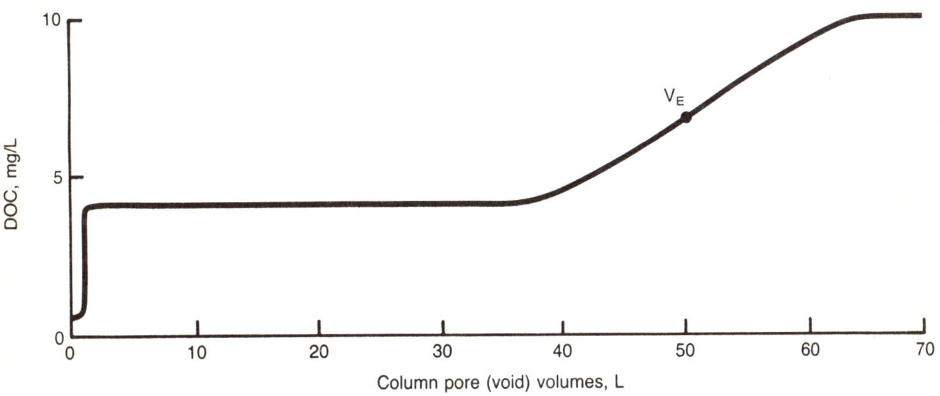

Figure 4 Idealized breakthrough curve for stream humic substances on XAD-8 resin (sample DOC of 10 mg/l, hydrophilic DOC of 4 mg/l and humic substances are the primary constituents of hydrophobic DOC. V_E is the sample elution volume).

and 4, and stored in an ice bath for later reconcentration. The "center-cut" (Part C), the 1.5 l Part B, and the 5.5 l Part D are designated as retained parts of the elution. The void volume during elution (Part A) prior to the retained parts and (Part E) until the eluate becomes strongly acidic (approximately between pH of 1 and 2) are combined, acidified to pH 2, and added to the 306 l filtered sample for the next column run.

After several column runs, the accumulated "other-than-center-cut" concentrated organic solutes (Parts B and D) are reconcentrated as a separate run on the large XAD-8 resin column. The final reconcentrations are conducted on smaller columns of XAD-8 resin. Virtually all of the humic substances are adsorbed upon reconcentration with little or no losses. During the initial concentration and elution of humic substances in some samples, the concentration of humic substances may be so low that the "center-cut" (Part C) of the elution is less than 500 mg C/l. In these instances, Part C should be combined with Parts B and D and treated accordingly.

Separation of Fulvic and Humic Acids

All the elution "center cuts" and the reconcentrated humic substances are mixed in a 45 l glass container and acidified to pH 1.0 with concentrated HCl. In the well-homogenized concentrate, a carbon concentration of 500 mg/l is minimal for rapid and complete precipitation of humic acids. The concentrated sample should be chilled to 2°C in an ice bath, the flocculated humic acids resuspended several times over a 12-hour period, and then allowed to settle. The precipitated humic acids are separated from the soluble fulvic acids by centrifugation.

Desalting, Hydrogen Saturation, and Freeze-Drying

The solution of fulvic acids contains high concentrations of sodium and HCl. The major part of these ions and salts is removed by desalting on an XAD-8 resin column. Fulvic acids at pH 1.0 should be pumped slowly (400 to 500 ml/min) onto a 9 l column of XAD-8 resin until the observed color of the sorbed fulvic acids extends approximately one-third down the length of the column; at this time, pumping of fulvic acid should cease and pumping of distilled water should begin. Specific-conductance monitoring of the effluent also is initiated using a flow-through cell. During leaching of the acidic salt solution from the void volume of the column, the specific conductance decreases rapidly. Concurrently with acid removal, the pH increases, and the fulvic acid begins to move slowly down the column. The column should be rinsed with distilled water until the specific conductance is decreased to 250 μS/cm. The XAD-8 resin column is then back-eluted with 4 column volumes of 0.1 M NaOH. Color, due to fulvic acids, is normally observed to elute from the column during rinsing when the specific conductance decreases to less than 700 to 800 μS/cm. These colored washings, collected until the specific conductance decreases to 250 μS/cm, should be acidified and added to the next desalting run.

The XAD-8 resin column should be back-eluted rapidly with 0.1 M NaOH, the dilute first part of the back eluate mixed with the initial concentrated basic part of the elution, and the solution passed rapidly through a small precolumn (for example, 60 cm by 3.5 cm) of hydrogen-saturated exchange resin. This limits the contact time of fulvic acids with base to only a few minutes. The last part of the base eluate of fulvic acids is introduced directly into the cation-exchange resin column. The purpose of the precolumn is not to completely hydrogen saturate the fulvic acids, but to remove most of the sodium and quickly neutralize the basic solution, which should prevent oxidation of fulvic acids in basic solution. The solution of fulvic acids will be below pH 4.5 after the pretreatment. The solution of fulvic acids is then passed at a fast, dropwise rate through another hydrogen-saturated resin exchange column (for example, 100 cm by 10 cm) for complete hydrogen saturation of the fulvic acids. After complete hydrogen saturation, the sodium concentration should be less than 0.1 ppm. The purpose of mixing the first dilute part of the elution with the concentrated "center cut" part is to dilute the fulvic acids to less than 300 mg C/l. Sometimes fulvic solutions of carbon concentrations in excess of 300 mg C/l will precipitate on the acidic precolumn and will not hydrogen saturate normally. If precipitation occurs on the hydrogen-saturated precolumn of cation exchange resin, the column must be unpacked, rinsed with 0.1 M NaOH, the entire fulvic acid solution diluted with distilled water, and then precolumn treatment repeated.

A non-macroporous cation exchange resin is preferred over the macroporous type. The cation exchange capacity of both types is almost the same, but the macroporous type allows penetration of the fulvic acids into the beads and the end of the saturation is protracted with a long elution tail from the cation exchange column. This excess dilute solution requires more rotovap preconcentration before freeze-drying.

The precipitated humic acids are kept moist with distilled water until they are desalted and hydrogen saturated. To desalt the humic acids, they are solubilized in dilute NaOH and the DOC concentration adjusted to 250 mg/l as carbon or less. The solution is then adjusted to pH 2.0 and pumped slowly onto a large XAD-8 resin column. At this point, the procedure used for desalting the humic acids is the same as that for fulvic acids. To accomplish hydrogen saturation of the humic acids without precipitation in the cation-exchange resin column, it is imperative that the humic-acid concentration in solution not exceed 250 mg/l as carbon during passage through the cation exchange resin. Concentrations of humic acids in excess of 250 mg/l frequently precipitate and clog the cation exchange resin, necessitating a repeat of the desalting procedure.

The hydrogen-saturated fulvic acids and the hydrogen-saturated humic acids should be separately concentrated by vacuum rotovap at room temperature to approximately 1500 mg C/l to facilitate freeze-drying of reasonable volumes of humic solution. Care should be taken to maintain a high vacuum during freeze-drying to insure a light fluffy product. A poor vacuum will result in dark, hard crystal-like formation due to thawing and reconcentration. Air drying or oven drying should be avoided due to extreme denaturation and possible alteration of the humic samples.

Chromatographic Considerations

Resin columns do not have an infinite capacity for solute sorption. This concept has been the most difficult for many users of the resin technique to realize. It is counterproductive to ignore all capacity considerations and pump large amounts of water at a rapid rate through small columns of resin and expect solute recoveries to be representative. The capacity factor (k'), volumes of XAD-8 resin, and amounts of water from which 95+ percent recovery of a given solute can be concentrated, and void volumes of the resin columns are given in Eqns (1) and (2). After the determination of the volume of XAD-8 to be used, a column size can be selected to contain the resin plus approximately 5 to 10 percent additional volume. The unfilled 5 to 10 percent column volume above the resin bed is desirable for efficient column operation. It permits small volumes of air to enter the column without effecting the water flow in the resin column during solute sorption, the observation of slow leaks in plumbing of the column, the time lag necessary to stop the water flow in the column before air enters the column when the pump tubing periodically ruptures, and permits small volume changes in the packing of the resin during column elution and regeneration. The void volume of XAD-8 resin is approximately 65 percent per unit volume as determined experimentally by the breakthrough curve for unretained inorganic ions.

The k' distribution for natural organic solutes dissolved in water onto XAD-8 resin has been found to be a bimodal; low k' values of less than 5 to 10 are termed hydrophilic and higher k' values of greater than 40 are termed hydrophobic. Because of the sparcity of solutes in natural waters with k' values in the 10 to 40 region, there is a natural resolution between hydrophilic and hydrophobic solutes according to sorption on XAD-8 resin. The use of the terms hydrophilic and hydrophobic according to XAD-8 resin sorption can be somewhat ambiguous to the natural water system because all the dissolved organic solutes are hydrophilic in natural waters.

Humic substances have high k' values and are acidic in nature; therefore, they are designated as hydrophobic acids according to XAD-8 resin isolation. The k' values of all the colored, non-specific, carbonaceous organic substances in water which comprise the range of compounds called humic and fulvic acids in water are pH dependent with 99+ percent of these substances having a k' greater than 50 at pH 2 with less than 10 percent having a k' of 50 at pH 7. The components with a k' of 50 or greater at a pH of 7 are termed weak hydrophobic acids and are typically high in phenolic character. A strong hydrophilic acid fraction which remains ionized at pH 2 and will not sorb onto XAD-8 is usually very small (less than 1 to 2 percent of the total humic substances).

The k' of humic substances is slightly affected by concentration. The k' of humic substances in dilute solution and up to approximately 10 mg/l is essentially constant at 50 or greater. This k' is applicable for all uncolored waters and for colored water with a DOC up to approximately 15 mg/l. At higher DOC values and humic substance concentrations, the k' of humic substances (primarily fulvic acids) decreases. At DOC values of near 30 and 50 mg/l, the k' of fulvic acid decreases by approximately 20 and 50 percent, respectively. The k' for humic substances on any size of column can easily be determined experimentally by plotting DOC versus the column pore (void) volumes of water sample passed through the resin column. After 1 column pore volume of sample input, there will be a constant breakthrough of hydrophilic sample DOC which will be approximately 40 percent of a surface water DOC value or approximately 70 to 80 percent of a groundwater DOC value. After approximately 40 column void volumes, the hydrophobic constituents of the DOC (primarily humic substances) will initially breakthrough with an increase in the breakthrough DOC. The humic substances breakthrough will increase as a typical S-shaped breakthrough curve until the DOC of the sample effluent is equal to the DOC of the sample influent. The breakthrough volume (V_E) at the midpoint of the S-shaped breakthrough curve is the value used to calculate the k' of humic substances in a given sample (Eqn 1). A curve of ultraviolet absorbance, fluorescence intensity, or any other quantitative parameters of humic substances versus column pore volumes can be used to determine the k' of humic substances. An idealized breakthrough curve for humic substances in water is shown in Fig. 4, in which the initial sample DOC is 10 mg/l, hydrophilic DOC of the sample is 4 mg/l, and the resin column pore (void) volume is 66 percent of the packed column volume.

The column capacity factor, k, is greatly affected by the flow rate of water through the resin column. Ideal sample flow rates have been determined to be between 5 and 10 column volumes per hour. For most organic solutes, k' decreases sharply with flow rates in excess of 10 to 15 column volumes per hour. At temperatures near room temperature (20°C), there are very small changes in k' with temperature fluctuations.

Column performance is seriously affected by fine particles of resin less than 60-mesh size. With time and usage of resin columns, the fine resin particles have a tendency to accumulate into a thin layer within the column. As this layer becomes thicker, it becomes more difficult to maintain adequate column flow rates without development of backpressure and subsequent leaks from column fittings. Thorough initial washing and decanting of resin fines is critical to efficient column operation.

Precautions in XAD-8 Resin Usage and Helpful Hints in Procedure Usage

As with any analytical method, certain precautions must be taken to assure quality results when using any resin including XAD-8. The magnitude of most of the potential problems of resin usage can be minimized or prevented if the resin is thoroughly cleaned initially. Inadequately cleaned resin will lead to problems in resin usage and in the organic isolates obtained by the resin procedure.

Long-term resin storage in solvents such as methanol, acetonitrile, acetone, or even water will result in slight resin dissolution or bleed. Most dissolution products can be removed by rinsing with several column volumes of high-quality distilled water, followed by rinsing alternately in dilute HCl and dilute NaOH. All XAD-8 resin columns should be rinsed alternately with dilute HCl and dilute NaOH daily before usage. Storage of bulk resin or of packed resin columns in water or dilute acid should not occur for more than a few days in order to prevent microbial or algal growth on the resin. Immediately after the usage of a column of resin is completed, the column should be unpacked and the resin stored in methanol until it is cleaned by recycling in Soxhlet extractors. The time between resin use and recleaning should be minimal to prevent any microbial growth on or in the resin beads. The cleaned resin should be stored in methanol to prevent microbial growth, to prevent breaking of the beads during drying and storage, and to prevent the slacking of the beads (production of fines) upon drying and rehydration.

One precaution in XAD-8 resin usage must be adhered to, i.e., is do not store the resin in basic solution. The time periods during elution when the resin is in basic solution must be minimized to prevent resin hydrolysis which results in resin bleed and sample contamination. With continuous daily column use, do not leave the resin standing in basic solution overnight. After eluting the last sample concentration run of the day, the resin must be rinsed with dilute acid until the resin column effluent is acidic before leaving the column to stand overnight. XAD-8 resin is quite stable for short periods in basic solution, but hydrolysis of the resin increases with high pH (greater than 10) and with the length of time in basic solution.

To minimize any possibility of resin bleed in acidic solution, XAD-8 resin columns with sorbed humic acids should not be left standing overnight. The sorbed humic substances should be eluted and the column reacidified for overnight storage. Finally, recovery or elution of the sorbed organic solutes from the resin should not be attempted with any solvent that has not been used initially to clean the resin. All of the previous precautions apply for any special usage of XAD-2 or XAD-4 resins.

After the elution of adsorbed humic substances from the XAD resin column with 0.1 M NaOH at the end of each "column run", the NaOH is followed by rapid pumping of 0.1 M HCl until the column effluent is acidic to pH indicator paper. After stopping the pump, the column pumping is reversed for frontal or downward pumping through the resin column. With pumping ceased, the resin column will slowly repack itself from the bottom of the column upward. The repacking processes can be accelerated by gently shaking or jostling the glass column. During the repacking any small air bubbles or channels in the resin column are removed by upward flotation. After the resin column is repacked, one column void volume of 0.01 M HCl (a solution of pH 2) is pumped through the column before starting the next sample run.

In addition to saving time by reacidification of the resin column with 0.1 M HCl (pH 1) rather than 0.01 M HCl (pH 2), there is a tenfold savings of distilled water used to make acidic solutions. This savings of distilled water is especially useful in field operations. The next sample <u>should not be used for</u> acidification of the resin column because of possible losses of humic substances. The humic substances in the sample volume which neutralized the basic front from pH 13 to a low pH would not sorb on the column and would be lost. Also some humic acids in water, especially in colored waters of high DOC (20 to 75 mg/l), have a tendency to be sensitive to flocculation, even at low concentrations and at a pH of lower than 2. Such flocs of humic acid do not sorb onto the XAD resins or will partially sorb. Such possible flocculation causes potential problems and should be avoided.

The AG 50W-X8, non-macroporous cation exchange resin used for hydrogen saturating humic substances must be Soxhlet cleaned in the same sequence of solvents as XAD resins. Slurry pack the Soxhlet-cleaned resin into a large glass preparatory column and rinse with distilled water to free the resin of methanol. Hydrogen saturate the resin by slowly passing 15 column volumes of 10% HCl by weight (a volume/volume ratio of 1 to 3 consisting of 1 volume of 36 percent concentrated HCl reagent and 3 volumes of distilled water) through the resin column. The resin should remain in 10 % HCl for an overnight period. Rinse the resin free of excess HCl with distilled water until the conductivity of the effluent is the same as the influent distilled water. Transfer the hydrogen-saturated cation exchange resin to smaller glass precolumns and final saturation columns as specified in the Materials and Methods Section under glassware.

The cation exchange resin has a tendency to bleed slowly when standing in distilled water. It is essential to adequately rinse the cation exchange resin with distilled water before each usage until its hydrogen exchange capacity is exhausted. When the resin becomes sodium-saturated, the resin becomes darker and the color band between sodium-saturated and hydrogen-saturated resin is clearly visible in the column of resin. Upon sodium-saturation of three-fourths of the resin column, remove the resin from the analytical column and resaturate with hydrogen in a preparative column.

Scale, Cost, and Planning for Humic Substances Isolation

The XAD-8 resin procedure for the concentration of humic substances from water can be scaled to accomplish the isolation of mg to 100 g quantities of humic substances. It is only a matter of planning, equipment cost, and time. Time may not be a major factor in some research situations, but because time is cost in most operations, the major factor is the balance of equipment, cost, and time.

For the example given previously, to isolate 400 mg of fulvic acids from a surface water with a DOC of 4.5 mg/l requires the processing of approximately 275 l of water. If the sample were processed on a 2 l column of resin at a k' of 50, 68 l of sample could be processed during each run before elution and regeneration of the resin column. The concentration of the sample could be completed in four sample runs with each run requiring 8 hours (a normal workday) for sample sorption and elution. For completion of the sample in 5 workdays, 55 l of water must be filtered each day (in a 24-hour period because sample filtration can proceed unattended during the night).

Production of 55 l of filtered surface water can usually be accomplished with four filter units of 142-mm size or 1 filter unit of 293-mm size. Equipment cost for pressure filtration equipment, a 2.5 l glass column, four 45 l glass jugs, and pump accessories for such an experiment is approximately $4,000. The cleaning of 2 l of XAD-8 resin would require a 2-month period prior to the resin use, approximately two man weeks of labor, and an initial investment of $3,000 for a large Soxhlet with a thimble of 2 l capacity. After the one week concentration period on XAD-8, two more man-weeks of time are required for further laboratory processing (humic/fulvic separation, desalting, hydrogen saturation, and freeze-drying) to attain the final hydrogen-saturated, freeze-dried fulvic acid.

In order to concentrate 400 mg of humic acids from the same source of surface water would require 9 weeks of sample concentration time on 2 l, XAD-8 resin columns to process approximately 2250 l of water. To avoid expensive time and labor costs, the concentration process could be scaled-up to be accomplished in one week with the purchase of a 10 l glass resin column, 21 additional small 142-mm plate filter units or 5 additional large 293-mm plate filters, 10 additional 45 l glass jugs, and 2 additional Soxhlet units at a total cost of $30,000. The production of 300 l of filtered surface water per day can be processed on a 9 l resin column during a normal 8-hour workday. Additional sample processing in the laboratory to achieve a final product of 400 mg of humic acids and 3600 mg of fulvic acids, would require an additional 5 man-weeks in the laboratory.

The isolation of 400 mg of fulvic acids from groundwaters and seawaters requires another order of magnitude of time and effort than freshwater stream sampling. Because these waters are relatively sediment-free, the filtration process is not limiting, but the processing of the water sample on the XAD-8 resin columns is time limiting. To process 25,000 l of groundwater or surface water requires approximately 85 workdays (more than 4 months of time) with one 9 l resin column at one run per 8-hour workday. By running two 9 l resin columns simultaneously for 18 hours per day (2 work shifts of 9 hours per shift) the sample concentration time can be reduced to 20 workdays. The capital investment in additional 9 l glass resin columns is minimal, but the time and effort in cleaning an additional 20 l of XAD-8 resin is significant.

From the previous discussion, one can readily understand that to isolate larger gram quantities of humic substances from water or to isolate several 100-mg quantities of humic substances simultaneously from different sources or different depths within a lake requires the appropriate scale factors for equipment, time, and resin cleaning. For the processing of large volumes of surface water, the time limiting factor is usually the rate of water filtration; therefore, the number and size of the plate filter units are critical. Some investigators have used cartridge-depth filters, such as a Balston filter unit, to produce filtered water at a faster rate than can be produced with several membrane plate filter units. The cost of such depth filter units is also much less than for plate filter units. Unfortunately, the filtration process appears to be less efficient in removing suspended-sediment particles and usually higher ash contents are found in humic isolates when using the depth filter cartridges.

Logistics of Water Handling and Processing of XAD-8 Resin Concentrates

The isolation of humic substances from surface waters and groundwaters close to one's own research laboratories usually presents very few problems. Freshwater from the sampling site can be transported daily in stainless-steel or glass containers to the laboratory for processing. When the sampling site is 100 miles or less from the laboratory it is usually more efficient to transport the water to the laboratory than to transport the sampling equipment to the sampling site unless a large, self-contained laboratory trailer is available. When sampling humic substances from surface waters and groundwaters at a great distance from the laboratory, it is usually desirable to transport the sample filtration and resin concentration columns to the sampling site. The combined humic substance concentrates after resin concentration and elution are kept chilled and transported back to the laboratory for further processing.

For concentrating humic substances from seawaters, the author processed the sample on land nearest to the deep ocean sampling site. The resin concentration of humic substances in an onshore laboratory rather than on ship is advocated to decrease the enormous cost of ship time, to preclude the necessity of making all the equipment seaworthy, to avoid seasickness during on-board processing, and to provide for a more normal laboratory working environment. Seawater must be transported from the sampling site to the on-shore field laboratory in glass, glass-lined, or Teflon-coated containers. Transport of seawater in stainless-steel containers is unacceptable due to the corrosive nature of seawater on stainless-steel. The humic substances concentrated after elution from the XAD-8 resin should be transported from the field laboratory to the analytical laboratory for further processing to the point of freeze-drying.

Characterization of the Isolated Humic Substances

Fulvic acids or humic acids are not identified in the classical organic chemical sense; that is, in the same way an unknown compound would be qualitatively identified in qualitative organic chemical analysis. Humic substances are only generally identified by characterization of general properties rather than specifically identified, as are specific organic compounds. Identification by characterization is necessitated because humic substances (either fulvic acids or humic acids) are a group of closely related compounds for which the chemical formula or structure, for all or any one of the components, is not known. The characterization of humic substances may be categorized into three types of analyses: traditional, new, and general characterization methods.

The traditional methods for characterization of humic substances include elemental analysis, titration analysis for carboxyl and phenolic functional groups, E_4/E_6 ratio, ultraviolet spectrographic analysis, infrared analysis, and molecular weight by sephadex chromatography or vapor pressure osmometry. Most of these analyses indicate that humic substances have a wide range of chemical properties. Newer methods of characterization include solid- and liquid-state ^{13}C NMR spectroscopy, solid- and liquid-state ^{1}H NMR spectroscopy, pyrolysis/mass spectroscopic analysis, δ ^{13}C value, ^{14}C-age, amino acid analysis, saccharide analysis, density, color intensity per unit of carbon, synchronous fluorescence emission, low-angle X-ray scattering, and molecular weight by ultracentrifugation. Several of these methods have been available for sometime, but only recently been applied to humic substances characterization. The general characterizations are for those properties that characterized humic substances on a specific research basis, but are not routinely performed. These include specific metal interactions, differential thermal analysis, various chromatographic separations, chlorine and bromine analysis, methoxyl group content, mass spectroscopy, electron spin resonance (ESR), and many other specific research characterizations.

Even though the traditional characterization methods are generally not very specific for humic substances nor is any characterization a written requirement for characterization of humic substances with the exception that humic substances must be colored; elemental analysis, carboxyl functional group content, and phenolic functional group content are the closest to being generally required characterization parameters. Many of the new characterization methods, such as ^{13}C-NMR and ^{1}H-NMR spectroscopy, are much more definitive for identification of humic substances and are gradually becoming more of a required characterization. For some groups of humic acid scientists, ultraviolet spectroscopy and E_4/E_6 ratios are generally required characterizations, but the relative non-definitiveness of this characterization has been seriously questioned. Because of their definitive nature, the characterization of humic substances by pyrolysis/mass spectroscopy, amino acid analysis, and synchronous fluorescence spectroscopy are expected to become standard characterizations in the near future.

The order in which the various characterizations and the number of characterizations conducted on humic substances is a special consideration in some studies, especially of aquatic humic substances, because of the small amounts of humic substances usually isolated. As presented in the isolation section of this paper, the isolation of large quantities (1 gram or more) of aquatic fulvic or humic acids is not common and is only achieved with considerable effort and expense. With our modern technology, the minimum amount of fulvic or humic acids that can be adequately characterized is 50 mg. The solid state ^{13}C-NMR spectroscopy must be run first because 50 mg is the absolute minimum sample size for high-spectral resolution in a period of 2 to 3 days of spectrometer time. The sample is quantitative recovered and uncontaminated after solid state ^{13}C-NMR analysis. The sample can now be used for various destructive characterization analyses. CHONSP elemental analysis and ash content can be accomplished on 20 mg, titration analysis for carboxyl, phenolic, and ester functional groups on 10 mg, pyrolysis/mass spectroscopy analysis on 1 mg, amino acid analysis on 2 mg, saccharide analysis on 2 mg, infrared analysis on 1 mg, δ ^{13}C value and ^{14}C-age by tandem mass spectroscopy on 2 mg, and combined color intensity per unit of carbon, ultraviolet scan, synchronous fluorescence scan, E_4/E_6 ratio, low angle X-ray scattering, density, and molecular weight by high-speed ultracentrifugation can all be determined sequentially on the same 7 to 10 mg of sample. A total of 17 characterizations can be accomplished on as little as 50 mg of fulvic or humic acids.

An additional 50 mg or more of humic sample is needed for liquid state ^{13}C-NMR and ^{1}H-NMR spectroscopy. Due to various reagents and solvents used in liquid NMR, it is often difficult to recover the fulvic or humic acids sample in an uncontaminated state.

Summary and Conclusions

Due to low concentrations of humic substances (<2 mg/l) in water and the many interferences from inorganic and other organic constituents, few direct or in situ experiments can be made on humic substances. Also, experiments conducted on the whole DOC of water are usually not representative of humic substances, because humic substances are only approximately one-half of the DOC. For these and other reasons, humic substances must be isolated from water in order to study most aspects of their nature and reactivity. The XAD-8 resin isolation procedure discussed in this paper is one of the best methods of isolating humic substances from water for research purposes.

The success of XAD-8 procedure has been due to many systematic experiments which were conducted to understand the sorptive and desorptive processes on the resin, to incorporate modifications in the procedure which overcome limitations in other isolation procedures for humic substances, the strict adherence to general chemical and chromatographic details, and the incorporation of quantitative chromatographic expressions into the procedure to define the resin sorptive capacity, k', for sorption of humic substances from water on a reproducible basis. The XAD-8 resin isolation method, which has been used successfully for the last few years by a few research groups, is rapidly expanding in usage. A detailed procedure for use of the procedure has been presented in this paper so that it can be used easily and successfully by both experienced chemists and relatively unexperienced scientists in organic chemistry. For the successful use of the method and to obtain uncontaminated humic and fulvic acid isolates, it is imperative that the user give heed to careful and complete cleaning of the resin before usage and to rigorously adhere to all the details and precautions presented concerning the method.

The XAD-8 resin procedure for isolating humic and fulvic acids from water is a time consuming and labor intensive process, but the humic isolates obtained are free from the many limitations which have been encountered in previous efforts to isolate aquatic humic substances. The XAD-8 resin procedure is quantitative for the removal and concentration of humic substances from water. The resin adsorbs 95+ percent of humic substances from water with 100 percent of the adsorbed humic substances being desorbed from the resin. There are no specific interactions between the resin and humic solutes resulting in irreversible sorption of humic substances which could result in low recoveries of humic substances from the resin. Because the isolation procedure involves pressure filtration, essentially all the clay minerals are removed from the humic substances. The combined treatment of pressure filtration and hydrogen saturation by resin exchange results in humic isolates with low ash contents. The saccharides, low-molecular-weight specific acids, and other hydrophilic low-molecular-weight non-humified compounds in water pass through the resin column and are not adsorbed with the humic substances. Elution of the adsorbed humic substances in dilute base (0.1 M NaOH) results in the elution of only humic substances without admixture with the hydrophobic neutral compounds which remain adsorbed on the XAD-8 resin column. Because the humic isolates obtained by the XAD-8 resin isolation procedure are free from inorganic salts and non-humified specific organic compounds, the author believes and data support the contention that they are truly fulvic and humic acids and not fulvic and humic acid fractions (admixed with saccharides) as in the past.

The sound chemical chromatographic basis of the XAD-8 resin isolation method is a major strength and advantage of the procedure. The use of k' and Equation 2 enables the quantitative and reproducible usage of the method for isolating humic substances from all aqueous environments. The humic isolates from different aquatic environments isolated by this procedure can be compared quantitatively for differences in composition and reactivity. The same criteria used to determine the quantitative and reproducible acceptability of this XAD-8 isolation procedure should be used for any other method developed or existing to isolate humic substances from water.

The XAD-8 resin procedure can be used to isolate mg to 100 g quantities of humic substances from water; it is a matter of scale, time, and equipment cost. Even the isolation of mg quantities of humic substances from water requires a period of 3 to 4 months; a major part of this time is spent cleaning the XAD-8 resin before use. The isolation of 100-g quantities by this method usually requires a minimal equipment investment of $50,000. Even though this method of isolating humic substances from water is time consuming an expensive, it appears to be the method of choice for obtaining quantitative yields of unfractionated humic substances from water.

Humic substances may be characterized in many ways by both traditional and new methods. The new characterizations of ^{1}H-NMR, ^{13}C-NMR, $\delta\ ^{13}$C content, amino acid analysis, saccharide analysis, and pyrolysis/mass spectroscopy are usually more definitive than older traditional methods. These new methods are rapidly becoming more common and will soon be considered as essential characterizations.

The older traditional characterizations are useful, seldom definitive, and generally augment the newer more definitive characterization methods. The more useful traditional methods include elemental analysis, ash content, titration analysis for carboxyl and phenolic content, functional group analysis by infrared spectroscopy, and molecular weight by various techniques. Some of the traditional characterization methods such as E_4/E_6 ratio, size analysis by sephadex chromatography, and ultraviolet spectroscopy are of limited value.

References

1. Packham, R.F. Proc. Soc. Water Treatment Exam. **13**:316 (1964).

2. Leenheer, J.A. and R.L. Malcolm. U.S. Geological Survey Water Supply Paper No. 1817-E (1973).

3. Burnham, A.K., G.V. Calder, J.S. Fritz, G.A. Junk, H.J. Svec, and R.Willis. Anal. Chem. **44**:139 (1972).

4. Grieser, M.D. and D.J. Pietrzyk. Anal. Chem. **45**:1348

5. Junk, G.A., J.J. Richard, M.D. Grieser, D. Witiak, J.L. Witiak, M.D.Ariguello, R. Vick, H.J. Svec, J.S. Fritz and G.V. Calder. J. Chromatogr. **99**:745 (1974).

6. Riley, J.P. and D. Taylor, Anal. Chim. Acta **46**:307 (1969).

7. Mantoura, R.F.C. and J.P. Riley. Anal. Chim. Acta **76**:97 (1975).

8. Leenheer, J.A. and E.W.D. Huffman. J. Research U.S. Geological Survey **4**:737 (1976).

9. Malcolm, R.L., E.M. Thurman and G.R. Aiken. In D.D. Hemphill, Ed., Trace Substances in Environmental Health-XI, pp. 307-314 (University of Missouri, Columbia, 1977).

10. Thurman, E.M., R.L. Malcolm and G.R. Aiken. Anal. Chem. **50**:775 (1978).

11. Thurman, E.M., G.R. Aiken and R.L. Malcolm. Proc. of the 4th Joint Conference on Sensing of Environmental Pollutants, pp. 630-634 (New Orleans: American Chemical Society, 1978).

12. Aiken, G.R., E.M. Thurman and R.L. Malcolm. Anal. Chem. **51**:1799 (1979).

13. Malcolm, R.L., G.R. Aiken and E.M. Thurman. American Soc. of Agronomy Abstr., p. 161 (1979).

14. Leenheer, J.A. and T.I. Noyes. U.S. Geological Survey Water-Supply Paper No. 2230 (1984).

15. Miles, C.J., J.R. Truschall and P.L.. Brezonik. Anal. Chem. **55**:410 (1983).

16. Malcolm, R.L and W.H. Durum. U.S. Geological Survey Water-Supply Paper 1817-G:1-20 (1976).

17. Leenheer, J.A. and E.W.D. Huffman. U.S. Geological Survey Water Resources Investigations **79**-4:1-16 (1979).

18. Leenheer, J.A., R.L. Malcolm, P.W. McKinley and L.A. Eccles, U.S. Geological Survey Journal of Research **2**:361 (1974).

19. Williams, P.J., H. Oeschger, and P. Kinney. Nature **224**:256 (1969).

20. Williams, P.M. and E.R.M. Druffel. Nature **330**:246 (1987).

21. Sugimura, Y. and S. Yoshimi. Marine Chem. **24**:105 (1988).

22. Black, A.P. and R.F. Christman. J. Am. Water Works Assoc. **55**:753 (1963).

23. Kerr, R.A. and J.G. Quinn. Deep-Sea Research **22**:107 (1975).

24. Jeffrey, I.M. and D.W. Hood. J. Marine Res. **17**:247 (1958).

25. Klöcking, R. and D. Mucke. Z. Chem. **9**:453 (1969).

26. Sridharan, N. and G.F. Lee. Environ. Sci. Technol. **6**:1031 (1972).

27. Leenheer, J.A. Water Analysis: Volume III, pp. 83-165 (San Diego: Academic Press, Inc., 1984).

28. Kunin, R. Polym. Eng. Sci. **17**:58 (1977).

29. Leenheer, J.A. Environ. Sci. and Technol. **15**:578 (1981).

Comparison of Aquatic Humic Substances of Different Origin

Fritz H. Frimmel and Gudrun Abbt-Braun
Engler-Bunte-Institut
University of Karlsruhe, D-7500 Karlsruhe, FRG

Abstract

Fulvic acids isolated by the XAD-method from anaerobic and aerobic landfill leachates are compared with samples isolated from a soil extract and from bog lakes. Elemental composition, functional groups, and spectroscopic and chromatographic data suggest that there is an "ageing" effect from the leachate samples to the soil and brown-water samples.

Introduction

Humic substances (HS) are a class of biogenic, heterogeneous and refractory organic compounds. They constitute the major part of organic carbon in soils and aquatic systems [1,2]. Due to the lack of knowledge about the detailed chemical structure of humic substances, it is difficult to understand their role in the environment [3].

Promising approaches focus on the spatial and temporal variation of samples. Well defined operations are necessary to lay a sound basis for the interpretation of the results [4]. There is no question that the comparison of samples isolated from different sources is the most feasible way to study the genesis of HS.

The goal of the current work was:

1. to isolate the FA fractions from brown-water, soil extracts and landfill leachates,
2. to characterize the samples by chromatographic, spectroscopic and electrochemical methods, and
3. to compare the results and draw some conclusions about the genesis of the HS.

Material and Methods

Origin of samples

The brown-water samples were taken from a bog lake in South Bavaria (Brunnensee-moor; BM 12; Sep 87 [5]) and from a lake in the Black Forest (Hohlohsee; HO 1-2; Aug 88).

Leachates were taken from a landfill near Braunschweig, FRG (BR 5 (O); June 87; BR 6 (A); Aug 87 [6]). BR (A) stands for anaerobic and BR (O) for aerobic water.

Soil extracts were taken from a cultosole (Ah horizon of a cultivated rendzina) of the Munich Gravel Plain, FRG, with 0.1 M pyrophosphate (MUC 2-2; Mar 87; [18]).

Isolation procedure

All samples were isolated according to the XAD-2 method, described elsewhere [8] and outlined in Fig. 1.

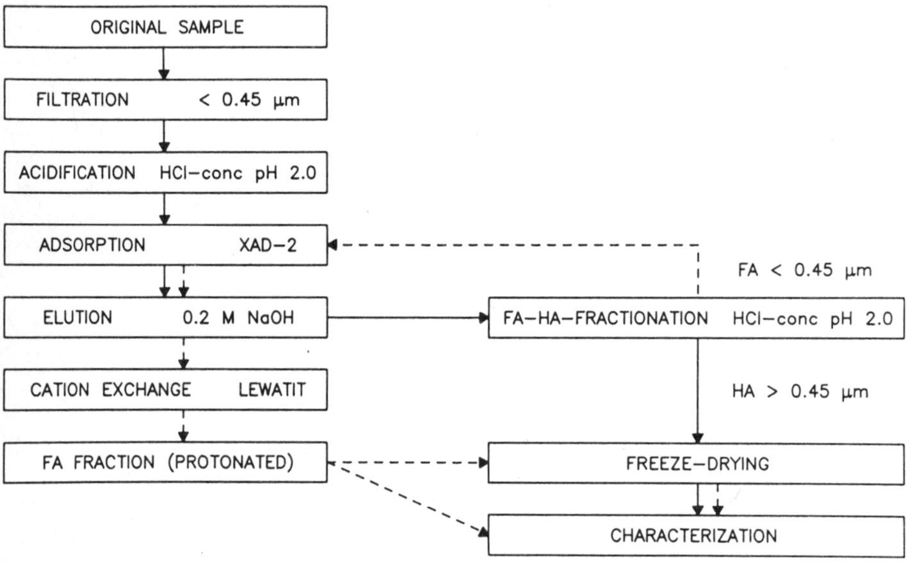

Figure 1 Isolation procedure for the fulvic acid (FA) fraction from aqueous samples.

The original water samples from the bog lakes and the landfill areas were filtered (0.45 µm) before the isolation procedure. The soil sample [18] was extracted at ambient temperature for 70 h with 0.1 M $Na_4P_2O_7$ (5 l per 2 kg soil) at pH 7.0, and the extract was treated like the other aqueous samples.

The freeze dried fulvic acid (FA) fractions were vacuum (10^{-3} mm Hg) dried at 70°C for elemental analysis and IR-measurement.

Aqueous solutions of defined mass concentration were prepared with double distilled water for all other investigations.

Analytical determinations

The determination of C, H, O, N and S was done as previously described [7]. Dissolved organic carbon (DOC) concentrations were determined in membrane (0.45 µm) filtered solutions using a UV/DOC analyzer MAI 3 (MAIHAK).

The proton capacity was determined by titration of deaerated solutions of pH from 3.0 to 11.0, using an autotitrator (DL 25; METTLER) connected to a printer plotter (FX 800; EPSON) [7].

The complexation capacity (CC) for Cu(II) was measured by differential pulse polarography (DPP) as previously described [8] and by the fluorescence quench method [9].

The molecular weight distribution was determined by gel permeation chromatography according to the method described by Fuchs [10]; a detailed protocol has been presented by Weis et al. [7]. Briefly, 1 ml of the concentrated sample (ca. 150 mg DOC/l) was applied to the TSK column and chromatographed with 0.15 M phosphate (pH 7.0) as mobile phase.

IR-spectra were measured using the KBr technique with a FT-IR spectrophotometer FTS 50 (DIGILAB).

The spectral absorbance at 254 nm and 436 nm was measured with a spectrophotometer (LAMDA 5, PERKIN ELMER and PYE UNICAM SP 8-100, PHILIPS) using 1 cm and 5 cm quartz cells. The pH of the solution was 11.0, and double distilled water was used as reference.

Results

The elemental composition of the FA samples is given in Table 1.

It is obvious that the FA fraction from the anaerobic leachate BR 6 (A) has by far the highest content of C, H and S. The C/H ratio is small and the C/O ratio is high compared to the other samples. There is not much difference in the other values, which are in good agreement with literature data [11]. It is interesting to note, that the sample from the anaerobic leachate (BR6 (A)) fits better into the group of humic acids (HA) than into the family of FAs [12].

Table 1 Elemental composition (%, dry weight) and elemental ratios of the isolated FA-fractions.

	Bog lake **(BM 12)**	Brown water **(HO 1-2)**	Leachate **(BR 5 (O))**	Leachate **(BR 6 (A))**	Soil extract **(MUC 2-2)**
C	53.2	51.6	51.9	57.9	49.4
H	3.4	3.9	4.3	5.9	3.8
N	2.2	3.2	3.7	2.3	3.9
O	40.2	37.2	36.0	30.4	40.5
S	0.9	0.9	1.8	2.3	0.9
Ash	0.1	3.2	2.3	1.2	1.5
C/H	1.3	1.1	1.0	0.8	1.1
C/O	1.8	1.8	1.9	2.5	1.6

The spectroscopic behaviour in the UV and visible ranges leads to further differentiation (Table 2). The FAs from the soil and the bog lakes absorb similarly, however, in the case of the FAs from the leachates, absorption is less intense. The anaerobic sample has by far the weakest spectral absorbance. In addition, the absorption at 254 nm for the leachate FAs is relatively strong compared to the yellow colour at 436 nm.

Table 2 Specific spectral absorbance at λ= 254 nm (A 254) and at λ= 436 nm (A 436) of dissolved FAs (l/(m x mg DOC); pH 11.0).

	Bog lake (BM 12)	Brown water (HO 1-2)	Leachate (BR 5 (0))	Leachate (BR 6 (A))	Soil extract (MUC 2-2)
A 254	6.5	5.7	3.4	1.0	5.3
A 436	0.62	0.64	0.26	0.09	0.60
A 254/A 436	10.3	8.9	12.9	11.2	8.6

Ligand functional groups contribute to the typical properties of FAs. Proton capacities and complexation capacities are well suited for comparison, even though the values are operationally defined. Table 3 shows that all FA samples have a total proton capacity (Σ) of around 15 µmol/mg DOC, with acidic groups (pH $\leq$7) contributing about 11 µmol/mg DOC and less acidic ones (pH > 7) contributing about 4 µmol/mg DOC.

Table 3 Proton capacities (µmol/mg DOC) of isolated FAs.

	BM 12	HO 1-2	BR 5 (0)	BR 6 (A)	MUC 2-2
≤ pH7	11.2	9.8	10.7	11.5	12.4
> pH7	5.8	2.0	3.0	3.5	4.2
Σ	17.0	11.8	13.7	15.0	16.6
≤ / > pH7	1.9	4.9	3.6	3.3	3.0

The anaerobic sample, in particular, has a very small complexation capacity (Table 4). The polarographically determined values (POL) are significantly higher than the ones measured by the fluorescence quench method (FLQ). Although this frequently observed effect [9] needs further investigation, it is attractive to assume that, in the case of polarography, adsorption effects add to the complexation reaction, whereas the fluorescent regions might not all be quenched by complexed paramagnetic metal ions.

Table 4 Cu(II)-complexation capacities (µmol/mg DOC) of isolated FAs.

	BM 12	HO 1-2	BR 5 (0)	BR 6 (A)	MUC 2-2
POL	2.7	2.3	1.8	0.8	3.3
FLQ	1.6	1.2	1.2	0.2	2.6

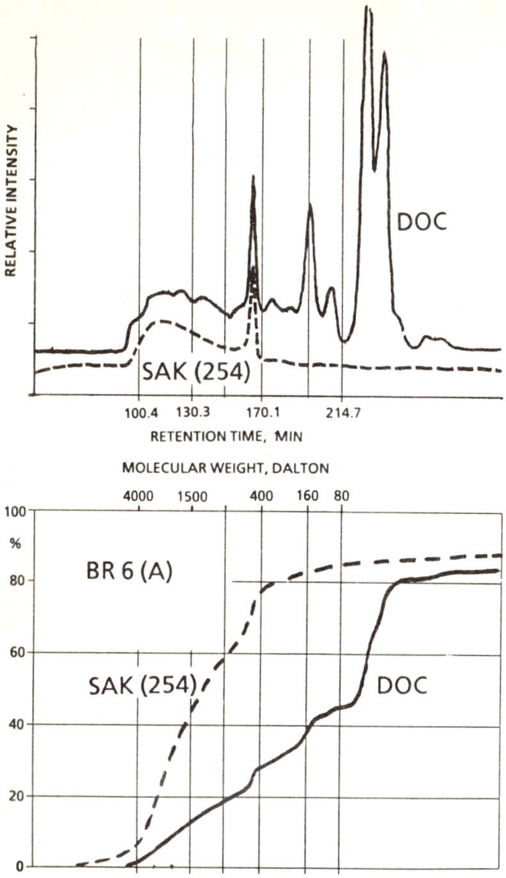

Figure 2 Gel permeation chromatogram and molecular weight distribution for BR 6 (A) FA; correction for retention time UV-DOC = 300 s.

The molecular structure of humic substances cannot be described in a non-degradative way, and it is therefore desirable to learn more about the general physical and chemical characteristics of the individual samples. Liquid chromatograms are well suited to give information on polarity, molecular weight distribution, etc. [13], even though there can be serious drawbacks [14]. Fig. 2-5 show the chromatograms gained by gel permeation. The chromatograms were run with simultaneous detection of the DOC and the spectral absorbance at 254 nm (SAK 254). The areas under both chromatograms were integrated to give the molecular weight distribution, shown below the chromatograms.

It has to be kept in mind, that the molecular weight numbers are based on the calibration with various compounds listed in the experimental section. The lack of identity with FAs is unsatisfactory, therefore the molecular weight distribution can only be seen as an estimate and is referred to as an apparent molecular weight distribution.

The sample from the anaerobic leachate (Fig. 2) shows a relatively small amount of DOC in the higher molecular weight range. The difference between the DOC line and the spectroscopic line is striking, especially in comparison to the other samples. The aerobic

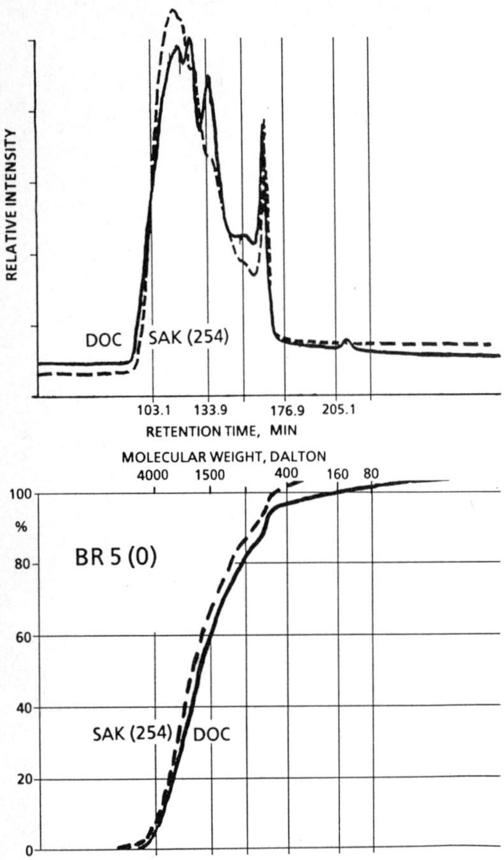

Figure 3 Gel permeation chromatogram and molecular weight distribution for BR 5 (O) FA; correction for retention time UV-DOC = 300 s.

leachate sample (Fig. 3) has more than 50% of the matter in the range of a few thousand Dalton. Characteristic peaks occur around 1500 and around 600 Dalton, although the peak at 1500 Dalton has a relatively low UV absorbance. The FA sample from the soil (Fig. 4) contains the highest molecular weight substances, and nearly 80% of the sample has a nominal weight higher than 1500 Dalton. A smaller fraction appears at about 400 Dalton. The chromatogram and distribution for the FAs from the brown water (Fig. 5) are similar to the chromatogram for the FAs from the bog lake. They are least structured and closest to a normal distribution with a mean between 2500 and 1500 Dalton. The similarity to the main fractions of the sample from the aerobic seepage water is obvious. Most of the soil FAs show higher and most of the anaerobic leachate FAs show lower molecular weights. The high molecular weight fraction of BR 6 (A) behaves chromatographically similar to the brown water FAs, indicating that a general maturing of aquatic FAs can be derived.

A survey of the dominant bonds and functional groups of FAs can be obtained with infrared spectra (IR). Assignments for the absorption bands of humic substances can be found in the literature [1,15,16]. Absorption bands of defined substances can be compa-

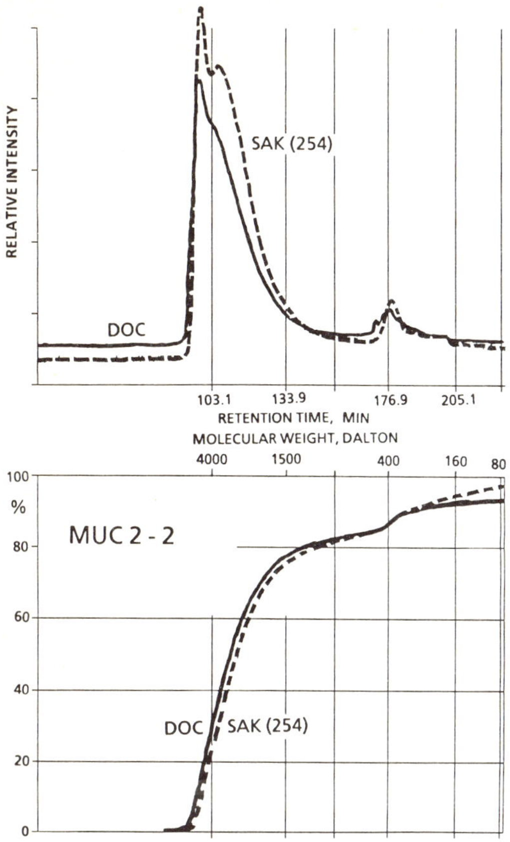

Figure 4 Gel permeation chromatogram and molecular weight distribution for MUC 2-2 FA; correction for retention time UV-DOC = 300 s.

red with the data collected by Bellamy [17]. Especially the Fourier transformed (FT) data reveal typical bands for the protonated samples (Fig. 6). The broad band around 3400 cm^{-1} and 3200 cm^{-1} are due to O-H stretching bonds of phenolic structures or alcohols. There is strong evidence for intermolecular and intramolecular hydrogen bonds. The band at 3530 cm^{-1} of the soil derived sample can be explained by single bridged O-H stretching bonds. The relatively sharp bands at 2967 and 2940 cm^{-1} are due to aliphatic C-H stretching bands. They are most obvious in the anaerobic leachate FAs and fit well to low C/H ratio. The shoulder at 2600 cm^{-1} can be assigned to the hydrogen bonded O-H stretching vibration of carboxylic acid groups. The dominant band at 1720 cm^{-1} is very likely caused by the C=O stretch vibration of acids, aldehyds and ketones. Typical for the brown water FAs is the second band at 1640 cm^{-1}, which is due to C=C stretching vibration of unsaturated and aromatic structures. Conjugation of C=O with C=C can also contribute to the absorption. The strong bands in the HO 1-2 sample and in the BM 12 sample (not shown) parallel the high spectral absorbance at 254 nm (A 254, Table 2). It is also tempting to assume that the IR band is connected with keto-enol-tautomerism, which leads to stable metal complexes by chelation. A high complexation capacity for Cu(II) (Table 4) favours this hypothesis. The O-H bending vibration of alcohols and carboxylic acids at 1440 cm^{-1} is also strongest in the brown water FAs, whereas the

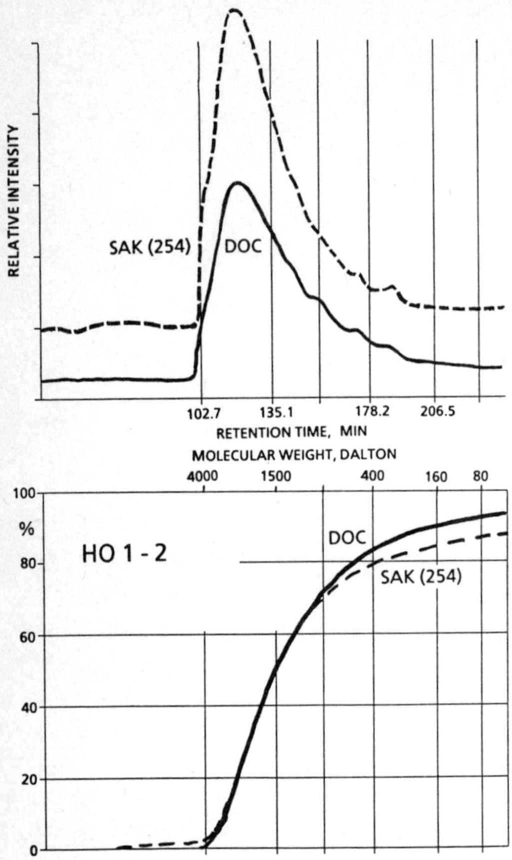

Figure 5 Gel permeation chromatogram and molecular weight distribution for HO 1-2 FA; correction for retention time UV-DOC = 300 s.

combination of the O-H deformation band and the C-O stretching band of ethers, esters and phenols at 1215 cm^{-1} is dominant in all other samples. Whether the relative intensity of the band around 1440 cm^{-1} and 1215 cm^{-1} are typical for the genesis of FAs or not must be further investigated.

Conclusions

Meaningful characterization of humic substances must include a clean definition of the origin and clearly described isolation procedures. On that basis samples from different origins can be compared with one another. A multiple method approach is suited to characterize the samples.

The results for FAs isolated from anaerobic and aerobic landfill leachates, from a soil extract and from bog lakes, lead to the following conclusions:

- The sample from the anaerobic leachate differs most from all other samples. The C/H ratio is low and the C/O ratio is high.

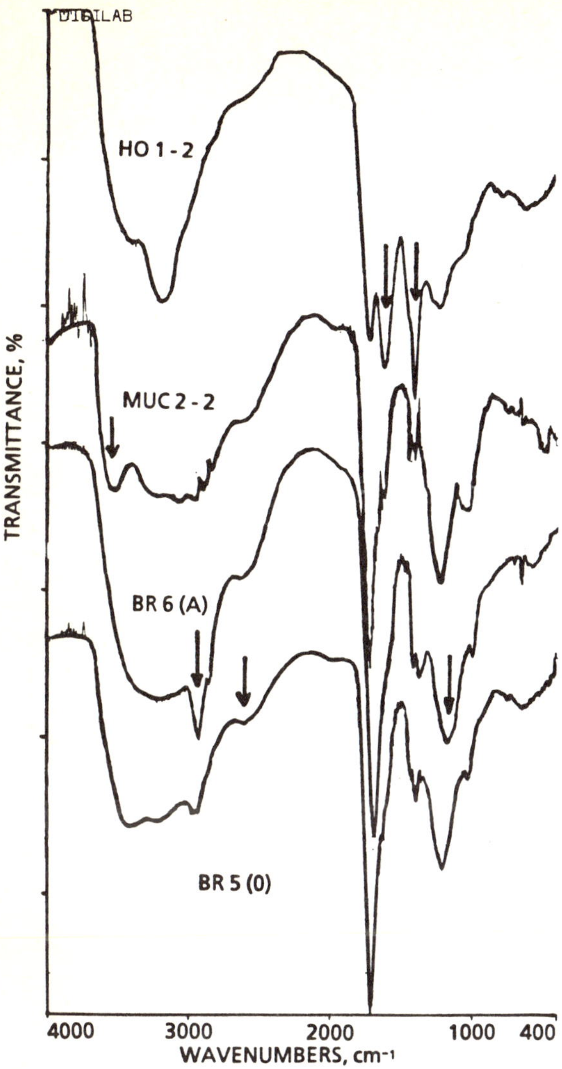

Figure 6 FT-IR spectra for the isolated FAs (KBr disk; resolution 2.0 cm^{-1}; 64 (MUC) and 256 (all others) scans).

- A relatively weak spectral absorbance and the low amount of functional groups suggest that the anaerobic FAs are fairly young.

- Ageing leads to products more like the aerobic FAs or the ones isolated from soil and brown-water.

- A large portion of the substances have an apparent molecular weight of about 2000 Dalton, and there is a larger amount of lower molecular weight material in case of the anaerobic FAs.

- The dominance of the band at 1640 cm^{-1} in the IR spectra seems to be typical for FAs from aqueous systems stabilised by ageing.

Acknowledgements

The leachate FAs were supplied by M. Weis. We gratefully acknowledge the fine experimental work of H. Niedermann and the analytical determinations of E. Kordik. The work was financially supported by Deutsche Forschungsgemeinschaft, Bonn-Bad Godesberg (grant FR 536/6 and FR 536/9).

References

1. Stevenson, F.J. Humus Chemistry (John Wiley & Sons, New York, 1982).
2. Christman, R.F. and E.T. Gjessing Eds, Aquatic and Terrestrial Humic Materials (Ann Arbor: Ann Arbor Science, 1983).
3. Frimmel, F.H. and R.F. Christman Eds, Humic Substances and their Role in the Environment (Chichester: John Wiley & Sons, 1988).
4. Aiken, G. R. a.o. Eds, Humic Substances in Soil, Sediment, and Water (New York: John Wiley & Sons, 1985).
5. Frimmel, F.H. and H. Niedermann. Z. Wasser Abwasser Forsch. **13**:119 (1980).
6. Frimmel, F.H. and M. Weis. Vom Wasser **71**:255 (1988).
7. Weis, M., G. Abbt-Braun and F.H. Frimmel. Sci. Tot. Environ. **81/82**:343 (1989).
8. Frimmel, F.H. and J. Geywitz. Fresenius Z. Anal. Chem. **316**:582 (1983).
9. Frimmel, F.H. and W. Hopp. Fresenius Z. Anal. Chem. **325**:68 (1986).
10. Fuchs, F. Vom Wasser **64**:129 (1985); **65**:95 (1985); **66**:127 (1986).
11. Schnitzer, M. and S.U. Khan Eds, Soil Organic Matter. (New York: Elsevier, 1978).
12. Steelink, C. In [4] p. 457.
13. Swift, R.S. in [4] p. 387.
14. Hine, P.T. and D.B. Bursill. Water Res. **18**:1461 (1984).
15. Ertel, J.R. and J.I. Hedges in [2], p. 143.
16. Mac Carthy, P. and J.A. Rice in [4], p. 527.
17. Bellamy, L.J. The Infrared Spectra of Complex Molecules (London: Chapman and Hall, 1975).
18. Frimmel, F.H. in Matthess, G.a.o. Eds, Progress in Hydrogeochemistry-Organics, Carbonate Systems, Silicate Systems, Microbiology, Models (Springer, Berlin, 1990) in press.

Structural Features of Aquatic Fulvic Acids by Analytical and Preparative HPLC Followed by Spectroscopic Characterization

Farida Y. Saleh, Wenching A. Ong, Inyoung Kim, and Qasem Haj-Mahmoud
Institute of Applied Sciences and Department of Chemistry,
University of North Texas, Denton, Texas 76203, USA

Abstract

Analytical and preparative reversed-phase liquid chromatography were used to separate constituents of Suwannee River aquatic fulvic acid. Photodiode array UV-Vis, and fluorescence detectors were used to monitor the HPLC separation. Preparative FA fractions were subjected to ^{1}H and ^{13}C solid state NMR, FT-IR, and ESR. The preparative and analytical RP-HPLC separations have shown that the FA macromolecule consists of at least two polymeric fractions, each of which can further be separated into individual constituents by analytical RP-HPLC. Vanillic acid structures were identified in the analytical chromatograms of the total FA sample, but not in the fractions. The first fraction included the hydrophilic constituents of FA and represented about 40 percent of the total sample. It could be resolved into six constituents by analytical RP-HPLC. Their t_R and UV-Vis scans were characteristic of carboxylic acids in rigid structures which exhibit strong hydrogen bonding. The second fraction included the hydrophobic constituents and represented about 30 percent of the total sample. It could be resolved into twelve constituents. Repeated structurally related units were separated and their UV-Vis scans were characteristic of conjugated ketone and phenolic structures. Free radical structures were present in both hydrophilic and hydrophobic fractions.

Introduction

The structures of many naturally occurring compounds have recently been elucidated through the combined use of preparative HPLC followed by NMR, FT-IR or mass spectrometry [1-4]. For example, carotenoid pigments have been sucessfully separated by HPLC. The ^{1}H and ^{13}C spectra of as little as 2 mg of a pure compound were interpretable [5]. In the case of FA, the complexity and heterogeneity of the macromolecules, even after elaborate isolation and purification, continue to limit the abilities to develop detailed structural information. Several recent ^{1}H- and ^{13}C-NMR and FT-IR studies [6-9] have provided useful information but so far, no real breakthrough has been developed.

Recent ^{13}C NMR studies on FA and humic acids were centered around the aromaticity and the ratio of aliphatic to aromatic carbons. New techniques such as ^{13}C - ^{1}H

dipolar dephasing [10], spin-echo and broad band decoupled ^{13}C CP/MAS [11] were utilized in humic substances research. A recent spin counting experiment [12], indicated that 97% of the carbons in Suwannee River fulvic acid, (SR-FA) can be detected by ^{13}C CP/MAS. Hydroxyl and aromatic ketone groups in FA and HA were examined with ^{13}C NMR for chemically derivatized samples [13,14]. With CH_3I/NaI permethylation, the ratio of carboxylic acid hydroxyl to total hydroxyl content could be determined. Products of side reactions by derivatization were identified. It was estimated that about one ketone group per monocyclic aromatic ring exists in both humic and fulvic acids. The pH effect on the dissociation of carboxyl and phenolic groups of HA were observed by ^{13}C-NMR spectroscopy [15].

It seems that the inability to develop detailed structural information on FA can be attributed to two factors. One is the inability to unfold or fractionate the constituents of FA into small molecular constituents. The other is the lack of detection of orderly repeated structural units in the FA macrostructure. Until now the polymeric nature of FA has not been unequivocally verified [6].

This paper presents the results of an investigation of the structural features of aquatic fulvic acids and HPLC fractions by solid state ^{1}H and ^{13}C NMR, FT-IR and ESR. Detailed information on the analytical and preparative separations have recently been published [16, 17]. The overall approach in this research was to utilize purified FA as starting material, to carefully evaluate the preparative and analytical separations, and to utilize combined techniques to develop structural information.

Materials and Methods

Samples

Total samples included reference and standard Suwannee River fulvic acids, purchased from the International Humic Substances Society [18]. Preparative HPLC fractions included the hydrophilic (A-1) and the hydrophobic (B-1) fractions, separated by preparative RP-HPLC as described [16]. Table 1 shows a summary of the mobile phases, columns and gradient programs.

Instrumentation

A Waters model 201 HPLC, equipped with a fixed UV detector (Beckman Model 160) and a fluorescence detector (Schoeffel Model 970), was used in the first preparative experiment. A Hewlett Packard HPLC model 1090 with a model DR-5 pump and a UV-Vis photodiode array Detector (DAD) was used in all other experiments.

NMR experiments included ^{13}C cross polarization magic angle spinning (CP/MAS) and ^{1}H combined rotation and multiple pulse spectroscopy (CRAMPS) on total reference fulvic acids FA-T and on the two HPLC factions. ^{13}C NMR spectra were obtained with a Nicolet NT-150 spectrometer at 37.7 MHz using a home built CP/MAS modification including the probe. The cross polarization contact time was 1 ms and the pulse repetition time was 1 s. The ^{1}H irradiation field was 11G and 1K data points were zero filled to 2K points in the spectra. Chemical shifts were measured with respect to tetramethylsilane via hexamethylbenzene as a secondary substitution reference (aromatic peak at 132.3 ppm).

Table 1 Continuous gradient programs and types of columns and samples used.

Gradient Program	Mobile Phase	pH at 25°C	Gradient	%A	%B	Column Types	Samples
I	A: H_2O + 0.01% AcH B: CH_3CN	4.0	t_0 min t_2 min t_{16} min t_{20} min	99 70 15 15	1 30 85 85	-Novapak C 18 5 µm 100 mm L x 3.9 mm ID Flow 0.5 ml/min -Hypersil ODS C18 200 mm x 2.1 mm ID Flow 0.3 ml/min	SR-FA, HPLC fractions, model compounds SR-FA, HPLC fractions
II	A: H_2O (He Purged) B: CH_3 CN	7.0	t0 min t_2 min t_{16} min t_{20} min	99 70 15 15	1 30 85 85	-Novapak C18 5 µm 100 mm L x 3.9 m ID Flow 0.5 ml/min -Hypersil ODS C18 200 mm x 2.1 mm ID Flow 0.3 ml/min	SR-FA, HPLC Fractions, Model Compounds SR-FA, HPLC Fractions
III	A: H_2O (He Purged) B: CH_3OH	7	t_0 min t_{30} min t_{35} min t_{45} min t_{50} min	99 99 15 15 15	1 1 85 85 85	-Custom made Novapak C18 4 µm 300 mm L x 7.8 mm ID Flow 1.5 ml/min -Novapak C18 5 µm 150 mm L x 4 mm ID	SR-FA Second Preparative Experiment SR-FA, HPLC Fractions

Usually 13000-50000 scans were accumulated. Bullet-shaped spinners were used with a sample volume of 0.4 ml and were spun at about 3.8 kHz. The ^1H CRAMPS spectra were recorded on a modified NT-200 at a proton Lermor frequency of 187 MHz using the BR-24 sequence. The Br-24 cycle time was 108 ms corresponding to $\tau = 3.0$ µs. The RF field strength, V 1H was 52.5 G. Chemical shifts are reported relative to the proton resonance of TMS and are accurate to $\pm$ 0.3 ppm. The FT-IR instrument was a Nicolet Model 60 SAB equipped with Spectra-Tech DRIFTS attachment. The operating conditions were: 265 scans, KCl sample background, resolution 1 cm^{-1}, DGTS detector. The ESR spectrometer was a Varian model 4502. ESR spectra were recorded on the solid samples in a quartz tube of 14.6 cm (L) x 4 mm (ID) at room temperature (24 $\pm$ 1°C). The measurements were made under the following conditions: Frequency, 9.5 GHz; modulation frequency, 100 kilocycle (Kc); magnetic field setting, 3,350 to 3,370 gauss; standard for solid state, 2,2-diphenyl-1-picrylhydrazyl (DPPH) diluted with KC1; and standard for liquid phase, DPPH in benzene.

Table 2 Preparative experiments 1 and 2. Percent recoveries of fractions.

	Initial total wgt in mg	Wgt of A-1 fraction mg	Wgt of B-1 fraction mg	% Total recovery A + B	% of A-1 recovery from total FA	% of B-1 recovery from total FA
Preparative Experiment 1						
Combined batches (1-5)	50	19.0	11.6	61.2	38.0	23.2
Combined batches (6-8)	30	11.20	9.8	70	37.3	32.7
Total from experiment 1	80	30.20	21.4	64.5	37.8	26.8
Preparative Experiment 2						
Combined batches (1-2)	20	7.91	6.51	72.1	39.6	32.6
Combined batches (3-4)	20	9.29	5.98	76.4	46	29.9
Total from experiment 2	40	17.20	12.49	74.2	43.0	31.2
Mean recoveries experiments 1 and 2	-	-	-	69.4	40.4	29

Results and Discussions

Preparative and Analytical HPLC

The total recoveries of FA and the contribution of each fraction are shown in Table 2.

Gradient Program I - Novapak Column

Fig. 1 shows the analytical chromatogram and the UV-Vis scans of a total SR-FA sample dissolved in a pH 7 buffer. Six peaks are resolved. The scans of the first two peaks are featureless and showed notable decrease of absorbance with increase of wavelength. Peak 3 is well defined and symmetrical and its scan is featureless and closely resembles some of the published UV-Vis spectra of aquatic FA [19]. Peak no. 4 is a stronger and broader one. However its spectrum is very similar to peak 3. This is rather interesting in view of the ideal peak shape and symmetry of peak no. 3. While both peaks, no. 5 and 6, differ by one minute in t_R, the UV-Vis scans are almost the same. Both spectra show the distinct absorption lines at 260 and 290 nm, characteristic of vanillic acid structures. The t_R and uv scans of standard vanillic acid are similar to those of peak no. 5. This may be one of the first detections of vanillic acid structural units in FA, via HPLC analysis. Additional weak UV and visible absorption lines were detected in the scans indicating that the VA structures are not present in the pure acid form. The analogy between the UV-Vis scans of vanillic acid (VA) and those detected in aqueous solutions of FA leaves little doubt regarding the presence of vanillic acid structural units in the FA macro-molecule. As discussed below, it is interesting to note that the vanillic acid structures are not detected in the chromatograms of either fraction A-1 or B-1. A possible explanation is that vanillic acids are only intermediates formed by the slow degradation from lignin precursors. Once the macromolecule is fractured during the HPLC seperation, VA may undergo further degradation to simpler carboxylic or phenolic compounds.

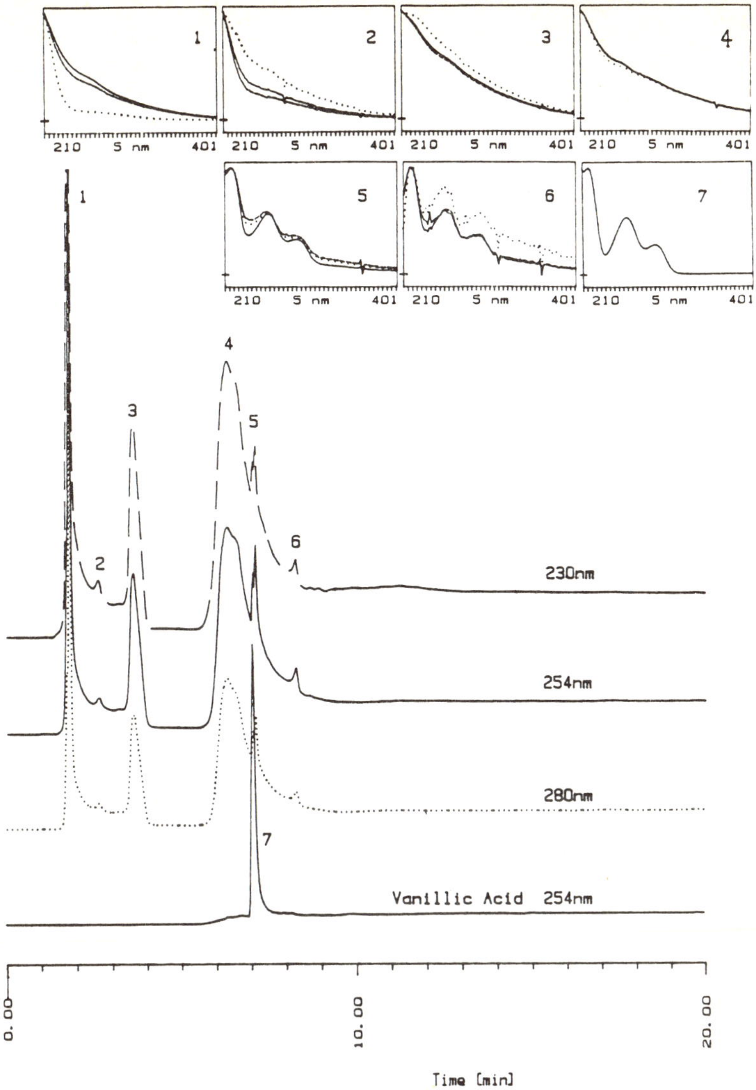

Figure 1 Chromatogram and UV-Vis scans of 50 µl 0.02 % SR-FA and 50 µl 0.02 % vanillic acid in pH 7 buffer. Gradient program I.

Gradient Program I - Hypersil C_{18} Column

Fig. 2 shows the chromatogram and UV-Vis spectra of 50 µl of an aqueous 0.02% SR-FA solution and the hydrophilic fraction A-1. The UV-Vis scans of the peaks 1-3 are all featureless and similar to the scans of the early eluting peaks with the Novapak column. Peaks 4 and 5 are also featureless, but show some absorption lines in the visible region indicating the presence of chromophores and auxochromes. This scan represents the predominant polymeric structure in FA.

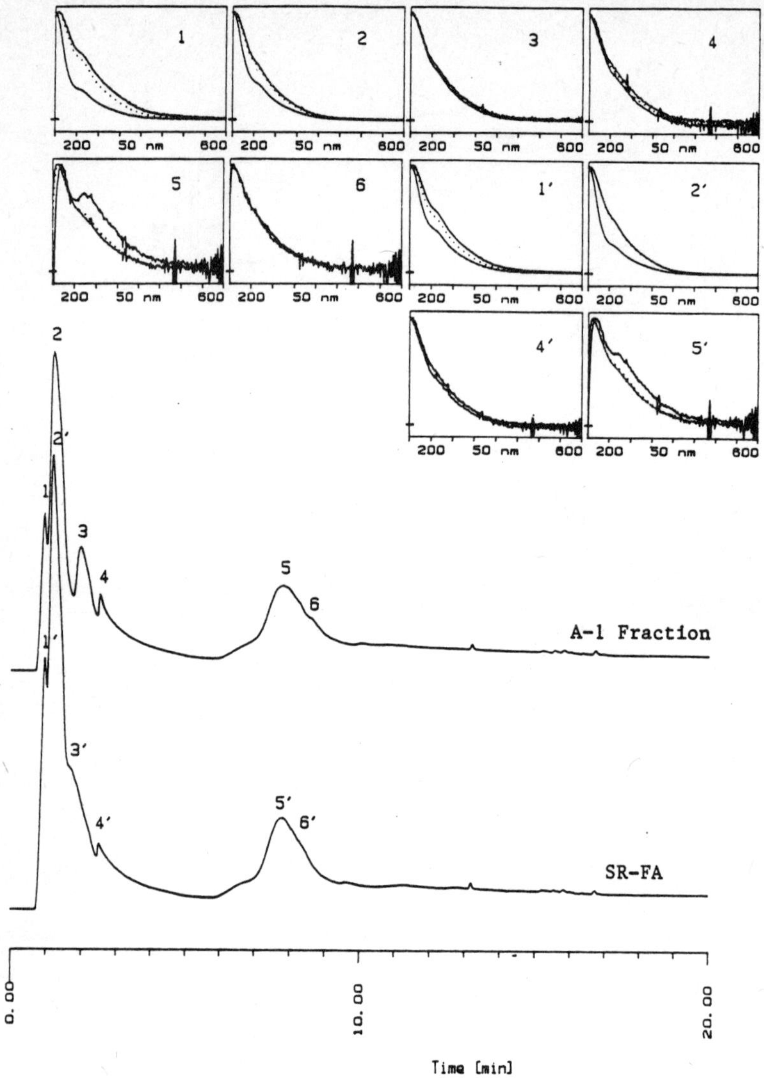

Figure 2 Chromatogram and UV-Vis scans of 50 µl 0.02 % SR-FA and 50 µl 0.02 % A-1 fraction dissolved in MilliQ water. Gradient program II. Hypersil ODS column.

Fig. 3 shows the chromatogram and scans of the hydrophobic fraction B-1. The chromatogram was monitored at $\lambda = 230$ nm. The scans of the early eluting peaks in the first two minutes are similar to those present in the total sample and hydrophilic fractions. Peaks 2-12 were eluted between 6 and 18 minutes with only peak 2 indicating a broad signal with a scan characteristic of the predominant colored fragment of FA. Peaks 3, 5, 6, 7, and 8 showed distinct absorption line at $\lambda = 226$ nm and a weak line at 265 nm. The similarities and intensities of these peaks suggest the presence of repeated structural units in the FA macromolecule. It should be remembered that these peaks occurred between t_R

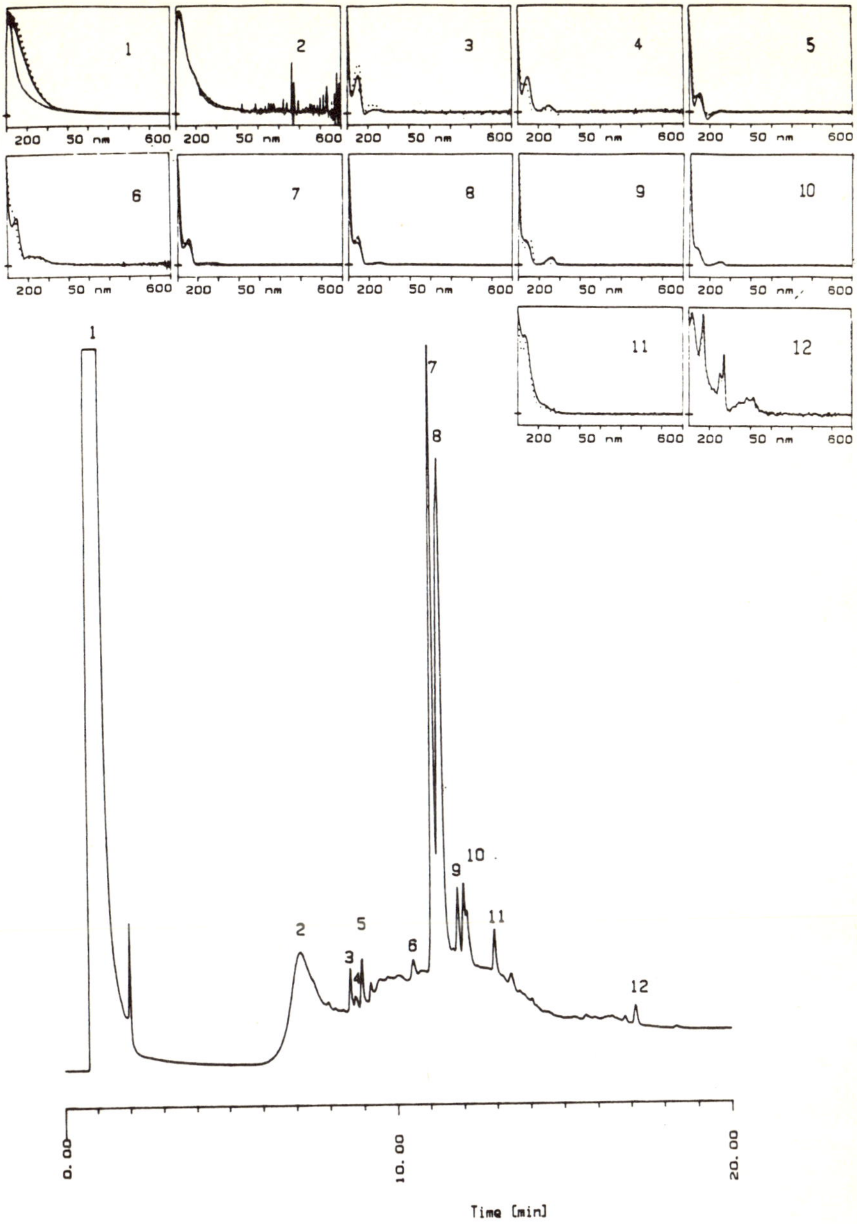

Figure 3 Chromatogram and UV-Vis scans of 50 μl 0.02 % aqueous hydrophobic fraction B-1 dissolved in methanol. Preparative experiment 2. Gradient program I. Hypersil ODS Column.

8.62 and 11.42 minutes i.e. within 2.8 minutes in the hydrophobic region. It is suspected that these peaks correspond to conjugated ketone structures. Scans of peaks 4, 9 and 10 are all similar and contain a defined additional absorption line at 275 nm. Such UV spectra are characteristic of phenolic compounds. In peak 11 the 275 nm line is barely detectable but a well defined line at 221 nm is noticeable. Peak 12 occurred at 17.2 minutes and t_R show several absorption lines at 209, 235, 260, 275, and 285 nm. These lines are characteristic to polynuclear aromatic structures. The chromatogram of Fraction B-1 on the Hypersil column represents one of the most successful chromatographic separations of FA. Their UV-Vis scans provide the first evidences of the presence of structurally similar units in FA. The effect of pH on the B-1 chromatogram was evaluated by injecting the same sample using mobile phase II at pH 7. The chromatogram showed essentially the same resolution and UV-Vis scans. The effect of storage of the B-1 solution was evaluated by injecting the same sample solution under the same conditions, after storage at 4°C for 5 months. The chromatogram showed essentially the same resolution and UV-Vis scans. The only detecable difference was the increase of the magnitude of the phenolic peaks and the decrease in the magnitude of the conjugated ketone peaks.

Solid State Proton CRAMPS NMR

Fig. 4 shows a comparative ^{1}H CRAMPS Spectrum of total FA and the two HPLC fractions. Notable improvement is shown in the spectrum of the B-1 fraction. The spectra of Total FA and A-1 fraction show broad overlapping lines, representing resonances at $\delta = 0.5, 1.8, 3.00, 7.00$ and 10.5 ppm. Protons can be assigned to the following regions: i) the alkane protons between 1-2 ppm, ii) the hydroxyl protons between 3-4 ppm; iii) the aromatic protons between 7-9 ppm and iv) the acidic protons between 10-12 ppm. Detection of the acidic protons in a 24 h air dried sample verifies the assignment of acidic protons. It is interesting to note that the data on aromaticity of FA by ^{13}C and ^{1}H NMR represented 14.1 and 13.5%, respectively. These results are in good agreement with published data on SR-FA.

The ^{1}H CRAMPS spectrum of B-1 shows well resolved resonance lines corresponding to the alkyl, O-alkyl and aromatic protons. The broad signal centered at about $\delta = 2.5$ ppm corresponds to labile protons. Most of the intensity in the FA-B spectrum is in the aliphatic region and is remarkably well resolved. The most upfield line is at $\delta = 0.25$ ppm. The line width in Fig. 4 ranges from 2-4 ppm which is considered excellent for a solid polymeric materials such as FA. Considering the hydrophobic nature of the B-1 fraction, the proton distribution reflects an essentially aliphatic structure where at least 50 percent of the protons are labile and correspond to orderly arranged structures. Combined with evidence from the UV-Vis scans and the FT-IR, conjugated ketones are the major constituents of this fraction. The percentage of aromatic protons are 5.6, which is consistent with the information derived from the HPLC results.

Solid State CP/MAS - ^{13}C NMR

Fig. 5 shows the ^{13}C spectra of total SR-FA and the hydrophobic fraction B-1. The major ^{13}C resonance lines in the total sample are very comparable to those reported in the literature on SR-FA. It is interesting to note that our calculated percentages of aromatic and carboxylic carbons are 14 and 18 percent, respectively. These results are in good agreement with the data derived from the ^{1}H-CRAMPS as well as the published data on SR-FA [13].

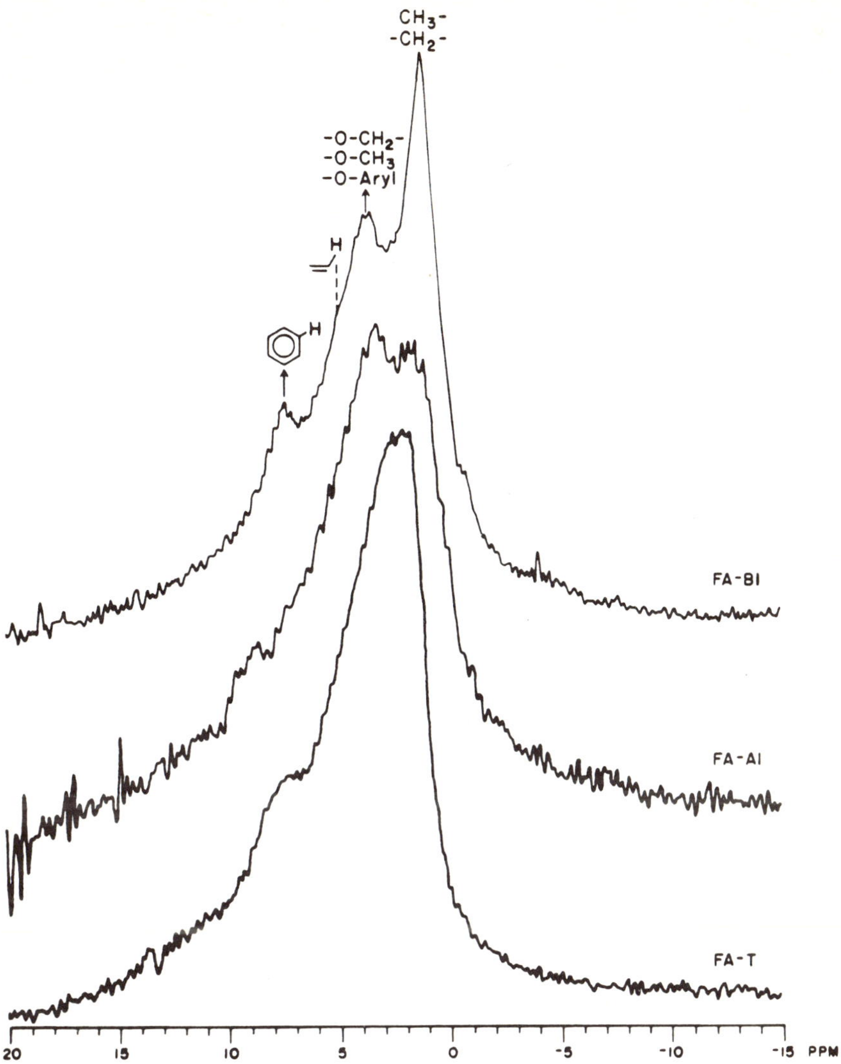

Figure 4 Solid state ^{1}H CRAMPS of fulvic acids and two HPLC fractions.

The CP/MAS ^{13}C spectrum of the hydrophilic fraction A-1 did not show any measurable signal. This fraction weighed ~ 30 mg and was expected to produce better signals than the B-1 fraction. The inability to obtain a good ^{13}C spectrum on this sample may be due to the presence of trace amounts of metallic paramagnetic contaminants. As discussed in the ESR section, this sample shows several ESR absorption lines with coupling constants in the range of 60-70 gauss.

The hydrophobic fraction B-1 weighed 21 mg and it was suspected that a useful ^{13}C spectrum could be obtained from such as small amount. Indeed it was difficult to obtain a spectrum of this sample. As shown in Fig. 5, the ^{13}C CP/MAS of the FA-B-1 is very

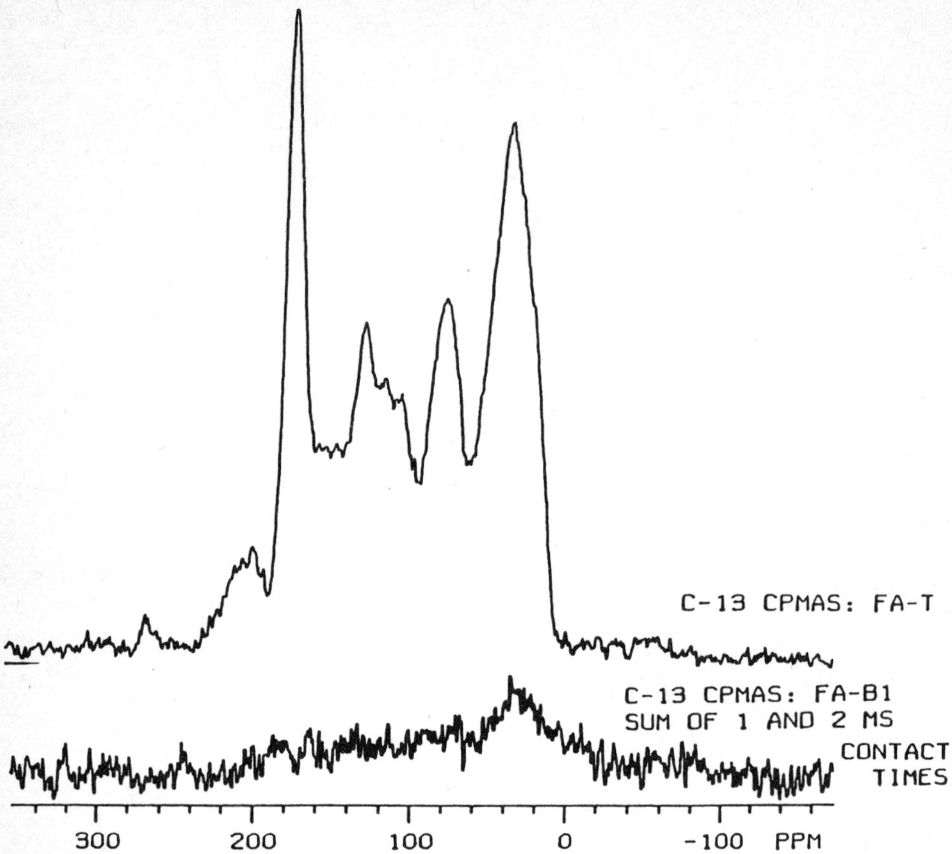

Figure 5 Solid state ^{13}C CP/MAS spectrum of the FA-T and B-1 fraction.

weak. The only notable feature of the specturm is the predominance of aliphatic structures.

Fourier Transform IR

Fig. 6 shows the FT-IR spectra of total SR-FA and fractions A-1 and B-1. Four IR absorption regions are discussed.

Absorption in the 3400 cm^{-1} region is due to the OH stretching and its broadness is usually attributed to hydrogen bonding. The feature in this region is broad and extends between 3300-3700 cm^{-1} indicating the presence of OH groups vibrating over a wide range of energies. This range may cover various types of OH stretching. For example, aliphatic alcohol absorb at 3640-3600 cm^{-1}, phenolic OH at 3612-3593 cm^{-1}, peroxide OH at 3550 cm^{-1}, polymeric hydrogen bonded OH at 3400-3200 cm^{-1} and chelated OH at 3200-2500 cm^{-1}. The flat broad band at 2600 cm^{-1} can be attributed to OH stretching vibrations of the carboxyl groups.

Examination of Fig. 6 shows that the total FA and the hydrophilic fraction, have similar bands in this region. This is expected since both samples contain substantial OH groups from carboxylic acids and phenols. In the hydrophobic fraction B-1 this region is significantly reduced, and extends over a narrower range between 3600-3100 cm^{-1}. It seems better resolved from the next region which extends between 3000-2800 cm^{-1}. Since this fraction contains only hydrophobic constituents, polymeric hydrogen bonded OH may be the major contributor to the absorption in this region.

Absorption in the 3000-2750 cm^{-1} region is caused by aliphatic CH stretch and is expected to be enhanced by methylation resulting from introduction of CH_3 group. The aromatic stretching mode occurs at 3055 cm^{-1} and the aliphatic CH stretch frequencies at 2970 and 2870 cm^{-1} (CH_3) and 2920 and 2850 (CH_2).

Fig. 6 shows that the total sample has two poorly resolved bands in this region at 2960 and 2920 cm^{-1}, which can be assigned to the symmetric and antisymmetric stretching vibrations of aliphatic CH bands of CH_3 or CH_2 groups. Methylation would increase absorption in this region. In the spectrum of the hydrophilic fraction, this region is slightly more resolved than in the total sample. In the spectrum of the hydrophobic fraction, this regions is well enhanced and is resolved into four distinct but overlapping bands at 3010, 2950, 2880 and 2850 cm^{-1}. The enhancement of this region is due to marked decrease in the OH stretching region and the mild methylation due to the use of 85% CH_3OH as a mobile phase, during the HPLC separation. It is interesting to note the appearance of a shoulder at 3010 cm^{-1} indicating some aromatic C-H stretch that was masked in the total sample and hydrophilic fraction. A distinct but small shoulder appears at 2720 cm^{-1} which may indicate C-H of an aldehyde.

Within the 2700-1950 cm^{-1}region the broad band at 2600 cm^{-1} in the total sample can be attributed to the broad hydrogen bonded COOH. The band is less pronounced in the hydrophilic fraction. The total sample shows a minor shoulder at 2380 cm^{-1}. The hydrophilic fraction shows a well defined band at 2350 cm^{-1}. The hydrophobic fraction has two absorption bands in this region, one is centered at 2550 cm^{-1} and the other is centered at 2120 cm^{-1}. Absorption in this region corresponds to triple bond vibrations.

The major part of absorbances in the 1720-1450 cm^{-1} region is caused by C=O stretching of COOH group which absorb at 1725 cm^{-1}. This band often shifts to a slightly higher frequency upon methylation. This is attributed to the fact that C=O of esters generally absorb at higher frequency than the C=O of the corresponding acid. The COO^- anion has two bands at 1550 and 1420 cm^{-1}. Ketones and aldehydes also absorb at 1720 cm^{-1}. Quinone C=O absorbs at 1650 cm^{-1}, C=C of aromatic ring at 1600 cm^{-1} and 1510 cm^{-1} (ring breathing bands) and aryl esters absorb at 1720 cm^{-1}.

The total FA sample show a strong band at 1750 cm^{-1} which can be attributed to the C=O of COOH, ketones and aldehydes. Absorption bands at 1550 and 1420 cm^{-1} corresponding to the COO^- are not likely to be strong in the total sample since the spectrum was recorded on the solid FA which is known to contain very little ash or mineral constituents. A distinct shoulder is apparent between 1620-1640 cm^{-1} and this has usually been attributed to C=C vibrations of aromatic structures. There are some doubts regarding this assignment. First the 1510 cm^{-1} band characteristic of ring breaking

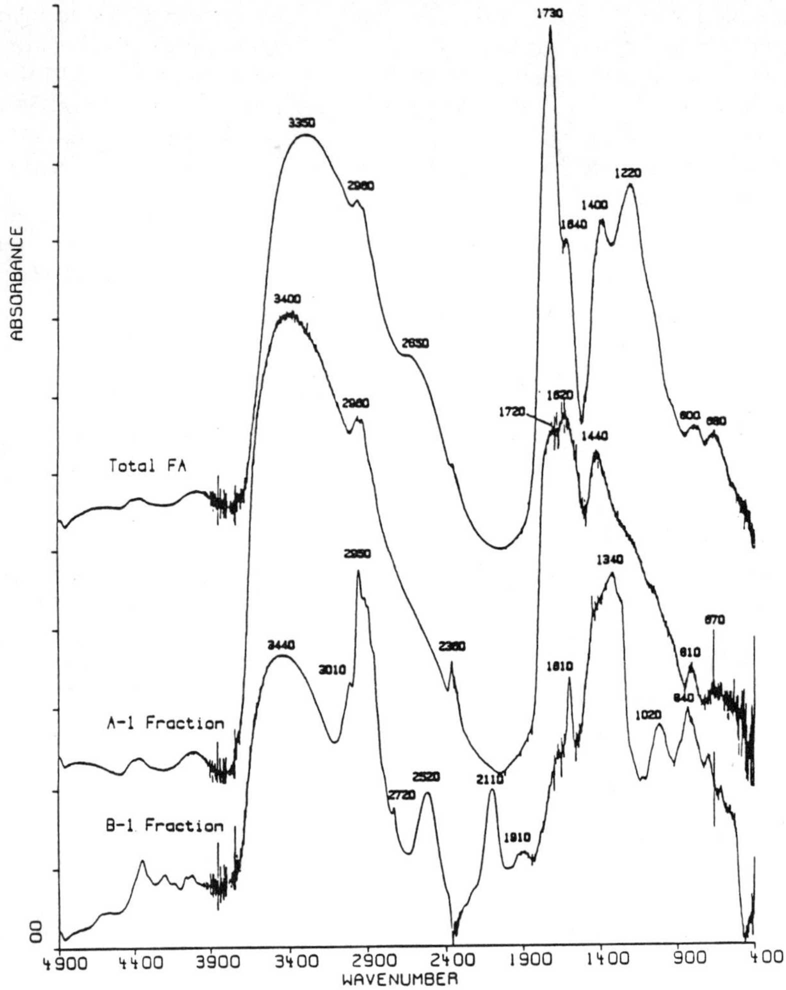

Figure 6 Solid state FT-IR spectra of solid total FA, hydrophilic fraction A-1 and hydrophobic fraction B-1.

is lacking. Second the band at 3030 cm^{-1} characteristic of CH stretch is not detected in the total sample.

The hydrophilic fraction show a weaker and broader band at 1720 cm^{-1}. Meanwhile the COO^- bands at 1560 and 1400 cm^{-1} are present. This may be attributed to the COO^- formed by interaction between the hydrophilic fraction and trace metals released during the HPLC separation.

The hydrophobic fraction shows a slight shoulder at 1720 cm^{-1} well defined sharp band at 1600 cm^{-1} and a slight shoulder at 1510 cm^{-1}. The 1600 cm^{-1} band can be assigned to α, β or α', β'- unsaturated ketones which are known to absorb in this region. There could be a minor contribution from aromatic C=C vibrations since the 1510 and 3010

cm^{-1} bands are only minor. The ^{13}C NMR, ^{1}H CRAMPS and UV -Vis scans of components of this fraction are in full agreement with this conclusion. Contribution of H-bonded conjugated C=O groups, to the band near 1610 cm^{-1} has been stressed by Theng and Posner [20].

Carbohydrate bands at 1050 cm^{-1} appear as a slight shoulder in the total sample and hydrophilic fraction. By contrast the hydrophobic fraction shows a well defined band at 1100 cm^{-1} which can be assigned to aliphatic alcohols, polysaccharides or to Si-O impurities.

ESR of Total FA and HPLC Fractions A-1 and B-1

ESR of the total sample showed a single symmetrical absorption line at g ~ 2.00 and spin content of 7.9 x 10^{17} [S] /g. The ESR spectra of the HPLC fractions A-1 and B-1 are shown in Fig. 7. The ESR spectrum of fraction A-1 shows several absorption lines with coupling constants of 71.4, 71.4, 59.6, 67.5, 65.6 and 71.4 gauss. This magnitude of coupling constants corresponds to field splitting by metallic protons. Line intensities do not follow a regular pattern which makes it difficult to identify exactly. The major absorption line correspond to the organic free radical and has a g value of ~ 2.0. These results suggest that the A-1 sample contains metallic paramagnetic ions possibly due to contamination from the HPLC system. Such contaminations are frequently reported in the literature. The inability to obtain a satisfactory ^{13}C CP/MAS spectrum for the A-1 sample, even though its weight is larger than that of the B-1 sample, is a further proof of the existence of paramagnetic contamination.

The ESR spectrum of the B-1 fraction shows a well defined absorption line of g ~ 2.00, due to organic free radicals. Additional weak absorption lines with coupling constants of 15 to 40 gauss are noted. It is not known whether these lines are caused by structures initially present in the B-1 sample or due to HPLC system contamination. It is

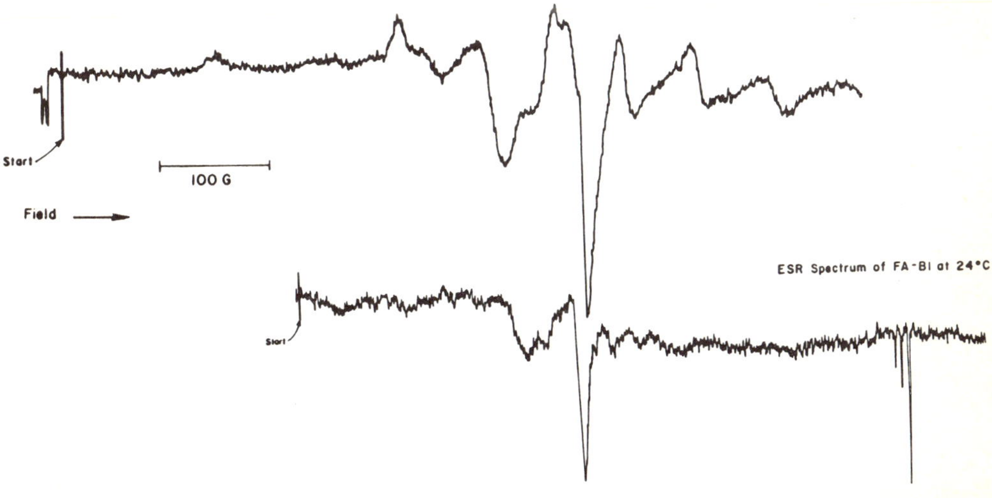

Figure 7 Solid state ESR spectra of hydrophilic fraction A-1 and hydrophobic fraction B-1 at 24 ±1°C.

remembered, however, that this fraction was collected using 85% CH_3OH as a mobile phase and the extent of metal contamination is expected to be less than in the A-1 fraction. The presence of organic free radical in this sample is also evident. Regardless of the paramagnetic contamination problem, the ESR spectra of both HPLC fractions indicated the presence of free radical structures. These results imply the presence of more than one type of organic free radicals in FA.

Summary and Conclusions

The yield from the preparative HPLC experiments of SR-FA was almost 70 percent. The hydrophilic constituents of FA represent ~ 40% of the total sample and can be resolved into six constituents each exhibiting featureless UV-Vis scans characteristic of aliphatic carboxylic acids and a few aromatic acids. The ^{1}H CRAMPS experiments confirmed the presence of acidic protons, strong hydrogen bonding and rigid structures. The presence of paramagnetic metallic contamination was indicated by the ESR and FT-IR spectra, and by the inability to obtain successful solid state ^{13}C CP/MAS spectrum. However, the ESR spectra indicated the presence of an organic free radical in this fraction.

The hydrophobic constituents of the FA represent ~ 30% of the total sample and can be resolved into 12 constituents as illustrated by the analytical chromatograms of this fraction. The UV-Vis scans of the peaks indicated the presence of repeated structurally related units. This is one of the first resolution of individual constituents of FA without chemical or thermal degradation. The UV-Vis scans of the repeated units were characteristic of conjugated aliphatic ketones and phenolic compounds. The ^{1}H CRAMPS and ^{13}C CP/MAS NMR spectra confirmed the predominant aliphatic structures in this fraction. The resolution in the ^{1}H CRAMPS spectra were comparable to those of model compounds. The reproductibility of proton distribution was excellent and indicated about 6% aromatic protons and 50% labile protons. Combined with evidence from the UV-Vis scans and the FT-IR, conjugated aliphatic ketones are likely to be significant constituents of this fraction. Contribution of H-bonded conjugated C=O groups in the FT-IR spectrum of this fraction provided further evidences to this conclusion. Free radical structures were also detectable in this fraction. Conjugated ketones and/or quinone structures are likely responsible for the presence of organic free radicals in this fraction.

The overall structure of SR-FA can be visualized as a mixed polymeric structure consisting of hydrophilic and hydrophobic subunits linked by weak interactions. The detection of several structurally related or repeated units in both FA fractions strongly suggests that FA is polymeric. Vanillic acid structures seemed to be present only in the FA macrostructure.

The forementioned picture of FA has some features that are in common with the published proposed models and formation pathways. However, it is not in full agreement with any single model. It is important to visualize the structure of FA only as a representation of a dynamic equilibrium at a given time. While the major structural units exhibit the same general properties, the contribution of each group to the macromolecule may vary. This approach would allow us to explain areas of homogeneity and heterogeneity in FA macromolecular structure.

Acknowledgement

Research was supported by the U.S. Geological Survey, Department of Interior Assistance Award No. 14-08-00001-G1146.

The views and conclusions contained in this document are those of the authors and should not be interpreted as necessarily representing the official policies, either expressed or implied of the U.S. Government.

We express our appreciation to scientists at Colorado State University Regional NMR Center, funded by the National Science Foundation, Grant No. CHE 78-18581, for running the NMR spectra.

References

1. Tanaka, Y. In: J.C. Randall, Ed., NMR and Macromolecules. ACS Series 247 pp 233-44 (Washington D.C:1984).
2. Hikichi, K., T. Kiroki, S. Takemura, M. Ohuchi and A. Nishioka, ibid, pp 119-130 (1984).
3. Hunt, D.F., J. Sharbanowitz, T.M. Harvey and M.L. Coates. J. Chromatogr. **271**:93 (1983).
4. Williams, D.H. In: P.A. Lyon, Ed.,Desorption Mass Spectrometry, ASC Symp. Ser. No. 219 pp 217-26 (Washington D.C: 1985).
5. Muller, R.K., K. Bernhard and M. Vecchi, In: Carotenoid Chemistry and Biochemistry, G. Britton and T.W. Goodwin, Eds., Pergamon Press pp 27-54 (1981).
6. MacCarthy, P. and J.A. Rice, In: G.R. Aiken, D.M. McKnight and R. Wershaw, Eds, Humic Substances in Soil, Sediment and Water. pp 528-560 (New York: John Wiley and Sons, 1985).
7. Saiz-Jimenez, C., B.L. Hawkins and G.E. Maciel. Org. Geochem. **9**:277 (1986).
8. Thorn, K.A., J. Rice, R. Wershaw and P. MacCarthy. Sci. Tot. Environ. **62**:185 (1987).
9. Norwood, D.L., R.F. Christman and P.G. Hatcher, Environ. Sci. Technol. **21**:791 (1987).
10. Gillam, A.H., M.A. Wilson and P.J. Collin. Org. Geochem, **11**:1 (1987).
11. Steelink, C. and A. Petsom, Sci. Tot. Environ. **62**:165 (1987).
12. Vassallo, A.M., M.A. Wilson, P.J. Collin, J.M. Oades and A.G. Waters. Anal. Chem. **59**:558 (1987).
13. Thorn, K.A., C. Steelink and R.L. Wershaw, Org. Geochem. **11**:123 (1987).
14. Leenheer, J.A., M.A. Wilson and R.L. Malcolm. Org. Geochem. **11**:237 (1987).
15. Preston, C.M. and M. Schnitzer. pH and Solvent Effects. J. Soil Sci. **38**:667 (1987).
16. Saleh, F.Y., W.A. Ong, I. Kim, Q.H. Mahoud and D.Y. Chang, U.S. Geological Survey, Final Report USGS Grant No. 14-08-0001-G1146, pp. 332 (March (1989). NTIS PB89-193429.
17. Saleh, F.Y., W.A. Ong and D.Y. Chang. Anal. Chem. **61**:2786 (1989).
18. Internationl Humic Sbustances Society (IHSS), Standard and Reference Humic Substances, Boulder, Colorado, USA. Personal communication 1986.
19. Stevenson, F.J. Humus Chemistry, pp. 264-267 (New York: John Wiley and Sons Publ. 1982).
20. Theng, B.K. and A.M. Posner. Soil Sci. **104**:191 (1967).

Fluorescence Spectroscopy as a Means of Distinguishing Fulvic and Humic Acids from Dissolved and Sedimentary Aquatic Sources and Terrestrial Sources

Nicola Senesi[1], Teodoro M. Miano[2] and Maria Rosaria Provenzano[1]

[1] Istituto di Chimica Agraria, Università di Bari, Bari-70126, Italy
[2] Istituto di Chimica Agraria e Forestale, Università della Basilicata, Potenza-85100, Italy

Abstract

Thirteen fulvic acids (FA) and humic acids (HA) isolated from river waters and sediment, marine sediments, leonardite, soils, and paleosol, have been investigated by fluorescence spectroscopy in the emission, excitation and, partly, synchronous scan excitation modes. Emission spectra are generally characterized by a unique broad band, whereas excitation spectra exhibit a variable number of peaks or shoulders of various intensity; these peaks are particularly well-resolved for sedimentary HA samples. A decrease in the relative intensity of fluorescence, which is associated with a red-shift (longer wavelengths) of both the emission maximum and the main excitation peaks, is observed when passing from dissolved aquatic and soil FA to river and marine sedimentary HA, to leonardite and soil HA, and, finally, to paleosol HA. Evident differences are shown in the relative intensity and wavelength maxima, measured in any mode, between soil FA and HA from the same source. For FA and HA of various nature and origin, the fluorescence is suggested to be caused by chemically different structural units. These units fluoresce from the blue-violet to the green and consist of variously extended, condensed, aromatic and/or heterocyclic ring systems, with a high degree of electronic conjugation and bearing suitable hydroxyl, alkoxyl and carbonyl groups (e.g salicyl, cinnamic and hydroxybenzoic derivatives, naphtols, naphtoquinones, coumarin), and quinoline-derivatives, flavonoids and Schiff-base derivatives. Fluorescence properties of humic substances may represent an additional diagnostic criterium useful in distinguishing between FA and HA from the same or various natural sources.

Introduction

In the last two decades fluorescence spectroscopy has been applied extensively as a useful means of general chemical characterization and, in particular, identification of certain structural and functional constituents (fluorophores) in natural, artificial and synthetic humic substances of various origin. With this method, the effects of concentration in solution and pH can also be taken into account. A comprehensive review on fluorescence of HS has been recently published [1]. However, only a few comparative studies are available, which discuss the potential of this technique to furnish spectra that can enable fulvic acid (FA) and humic acid (HA) fractions from the same source to be unambiguosly distinguished, and that can allow differentiation of FA or HA of various origin and genesis. In one of these studies, a marine sedimentary HA was found to fluoresce more intensively and showed emission and excitation fluorescence maxima at wavelengths longer than FA from the same source [2]. In another investigation, fluorescence excitation spectral shapes and maxima of FA and HA couples extracted from a large number of soils of widely differing geographic and pedologic environments, revealed apparent differences typically related to the soil nature and properties of the soil [3]. Sedimentary marine HA has been found to be characterized by excitation wavelength maxima lower than terrestrial HA, and both fluoresce less intensively than dissolved organic matter [4]. Emission fluorescence spectra corrected for Raman and Tyndall scattering have been presented for FA and HA extracted from estuarine particles. These spectra showed wavelength maxima higher than those measured for their counterparts in autochthonous marine sediments [5]. Recently, evidence has also been obtained of typical differences in fluorescence intensity and spectral shape and maxima between FA and HA and among humic substances of various sources, including soil, peat, leonardite and river water [6]. Even more recently, results of two comparative fluorescence studies have shown that terrestrial HA had the most pronounced emission and excitation maxima at the longest wavelength (green), followed by HA from lake and deep marine sediments, river and marine aquatic HA, river and marine surface sedimentary HA, and, finally, by terrestrial and aquatic FA, which showed fluorescence maxima at the shortest wavelength (blue-violet). When comparing terrestrial FA and HA, it has also been found that fluorescence spectra are characteristically well differentiated in shape, with maxima at wavelengths distinctly higher for HA than for correspondent FA. When comparing aquatic HA and FA, however, it was observed that spectra were less differentiated, with maxima centered at much closer wavelengths [7,8].

The principal objective of the present study was to furnish further evidence of the possibility of applying fluorescence spectroscopy to differentiate humic materials according to their origin and to distinguish between FA and HA from the same source. For this purpose, the fluorescence spectra of a number of FA and HA samples, isolated from various aquatic, sedimentary and terrestrial sources and previously characterized extensively for chemical and spectroscopic properties, have been comparatively analyzed and discussed.

Materials and Methods

The identification and origin of the thirteen fulvic and humic acid samples investigated in this study are furnished in Table 1. The samples were supplied by various colleagues: the river aquatic FA (R29 to R32) by Dr. R.L. Malcolm (U.S.G.S., Denver, CO, USA); the leonardite HA (M20) and the Chino soil HA (M23) by Dr. J.P. Martin (Univ. of California, Riverside, CA, USA); the gley soil HA (NG5) by Dr. K. Yonebayashi (Prefectural Univ., Shimogamo, Kyoto, Japan); and the paleosol HA (P5) by Dr. G. Calderoni (Univ. La Sapienza, Roma, Italy).

Table 1 Identification and origin of fulvic acids (FA) and humic acids (HA).

Sample Identity	Origin	Sampling Place	Sampling Depth	Additional Information
R29, FA	Missouri River, water	Sioux City, Iowa,U.S.A.	surface	3.2[a]
R30, FA	Ohio River, water	Cincinnati, Ohio,U.S.A.	surface	3.2[a]
R31, FA	Ogeechee Stream, water	Grange, Georgia,U.S.A.	surface	8.1[a]
R32, FA	Yampa Stream water	Yampa, Colo.,U.S.A.	surface	l.0[a]
MS7, HA	Filiouris River, sediment	Delta area at Xilagani,Greece	0-40 cm	mud[b]
MS3, HA	Strymonikos Plateau, marine sediment	North Aegean Sea,Greece	0-60 cm	medium sand, gravel[b]
MS1, HA	Samothraki Plateau, marine sediment	North Aegean Sea,Greece	0-170 cm	silt, silty mud[b]
M20, HA	Leonardite deposit	Wyoming,U.S.A.	-	-
NG5, HA	Gley soil	Nagaoka, Niigata,Japan	0-15 cm	paddy[b]
M23, HA	Chino soil	Chino, Cal.,U.S.A.	0-15 cm	clay loam[b]
P5, HA[b]	Volcanic paleosol, exposed cliff	Procida Island, Thyrr.Sea,Italy	30 m	c
F4, FA	Brown Mediterran. soil	Sassari, Sardinia,Italy	0-15 cm	loamy sand[b]
H4, HA	Brown Mediterran. soil	Sassari, Sardinia,Italy	0-15 cm	loamy sand[b]

[a]DOC: Dissolved organic carbon (mg C/l) [9]. [b]Texture.
[c]Interbedded within pyroclastic rocks; age, 28,850±860 yr BP [13].

Methods of extraction and analytical properties of the FA and HA samples used have been previously described [8-14]. Some analytical data are presented in Table 2.

Sample solutions were prepared by dissolving each FA or HA in deionized water at a concentration of 100 mg/l. After overnight equilibration at room temperature (RT), solutions were filtered through Whatman No. 2 paper and adjusted to pH 8.0 with 0.05 N NaOH.

Table 2 Major elemental atomic ratios and some analytical data for fulvic acids (FA) and humic acids (HA).

Sample	C/H	C/N	O/C[a]	Ash	E_4/E_6	Free radical	Ref.
	Atomic ratios[b]			%	ratio	concent[c]	
R29, FA	0.91	54.82	0.56	<1.00	23.7	1.22	[10]
R30, FA	0.92	43.94	0.59	<1.00	15.9	4.79	[10]
R31, FA	0.98	47.50	0.61	<1.00	17.1	3.42	[10]
R32, FA	1.00	53.86	0.62	<1.00	18.8	1.06	[10]
MS7, HA	0.90	18.30	0.61	12.30	4.5	7.26	[8]
MS3, HA	0.65	10.10	0.46	18.20	5.3	1.18	[8]
MS1, HA	0.94	16.60	0.38	11.40	5.1	3.53	[8]
M20, HA	1.32	20.25	0.41	3.10	5.4	2.23	d
NG5, HA	0.85	12.79	0.48	<1.00[d]	5.0[d]	1.50[d]	[12]
M23, HA	1.01	13.89	0.40	2.30	5.2[d]	2.21[d]	[11]
P5, HA	2.38	33.33	0.42	2.20	5.0	7.03	[13]
F4, FA	0.49	16.24	0.95	14.20	6.2[d]	1.84	[14]
H4, HA	1.01	17.77	0.49	1.40	4.6[d]	16.55	[14]

[a]Oxygen calculated by difference. [b]Calculated on an ash- and moisture-free basis.
[c]Spins/g x 10^{-17}. [d]Present study.

Fluorescence spectra were recorded at RT on a Perkin Elmer LS-5 luminescence spectrophotometer equipped with a Perkin Elmer Data Station 3600. Data were generated and processed by Perkin Elmer PECLS programs. The emission and excitation slits were set at 5-nm band width and a 120-nm/min scan speed for both monochromators selected. The fixed scale setting was chosen according to the fluorescence intensity of each sample, so that the highest peak in the spectrum could approximate 90% relative intensity. Comparable values of relative intensity were then calculated for the various samples. Emission spectra were recorded over the range 380 to 550 nm at a constant excitation wavelength of 360 nm. Excitation spectra were obtained over a scan range of 270 to 500 nm, by measuring the emission radiation at a fixed wavelength of 520 nm. Synchronous-scan excitation spectra were measured only for brown soil FA and HA, by scanning simultaneously both the excitation and emission wavelengths. The excitation wavelength was varied from 290 to 550 nm during the scans, while maintaining a constant wavelength difference of $\Delta\lambda = \lambda_{em} - \lambda_{exc} = 18$ nm, which was found to produce optimally resolved spectra.

Results and Discussion

Wavelengths of fluorescence maxima (peaks and shoulders) obtained in the emission and excitation modes and relative fluorescence intensities for FA and HA examined are listed in Table 3. Fig. 1 shows the fluorescence emission and excitation spectra of most examined FA and HA samples. For simplicity, the spectra of only one (the Ohio River FA) of the four river aquatic FAs is shown, because all four stream fulvic acids are very similar. Fluorescence emission, excitation and synchronous-scan spectra of FA and HA isolated from the brown soil are presented, in Figs 2 and 3, respectively, to allow an immediate visual comparison.

Fluorescence emission spectra measured at an excitation wavelength of 360 nm (Figs 1 and 2, Table 3) generally feature a broad band of relative intensity which decreases and a wavelength maximum which increases in the order: stream and soil FA, stream and marine sediment HA, soil and leonardite HA, paleosol HA. Sedimentary HA are characterized by flat fluorescence maxima with tails toward higher or lower wavelengths, whereas leonardite and soil HA show shoulders at lower wavelengths. Therefore, these results, together with comparable data reported in the current literature [1,4,6-8,15], permit the establishment of two first-evident parameters of distinction between FA and/or HA of various natural sources, based on the values of relative intensity and wavelength of the fluorescence emission maximum and supposing that the same excitation wavelength is maintained in all experiments.

Fluorescence excitation spectra (Figs 1 and 2) are apparently better structured than the corresponding emission spectra and confirm the trends previously described. FA samples show, in any case, a typical excitation peak at 390 nm (Figs 1a and 2a, Table 3). Comparable wavelengths have been measured by some authors [6,7], whereas others have obtained few lower values for FA of various origin [1,2,16,17]. In the case of soil FA, the peak at 390 nm is accompanied by an equally intense fluorescence at lower wavelengths and by additional secondary peaks and shoulders at higher wavelengths (Fig. 2, Table 3). This behaviour, however, does not seem to be of general validity for other soil and peat FAs, which often show, in addition to the previous peak, one or more intense peak(s) at higher wavelengths [3,6,7,16]. The excitation spectral shape of stream FA (Fig. 1a) evidently differs from that of soil FA in that it always shows a unique peak at around 390 nm. This peak is associated with shoulders on the low wavelength side [6,7].

In general the excitation spectra of sediment HA usually have three (or four) peaks of similar intensity in the low, intermediate and high wavelength ranges, although the HA from deep marine sediment features more intense peaks at high wavelengths (Figs 1b-d; Table 3). Leonardite and soil HA generally show two close excitation peaks at high wavelengths, with a shoulder between 390 and 400 nm (Figs 1e, 1g, 2; Table 3). Gley soil HA, however, features an additional intense peak at 393 nm and weak shoulders in the low wavelength range (Fig. 1f). Paleosol HA is characterized by a unique excitation maximum centered at the highest wavelength (471 nm) among examined samples, with shoulders (Fig. 1h). These results, which generally agree with data reported in the literature for FA and HA of similar origin [2-4,6-8,15-17], make it possible to further distinguish between FA and/or HA on the basis of the shape and wavelength maxima of excitation fluorescence spectra.

Table 3 Fluorescence wavelengths (nm) of emission and excitation maxima and relative intensity (R.I., arbitrary units) measured for fulvic acids (FA) and humic acids (HA).

Sample	Emission peak		shoulders/	Excitation peaks		
	λem	R.I.	tail (--->)	Main[a]	Secondary[a]	Shoulders[a]
R29, FA	453	57	-	390	-	342,303
R30, FA	456	69	-	390	-	360,340
R31, FA	458	57	-	389	-	356,344
R32, FA	459	47	-	389	-	357,342
MS7, HA	462	14	--->515	306,442,391	-	462
MS3, HA	472	8	--->515	312,444,390	-	460
MSI, HA	505	10	--->460	448,463	391,330	310
M20, HA	510	2	490,470	465,451	-	396
NG5, HA	507	5	--->470	446,464	393	438,360,338
M23, HA	514	2	470	464	451	393
P5, HA	522	1	-	471	-	453,394
F4, FA	451	14	-	388,336,301	438,457	493
H4, HA	514	5	-	464,450	-	392

[a]Listed in decreasing order of relative intensity in the spectrum of each sample.

Comparison of FA and HA from the brown soil in regard to fluorescence emission and excitation spectral shape, relative intensity and wavelength (nm) maxima, (Fig. 2; Table 3), permits the two humic fractions to be easily distinguished. This difference in emission and excitation wavelength maxima is also apparent for other FA and HA couples of terrestrial origin, but it is not so evident when comparing river or marine aquatic FA versus HA [5,7]. This is not surprising if one considers that the molecular, chemical and genetical characteristics of FA and HA from the same aquatic source are more similar than the same characters of their terrestrial counterparts. Synchronous-scan excitation spectra of soil FA versus HA (Fig. 3) feature even more differentiated spectra than those previously examined. This type of fluorescence spectra also appears to be promising for distinguishing between FA and HA of aquatic origin [7].

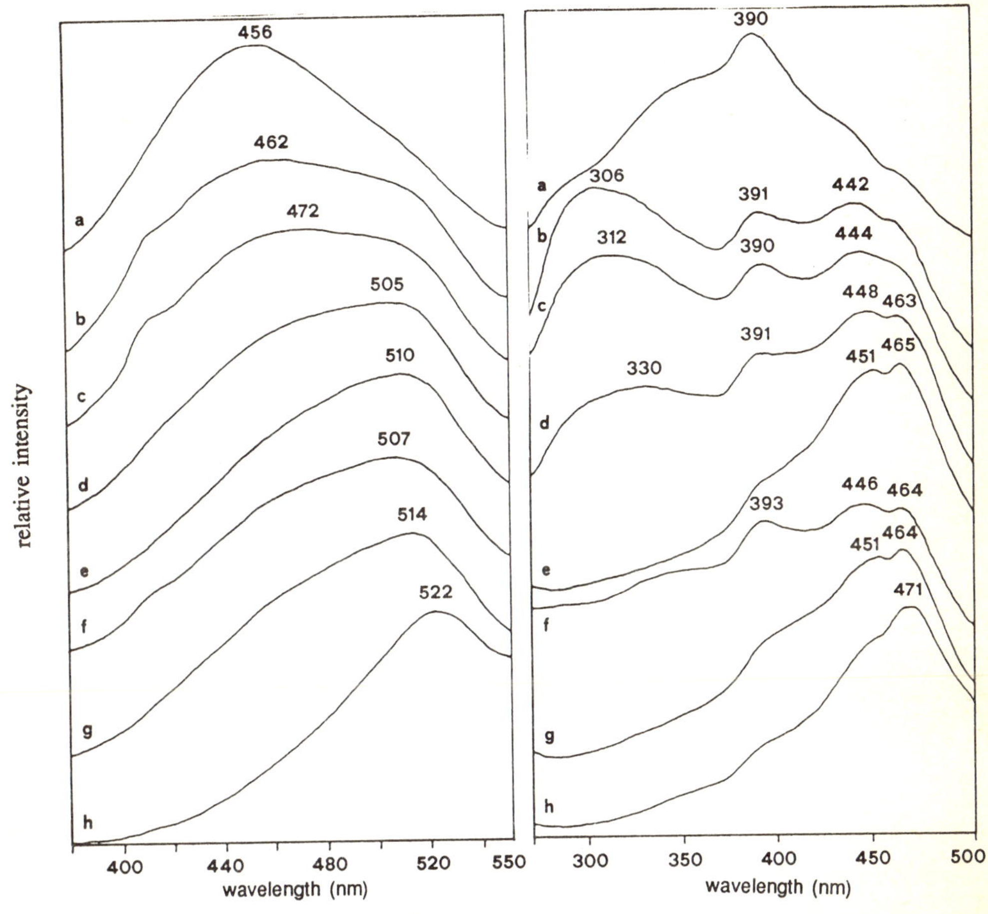

Figure 1 Fluorescence emission (left) and excitation (right) spectra. Ohio River FA (a); sediment HA from the Filiouris River (b), the Strymonikos Plateau (c) and the Samothraki Plateau (d); leonardite HA (e); gley soil HA (f), Chino soil HA (g), and paleosol HA (h).

Fluorescence, as is well-known, depends on the availability of delocalized electrons. For example, fluorescence occurs in the presence of condensed aromatic ring systems bearing, in suitable positions, electron-donating or accepting substituent groups, such as hydroxyls, alkoxides and carbonyls of various nature. It can also occur in the presence of conjugated unsaturated systems capable of a high degree of resonance, i.e. of electron delocalization [18,19]. Because of the well-known chemical and structural complexity and heterogeneity of humic materials, only hypothetical identification of the relevant molecular constituents responsible for fluorescence in FA and HA may be made by comparison with fluorescence properties of chemically-defined molecules. This should possibly be carried out with the aid of other available analytical, spectroscopic and genetical information. In Table 4 a number of pure compounds and their fluorescence maxima are listed. These compunds may be expected to be potential contributors to the fluorescence of natural FA and HA, on the basis of similar structural units identified as products of hydrolysis and/or degradation of humic materials or units hypothesized as components of molecular structures proposed for them [20,21].

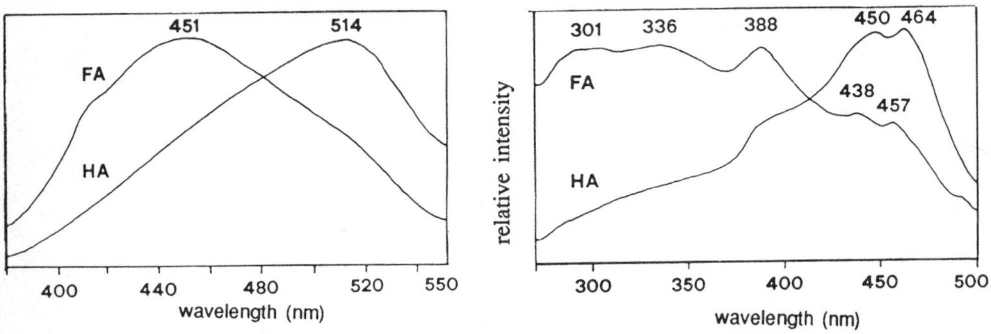

Figure 2 Fluorescence emission (left) and excitation (right) spectra of brown soil FA versus HA.

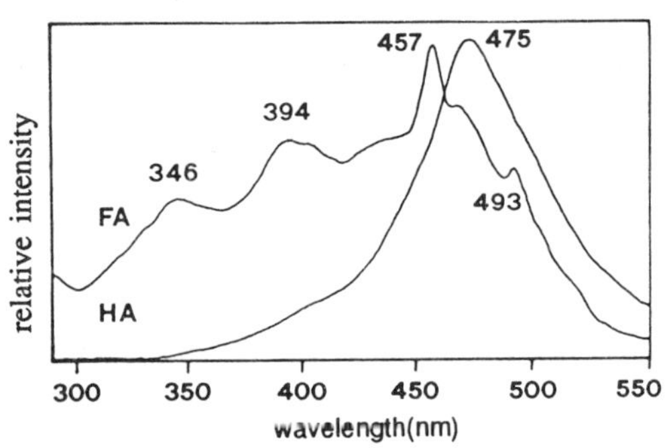

Figure 3 Synchronous-scan excitation spectra of brown soil FA versus HA.

As the number of aromatic rings condensed in a straight chain and the number of conjugated double bonds increase, the fluorescence emission wavelength increases [19,22]. Functional groups such as hydroxyls, alkoxides, carboxyls, aldehydes, ketones and esters, are also known to produce a red-shift in the emission wavelength [19,22]. This would explain the long fluorescence wavelength which characterizes terrestrial HA. Feasible contributors to fluorescence of terrestrial HA (Table 4) include: structural units such as aromatic rings condensed in long chains, methylnaphto-

Table 4 Fluorescence emission and excitation maxima of potential contributors, as structural units, to the fluorescence of fulvic acids (FA) and humic acids (HA) from soil, stream and marine environments.

Compound Name	Structural formula	λem max (nm)	λexc max (nm)	Ref.
1,2-Benzopyrene		400-500		[22]
Perylene		440,470		[22]
Methyl salicylate		448	302,366	[24]
Salicylaldheyde		525	327	[24]
Protocatechuic acid	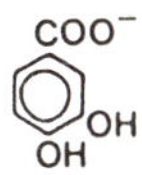	455	340-370	[18]
2-Hydroxycinnamic acid	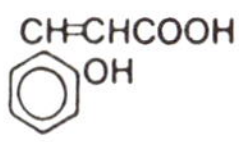	500	360	[24]
Caffeic acid	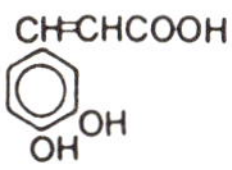	450	365	[24]
2-Methyl-1,4-naphtoquinone	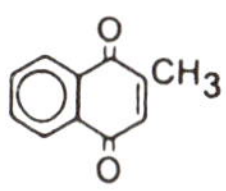	480	-	[18]
Coumarins:				
unsubstituted	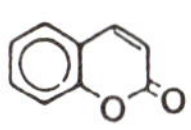	454	376	[22]
hydroxy- and methoxy-		450-480	320-365	[24]
esculetin (6,7-dihydroxy-)		475	390	[24]
scopoletin (7-hydroxy-6-methoxy-)		460	390	[24]
other disubstituted		449-515	367-420	[22,24]
Chromone-derivatives	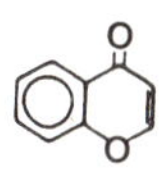	445-490	320-336	[24]
Xanthone	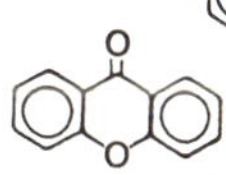	456	410	[24]
3-Hydroxyxanthone		465	343,365	[24]
Flavones and Flavonoids	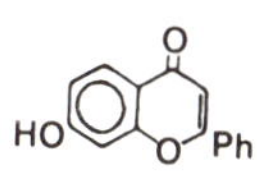	465-525	339-365	[24]
Schiff-base derivatives -N=C-C=C-N-		470	360-390	[23]
Hydroxy-quinolines	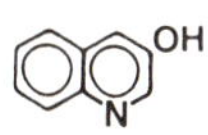	450-510	350-360	[18]

quinones, hydroxyquinolines and, particularly, coumarin derivatives and flavonoids, all derived from partial degradation of plant constituents and incorporated in the humic polymer. A blue shift in the fluorescence wavelength is observed when the aromatic system is branched or the straight chain is shorter [22]. Lignin degradation products (e.g coumarin derivatives, like esculetin, flavones, xanthones, and naphtoquinones) appear to fluoresce at wavelength maxima comparable to those of sedimentary HA (Table 4). The fluorescent chromophore group derived from browning reactions that are polycondensation of carbohydrates and amino acids to melanoidins (the Schiff base derivative N=C-C=C-N), has also been suggested to be an important contributor to fluorescence of marine FA and HA [23]. These considerations support the idea that sedimentary humic materials have a mixed terrestrial and autochthonous origin.

Fluorescence of both terrestrial and stream FA may be attributed to simpler molecular units of low degree of aromatic polycondensation and rich in oxygenated functional groups, such as methylsalicylate units, e.g dihydroxybenzoic acids like protocatechuic acid, chromones, xanthones, and hydroxyquinolines, which fluoresce at short wavelengths (Table 4). The possible high contribution to fluorescence by coumarin and its derivatives (Table 4) supports the suggestion that both types of FA originate from plant materials [17]. No correlations can be found between fluorescence data and compositional or other analytical properties presented in Table 2.

Conclusions

Fluorescence spectral data presented in this paper appear adequate for distinguishing between FA and HA (particularly terrestrial) of common origin, and for distinguishing HA of various natural sources. The emission spectra appear featureless, but exhibit relative intensity and wavelength maxima different enough to provide a clear distinction among the various types of humic materials. The excitation spectra are more distinctive than the emission spectra, whereas synchronous-scan excitation spectra are particularly promising as diagnostic criteria of distinction.

Fluorescence properties of humic materials, in terms of structural units responsible for fluorescence, may be tentatively interpreted. This can be done by comparison with analoguous data available for pure compounds which have been identified, or hypothesized, as structural constituents of the humic molecules.

However, further studies, associated to information from other chemical and spectroscopic techniques, such as NMR, IR, GC/MS and ESR, are necessary in order to confirm and extend the preliminary classification of humic substances on the basis of fluorescence properties, as proposed in this paper.Fluorescence analysis is expected to provide important implications in the evaluation of the genesis and environmental behaviour and functions of humic substances.

References

1. Choudry, G.G. Residue Rev. **92**:59 (1984).

2. Hayase, K. and H. Tsubota. Geochim. Cosmochim. Acta **49**:159 (1985).

3. Bachelier, G. Cah. O.R.S.T.O.M., ser. Pedol., **18** (**2**):129 (1980-81).

4. Ertel, J.R. and J.I. Hedges. In: R. F. Christman and E. T. Gjessing, Eds, Aquatic and Terrestrial Humic Materials, pp. 143-163 (Ann Arbor: Ann Arbor Science, 1983)

5. Ewald, M., C. Belin, P. Berger, and H. Etcheber. In: R.F. Christman and E.T Gjessing, Eds, Aquatic and Terrestrial Humic Materials, pp. 461-466 (Ann Arbor: Ann Arbor Science, 1983).

6. Miano, T.M., G. Sposito, and J.P. Martin. Soil Sci. Soc. Am. J. **52**:1016 (1988).

7. Senesi, N., T.M. Miano, M.R. Provenzano, and G. Brunetti. Sci. Total Environ. **81/82**:143 (1989).

8. Senesi, N. and F. Sakellariadou. Proc. 9th Int. Symp. Environ. Biogeochemistry, Moscow, SSSR, 3-8 September 1989 (In press).

9. McKnight, D.M., G.L. Feder, E.M. Thurman, R.L. Wershaw, and J.C. Westall. In: R.E. Wildung and E.A. Jenne, Eds, Biological Availability of Trace Metals, pp. 65-76 (New York: Elsevier, 1983).

10. Senesi, N. In: P. Hills. Ed. Pollution in the Urban Environment, Polmet 88, Vol. 2, pp. 6 0 7 - 612 (Hong Kong: Vincent Blue Copy Co. Ltd, 1988).

11. Senesi, N., G. Sposito, and J.P. Martin. In: T.D. Lekkas, Ed., Heavy Metals in the Environment, pp. 478-480 (Edinburgh: CEP Consultants Ltd, 1985).

12. Yonebayashy, K. Priv. Comm. (1988).

13. Calderoni, G. and M. Schnitzer. Geochim. Cosmochim. Acta **48**:2045 (1984).

14. Chen, Y., N. Senesi, and M. Schnitzer. Geoderma **20**:87 (1978).

15. Visser, S.A. In: R.F. Christman and E.T. Gjessing, Eds, Aquatic and Terrestrial Humic Materials, pp. 183-202 (Ann Arbor: Ann Arbor Science, 1983).

16. Ghosh, K. and M. Schnitzer. Can. J. Soil Sci. **60**:373 (1980).

17. Larson, R.A. and A.L. Rockwell. Archives Hydrobiol. **89**:416 (1980).

18. Williams, R.T.J. Royal Inst. Chem. **83**:611 (1959).

19. Seitz, W.R. In: P.J. Elving, Ed., Treatise on Analytical Chemistry, Part 1, Theory and Practice, Vol. 7, Sect. H, Optical Methods of Analysis, pp. 159-248 (New York: Wiley, 1981).

20. Schnitzer, M. and S.U. Khan. Humic Substances in the Environment (New York: Dekker, 1972).

21. Stevenson, F. J. Humus Chemistry (New York: Wiley, 1982).

22. Guibalt, G.G. Practical Fluorescence. Theory, Methods and Techniques (New York: Dekker, 1973).

23. Laane, R.W.P.M. Marine Chem. **15**:85 (1984).

24. Wolfbeis, O.S. In: P.J. and J.P. Winefordner, Eds, Chemical Analysis, Vol. 77, Molecular Luminescence Spectroscopy, Methods and Applications: Part 1, pp. 167-370 (New York: Wiley, 1983).

Dielectric Spectroscopy of Aqueous Solutions of Fulvic Acids

Francisco J. Gonzalez-Vila[1], Udo Kaatze[2], Harro Lentz[3], Francisco Martin[1] and Reinhard Pottel[2]

[1] Instituto de Recursos Naturales y Agrobiologica, C.S.I.C., Apartado 1052, E-41080 Sevilla, Spain
[2] Drittes Physikalisches Institut, Universität Göttingen, Bürgerstrasse 42-44, D-3400 Gottingen, F.R. Germany
[3] Fachbereich 8, Universität-Gesamthochschule Siegen, Postfach 101240, D-5900 Siegen, F.R. Germany

Abstract

The complex dielectric spectrum of aqueous solutions (10% w/w) of fulvic and polymaleic acids has been measured at various frequencies between 1 MHz and 40 GHz. Similar to pure water a dispersion/dielectric loss region emerges in the range above 1 GHz. The measured spectra have been analytically represented by the empirical Cole-Cole relaxation spectral function to yield values for the extrapolated high- and low-frequency permittivity, the principal dielectric relaxation time, the relaxation time distribution parameter and the specific electric d.c. conductivity. In correspondence with aqueous solutions of synthetic organic molecules, the low-frequency permittivity of the fulvic acid/water mixtures is reduced, and the principal dielectric relaxation time is enhanced with respect to the pure water value. As with solutions of polyacrylic acid no solute contributions to the real part of the dielectric spectrum are found.

Introduction

As an important component of natural waters, humic substances play a significant role in geochemical and ecological processes. These acids act as complexing agents and, by this means, mobilize metal ions and organic pollutants in an aquatic environment [1,2]. Though some information exists on the physico-chemical and structural characteristics of humic substances in solution [3], many physical properties of these systems are still unclear. An open problem is the tertiary structure of humic substances in water. Other than with aqueous solutions of synthetic polymers, micellar or membrane-like aggregates have been recently proposed for the humic acid/water mixtures [4].

Aiming at information on specific solute-solvent interactions which might be associated with the tertiary structure of humic substances in water, we performed a dielectric relaxation study on aqueous solutions of fulvic acids (FA) and polymaleic acid (PA). The latter, a recently characterized [5-7] polycarboxylic acid formed by hydrolysis of base-catalyzed homopolymerized maleic anhydride, is assumed to be an interesting model of FA [8,9]. The microwave dielectric spectrum of aqueous solutions clearly reflects hydration properties of the solutes [10-18]. The evaluation of the measured spectra in terms of molecular models, however, is not unambiguously possible. We therefore restrict ourselves to a comparative discussion of the present data and of results for aqueous solutions of synthetic polymers, as well as of low-molecular solutes.

Materials and Methods

Samples

Two fulvic acids were isolated from the Bh horizons of Humic Haplorthod (podzol) located in Vollbüttel, West Germany (FA-PV) and Armadale, Canada (FA-PC), respectively. A third fulvic acid was isolated from water of a lake in Huelva, Spain (FA-T). This lake is surrounded by peatland. Details of the extraction and purification procedures have been published elsewhere [19-21].

The polymaleic acid (PMA-I) was prepared according to Braun and Pomakis [22] and subsequently acidified with a strong cation exchange resin in the protonated form. A purified PMA sample (free of pyridine and other aromatics) was obtained by redissolution of freeze-dried PMA-I in acetone and coagulation by addition of chloroform in excess. The coagulate was removed by filtration and dried at room temperature afterwards (PMA-II). The results of an elementary composition and a functional group analysis of the samples is presented in Table 1. C and H were determined by dry combustion, N by the automated Dumas method, and O was estimated from the difference. The total acidity was determined by the barium hydroxide method and the carboxyl groups by calcium acetate exchange, as previously described by Schnitzer and Khan [23]. Phenolic hydroxyls were considered to be equal to the difference between the total acidity and the carboxyl groups.

Table 1 Elementary composition (% w/w) and concentration of oxygen-containing functional groups (meq/g) of the humic acid samples. All values are calculated on a dry, ash-free humic acid.

Sample	C	H	N	0 + S	Total Acidity	COOH	Phen. OH
FA- PV	46.2	3.4	0.8	49.6	11.2	8.8	2.4
FA- PC	50.9	3.3	0.7	45.1	11.5	9.1	2.4
FA- T	43.4	4.2	1.3	48.9	12.7	7.6	5.1
PMA	45.9	4.4	0.6	49.1	12.0	8.8	3.2

Table 2 Concentration data (c_s, weight of solute per volume of solution; s, weight per cent), density ρ, molarity c_w of solvent, volume fraction v of solute, and pH value of the humic acid solutions at 25°C.

Solute	c_s g/cm³ ± 0.0002	s % w/w ± 0.02	ρ g/cm³ ± 0.004	c_w mol/l ± 2	v ± 0.004	pH ± 0.05
FA- PV	0.0990	9.59	1.033	51.8	0.065	1.80
FA-PC	<0.015	<1.5	≈1.002	>54.8	≈0.008	-
FA-T	0.0954	9.16	1.041	52.5	0.052	2.25
PMA-I	0.1000	9.70	1.031	51.7	0.066	2.75
PMA-II	0.1000	9.77	1.023	51.2	0.074	-

Aqueous solutions with a concentration of about 10 % w/w have been prepared of FA-PV, FA-T, PMA-I, and PMA-II by adding bidistilled water to a preweighed amount of the respective fulvic acid. FA-PC was less soluble (< 1.5 % w/w). Concentration data, density, and pH value of the solutions, as well as the volume fraction of solute, are given in Table 2.

Complex Permittivity Measurements

The complex (relative) electric permittivity (i, imaginary unit)

$$\varepsilon(\nu) = \varepsilon'(\nu) - i\varepsilon''(\nu) \qquad (1)$$

of the sample liquids has been determined as a function of frequency ν (Fig. 1) by frequency domain measurements [24]. The liquids have been exposed to harmonically alternating, weak electric fields to observe the responding dielectric polarization. Three different methods, which had been successfully applied in various previous studies on aqueous solutions, have been used to cover the frequency range between 1 MHz and 40 GHz.

From 1 to 100 MHz we utilized a sensitive rf-admittance bridge (Boonton 33D/l) to perform at seven fixed frequencies input impedance measurements on a small specimen cell. This cell contains the sample in a short piece of a circular waveguide, which is excited far below its cut-off frequency [25]. Modal analysis of the transition between the coaxial line feeder and the waveguide-below-cut-off section has been performed [26]. It was found that the cell can be represented by a simple equivalent network [26]. For the liquids under consideration (ε'>60) this network simply consists of two capacitors, the capacitances of which have been determined by calibration measurements with the empty cell and with water as reference liquid.

Between 1 MHz and 1 GHz a vector voltmeter (Rohde & Schwarz ZPU) has been additionally used to measure the transmission coefficient of a cell in which the sample is also placed in a waveguide-below-cut-off section [26]. This cell essentially consists of a coaxial line, the inner conductor of which is interrupted for a certain distance to form a small piece of circular cylindrical waveguide. This waveguide contains the liquid sample between two dielectric windows. This "transmission" cell, in analogy to the "reflection" cell, has been likewise treated by modal analysis [26]. A π-network representation follows thereby. The values of the capacitors of this lumped-element circuit have also been found by calibration measurements.

At frequencies above 1 GHz a travelling-wave method has been applied. The wave transmitted through a liquid-filled circular cylindrical waveguide was balanced against a reference wave by a doublebeam interferometer technique [27-29]. Five small-band microwave bridges, consisting of standard coaxial line components or waveguide devices, were used to cover the frequency range.

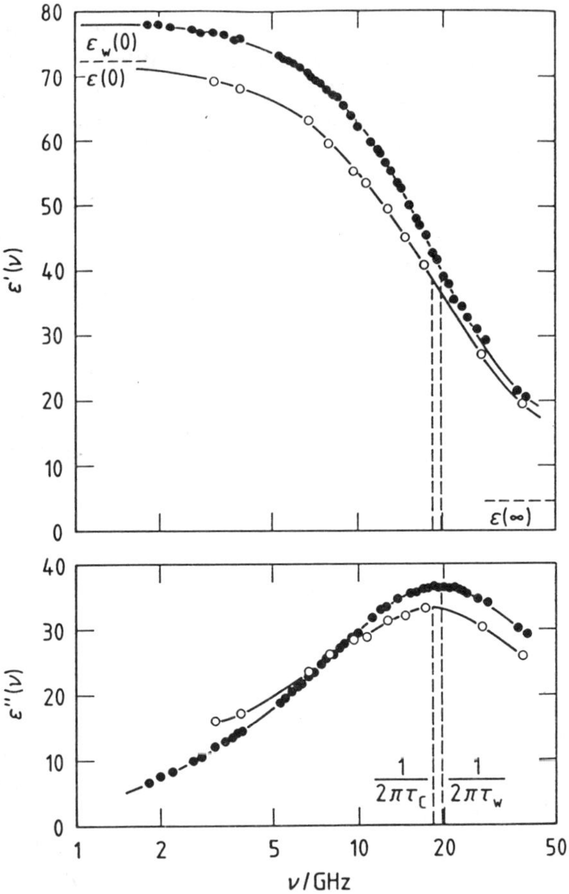

Figure 1 Real part $\varepsilon'(\nu)$ and negative imaginary part $\varepsilon''(\nu)$ of the complex (electric) permittivity plotted versus frequency ν for pure water (• [30]) and for the aqueous PMA-I solution (o) at 25°C. The curves represent the relaxation spectral functions eq. (3) and eq. (4), respectively, with the parameter values given in Table 3.

Experimental Errors

The experimental error in the complex permittivity data depends on the frequency of measurement and also on the applied method. Above 200 MHz it may be globally characterized by an uncertainty of ±2% for both ε' and ε''. Due to the high d.c. conductivity of the samples, measurements below 200 MHz were substantially affected by electrode polarization effects. We therefore restrict the following discussion to the microwave part (ν>200 MHz) of the dielectric spectra. Errors in the determination of the frequency ν were negligibly small. The temperature was controlled to within ±0.1 K during the measurements.

Results and Treatment of Data

In Fig. 1 the real part ε' and the negative imaginary part ε'' of the complex permittivity are displayed as a function of frequency ν for the PMA-I solution at 25°C. Also shown for comparison is the complex dielectric spectrum of pure water at the same temperature. The curves for the mixture resemble the corresponding curves for the solvent. Some differences, however, emerge, which are characteristic for all measured spectra of humic acid solutions. The extrapolated static permittivity $\varepsilon(0)$ of each solution is smaller than that of water, $\varepsilon_w(0)$. We shall comment on this effect below. The frequency $\nu_C = (2\pi\tau_c)^{-1}$ at which $\varepsilon''(\nu)$ adopts its relative maximum $(d\varepsilon''(\nu_C)/d\nu = 0,\ d^2\varepsilon''(\nu_C)/d\nu^2 < 0)$ is different from that of water, $(2\pi\tau_w)^{-1}$. The dispersion region $(d\varepsilon'(\nu)/d\nu < 0)$ of the mixture extends over a slightly broader frequency band. Finally, at low frequencies (ν<7 GHz) the $\varepsilon''(\nu)$ values of the solution exceed those of the pure solvent. This finding is an indication that the total loss is a sum

$$\varepsilon''(\nu) = \varepsilon_d''(\nu) + \sigma/(\varepsilon_0\omega) \qquad (2)$$

of a contribution $\varepsilon_d''(\nu)$ originating in dielectric relaxation processes and of another one resulting from the specific electric conductivity σ. In eq. (2), ε_0 denotes the electrical field constant and $\omega = 2\pi\nu$ the angular frequency.

Within the limits of experimental error, the microwave dielectric spectrum of pure water is characterized by a relaxation with one discrete relaxation time τ_w [30,31]. It can thus be represented by the Debye-type relaxation function [32] given by

$$\varepsilon(\nu) = \varepsilon_w(\infty) + [\varepsilon_w(0) - \varepsilon_w(\infty)] \,/\, (1 + i\omega\tau_w) \qquad (3)$$

The values of the parameters of eq. (3) at 25°C are given in Table 3.

Empirically, the dielectric spectra of the solutions can be described by the Cole-Cole relaxation spectral function [33]. The function

$$\varepsilon(\nu) = \varepsilon(\infty) + [\varepsilon(0) - \varepsilon(\infty)] \,/\, [1 + (i\omega\tau_C)^{(1-h)}] - i\sigma/\varepsilon_0\omega \qquad (4)$$

is therefore appropriate to analytically represent the frequency-dependent complex permittivity data. In this relation parameter h is a measure of the width of the underlying relaxation time distribution. The values of the parameters $\varepsilon(\infty)$, $\varepsilon(0)$, τ_C, h, and σ have been found by fitting eq. (4) to the measured spectra using a nonlinear least-squares regression analysis. The results obtained by this fitting procedure are also presented in Table 3. Also included in this table for comparison are data for a nearly l-molar aqueous solution of polyacrylic acid, PAA.

The PMA-II solution has been measured below 1 GHz only. In correspondence with the other samples no dispersion/dielectric loss region has been found in this frequency range. This finding is in accordance with previous results for aqueous solutions of PAA [34]. In contrast, solutions of salts of polyacrylic acid, like NaPA, at frequencies below about 1 GHz, clearly exhibit solute contributions in their dielectric spectra [34].

The solution of FA-PC has been measured at one frequency only (ν = 6.0 GHz). For this reason parameters $\varepsilon(\infty)$ and h have been fixed at reasonable values in the above fitting procedure. By that means estimates for the extrapolated static permittivity $\varepsilon(0)$ and the principal dielectric relaxation time τ_C have been obtained.

Table 3 Parameters of the relaxation spectral functions for water (eq. 3) and for aqueous solutions (eq. 4) of fulvic acids, polymaleic acid, and polyacrylic acid at 25°C. Values marked by an asterisk (*) have been fixed during the fitting procedure.

Sample	s	$\varepsilon(\infty)$	$\varepsilon(0)$	τ	h	σ
	% w/w			ps		S/m
Water [29]	-	5.16±0.08	78.36±0.05	8.27±0.02	-	-
FA-PV	9.59	5.0±0.5	72.3±0.5	8.50±0.05	0.03±0.03	1.25±0
FA-PC	<1.5	4.7*	76.9±0.5	8.34±0.1	0*	0.5±0
FA-T	9.16	4.5±0.5	72.1±0.5	8.41±0.15	0.05±0.05	2.52±0
PMA-I	9.70	4.5±0.5	72.5±0.5	8.64±0.15	0.05±0.05	0.80±0
PMA-II	9.77	-	70.0±1	-	-	0.78±0
PAA [33]	6.95	5.5±0.5	73.0±0.5	8.69±0.15	0.02±0.02	0.17±0

Discussion

To look for potential special properties of the fulvic acid/water mixtures it is useful to compare the parameter values (Table 3) of the relaxation spectral function (eq. 4) with data for aqueous solutions of other organic solutes. For this purpose, the extrapolated static permittivity values of the fulvic acid solutions are plotted in Fig. 2 as a function of volume fraction v of the solute. Also presented in that diagram are data for aqueous solutions of non-dipolar small organic molecules, of PAA, and of other synthetic polymers. The latter solutes are dipolar themselves. The contributions of the solvents to the static permittivity are therefore used here as $\varepsilon(O)$ value. This value should compare with the static permittivity of aqueous solutions of non-dipolar solutes.

As a result of the increasing dilution of the dipolar solvent by solute molecules there is a general tendency in the $\varepsilon(0)$ values to decrease with v. The static permittivity data of the fulvic acid and PAA solutions nicely fit into this trend (Fig. 2). There are no indications of special effects in the fulvic acid solutions. Such effects could be due to a particular shape of the solute particles or to uncommon hydration properties.

The principal dielectric relaxation time τ_C of the fulvic acid solutions is enhanced with respect to the pure water value τ_w. Such an enhancement, however, is a common feature of aqueous solutions of organic solutes [10-15, 17, 18, 35-38]. The shift τ_C-τ_w in the dielectric relaxation time of the present solutions has reasonable values. With the fulvic acid/water mixtures (10% w/w) it is even somewhat smaller than with the less concentrated PAA solution (7% w/w). We therefore conclude that the fulvic acids do not seem to induce unusual hydration water properties.

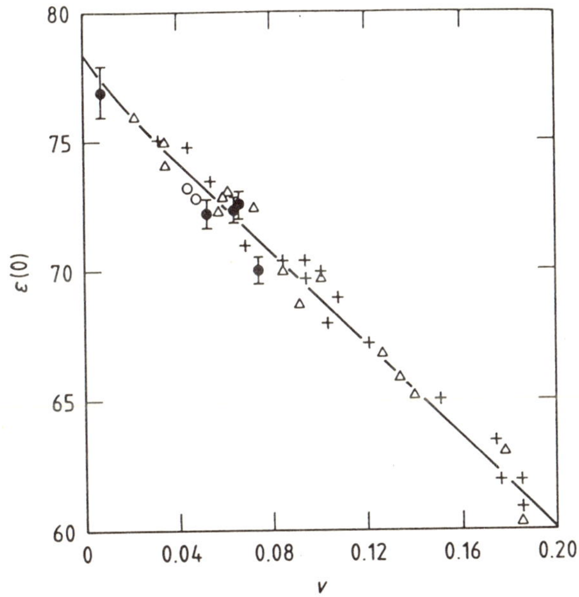

Figure 2 Plot of the extrapolated static permittivity $\varepsilon(0)$ as a function of the volume fraction v of solute for aqueous solutions of fulvic acids (•), polyacrylic acid (o) [34], nonionic synthetic polymers (Δ) [10,12,18] and small non-dipolar organic molecules (+) [35,39] at 25°C. With the solutions of dipolar synthetic polymers (Δ) only the solvent contribution to $\varepsilon(0)$ is considered here. The curve is hand drawn to indicate the trend in the data.

The specific electric d.c. conductivity σ of the fulvic acid solutions is higher than expected from the pH values. This enhanced conductivity is most probably due to the presence of small amounts of inorganic salts which, possibly, were not removed by the sample purification and preparation procedure.

References

1. Oliver, B.G., E.M. Thurman and R.L. Malcolm. Geochim. Cosmochim. Acta **47**:2031 (1983).

2. Aiken, G.R., D.M. McKnight, R.L. Wershaw and P. McCarthy. Humic Substances in Soil, Sediment and Water. Geochemistry, Isolation and Characterization (New York: John Wiley, 1985).

3. Ziechmann, W. Huminstoffe (Weinheim: Verlag Chemie, 1980).

4. Wershaw, R.L. J. Contaminant Hydrol. **1**:29 (1986).

5. Bracewell, J.M., G.W. Robertson and D.J. Welch. J. Anal. Appl. Pyrolysis **2**:239 (1980).

6. Spiteller, M. and M. Schnitzer. J. Soil Sci. **34**:525 (1983).

7. Martin, F., F.J. Gonzalez-Vila, and H.-D. Lüdemann. Z. Naturforsch. **39c**:244 (1984).

8. Anderson, H.A. and J.D. Russell. Nature **260**:597 (1976).

9. Young, S.D., B.W. Bache, D. Welch, and H.A. Anderson. J. Soil Sci. **11**:597 (1981).

10. Kaatze, U. Advan. Mol. Relaxation Processes **7**:71(1975).

11. Kaatze, U. and R. Pottel. Colloques internationaux du C.N.R.S. **246**:111(1976).

12. Kaatze, U., O. Göttmann, R. Podbielski, R. Pottel, and U. Terveer. J. Phys. Chem. **82**:112 (1978).

13. Pottel, R., U. Kaatze, and St. Müller. Ber. Bunsenges. Phys. Chem. **82**:1086 (1978).

14. Kaatze, U. Progr. Colloid Polymer Sci. **65**:214 (1978).

15. Pottel, R., K.-D. Göpel, R. Henze, U. Kaatze, and V. Uhlendorf. Biophys. Chem. **19**:233 (1984).

16. Kaatze, U., A. Dittrich, K.-D. Göpel, and R. Pottel. Chem. Phys. Lipids **35**:279 (1984).

17. Kaatze, U. and R. Pottel. J. Mol. Liquids **30**:115 (1985).

18. Kaatze, U., O. Göttmann, R. Podbielski, and R. Pottel. J. Mol. Liquids **37**:127 (1988).

19. Schnitzer, M. and S.I.M. Skinner. Soil Sci. **105**:392 (1968).

20. Martin, F. and F.J. Gonzalez-Vila. Z. Pflanzenernähr. Bodenk. **146**:409 (1983).

21. Martin, F. and F.J. Gonzalez-Vila. Chem. Geol. **67**:353 (1988).

22. Braun, D. and J. Pomakis. Makromol. Chem. **175**:1411 (1974).

23. Schnitzer, M. and S.V. Khan. Humic Substances in the Environment (New York: Marcel Dekker, 1972).

24. Kaatze, U. and K. Giese. J. Phys. E: Sci. Instrum. **13**:133 (1980).

25. Kaatze, U. and D. Woermann. Ber. Bunsenges. Phys. Chem. **86**:81 (1982).

26. Göttmann, 0. and A. Dittrich. J. Phys. E: Sci. Instrum. **17**:279 (1984).

27. Pottel, R. Ber. Bunsenges. Phys. Chem. **69**:363 (1965).

28. Kaatze, U. Mikrowellen Magazin **6**:46 1980).

29. Kaatze, U. and K. Giese. J. Mol. Liquics **36**:15 (1987).

30. Kaatze, U. and V. Uhlendorf. Z. Phys. Chem. N. F. **126**:151 (1981).

31. Kaatze, U. Chem. Phys. Lett. **132**:291 (1986).

32. Debye, P. Polare Molekeln (Leipzig: Hirzel, 1929).

33. Cole, K.S. and R.H. Cole. J. Chem. Phys. **9**: 341 (1941).

34. Uhlendorf, V. Messungen der dielektrischen Relaxation wässriger Polyelektro-lyt-Lösungen zwischen 5 MHz und 40 GHz, Diplom-Thesis (Göttingen: University, 1978).

35. Pottel, R. and U. Kaatze. Ber. Bunsenges. Phys. Chem. **73**:437 (1969).

36. Kaatze, U. and W.-Y. Wen. J. Phys. Chem. **82**:109 (1978).

37. Kaatze, U., H. Gerke and R. Pottel. J. Phys. Chem. **90**: 5464 (1986).

38. Kaatze, U., R. Pottel and P. Schmidt. J. Phys. Chem. **92**:3669 (1988).

39. Akerlöf, G. and A.O. Short. J. Am. Chem. Soc. **58**:1241 (1936).

Cobaltihexamine as an Index Cation for Measuring the Cation Exchange Capacity of Humic Acids.

André Maes, Francis Van Elewijck, Jacqueline Vancluysen,
Jan Tits and Adrien Cremers
K.U. Leuven, Laboratorium voor Colloidale Scheikunde
Kardinaal Mercierlaan 92, 3030 Heverlee, Belgium

Abstract

Cobaltihexamine (CoHM) is proposed as an index cation for measuring the cation exchange capacities (CEC) of "intact" and dissolved soil humic substances. Samples containing 2 to 5 mg humic substances were sufficient for each CoHM-CEC measurement. CoHM cation exchange capacity versus pH curves were better resolved than acid base titration curves in 0.1 M $NaClO_4$ and were slightly displaced towards lower pH values. Three adsorption maxima were obtained in the pH range 3 to 10, and these corresponded to functional group entities observed in $\delta pH/\delta ml$ curves from acid-base titration.

Introduction

One of the key problems in assessing the quantitative role of humic substances in transition metal ion binding is their quantitative measurement in terms of the diversity and the number of contributing species. Perdue [1] discussed the different methods for assessing the acidic functional group capacities of humic substances. Acid-base titrations are a classical means to measuring the potential number of complexing functional groups.

Methods based on the measurement of the metal ion complexation capacity may depend on the nature of the metal ion involved and generally lead to different complexation capacities [2]. The main disadvantage lies in the fact that these cations are hydrolysable, which leads to unreliable measurements in a broad pH range due to the fact that their adsorption mechanism is still a matter of discussion [3,4].

In previous contributions [5,6]) we reported on the use of silver(thiourea)$_n$ ($Ag(TU)_n$) for measuring the cation exchange capacity (CEC) of natural organic matter (NOM) in soil extracts. In the present study we examined the use of cobaltihexamine (CoHM) as an alternative means of measuring the CEC of the more hydrolyzable metal cations; this method takes advantage of the coagulating power of CoHM, which is similar to that of the $Ag(TU)_n$ complex.

The CoHM method was demonstrated for two different cases: a) for soil organic matter in its original form ("intact" or "in situ"), as it occurs in the Bh horizon of a Podzol soil, and b) for small concentrations (2 to 5 mg) of Podzol Bh soil extracts.

Experimental

Samples

Samples of the Bh horizon of a Podzol soil (Podzol Bh) were taken in Kalmthout (Belgium). Podzol Bh was essentially sand with an organic carbon content of 4.1±0.3% (10 determinations) [7] or 7% organic matter. The sample contained only traces of clay minerals. No amorphous or crystalline iron oxides were detected by either the NH_4-oxalate [8] or the Na-dithionate [9] method. The organic matter contained 58% humic acid, 12% fulvic acid and 30% humin [10]. The soil organic matter was considered as the dominant adsorption sink.

Extracts of Podzol Bh were obtained by repeated (3 times) two-hour mixing of 100 g Podzol with 1000 ml of 0.01 M NaOH, followed by 30 min centrifugation at 10000 rpm. The collected supernatants were dialysed using a Minitan ultrafiltration unit (Millipore), and the fraction $>10^5$ Dalton was used. The extract was salt-free and had a pH of 6 (Na^+-NOM22 and Na^+-NOM23). (The numbers refer to our laboratory classification). The E_4/E_6 ratio of the extract was 5, indicative of humic acids [11]. In another procedure the collected supernatants were used for the preparation of humic acid samples following the conventional humic acid precipitation in 0.01 M HNO_3. The humic acids were neutralized with NaOH prior to salt free dialysis (Na^+-HA21). Stock suspensions of about 1000 ppm of Na^+-NOM22 (82.5% NOM and Na^+-NOM23 (82.6% NOM) and Na^+-HA21 (83.4% NOM) were stored at 4°C in the dark.

Procedures

Cobaltihexamine (CoHM) spiked with ^{60}Co was used to monitor all CoHM adsorption equilibria. Since CoHM is an inert complex with an extremely low ligand exchange rate [12], it was labelled by adding 0.5 mCi ^{60}Co during its synthesis. The procedure of Bjerrum [13] was followed. A 0.01 M CoHM solution was stable in the pH range 2 to 12. Outside this range precipitation may occur.

All adsorption equilibria were run overnight at room temperature, followed by centrifugation (20 min. at 14000 rpm). Initial and supernatant solutions were radioassayed for ^{60}Co (Packard gamma scintillation counter), and the equilibrium pH was measured. The organic matter concentrated in the supernatant solutions was determined at 280 nm using an LKB Ultraspec K. The absorbances were corrected for the contribution of CoHM absorption to the optical density of the sample.

CoHM adsorption isotherms at constant pH were obtained by mixing 1 g of Podzol Bh with solutions containing known CoHM (spiked with ^{60}Co) concentrations and which were brought to a constant electrolyte concentration (0.05 M NH_4Ac or 0.1 M of either KNO_3, NH_4Ac or $NaNO_3$). The pH was maintained at predetermined values by repeated small additions of 0.1 M KOH or NaOH in order to limit exposure to pH values exceeding the equilibrium value by more than 0.5 pH units.

CoHM-CEC versus pH curves were obtained from single batch experiments under N_2 atmosphere and in the absence of background electrolyte. 1 g of Podzol Bh was mixed with 30 ml solutions of varying CoHM concentration (which corresponded to a 5 to 10 fold excess with respect to the final CoHM-CEC) and was held overnight (16 hours) at the desired pH values by using an automatic titrating unit (Radiometer) and 0.1 M NaOH or 0.1 M HCl. The amount of 0.1 M NaOH or HCl used was recorded. A similar titration procedure on separate 1 g batches of Podzol Bh in absence of CoHM served as a reference acid-base titration.

A CoHM adsorption isotherm on the humic acid fraction (Na^+-HA22) of a Podzol Bh extract was obtained by mixing 5 ml portions of the extract (≈5 mg) with CoHM solutions (spiked with ^{60}Co) of varying concentration. The final concentration of the equilibrium solutions was 0.05M NH_4Ac, and the pH was 6.85.

A stock suspension (≈1000 mg/l) of the considered Podzol Bh extract was slowly titrated with acid or base to different pH values. Separate 5 ml portions (containing approx. 5 mg NOM) were withdrawn at regular pH intervals and were equilibrated with CoHM solutions (brought to the same pH) containing a 5 to 10 fold excess of CoHM, with respect to the estimated maximum adsorption capacity at the considered pH. The CoHM-CEC-pH function was obtained in absence of supporting electrolyte on Na^+-NOM23.

Acid-base titrations were performed under N_2, on 100 ml solutions containing a known amount (48 mg) of Na^+-HA21 (MW > 10^5) in 0.1 M $NaClO_4$. These solutions (acidified to a start pH of 2.94) were titrated from acid to base and vice versa using 0.1 M NaOH or 0.1 M $HClO_4$, respectively. The titrating solutions were added as 0.05 ml increments; the pH readings were done when the change was smaller than 0.01 units/min. The whole titration took about 8 hours.

Results and Discussion

Adsorption Isotherms

Examples of CoHM adsorption isotherms on Podzol Bh and Podzol Bh extract obtained in 0.05 M NH_4Ac are shown in Fig. 1. The adsorption isotherms were plotted as the amount adsorbed versus the amount added. This was done to demonstrate the general feature that CoHM adsorption reached a maximum after adding a 3 to 5 fold excess, with respect to the maximum adsorption capacity. The maximum was identified with the CoHM cation exchange capacity (CoHM-CEC) at the specified pH.

Adsorption of CoHM resulted in the gradual coagulation of the organic matter. The organic matter concentrations remaining in the supernatant solution, after phase separation, are shown in Fig. 1 for Podzol Bh and Na^+-NOM22. These concentrations gradually decreased with increasing CoHM additions. Precipitation was complete upon addition of 1 symmetry value of CoHM. In the case of Podzol Bh, the organic matter content of the equilibrium solutions (0.5 mg at the lowest CoHM concentration) was negligible compared to the total organic matter content (70 mg), which made corrections unnecessary. Coagulation of organic matter extracts, however, gradually increased from zero to complete precipitation. The amount of CoHM adsorbed was therefore corrected for organic matter precipitation (using optical density) and was expressed per gram of precipitated organic matter. The CoHM-CEC measured in 0.05 M NH_4Ac equalled 3.76 meq/g ash-free NOM or 3.1 meq/g NOM (pH=6.90), which agreed with the value of 3 meq/g found in $Ag(TU)_n$ measurements [6] on a similar extract.

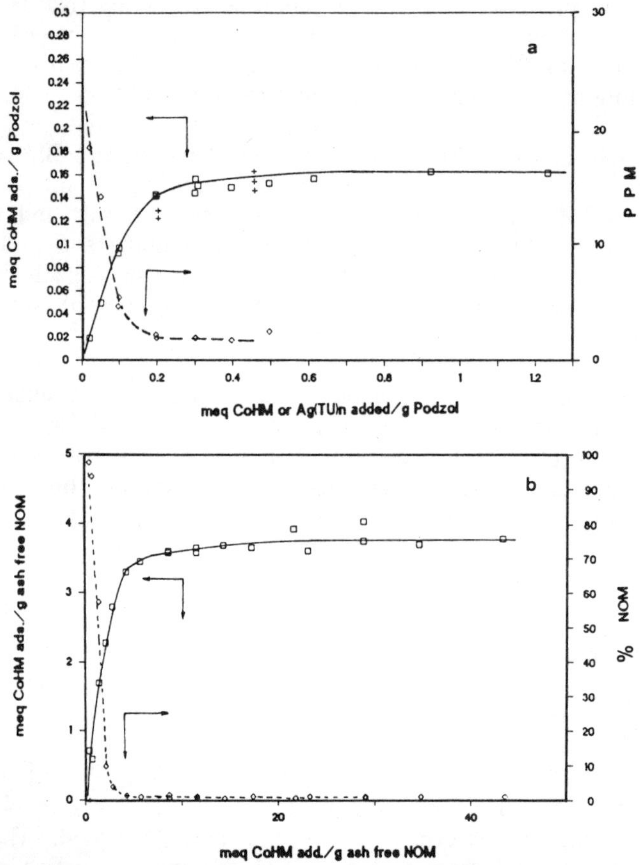

Figure 1 CoHM adsorption (meq/g) versus the amount of CoHM added (meq/g). a) Podzol Bh at pH 5.6 ± 0.1 (□). The concentration of organic matter (◇) in the equilibrium solution is also shown. Data for $Ag(TU)_n$ (pH 5.7±0.1) are included for comparison (+). b) Podzol Bh extract (Na^+-NOM 22) at pH 6.8±0.1. The % organic matter remaining in the equilibrium solution after phase separation is also shown. (Electrolyte concentration = 0.05 M NH_4Ac).

The CoHM-CEC for Podzol Bh equalled 0.162 meq/g at pH 5.6±0.1 and 0.05 M NH_4Ac. This was identical to the $Ag(TU)_n$ CEC determined under similar conditions of 0.05 M NH_4Ac and equilibrium pH (5.7±0.2) (see Fig. 1). Previously measured values for the $Ag(TU)_n$ CEC of a $CaCO_3$-amended Podzol (4 mg/g or ± 0.08 meq Ca/g) sample [6] in 0.1 M NH_4Ac buffer at pH=6, amounted to only 0.06 meq/g. This difference was easily explained by the much lower carbon content of the former sample (1.62%) as compared to the latter (4.06±0.2%).

CoHM isotherms were obtained on Podzol Bh at different pH values and 0.1 M electrolyte concentration (KNO_3 or NH_4Ac) (data not shown). The CoHM-CEC corresponding to the plateau of adsorption, increased with increasing pH as expected. The concentration of CoHM at equilibrium necessary to reach maximum adsorption (within 5%) ranged from $5x10^{-3}$ M at pH 3.5 to $1x10^{-3}$ M at pH 7.

It was sufficiently evident from the foregoing adsorption curves that CoHM-CEC values can be measured in both intact NOM of Podzol Bh and in NOM-extracts, when using a sufficient excess of CoHM (3 to 5 times the CEC). It should be stressed that CoHM-CEC measurements are possible on extracts containing about 2 to 5 mg organic matter, similar to observations made with both $Ag(TU)_n$ and CoHM on a commercial HA [6].

CoHM-CEC-pH Functions

CoHM-CEC values determined at different pH values using a sufficient CoHM excess are shown in Fig. 2 for Podzol Bh and a Podzol Bh extract, respectively. Fig. 2a shows CoHM adsorption data obtained under N_2 atmosphere in absence and presence of 0.1 M KNO_3 by overnight equilibration under careful pH control and using a 5 to 10 fold excess of CoHM. The adsorption maxima read from curves similar to Fig. 1 are also indicated. The CoHM-CEC values gradually increased from 0.060 meq/g at pH 2, to 0.225 meq/g at pH 7, and to 0.5 meq/g at pH 10. The CoHM-CEC-pH curves clearly showed different adsorption plateaus (see later).

The CoHM-CEC *vs* pH curve of the Podzol Bh extract in Fig. 2b shows adsorption plateaus similar to the observations made on Podzol Bh. CoHM-CEC values ranged from about 1.8 meq/g (at pH=3) to about 5.7 meq/g ash- free organic matter (at pH=9). Similar curves were obtained for other Podzol extracts.

Influence of Oxygen and pH History.

Fig. 2a demonstrates that under mild conditions of pH (pH 3 to 10), short term exposure of Podzol Bh to oxygen did not affect the CoHM-CEC, since the curves observed in a titration vessel under N_2 coincide with the data obtained in closed polycarbonate centrifuge tubes without special precautions to exclude air. All solutions were previously flushed with N_2. Such a result is in line with observations of Swift and Posner [14] and Borggaard [15], who found that no CEC variations occur, even under extreme conditions of pH, when working under N_2.

The pH history of the sample may, however, influence the available CEC by releasing blocked sites [16]. In an additional experiment it was indeed shown that the rapid addition of 10 ml 0.1 M NaOH to 1 g of Podzol Bh, equilibrated in a closed tube for 1 to 4 days, resulted in similar CoHM adsorption isotherms at pH 7. These isotherms, however, showed an enhanced CoHM-CEC equal to 0.390 meq/g.

The CoHM-CEC-pH functions of the soil extracts were not expected to change with the direction of the titration, since these extracts were obtained at high pH (12 to 13), and therefore no further solubilization and/or degradation and/or hydrolysis was expected in the considered pH range (3 to 10).

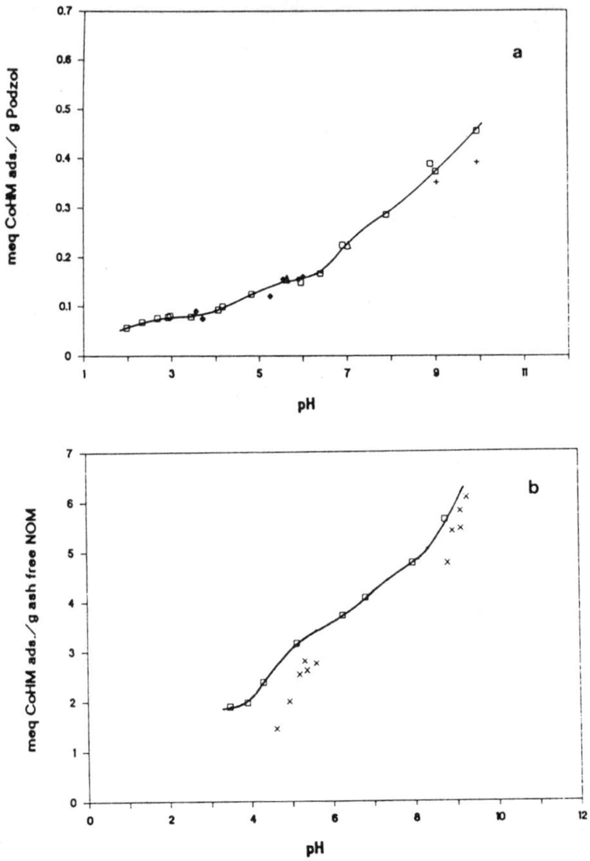

Figure 2 CoHM-CEC (meq/g) versus pH. a) Podzol Bh under N_2 atmosphere in absence (▫) and presence of electrolyte: 0.1 M KNO_3 (+) and 0.05 NH_4Ac (Δ). CoHM-CEC values from maxima of adsorption isotherms are also indicated: 0.1 M KNO_3 (◆) and 0.05 M NH_4Ac (▲). b) Podzol Bh extract (Na^+-NOM 23) in absence (▫) and presence (x) of increasing NaClO4 concentrations. From top to bottom: 0.0; 0.02; 0.05; 0.1; 0.25; 0.5 N (around pH 5) and 0.0; 0.02; 0.05; 0.1; 0.25 (around pH 9).

Adsorption Stoichiometry

Since CoHM was used as the chloride salt, the formation and adsorption of bivalent $Co(NH_3)_6Cl^{2+}$ complexes might influence the CoHM-CEC values. However,

CoHM-CEC values obtained in different $NaNO_3$/NaCl mixtures of 0.1 total molarity and pH 5.35 were identical, indicating that, essentially, $CoHM^{3+}$ is the adsorbing cation.

The mechanism of the CoHM adsorption process on Podzol Bh could be inferred from Fig. 3, which compares the CoHM-CEC function obtained under N_2 and in absence of background electrolyte, with the number of OH ions consumed to reach the different pH values in the presence and absence of COHM. The data refer to overnight equilibrium using a pH-stat in order to avoid variations due to the release of blocked sites. The starting pH of the system was 3.7. The Podzol Bh can be considered as being in the H^+-form. It was clear from Fig. 3 that the charge measured by acid-base titration in the presence of CoHM equalled or slightly exceeded the number of equivalents of CoHM adsorbed. The dissociated charge was therefore balanced by $CoHM^{3+}$ cations corresponding to a $3H^+$-$CoHM^{3+}$ stoichiometry in the considered pH range (2 to 10).

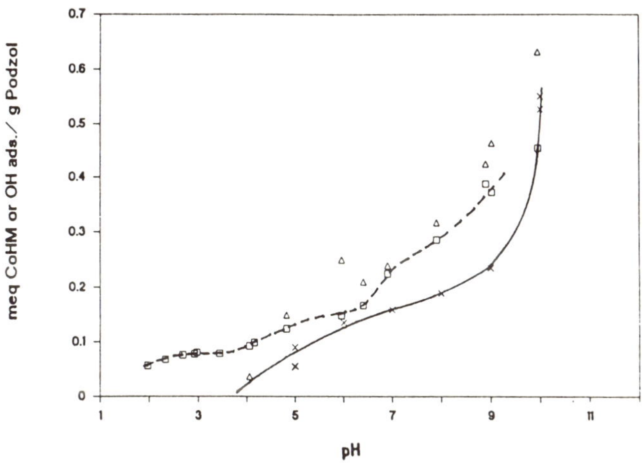

Figure 3 Comparison of CoHM-CEC-pH function (□) obtained by overnight equilibriation under N_2 with the number of OH-ions consumed to reach different pH values in absence (x) and in presence (△) of an excess of CoHM. Starting pH=3.7.

Influence of Ionic Strength and Counter Ion Effects

Inspection of Fig. 2a revealed that M^+-salt concentrations up to 0.1 M barely affected the CoHM-CEC values in the low pH range. Beyond pH 7 the CoHM-CEC decreased by about 8% upon increasing the salt concentration from zero to 0.1 M. This observation can be explained by the $CoHM^{3+}$-M^+ ion exchange behaviour of Podzol Bh. Indeed, using selectivity coefficients calculated for the different adsorption curves in 0.1 M M^+-salts at pH 3.55 (lnKc = 5) and pH 6 (lnKc = 9), one predicts a 6% decrease in CoHM occupancy upon increasing the M^+-salt concentration from 0.001 M to 0.10 M. Such differences amounted to 0.005 to 0.010 meq/g and were of a magnitude similar to the experimental uncertainties observed in this pH region.

The presence of about 0.005 meq Ca/l, which was of a magnitude similar to the Ca concentations present in soil solutions, had no effect on the CoHM-CEC, since the maximum adsorption capacity, determined at pH 6.95 in presence and absence of 0.005 meq/l $Ca(NO_3)_2$, equalled 0.225 meq/g.

A similar CoHM-CEC dependency on M^+-salt concentration was observed for NOM-extracts as shown in Fig. 2b. A decrease of about 8% in CoHM-CEC was again observed upon comparison of measurements in zero and 0.1 M $NaClO_4$.

In conlusion, it may be stated that CoHM-CEC values depended on ionic strength. The process could essentially be understood as an ion exchange competition.

CoHM probably did not displace Fe present in NOM. Indeed, interaction constants at low to intermediate CoHM occupancies on NOM extracts were estimated to equal $10^{4.5}$ at pH 7. Interaction constants for Fe^{3+} with NOM were estimated to be 10^{12}-10^{14} [17]. CoHM was therefore unable to displace Fe^{3+} from soluble and intact NOM. Comparison with interaction constants for Ca ($10^{3.5}$ at pH 5), Mg ($10^{2.15}$ at pH 4.5) [18] and Na and K ($\approx 10^{2.6}$) [19], also leads to the conclusion that equilibrium concentrations in the order of 5×10^{-3} M CoHM were more than sufficient to displace the alkali- and alkaline earth metal ions associated with NOM present in soil extract and intact soils. The CoHM method can therefore be used to obtain a first estimate of the available CEC.

Comparison with Acid-Base Titration

Careful inspection of the CoHM-CEC-pH curves of Podzol Bh and Podzol Bh extracts in Figs 2 a and b showed three regions of adsorption in both cases. Maxima were observed in the pH ranges 3.5 to 4, pH 5.5 to 6.5 and beyond pH 8. It is suggested that these regions correspond to the titration of two carboxylic type groups and one OH-type group, by analogy with generally accepted assignments.

The CoHM-CEC-pH functions were similar to conventional acid-base titration functions and are compared with them hereafter.

Comparison of the CoHM-CEC-pH function with the discontinuous titration curve of Podzol Bh in absence of CoHM in Fig. 3 revealed that the CoHM adsorption curve was displaced towards lower pH values. CoHM adsorption therefore occurred with the displacement of a number of protons. The data in Fig. 3 strongly suggest that the functional group capacity obtained from acid-base titration around pH 7 identifies with the CoHM-CEC value obtained at the plateau at pH 6.

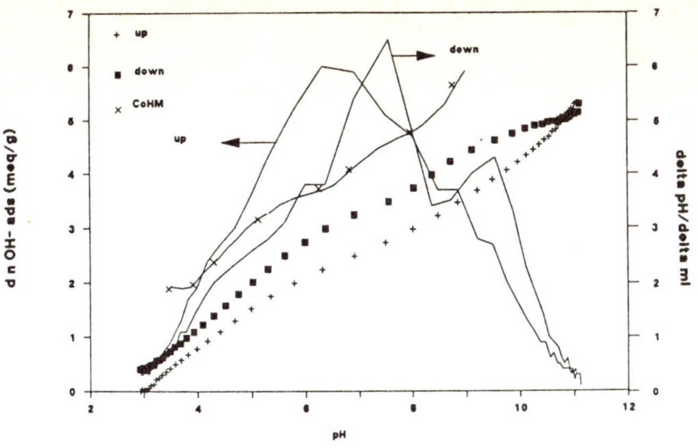

Figure 4 δn_{OH} (meq/g) and $\delta pH/\delta ml$ for the titration of Na^+-HA 21 in 0.1 M $NaClO_4$ background electrolyte. Starting pH=2.94. Forward (+) and backward (▪) titrations are shown. The CoHM-CEC (meq/g) versus pH curve (x) for Na^+-NOM 23 is shown for comparison.

Comparison of the CoHM-CEC-pH function with the continuous titration curves of Podzol Bh extract in Fig. 4 also revealed (by assuming zero charge at pH 3) that the CoHM adsorption curves were displaced towards lower pH values. The δpH/δml curves clearly show that three distinct functional groups with charges of 4.7 (pH 9.5), 3.5 (pH 7.5) and 1.6 meq/g (shoulder at pH 4.5), respectively were involved in the back titration. These data are comparable to the CoHM-CEC values of 5.0 meq/g (pH 8), 3.5 (pH 6) and 1.9 (pH 3.5).

The foregoing data strongly suggest that identical values of the various functional group capacities present in NOM were obtained from CoHM adsorption and acid-base titration but that equivalence was reached at lower pH values in the case of CoHM. Further research is necessary to confirm this observation.

The correspondance of the functional group capacity in Podzol Bh obtained from the Ca-acetate method (0.218 meq/g) and from the CoHM-CEC at pH 7 (0.225 meq/g) may be misleading. Indeed, comparison is in fact only allowed with CoHM-CEC values corresponding to the adsorption plateaus at pH 6 to 6.5 (CEC = 0.16 meq/g) or at pH 8 to 8.5 (CEC = 0.28 meq/g). Either choice fails to correspond to the Ca-acetate method.

Conclusions

CoHM adsorbs as a trivalent cation with dissolved and "intact" soil organic matter in the pH range 2 to 10. The adsorption plateau is reached at equilibrium concentrations of about 10^{-3} to $5x10^{-3}$ M CoHM and can be identified with the CoHM cation exchange capacity.

Dissolved soil humic acids are completely flocculated upon addition of 1 symmetry value of CoHM. This phenomenon allows easy phase separation and measurement of the adsorbed CoHM.

The CoHM-CEC-pH functions show the characteristic pattern of acid-base titration functions but are better resolved and are displaced towards lower pH values. At least three adsorption maxima are observed in NOM of intact Podzol Bh and in humic acids from a Podzol Bh extract. The nature of the functional groups involved cannot be identified and remains speculative, similar to the situation with acid-base titrations. The maximum of the CoHM-CEC-pH curve in the pH range 6 to 7 is suggested to be a good measure of the carboxylic functional group capacity. Functional group capacities obtained from CoHM-CEC-pH curves and acid-base titration data, appear to correspond, but should be critically evaluated in further research. Indeed, acid-base titration curves can depend on the NOM concentration and show hysteresis effects. Hydrogen bonding has been invoked [20] to explain these effects. CoHM-CEC-pH curves are independent on the NOM concentration. Increasing the ionic strength to 0.1 N monovalent salts leads to a 5 to 10% decrease in the CoHM-CEC, due to ion exchange competition. These changes are similar to observations made on acid-base titrations.

Although CoHM and $Ag(thiourea)_n$ adsorption lead to similar functional group capacities [6], CoHM is preferred over $Ag(thiourea)_n$. Indeed CoHM is a so-called inert highly stable complex, that can be considered as a non hydrolysable cation in the pH range of interest. In contrast, $Ag(thiourea)_n$ solutions are unstable beyond pH 8 and are subject to photochemical reactions. The advantage of using a non hydrolysable cation relates to the fact that methods based on the measurement of the complexing capacity using hydrolysable metal cations may depend a) on the nature of the cation and b) its ability to cause the release of a number of otherwise non titratable protons. CoHM on the contrary measures the charge imposed by the acid-base equilibrium alone.

The CoHM-CEC method is especially suited for determining small (2 to 5 mg), humic acid concentrations with a precision better than 5%. Samples containing less than 2 to 5 mg humic acid are less reliable, because the CoHM-CEC is determined as a difference of an initial and a final CoHM concentration, and because the necessary added CoHM excess needs to be sufficient for obtaining an equilibrium concentration of $\pm 5 \times 10^{-3}$ M CoHM to reach the adsorption plateau.

References

1. Perdue E.M. In: G. Aiken, D. McKnight, R. Wershaw and P. MacCarthy, Eds, Humic Substances in Soil, Sediment and Water, pp 493-526 (New York: John Wiley & Sons, 1985).

2. Alberts J.J. and J.P. Giesy. In: R.F. Christman and E.T. Gjessing, Eds., Aquatic and Terrestrial Humic Materials, pp 333-348 (Ann Arbor: Ann Arbor Sci. Publ., 1983).

3. Tipping E., C.A. Backes and M.A. Hurley. Water Res. **22**:597 (1988).

4. Cabanis S.E. and M.S. Shuman. Geochim. Cosmochim. Acta **52**:185 (1986).

5. Stalmans M., S. de Keijzer, A. Maes and A. Cremers. In: G. Desmet and C. Myttenaere, Eds., Technetium in the Environment, pp 155-167 (London: Elsevier Applied Science Publ., 1986a).

6. Stalmans M., S. de Keijzer, A. Maes and A. Cremers. In: T.H. Sibley and C. Myttenaere, Eds., Application of Distribution Coefficients to Radiological Assessment Models, pp 111-119 (London: Elsevier Applied Science Publ., 1986b).

7. Walkley A. and I.A. Black. Soil Sci. **37**:29 (1934).

8. Schwertmann U. Can. J. Soil Sci. **53**:244 (1973).

9. Mehra O.P. and M.L. Jackson. Proc. 7th Natl. Conf. on Clays and Clay Minerals, pp 317-327 (New York: Pergamon Press, 1960).

10. Van Loon L. Fixatie van Zink in Humusrijke Podzol Bodems, Thesis (K.U.Leuven, 1982).

11. Chen Y., N. Senessi and M. Schnitzer. Soil Sci. Soc. Am. J. **41**:352 (1977).

12. Basolo F. and R. Pearson. Mechanisms of Inorganic Reactions (New York: J. Wiley and Sons, 1967).

13. Bjerrum J. Metal Amine Formation in Aqueous Solution (Copenhagen: P. Haase and son, 1941)

14. Swift R.S. and A.M. Posner. J. Soil Sci. **23**:381 (1972).

15. Borggaard O.K. J. Soil Sci. **25**:189 (1974).

16. Stevenson F. Humus Chemistry (New York: Wiley Interscience, 1982).

17. Mantoura R.F.C. In: E.K. Duursma and R. Dawson, Eds., Marine Organic Chemistry, pp 179-223 (Amsterdam: Elsevier Scientific Publ., 1981).

18. Schnitzer M. and E.H. Hansen. Soil Sci. **109**:333 (1970).

19. Gamble D.S. Can. J. Chem. **51**:3217 (1973).

20. Sposito G., K.M. Holtzclaw and D.A. Keech. Soil Sci. Soc. Am. J. **41**:1119 (1977).

Investigation of Humic Acid Samples from Different Sources by Photon Correlation Spectroscopy

Marco S. Caceci and Valerie Moulin
IRDI-DERDCA-DRDD-SESD-SCPCS
Commissariat à l'Energie Atomique
B.P. 6, 92265 Fontenay-aux-Roses CEDEX, France

Abstract

Photon correlation spectroscopy (dynamic light scattering) indicated that relatively large scatterers (50-200 nm diameter) are present in a number of soil, lake and groundwater humic and fulvic acids, as well as in natural waters of high humic content, but absent in 'synthetic' humic acid. The influence of ionic strength, Ca(II), La(III), EDTA, fluoride ions, surfactants, and ultrasound on size and zeta potential of these scatterers has also been investigated.

Introduction

In order to predict the transport behavior of humic acids toward metal ions in porous media (clay and rocks) it is important to know the actual size of humic and humic/metal aggregates. Retention on rocks and clay may depend on the surface (zeta) potential of these aggregates.

The size of humic and fulvic acid (H/FA) molecules and/or aggregates has been the object of extensive investigations. Earlier studies were based on osmometry, viscosimetry, and ultracentrifugation [1]; size-exclusion chromatography [2], ultrafiltration [3], flow field-flow fractionation [4], small angle X-ray scattering [5,6] and time-dependent fluorescence depolarization [7] have been employed more recently. Published results for average molecular weights (M_w) range between 200 and 100,000 daltons. Many authors stress the extreme heterodispersity of humic acids, and the presence of a significant distribution tail toward larger molecular sizes has been consistently reported in size-exclusion chromatography.

Little is known of the influence of metal ions or other factors on the size and surface characteristics of H/FA molecules/aggregates. Whereas the idea of a colloidal [4] or micellar [8] nature of humic acid solutions is not new, its colloidal properties have been the object of only limited studies.

Photon correlation spectroscopy (PCS, also known as dynamic light scattering, quasi-elastic light scattering or laser doppler velocimetry) is a relatively new, powerful tool for the non-destructive investigation of particle size and shape in the 1 nm - 10 µm diameter range [9]. When coupled to electrophoresis, PCS allows the determination of surface (zeta) potentials [10]. A characteristic feature of PCS is that, since the signal intensity is proportional to the third power of particle size (z-weighting), it is more sensitive to the larger particles in a system.

Materials and Methods

A 3 W CW Argon ion laser was installed on a commercial PCS instrument (Malvern 4700) in order to achieve μg/l-level sensitivity (Fig. 1). The manifacturer's software was used to acquire the autocorrelation function of light scattered by the samples (128 channels, pseudo-logarithmically spaced). It is well known [11] that the autocorrelation function of coherent light scattered by a suspension of particles in Brownian motion can be described as a sum of exponential forms:

$$G(\tau) = A\,(1 + \beta\, \Sigma x_i\, e^{-2q^2 D_i \tau})$$

where A and β are, in practice, empirical constants, q is a function of wavelength and scattering angle, and x_i is the intensity of the signal due to particles of diffusion coefficient D_i. Particle hydrodynamic radius can be estimated from diffusion coefficients using the Stokes-Einstein relationship:

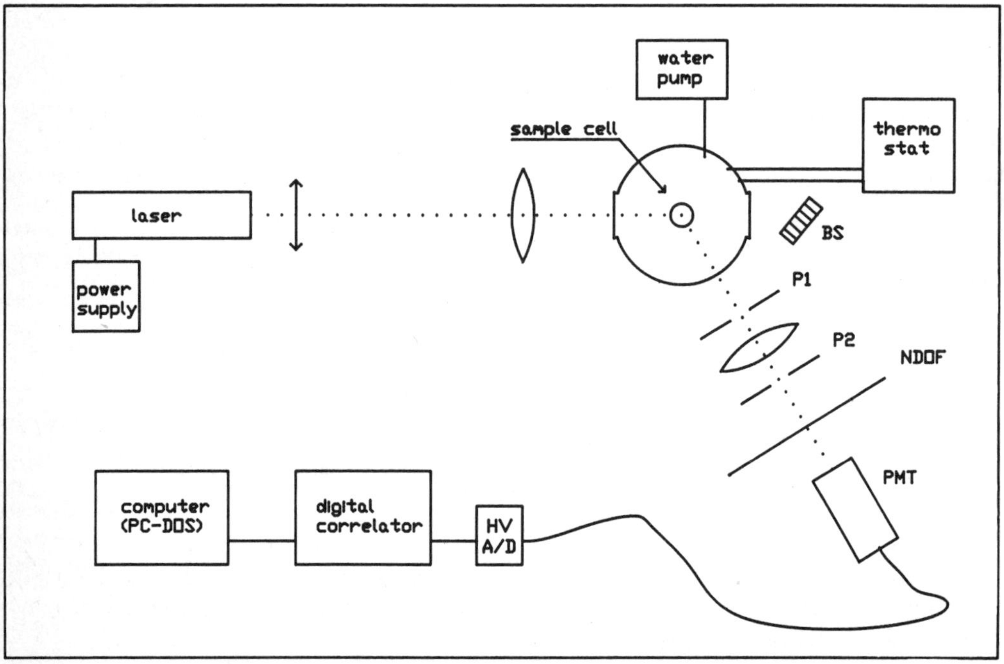

Figure 1 Set-up of photon correlation spectroscopy apparatus. Coherent light is focused by a lens inside a cylindrical glass cell. BS is a beam stop. Pinholes P1 and P2 define the detected area. A narrow band optical filter (NBOF) allows only scattered laser light to register on the photomultiplier (PMT). Pulses are amplified and discriminated (A/D) before entering, as TTL pulses, a 128-channel, 8 bit digital correlator.

$$r = k_B T / 6\pi \eta D$$

which only requires knowledge of k_B (Boltzmann's constant), T (the absolute temperature), and η (the viscosity of the fluid).

The autocorrelograms were deconvoluted to size distribution spectra by full positively constrained Inverse Laplace Transform using the Simplex [12] and the Maximum Entropy algorithm [13] as search and normalizing criteria, respectively. Exhaustive descriptions of the physical foundations and experimental set-ups for PCS experiments are given in the literature [11].

Zeta potentials were measured with a Malvern Zetasizer instrument (Fig. 2). This machine uses a 5 mW He-Ne laser, and it is, correspondingly, less sensitive.

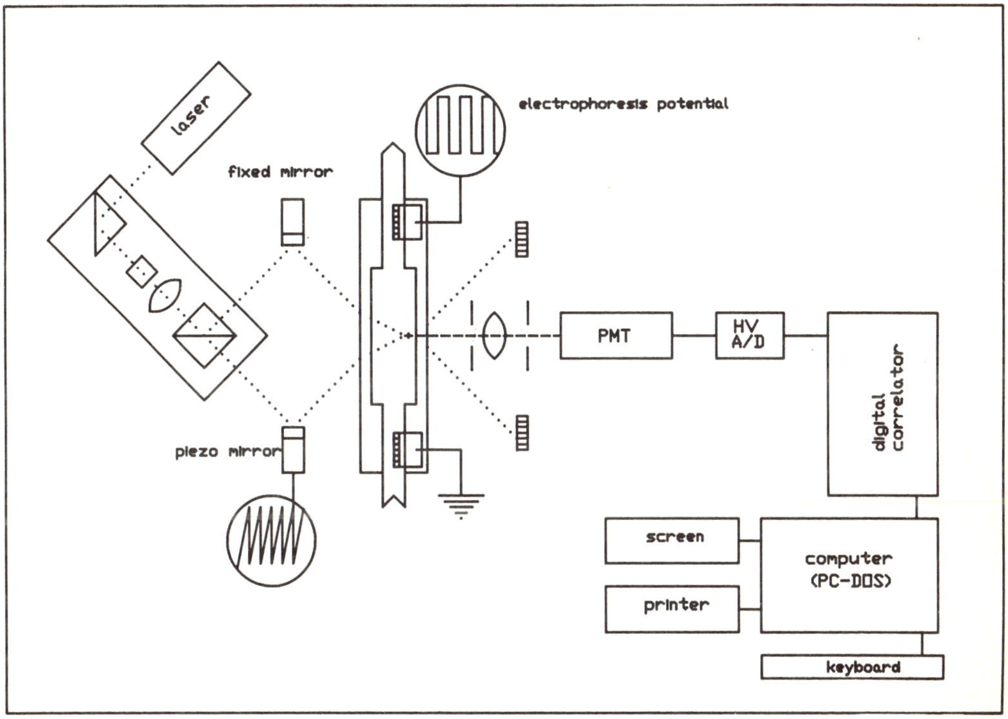

Figure 2 Set-up of the electrophoresis - photon correlation spectroscopy apparatus (Malvern Zetasizer IIc). Light from a He-Ne 5 mW laser is split in two beams, which recombine inside a 5 cm long, 0.4 mm diameter cell at the stationary layer. One of the mirrors vibrates at 200 Hz in order to impose a Doppler shift on the signal. An alternating potential (up to 200 V, 0.5 Hz) is applied through platinized platinum electrodes separated from the sample by semipermeable membranes. Actual potentials are read through sensing electrodes. The area at the crossing ot the two laser beams is observed by a PMT through two pinholes. After amplification and discrimination, pulses are sent to the hardwired digital correlator.

The samples had the following origins: Aldrich Na salt was purchased from Aldrich Chemical Co.; Aldrich purified (hydrogen form) and Gorleben (a groundwater humic acid) were from TUM, Garching (FRG) and Lake Bradford FSU, Tallahassee (Florida, USA) respectively; Fanay-Augeres (from a granitic groundwater) was purified in our laboratories [14]; podzol, rendzine (from soil) and 'synthetic' humic acid were from Centre de Pedologie Biologique, Nancy (France); sediment came from Laboratoire de Chimie Organique Structurale, Universit Pierre et Marie Curie, Paris (France); Mol (clay water) was from CEN/SCK, Mol (Belgium). 'Synthetic' humic acid was prepared by condensing catechol with triglycine [15].

Preparations were made using freshly ultrafiltered water (Millipore) at a constant temperature in a dark room and in a laminar flow hood in order to prevent dust contamination. Small amounts of NaN_3 were added to stock solutions where appropriate in order to prevent bacterial growth.

Results

The results for eight humic acid samples, one fulvic acid, one groundwater, and a sample of 'synthetic' humic acid are summarized in Table 1, giving the average particle diameter (z-weighted) and scattering intensity, the position of the maximum in the size distribution spectra, and the estimated fraction of humic matter responsible for the scattering signal.

All samples except Mol water were prepared from stock H/FAsolutions obtained by dissolving solid H/FA in excess 0.1 M NaOH. Composition of the samples was: [HA] = 100 mg/l, [$NaClO_4$] = 0.01 M, morpholino-ethane-sulphonic acid [MES] = 0.001 M (as buffer pH = 7.0), with the exception of Fanay fulvic acid (concentration: 1000 mg/l) and Mol water, which were collected, filtered and transported under rigorously anoxic conditions and measured as such. Mol water has a humic acid content of about 200 mg/l, an ionic strength equal to 0.02 M (mainly due to $NaHCO_3$), and a pH of 8.5.

Laser power was 0.5 W, with the exception of the Fanay fulvic acid (1.2 W) and the Mol water (0.15 W) sample. Reported values were normalized to 0.5 W. The observation angle was 90°. Electrolyte scatter (6.3 Kcounts/sec) was subtracted.

Scatterers in the size range 50 - 200 nm were observed in all samples except 'synthetic' humic acid. Fulvic acid contained a significantly lower amount of scatterers (one to two orders of magnitude less), as compared to humic acid from the same source and laboratory. Whether such a small amount of fulvic acid scatterers should be considered significant, or rather as an impurity left from humic acid separation, is open to debate. The existence of these scatterers was confirmed by filtering a solution of purified Aldrich HA (Amicon 15 nm nominal pore size) and observing the retentate by transmission electron microscopy.

Centrifugation in a bench centrifuge (7800 rpm nominal) was found to decrease scattering intensity but to only marginally affect the shape of the computed size distribution, and to have virtually no effect at all on the position of the maximum.

The fraction of humic matter causing the observed scattering, as reported in Table 1, is only a semi-quantitative estimate, computed by comparison with Triton X-100 micelles, on the assumption that refraction index is the same and that Rayleigh scattering

Table 1 Summary of results. Values in parenthesis are interpolated to account for different laser power and/or sample concentration.

Sample	Average Diameter (MZ) (nm) / Scattered Intensity (kcounts s^{-1}) non centrifuged	centrifuged 5'	centrifuged 30'	maximum	fraction (%)
Aldrich purified	123 / 492	122 / 495	110 / 495	100	3.9
Aldrich (Na salt)	129 / 533	141 / 711	123 / 556	120	2.5
Lake Bradford	94 / 127	94 / 121	83 / 102	80	2.2
Podzol	246 / 769	191 / 651	156 / 474	150	1.5
Sediment	216 / 469	164 / 367	138 / 295	120	1.4
Rendzine	340 / 172	330 / 116	125 / 58	120	0.3
Fanay HA	808 / 697	433 / 347	203 / 194	100	2.4
Gorleben	123 / 19	132 / 16	72 / 11	60	0.4
Synthetic			555 / 0.4	---	$3*10^{-5}$
	non filtered	centrifuged 5'	filtered 0.45 μm		
Fanay FA	--- / (29)	--- / (18)	--- / (11)	70	(0.03)
Mol Water	250 / (370)		143 / (147)	130	(1.0)

regime applies. A 10 g/l solution of Triton X-100 (9.6 nm diameter micelles) scatters 670 kcounts/sec under identical experimental conditions.

Further investigations on the size and charge of Lake Bradford and purified Aldrich humic acids indicated that:

- pH affects neither scatterers size (Fig. 3) nor zeta potential in the 4 to 9 pH range (samples at pH 8 and 9 contained, as a buffer, 4-(2-hydroxyethyl)-1-piperazine ethanesulfonic acid [HEPES]).
- Fluoride ion, EDTA, surfactants (Triton X-100, SDS) and ultrasound do not affect scatterers size. This indicates that the scatterers are of organic nature.
- Zeta potentials become less negative at higher ionic strength. The potential, within error, is a linear function of $\sqrt{I}$.
- In the presence of increasing amounts of Ca(II) or La(III), the size of scatterers increases abruptly at a definite ion concentration to a larger and still definite value (Fig. 4). Zeta potential, on the other hand, increases smoothly (Fig. 5). Critical concentrations for this Ca(II)- and La(III)-induced flocculation (for 40 μg/l Aldrich purified HA, 0.1 M $NaClO_4$, 0.001 M MES, pH 7) are about $5x10^{-3}$ and $7x10^{-5}$M, respectively, which is in good agreement with the Schulze-Hardy rule [16].

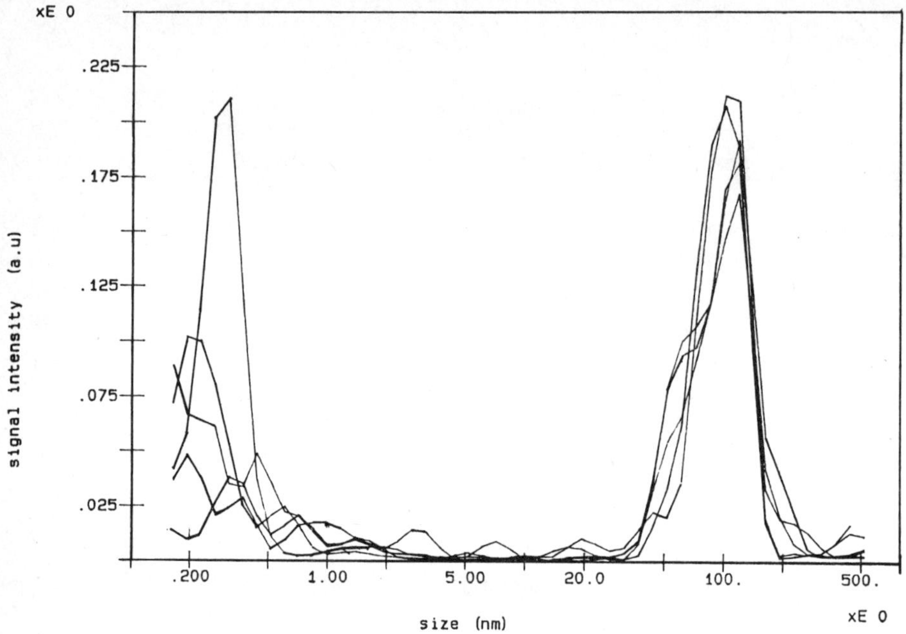

Figure 3 Size spectra of humic acid solutions. Lake Bradford HA, 42 mg/l, in 0.1 M $NaClO_4$, $1x10^{-4}$ M buffer (MES pH 4 to 7, HEPES pH 8 and 9). Signal around 0.3 nm is due to water.

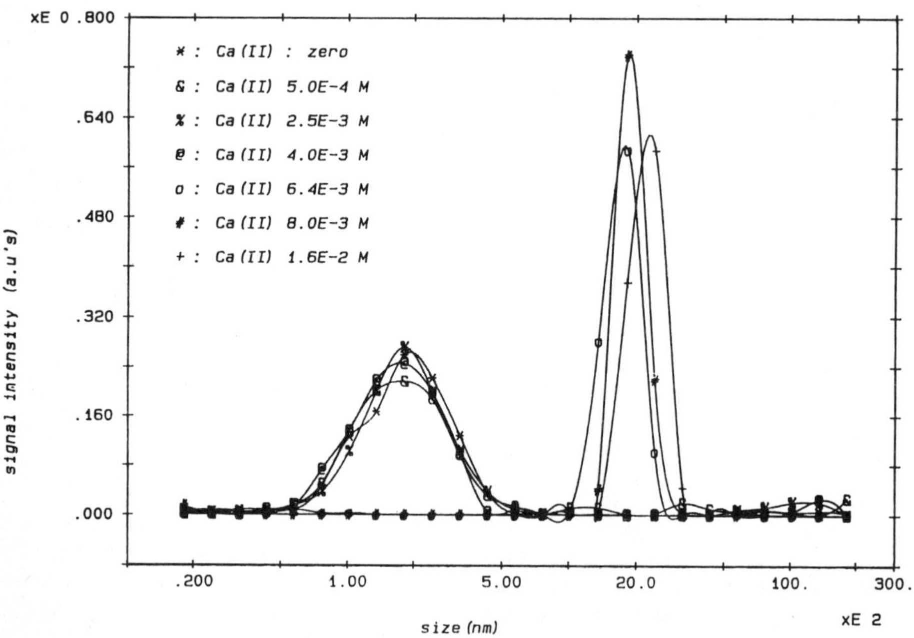

Figure 4 Intensity *vs* size distribution of scatterers in humic acid sample as a function of Ca(II) concentration. Aldrich purified HA, 40 mg/l, in 0.1 M $NaClO_4$, 0.001 M MES, pH 7.

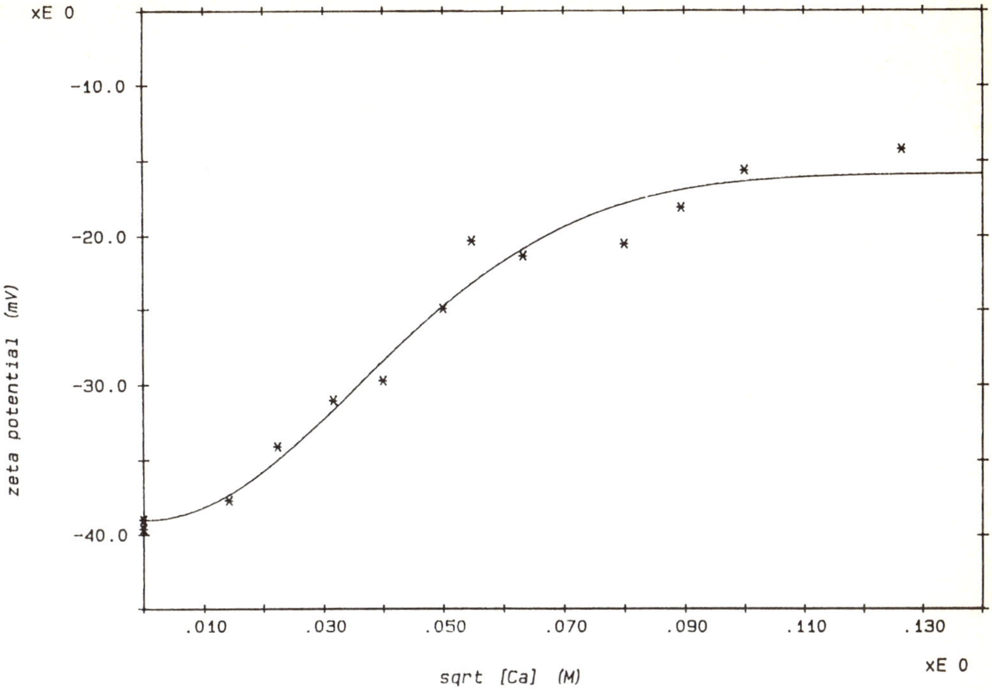

Figure 5 Zeta potential of scatterers in humic acid sample as a function of square root of Ca(II) concentration. Samples as in Fig. 4. Solid line is best least-square fit to equation $Z = A - (-39.3 - A) \exp(-C[Ca(II)])$ (mV), with $A = -15.9 \pm 1.4$ mV, and $C = 395 \pm 70\ M^{-1}$.

Conclusions

Large (60-200 nm), negatively charged, organic particles constitute a small (less than 4% in weight) but definite fraction of humic substances. Their existence may explain some results obtained by size-exclusion chromatography. Their possible role in the retention and mobilization of organics and metals in the environment, as catalysts for the hydrolysis of organics, and as nucleation centers for mineral precipitation, is open to discussion.

Acknowledgements

Humic acid samples were kindly provided by Gunnar Buckau, Gregory R. Choppin, and Paul Henrion. Part of this work was performed in a shared-cost program of the Commission of the European Communities under contract FI1W.

References

1. Schnitzer, M. and S.U. Khan. Humic Substances in the Environment (New York: Marcel Dekker, 1972).
2. Miles, C. J. and P. L. Brezonik. J. Chromatogr. **259**:499(1983).
3. Buffle, J., P. Deladoey and W. Haerdi. Anal. Chim. Acta **101**:339 (1978).
4. Beckett, R., Jue Zhang and J. C. Giddings. Environ. Sci. Technol. **21**:289 (1987).
5. Thurman, E. M., R. L. Wershaw, R. L. Malcolm and D. J. Pinckney. Org. Geochem. **4**: 27 (1982).
6 Wershaw, R. L. and D. J. Pinckney. J. Research U.S. Geol.Survey **1**:701 (1973).
7. Lochmuller, C. H. and S. S. Saavedra. Anal. Chem. **58**:1978(1986).
8. Perdue, E. M. In: R. F. Christman and E. T. Gjessing Eds, Aquatic and Terrestrial Humic Materials (Ann Arbor: Ann Arbor Science, 1983).
9. Barth, H.G., Ed., Modern Methods of Particle Size Analysis (New York: John Wiley & Sons, 1984).
10. Hunter, R.J. Zeta Potential in Colloidal Science (New York: Academic Press, 1981).
11. Berne, B.J. and R. Pecora. Dynamic Light Scattering (New York: John Wiley & Sons, 1976).
12. Caceci, M.S. and W.P. Cacheris. Byte Magazine **9**:340 (1984).
13. Livesey, A. K., M. Delaye, P. Licinio and J.-C. Brochon. Faraday Discuss. Chem. Soc. **83**:247 (1987).
14. Dellis, T. and V. Moulin. In: D.L. Miles, Ed. Proceedings of WRI-6, pp 197-201 (Rotterdam: Balkema, 1989).
15. Andreux, F., Thesis, Nancy University (France, 1978).
16. Vold, R. D., and M.J. Vold. Colloid and Interface Chemistry (Reading: Addison-Wesley, 1983).

The Behaviour of Diborane-Reduced Fulvic Acids in Flash Pyrolysis

Francisco Martin and Francisco J. Gonzàlez-Vila
Instituto de Recursos Naturales y Agrobiologia (C.S.I.C.)
Apartado 1052, 41080-Sevilla, Spain

Abstract

It is suggested that the striking structural changes introduced in humic substances by reduction with diborane may be useful in studying the role of carboxyl groups in the behaviour of these substances in flash pyrolysis. In the preliminary results shown in this communication, pyrograms of two fulvic acids of different origins and the corresponding diborane reduced substances are compared. It was found that the diborane reduction was responsible for both qualitative and quantitative changes in the pyrolytic patterns of the fulvic acids. These patterns reflect the changes in the reactivity and structural stability of the fulvic materials caused by the disappearance of the carboxyl groups.

Introduction

The progress in the application of analytical pyrolysis (Py) to the study of humic substances (HS) has been limited not only by technical factors [1], but mainly by the intrinsic complex chemical nature of these materials. This complexity has lead to serious difficulties in the interpretation of their complex pyrolytic patterns.

Recent reviews of the extense bibliography on this topic [2-5] show that Py, in combination with gas chromatography and mass spectrometry (Py-GC-MS, Py-MS), can be used as fingerprint techniques, illustrating the differences between humic fractions of various origin [6,7]. However, in contrast with the well known pathways of thermal breakdown of synthetic polymers [8], the mechanisms involved during the pyrolysis of HS are not completely understood.

The main structural information concerning the parent macromolecules obtained by Py is related to the identification of different components of the humic extracts. In fact, many typical Py-products of HS were recognized by comparison with those obtained by Py-GC-MS of well defined biopolymers [9-13].

Some approaches to the study of the Py mechanisms of HS have been carried out by examining the influence of chemical alterations of the humic molecules [14,15]. The specific modification of carboxyl groups (COOH), usually based on the preparation of ester derivatives, has not always been satisfactorily accomplished. In general, the esterification conditions are either too drastic (when methanol-HCl is used) or suffer from poor yields and non-specificity, when diazomethane is employed.

In a previous study [16] we were able to obtain stable FA preparations in the reduced state upon treatment with diborane in tetrahydrofuran (THF). The diborane treatment was found to be a useful method to obtain chemically transformed HS without carboxyl groups. This striking alteration is responsible for changes not only in the solubility and colloidal properties of this transformed HS, but probably also in the nature of the different intramolecular forces affecting the structural arrangement of the macromolecular constituents. Therefore, it is expected that the comparison by Py-GC-MS of original and reduced humic samples can provide information on the effects of carboxyl functionality upon the behaviour of HS against thermal degradation.

Materials and Methods

Two fulvic acids isolated from a podzol soil (FA-P) and from the lake water over a peatland (FA-W) were investigated. The processes for extraction and purification, as well as the physico-chemical characteristics of these samples have been reported elsewhere [17,18].

The treatment of the samples with diborane in THF to obtain the reduced preparations has been previously described [16]. Briefly, the samples were dissolved in a minimum amount of THF, and 5 ml of a 1M solution of diborane in THF (Aldrich) was slowly added under N_2 atmosphere at a temperature of 0°C. Additional amounts of diborane in THF were added periodically until hydrogen evolution was completed; the flask was stoppered under N_2 and heated to 50°C. After 20 days, the excess of diborane was destroyed by the addition of water; the tetrahydrofuran was evaporated under reduced pressure, and the boric acid formed was removed by the successive evaporation in the presence of methanol. The efficiency of the reaction was confirmed by the more or less complete disappearance of the carbonyl stretching absorption at 1720 cm^{-1} in the IR spectra, and the disappearance of the resonances for carboxyl carbons at about 175 ppm, in the ^{13}C NMR spectra.

Pyrolysis Conditions

The pyrolysis was carried out at 700°C in a CDS Pyroprobe 190 consisting of a Pt coil heated by an electric current at a rate of up to 20°C/$msec^{-1}$. The sample (2 mg) was placed in a quartz tube, using quartz wool for end plugs. The pyrolysis unit was mounted in the injection block of a HP 5730A gas chromatograph, or on a HP 5992B in the case of the Py-GC-MS system, through the resistively heated special device shown in Fig. 1, which allows the horizontal insertion of the pyrolyzer.

The volatile Py products were separated on a 25 m cross-linked fused silica column (i.d.=0.32 mm) coated with OV-101 (0.11 µm film thickness). Before starting the pyrolysis temperature program, a pre-heating of the pyrolysis unit at 250°C for 10 min allowed adhered lipid material to be removed, channeling any evolved material directly to the purge vent. The Py products were first concentrated in a loop of the column into a liquid N_2 cold trap and then the GC oven was heated from 40 to 300°C at a rate of 6°C/min. The Py products were identified by comparing their EI mass spectra with mass spectra libraries, and with mass spectra and GC retention times of standard compounds.

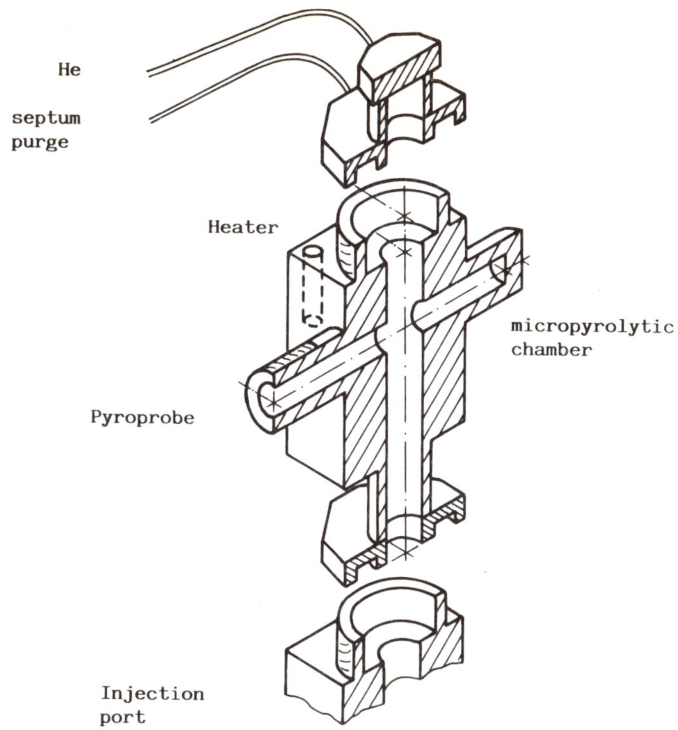

Figure 1 Expanded view of the pyrolytic accessory.

Results and Discussion

The great qualitative and quantitative differences between the pyrolytic patterns yielded by the original and diborane-reduced samples are evident (Figs 2 and 3). The original samples gave relatively simple pyrograms, as is usual for these humic fractions [17], whereas the reduced samples yielded complex chromatograms. With some exceptions identification of the Py compounds in the pyrograms was achieved (Table 1). There are many Py products, especially in the reduced samples, and Table 1 shows only the main series of compounds present, before and after diborane reduction. This exclusion is possible, since a detailed description of the identified compounds is not required in order to characterize the changes occurring upon reduction. Only significative peaks are labelled in the Figs 2 and 3 to illustrate the different nature of the Py products.

The Py products encountered in the original samples were very similar to those previously described [5, and references therein]. The podzol FA (FA-P), as other soil FA's, yielded mainly Py products from polysaccharides, together with some aromatic compounds from lignins and dialkyl phthalates. No nitrogen derivatives were identified. The FA-W gave a similar pattern of Py products, with a higher proportion of aromatic compounds, whereas dialkyl phthalates were not frequently encountered.

Table 1 Series of Py products identified in the original and reduced samples.

	FA-P	FA-P-R	FA-W	FA-W-R
Low b.p. compounds	+	+++	+	+++
n-alkanes (C8-C33)	+	++	+	++
Branched hydrocarbons	-	++	-	+
Furan,furfural,benzo-furan derivatives	+	+	+	+
Aromatic compounds from lignins	+	+	+	+
Hydroaromatics	-	++	-	++
Alkyl benzenes	-	+	+	+++
Alkyl naphthalenes	-	++	-	+

- not detected
+,++,+++ : low, medium and high presence

The reduced preparations (FA-P-R and FA-W-R) gave different types and amounts of Py products. In both cases the predominant components were series of multibranched aliphatic and aromatic hydrocarbons (labelled a and b in the chromatograms); the former were predominant in FA-P-R and the latter in the FA-W-R.

It is suggested that the loss of functionality leads to "more easily pyrolyzable samples", and these samples with a lower oxygen content yield higher relative amounts of Py products. These data are in agreement with the previous observations of Bracewell *et al.* [19] on the transformations occurring during the Py of organic matter from surface organic horizons. They found that, upon Py, raw humus originating from a retarded humification process involving the selective removal of functional groups containing oxygen and nitrogen, yielded high amounts of hydrocarbons. Py should also cause aromatic moieties from lignins and amino acids to lose functional groups and thereby yield aromatic hydrocarbons. However, aromatic hydrocarbons can also arise, by cyclisation of highly unsaturated chains, which can be formed during Py by elimination of electron-withdrawing side groups, such as hydroxyl and carboxyl groups [20].

Another explanation could be that the reductive treatment has progressed to complete the conversion of COOH into CH_3 groups. This would mean that after pyrolysis there would be more possibilities for alkyl rearrangements and reactions among alkyl radicals in the reduced samples. In fact, when appropriate electron-donating groups are present, complete reduction of COOH to a methyl group is possible [21]. Thus, studies of the diborane reduction of indole and pyrrole carbonyl derivatives, and of other "electron-rich" aromatic carbonyl compounds, showed that the carbonyl group was completely reduced to a methylene group [22].

From the studied samples it was not possible to select the most significant compounds to monitor the particular behaviour of some humus components. In general, all the

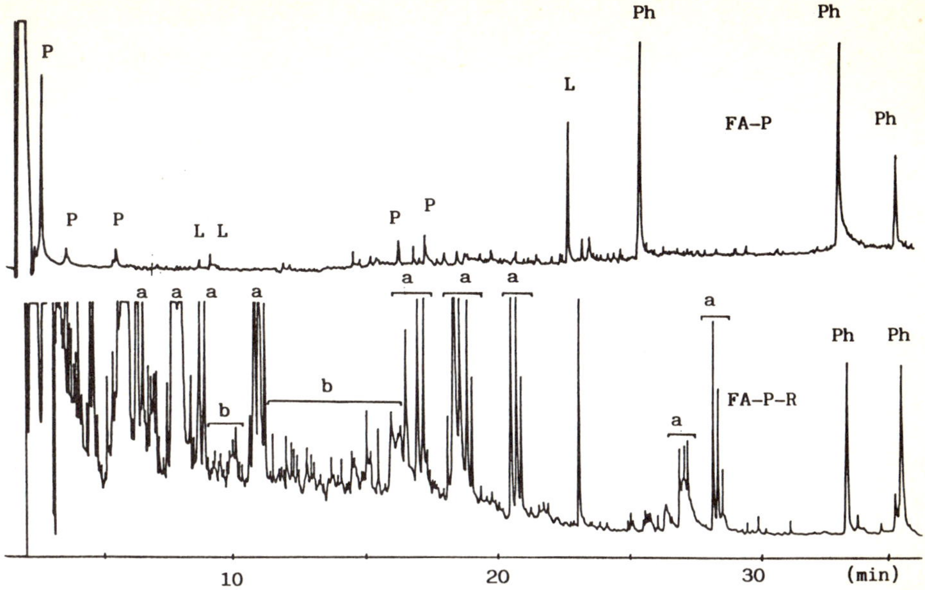

Figure 2 Pyrograms of the Podzol fulvic acid before (FA-P) and after (FA-P-R) the reductive treatment. Py-products characteristic for polysaccharides (P), lignin (L), dialkyl phthalates (Ph), multibranched aliphatic (a) and aromatic (b) hydrocarbons are indicated.

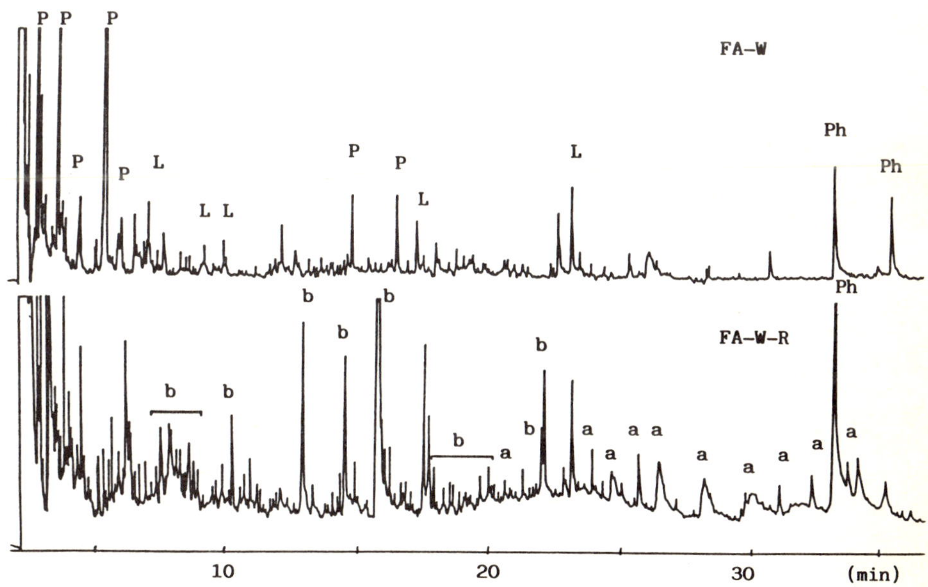

Figure 3 Pyrograms of the peat water fulvic acid before (FA-W) and after (FA-W-R) the reductive treatment. Labelled compounds as in Fig. 2.

identified compounds in the original samples were also present in the pyrograms of the reduced samples. Therefore, the diborane treatment does not apparently affect the polysaccharide moieties responsible for most of the Py products of the original samples. The present results should, however, be corroborated by further investigations with model compounds. Until now it has been demonstrated that the application of the hydroboration reaction to carbohydrates containing either a terminal or an endocyclic double bond yielded only hydration products [23].

The evolution of nitrogen compounds arising from Py of polypeptides would be very interesting, since several authors have shown that the treatment with diborane can be adapted for the specific reduction of carboxyl groups in amino acids, peptides and proteins [24, 25] without affecting the peptide bonds. Investigations on the particular behaviour of nitrogen-containing Py products are currently in progress.

References

1. Montaudo, G. J. Anal. Appl. Pyrol. **13**:1 (1988).
2. Irwing, W.J. Analytical Pyrolysis (New York: Marcel Dekker, 1982).
3. Meuzelaar, H.L.C., J. Haverkamp and F.D. Hileman. Pyrolysis Mass Spectrometry of Recent and Fossil Biomaterials (Amsterdam: Elsevier, 1982).
4. Larter, S.R. In: K.J. Voorkees, Ed., Analytical Pyrolysis. Techniques and Applications, p.212 (London: Butterworths, 1984).
5. Saiz-Jimenez, C. Origin and Chemical Nature of Soil Organic Matter. Ph.D. Diss. (Delft: University Press, 1988).
6. Wilson, M.A., R.P. Philp, A.H. Gillan, T.D. Gilbert and K.R. Tate. Geochim. Cosmochim. Acta **47**:497 (1983).
7. Saiz-Jimenez, C. and J.W. de Leeuw. J. Anal. Appl. Pyrol. **9**:99 (1986).
8. Albright, L.F., B.L. Crynes and W.H. Corcoran, Eds. Pyrolysis: Theory and Industrial Practice (New York : Academic Press, 1983).
9. Sigleo, A.C. Science **200**:1054 (1978).
10. Martin, F., C. Saiz-Jimenez and F.J. Gonzalez-Vila. Holzforschung **33**:210 (1979).
11. Obst, J.R. J. Wood Chem. Technol. **3**:377 (1983).
12. van der Kaaden, A., J. Kaverkamp, J.J. Boon and J.W. de Leeuw. J. Anal. Appl. Pyrol. **5**:199 (1983).
13. van der Kaaden, A., J.J. Boon, J.W. de Leeuw, F. de Lange, P.J.W. Schuyl, H.R. Schulten and U. Bahr. Anal. Chem **56**:2160 (1984).
14. Martin, F. and Gonzalez-Vila, F.J. Z. Pflanzenernhr. Bodenk. **146**:653 (1983).
15. Saiz-Jimenez, C. and J.W. de Leeuw. J. Anal. Appl. Pyrol. **11**:357 (1987).
16. Martin, F., F.J. Gonzalez-Vila and G. Almendros. Sci. Tot. Environ. **62**:121 (1987).
17. Martin, F. and F.J. Gonzalez-Vila. Z. Pflanzenernhr. Bodenk. **146**:409 (1983).
18. Martin, F. and F.J. Gonzalez-Vila. Chem. Geol. **67**:353 (1988).
19. Bracewell, J.M. and G.W. Robertson. Geoderma **40**:333 (1987).

20. Bracewell, J.M., G.W. Robertson and B.L. Williams. J. Anal. Appl. Pyrol. **2**:239 (1980).

21. Lane, C.F. Aldrichimica Acta **10**:41 (1977).

22. Biswas, K.M. and A.H. Jackson. Tetrahedron **24**:1145 (1968).

23. Rosenthal, A.F. and M. Sprinzl. Carbohyd. Res. **16**:337 (1971).

24. Rosenthal, A.F. and M.Z. Atassi. Biochim. Biophys. Acta **3**:410 (1967).

25. Atassi, M.Z. and A.F. Rosenthal. Biochem. J. **111**: 593 (1969) .

Some Aspects of the Characterization of Humic Substances in Lake Waters

Kalevi Pihlaja[1], Juhani Peuravuori[1], Pirjo Vainiotalo[2] and Bo Nordén[3]

[1] Department of Chemistry, University of Turku, SF-20500 Turku, Finland
[2] Department of Chemistry, University of Joensuu, SF-20500 Joensuu, Finland
[3] Department of Organic Chemistry, University of Umeå, S-90187 Umeå, Sweden

Abstract

Several different parameters, including elemental and functional group analyses, solid state ^{13}C NMR spectroscopy and E4/E6 ratios, were measured on the dissolved organic matter isolated and fractionated with ultrafiltration and XAD-8 techniques from two highly colored lakes in Finland. The results of the analyses are discussed in relation to the sample isolation (DOM, FA or HA). Multivariate analyses showed some trends in the average molecular masses, in the results of elemental analyses and, especially, in the characteristics of the ^{13}C NMR spectra so far obtained.

Introduction

In another paper [1] we described a systematic study on the isolation and fractionation of aquatic humus. The aim of the present report was to find common characteristics for the fractions isolated in the previous study by using UV-determination of E_4/E_6 ratios, elemental and functional group analyses, determination of average molecular masses (Mw, weight average), ^{13}C NMR, and multivariate analyses.

Materials and Methods

The DOM-fractions studied originate from the work presented in the paper by Peuravuori and Pihlaja [1]. Elemental analyses (C, H, N) of the freeze-dried materials were performed by various combustion techniques on a Carlo Erba Elemental Analyzer at the University of Joensuu. The concentrations of the acidic functional groups of the freeze-dried materials were estimated by titrating with NaOH [2]. The solid state ^{13}C NMR spectra (CP/MAS technique) of the freeze-dried materials were measured on a Bruker MSL-100 instrument at 25.178 MHz at the University of

Umeå. Experimental conditions: 1 ms contact time, 2.5 s repetition rate, 700 data points zero filled to 2K and 3.5 KHz spinning speed using the double air-bearing probe and Al_2O_3 rotors. The chemical shift scale is referenced to adamantine at 964 Hz (δCH_2 = 38.3 ppm). The E_4/E_6 ratios were measured conventionally [3] at 465 and 665 nm, respectively.

Results

Table 1a lists the amounts and distributions of the fractions isolated from the water samples from lakes Savojärvi (S) and Mekkojärvi (M1 and M2) [1].

Table 1a Amounts and distributions of the different ultrafiltration fractions of the water samples S (taken on February 15, 1988), M1 (taken on May 16, 1988) and M2 (taken on September 30, 1988) together with amounts of humic (HA) and fulvic acids (FA) and neutral substances (MeOH) separated with the XAD-8 technique (d-% = decrease-%).

Savojärvi (S)	mg/l	d-%	UF-%	Mekkojärvi (M1)	mg/l	d-%	UF-%	Mekkojärvi (M2)	mg/l	d-%	UF-%
UF.I	(nominal molecular mass cutoff of UF-membrane $>10^5$)										
(a)	4.4		7.9	(a)	1.3		2.9	(a)	6.9		9.6
(b)	4.2	4.5		(b)	1.2	7.7		(b)	6.6	4.3	
(c)	3.1	29.5		(c)	1.0	23.0		(c)	5.5	21.0	
UF.II	(nominal molecular mass cutoff of UF-membrane 10^4-10^5)										
(a)	10.7		19.3	(a)	15.2		34.0	(a)	25.2		34.8
(b)	8.7	18.7		(b)	12.8	16.0		(b)	22.1	12.2	
(c)	7.7	28.0		(c)	11.4	24.9		(c)	19.5	22.6	
UF.III	(nominal molecular mass cutoff of UF-membrane 10^3-10^4)										
(a)	40.3		72.8	(a)	28.1		63.1	(a)	40.4		55.7
(b)	22.7	43.7		(b)	22.1	21.5		(b)	32.9	18.6	
(c)	17.8	55.8		(c)	13.7	51.5		(c)	21.8	45.9	
UF.IV	(nominal molecular mass cutoff of UF-membrane $<10^3$)										
(c)	0.72			(c)	2.2			(c)	2.1		

Table 1b Amounts of the hydrophobic humic (HA+FA) and neutral (MeOH) substances separated with the XAD-8 technique from the minor subsamples.

	mg/l	HA/FA %		mg/l	HA/FA %		mg/l	HA/FA %
HA	4.3	17.8	HA	4.3	18.2	HA	7.7	21.2
FA	19.8	82.2	FA	19.3	81.8	FA	28.7	78.8
HA+FA	24.1		HA+FA	23.6		HA+FA	36.4	
(MeOH)	1.6		(MeOH)	1.1		(MeOH)	4.6	

Table 1b shows the amounts of humic and fulvic acids and neutral hydrophobic substances (MeOH) separated from the minor subsamples of S, M1, and M2 with the

XAD-8 technique [1]. The symbols a-c in Table 1 correspond to untreated, freeze-dried UF-concentrates (a); the same after a cation-exchange (b); or after XAD-8 treatment followed by cation-exchange (c) [1]. Experimental data for several parameters are collected in Table 2. The results of different multivariate analyses [4] based on the collected data are shown in Fig. 1-3 and discussed below.

Table 2 Results of different analyses for the water samples S, M1 and M2. The ^{13}C-ranges used were: aliphatics 0-50, carbohydrates 50-110 and aromatics 110-160 ppm.

Fraction	Mw	H/C	O/C	N/C	E_4/E_6	$COOH_{tot}$	Ar-OH	Aliph.	Carboh.	Arom.
		atomic ratio				meq/g	meq/g	%	%	%
SUF.Ia	150000	1.237	0.985	0.022	7.078	4.73	0.68	51.1	34.0	14.9
SUF.Ib	126900	1.424	1.182	0.026	7.250	4.65	0.97	46.6	41.5	11.8
SUF.Ic	43900	1.149	0.931	0.018	6.991	4.78	0.42	49.4	33.0	17.6
SUF.IIa	74400	1.283	1.092	0.018	8.868	4.32	0.97	45.8	41.5	12.7
SUF.IIb	21400	1.040	0.791	0.015	7.084	4.44	0.44	35.0	45.0	20.0
SUF.IIc	20900	1.018	0.789	0.015	6.630	4.02	0.40	37.2	40.4	22.4
SUF.IIIa	5400	1.546	1.743	0.020	9.725	4.73	0.76	44.9	32.4	22.7
SUF.IIIb	5700	1.420	1.425	0.014	9.787	4.86	0.48	33.7	37.9	28.5
SUF.IIIc	7000	1.070	0.751	0.012	9.186	5.16	0.42	34.9	36.4	28.7
SUF.IVc	3500	1.231	0.573	0.014	10.800	5.45	0.79	61.3	23.9	14.7
SHA	26200	0.992	0.719	0.019	6.097	4.56	1.44	35.2	36.2	28.7
SFA	8400	1.051	0.786	0.009	7.715	4.75	1.36	36.4	37.1	26.4
SMeOH	6400	1.245	0.746	0.012	10.833	0.72	0.01	52.4	30.0	17.6
M1UF.Ia	82700	1.389	0.876	0.038	7.213	3.85	0.87	41.9	46.8	11.3
M1UF.Ib	46400	1.270	0.763	0.032	7.170	4.33	0.74	39.7	44.8	15.5
M1UF.Ic	28500	1.172	0.918	0.032	6.468	4.93	0.53	42.8	40.1	17.1
M1UF.IIa	33000	1.155	0.983	0.017	8.542	4.84	0.98	43.9	40.4	15.7
M1UF.IIb	23200	1.007	0.819	0.018	6.018	4.44	0.72	36.1	40.4	23.5
M1UF.IIc	25000	0.983	0.776	0.019	8.396	5.01	0.37	34.9	40.3	24.7
M1UF.IIIa	7200	1.530	1.826	0.021	10.679	4.57	0.57	45.3	36.5	18.2
M1UF.IIIb	7200	1.473	1.552	0.018	9.976	5.11	0.40	34.6	38.1	27.4
M1UF.IIIc	7400	1.057	0.805	0.014	9.206	5.24	0.44	37.5	34.9	27.6
M1UF.IVc	5100	1.169	0.565	0.011	9.619	4.40	0.61	59.3	26.4	14.3
M1HA	25500	0.940	0.690	0.023	6.933	4.65	1.42	37.0	33.6	29.4
M1FA	11300	0.990	0.716	0.013	7.538	5.20	1.22	36.8	35.7	27.6
M1MeOH	6600	1.174	0.696	0.014	10.350	0.61	0.01	55.3	31.0	13.7
M2UF.Ia	102000	1.149	0.890	0.012	7.705	4.21	0.67			
M2UF.Ib	69500	1.026	0.696	0.014	6.857	3.81	0.40			
M2UF.Ic	67200	1.048	0.736	0.018	6.925	4.00	0.53			
M2UF.IIa	44400	1.137	0.964	0.016	9.167	4.37	0.88			
M2UF.IIb	27300	0.950	0.746	0.013	8.025	3.82	0.71			
M2UF.IIc	25900	0.951	0.747	0.015	7.652	4.59	0.35			
M2UF.IIIa	9700	1.236	1.278	0.012	10.833	4.66	0.85			
M2UF.IIIb	8900	1.072	0.933	0.011	9.556	5.88	0.42			
M2UF.IIIc	10000	0.937	0.740	0.010	9.333	4.29	0.39			
M2UF.IVc	6200	1.150	0.568	0.007	7.372	4.99	0.71			
M2HA	29400	0.935	0.649	0.018	7.690	4.52	1.22			
M2FA	12500	0.979	0.664	0.011	9.778	5.53	1.02			
M2MeOH	9000	1.877	0.667	0.009	9.867	0.50	0.01			
NoHA	26800	0.868	0.658	0.024	9.580	4.57	1.46	31.3	37.1	31.6
NoFA	8100	0.930	0.690	0.016	9.146	5.36	1.38	33.5	38.5	28.0

NoHA and NoFA are Nordic Reference Standards (humic and fulvic acid, respectively).

Discussion

E_4/E_6 Ratio

The correlation of the E_4/E_6 ratio (Table 2) with the average molecular masses (Mw, weight average) of the different fractions was not very good (r=-0.57, n=20), but nevertheless agrees with the postulation that the ratio should decrease with the increasing molecular size [3]. Even poorer correlations were found with the H/C ratio and aromaticity (Table 2), although the E_4/E_6 ratio appears roughly to increase with the increasing H/C ratio [5]. As a whole our results support the view that the E_4/E_6 ratio has no direct relationship to the nature of the water soluble humic substances nor to the content of condensed aromatic rings in HA and FA [3].

Functional Group Analyses

The titrimetric total concentration of the COOH-groups and that of the Ar-OH correlated reasonably well ($r \approx 0.7$) with the relative carbon contents (solid state NMR: 160-190 and 140-160 ppm, respectively). No other significant correlations were found, although the titrimetric contents of the functional groups (Table 2) are in accordance with those described in the literature [6]. However, our results did not allow a clear distinction between the FA- and HA-fractions.

Elemental Analyses

The H/C, O/C, and N/C ratios based on the average of three parallel C,H,N-analyses (the content of oxygen was taken as a difference from 100 %) are shown in Table 2. The H/C, O/C and N/C ratios are similar to the values of aquatic humic substances reported earlier [7]. Of the UF-concentrates, those obtained by method (c) gave the smallest H/C ratios, especially in the case of the UF.III-concentrates (Fig. 1). The O/C ratio behaved similarly but only in the UF.III category. This is not at all reflected by the carboxylic group contents and only moderately by the Ar-OH contents (Table II). The ability of the XAD-8 treatment to remove carbohydrates may, however, partly explain the decrease in the O/C ratio when applying method c. The N/C ratios alone do not allow any specific discussion.

The peculiar behavior of the UF.I-concentrates (Fig. 1) supports the postulation [8] that ultrafiltration can be subject to interactions which increase values of molecular masses, whereas pH-adjustment (method c) can result in lower molecular masses by disrupting interactions between the humic material and e.g. Fe and Al. Fig. 1d shows the H/C vs O/C Van Krevelen plot for all UF-fractions obtained by method (c) together with the HA- and FA-fractions isolated from the original water samples with the XAD-8-technique. It is also indicated that UF.IC-concentrates (cutoff $>10^5$) as a whole were not alike the other (c)-concentrates. Furthermore fractions UF.IVc (cutoff $<10^3$) formed another distinct cluster with high H/C values, which speaks for increased aromaticity [9]. All other fractions were practically clustered together with the humic substances (HA and FA). Here the HA-materials are always placed to the left of and below the FA homologues as suggested earlier

[9]. The M1 and M2 fractions do not fall closely together in plot 1d, which is obviously an indication of the difference between the spring and autumn sample.

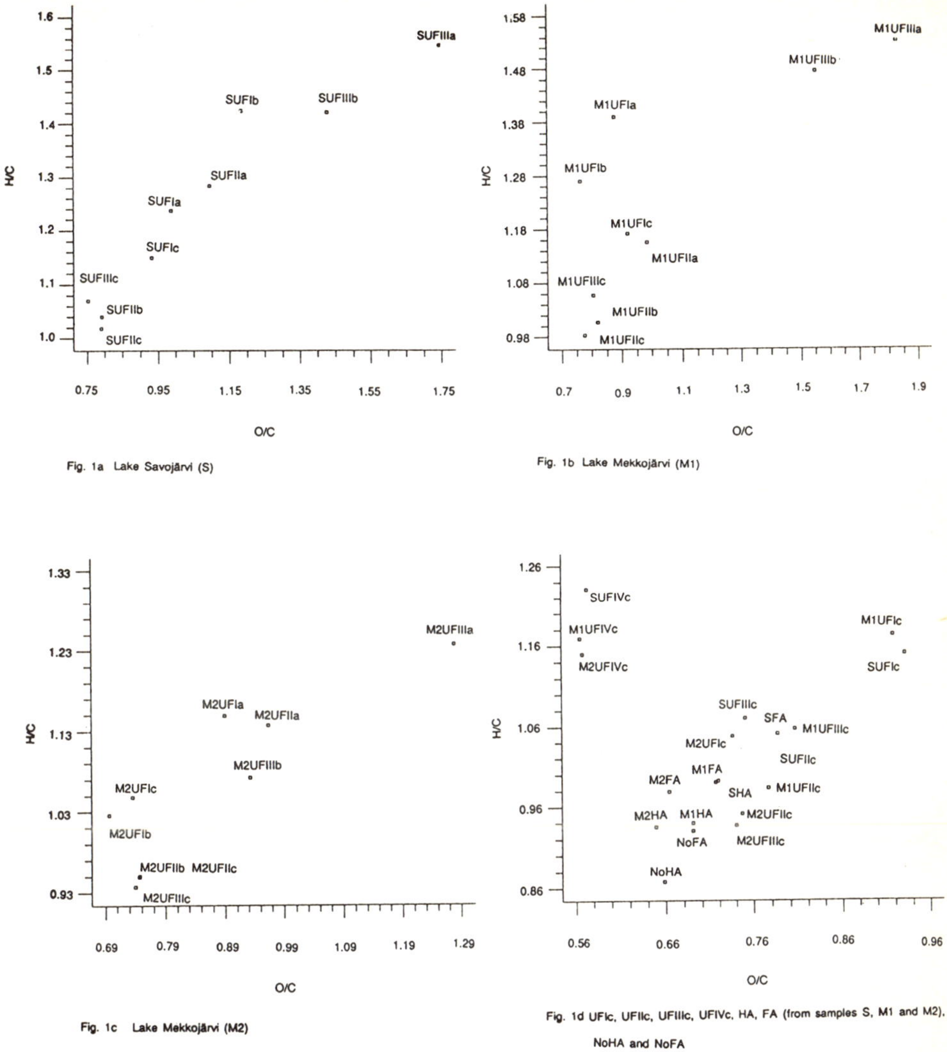

Figure 1 Van Krevelen plots for several groups of samples from the present study.

^{13}C NMR

Typical examples of the ^{13}C NMR spectra are shown in Fig. 2. According to the spectra the aromatic fraction between 110 and 140 ppm increased when carrying out the cation exchange (method b) on the original type a UF-concentrates. Method b decreased the relative amount of the aliphatic fraction (0-50 ppm) more than method c, especially in the cutoff ranges 10^4-10^5 and 10^3-10^4. Despite the fact that 97% of the carbon atoms present have been stated to appear in the solid state NMR spectra [10], it is not possible by inspection of the spectra alone to draw similar conclusions as, e.g., from Fig. 1d. However, certain deductions about the structural variations can be made quite easily on the basis of the percentage constitutions given in Table 2 [11].

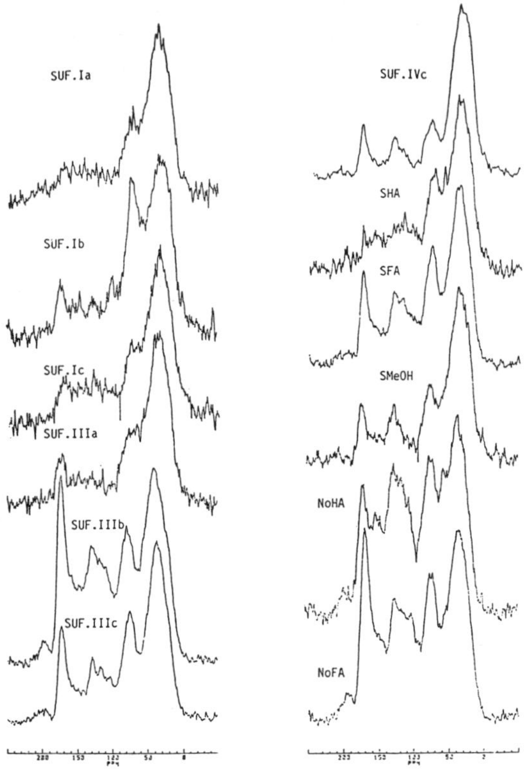

Figure 2 Typical examples of the ^{13}C CP/MAS NMR spectra.

Multivariate Analysis

It was not possible to find any single parameter specific for a certain group of fractions or to HA and FA. A multivariate analysis based on average molecular mass (Mw or Mn), H/C, O/C, N/C, aliphatic, aromatic, and carbohydrate contents, as well as on aromatic/aliphatic and aromatic/carbohydrate ratios, was proved to be the best approach. Principal component analysis (PCA) is a useful technique for reducing the number of variables in a data set by finding linear combinations of those variables that explain most of the variability [4]. In addition to the E_4/E_6 ratios [12] the functional group contents cannot be applied unambiguously to resolve different sample categories. Fig. 3 shows two example plots against the first two principal

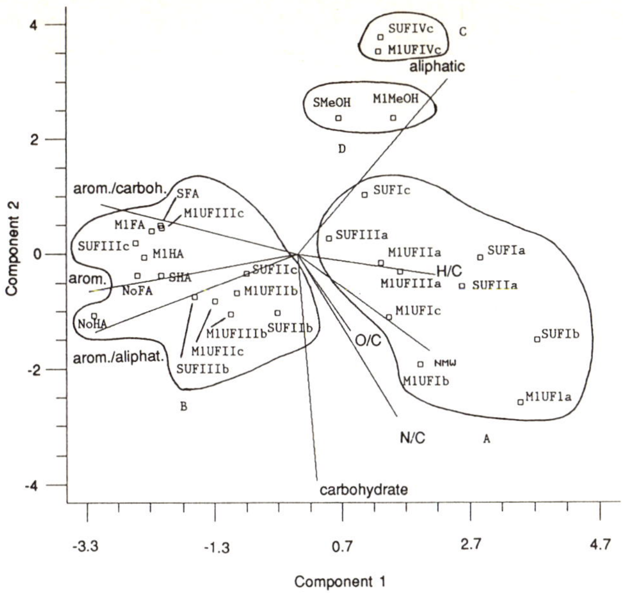

Fig. 3a Biplot of first two principal components

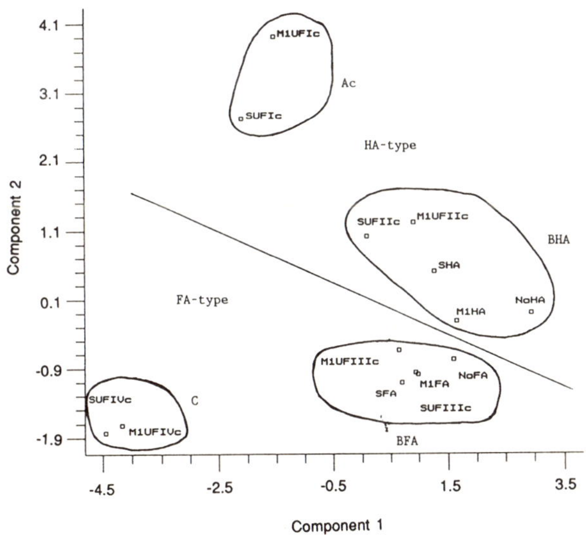

Fig. 3b Scatterplot of first two principal components

Figure 3 PCA plots for the materials studied.

components of the above variables. Since the NMR data were not yet available for the M2 material, the analysis consisted only of the data for samples S and M1.

In the biplot (Fig. 3a) the first component accounts for 46.4% of the total variability and the second component increased the statement level to a fair 70.9%. Four clearly separated clusters (A-D) can be seen in Fig. 3. Cluster A contains all UF.I-concentrates (types a, b and c) together with all other original UF-concentrates (only type a). All of the other UF-concentrates together with the HA-and FA-substances from the original water samples and with the Nordic Reference Standards (NoHA and NoFA; Table 2), fell into cluster B. As expected, two minor clusters were formed by materials of the nominal molecular mass cutoff $<10^3$ (C, Tables 1 and 2) and MeOH-extracts (D, Tables 1 and 2). Cluster C can also be seen in Fig. 3b (levels of statement 53.3 and 82.1% for the first and second PC, respectively), which actually demonstrates the clustering of HA- and FA-type substances (BHA and BFA, respectively). It also shows that clusters Ac and C, containing the UF.I-concentrates and UF.IV-filtrates, respectively, treated by method c are very well separated from clusters BHA and BFA.

To conclude, it can be stated that the cutoff range 10^3-10^4 in the ultrafiltration (UF.III-concentrates) is quantitatively largest, and the separated materials closest resemble fulvic acids.

Acknowledgement

The authors wish to thank the Maj and Tor Nessling Foundation and the Academy of Finland, the Research Council for Natural Sciences, for financial support.

References

1. Peuravuori, J. and K. Pihlaja. - Presented in this publication.

2. (a) Borggaard, O. K. Acta Chem. Scand. **A28**:121 (1974); (b) Perdue, E. M. Geochim. Cosmochim. Acta **42**:1351 (1978).

3. Chen, Y., N. Senesi and M. Schnitzer. Soil. Sci. Am. J. **41**:352 (1977).

4. Sharaf, M. A., D. L. Illman and B. R. Kowalski. Chemometrics (NewYork: John Wiley & Sons, 1986).

5. Ertel, J. R. and J. I. Hedges. In: R. F. Christman and E. T. Gjessing, Eds., Aquatic and Terrestrial Humic Materials, pp. 143-163 (Ann Arbor: Ann Arbor Sci. Press, 1983).

6. Thurman, E. M. Organic Geochemistry of Natural Waters (Dordrecht: Martinus Nijhoff/Dr. W. Junk Publishers, 1985).

7. Steelink, C. In: G. R. Aiken, D. M. McKnight, R. L. Wershaw and P. MacCarthy, Ed., Humic Substances in Soil, Sediment and Water, pp. 457-476 (New York: John Wiley & Sons, 1985).

8. Aiken, G. R. and R. L. Malcolm. Geochim. Cosmochim. Acta **51**:2177 (1987).

9. Visser, S. A. Environ. Sci. Technol. **17**:412 (1983).

10. Vassalo, A. M., M. A. Wilson, P. J. Collin, J. M. Oades, A. G. Waters and R. L. Malcolm. Anal. Chem. **59**:558 (1987).

11. Perdue, A. M. Geochim. Cosmochim. Acta **48**:1435 (1984).

12. MacCarthy, P., S. J. DeLuca, K. J. Voorhees, R. L. Malcolm, and E. M. Thurman. Geochim. Cosmochim. Acta **49**:2091 (1985).

Isolation and Fractionation of Humic Substances in Lake Waters

Juhani Peuravuori and Kalevi Pihlaja

Department of Chemistry,
University of Turku, SF-20500 Turku, Finland

Abstract

Dissolved organic matter (DOM) in water from two lakes (spring and autumn samples) was isolated and fractionated by ultrafiltration (UF) and XAD-8 techniques. The difference between the spring and autumn samples was greatest for the colour, COD_{Mn} and DOC values. High-performance size-exclusion chromatography gave useful information on the molecular size distribution of the DOM. The two largest fractions of DOM were within the nominal molecular mass cutoffs 10^3-10^4 and 10^4-10^5, respectively. Decreasing of DOM of UF-concentrates in the XAD-8 treatment was the greatest for the former cutoff. Some similarity of molecular masses (weight and number averages) could be found between the DOM-fractions within the 10^3-10^4 cutoff isolated by UF and fulvic acids isolated directly from the original water samples by the XAD-8 technique. The total amounts of humic substances separated from the UF-concentrates were about 25% higher than those obtained directly from the original water samples. The results show that the usefulness of a given reference humus depends strongly on its origin.

Introduction

Aquatic humic substances form the major fraction of the dissolved natural organic matter (DOM) in water, typically accounting for 40-80% of the dissolved organic carbon (DOC) [1-3]. Humic substances are generally characterized as high molecular mass, yellow-coloured organic acids that are refractory end products in the degradation of plant and microbial organic material. However, the processes by which aquatic humic substances are formed from precursor organic material are a subject of much speculation (allochtonous or autochtonous origin). Historically, there has been more investigation of the biogeochemical roles of DOC in marine ecosystems than in freshwater ecosystems. Because freshwater generally has higher DOC concentrations and lower concentrations of inorganic ions than seawater, DOC may be more important in freshwater ecosystems [4].

During the past ten years important advances have been made in the isolation and characterization of DOM [1,2,5-9]. For instance, DOC in river waters has been described to consist of on the average 50% hydrophobic acids (40% fulvic and 10% humic acids), 30% hydrophilic acids and 20% carbohydrates, carboxylic acids, amino acids and hydrocarbons [2].

The ultimate goals of our research are: (1) to compare different isolation and fractionation procedures for freshwater DOM; (2) to examine the chromatographic behavior of the original water samples and of the different isolated DOM-fractions; (3) to clarify the significance of a given reference humus; (4) to test the empirical conclusion [10] about the interdependence of the E2/E3-ratio and the average molecular mass (Mw, weight average) of the water-soluble humic substances; and (5) to estimate the influences of prefiltration and the sampling season on the water quality.

Materials and Methods

Sample Collection

Natural humic water samples (approx. 350 l) were collected from two lakes: Savojärvi, situated in a marsh region in the southwestern part of Finland [11]; and Mekkojärvi, a small, forest lake situated in the Evo district of Lammi in southern Finland [12]. A small water sample from Kevojärvi - an extension of the Utsjoki River situated in Utsjoki, the northernmost municipality in Finnish Lapland - served as a reference (11). Both Lake Savojärvi and Lake Mekkojärvi have highly coloured waters, whereas the water in the Utsjoki River is fairly faintly coloured. The water samples were collected 1 m below the surface into glass bottles. Samples were taken from Savojärvi 15.02.1988 (S), from Mekkojärvi 16.05.1988 (M1), 30.09.1988 (M2) and 24.05.1989 (M3) and from Kevojärvi 01.09.1988. The samples were prefiltrated (0.2 µm Nuclepore polycarbonate filter cartridge, no. 611101) [12] as soon as possible after sampling and thereafter stored in the dark at 4°C.

Characterization

Organic carbon concentrations were determined by combustion techniques (Ionics Model 555 Carbon Analyzer). The spectrophotometric measurements were made by use of a computer programmed spectrophotometer (LKB Ultrospec II). The concentrations of inorganic cations, as well as some other characteristic properties for the water samples, were determined at the Turku Water Works (Table 1). For gel chromatography, serially connected dextran gels, Sephadex G-100, G-75 and G-25SF columns (2.5x25 cm), and NaN_3 (0.02%, pH 7.8) as an eluent, were employed to attain the best possible resolution [12,13]. The ionic strength of the eluent was then fairly close to the ionic strength of the original water samples, and, therefore, it should not have caused any major structural changes in the humic substances. High-performance size exclusion chromatography (HPSEC) with a macroporous silicic-particle TSK G3000SW column (7.5x300 mm and a 7.5x75 mm precolumn;

0.01 M sodium acetate, adjusted to pH 7 with acetic acid, as an eluent) was carried out as described elsewhere [14]. This was done to obtain good resolution and reproducibility for DOM, e.g. when studying relative molecular mass (M_r) distributions.

Table 1 Some characteristic properties of water samples before (*) and after (**) prefiltration.

Physico-chemical property		Kevojärvi *	**	(S) Savojärvi *	**	(M1) Mekkojärvi *	**	(M2)Mekkojärvi *	**	(M3)Mekkojärvi *	**
Temperature (water)	°C	12.1		1.3		8.2		11.5		7.0	
Temperature (air)	°C	15.4		2.2		14.3		13.7		22.4	
Oxygen	mg/l					5.5				3.9	
Conductivity 25°C	mS/m			6.0	6.0	4.0	4.0	4.5	4.5	4.5	4.5
pH 20°C		6.7	6.7	5.8	5.8	5.5	5.5	5.7	5.7	5.5	5.4
Colour	mg Pt/l	20	20	225	200	175	175	300	250	180	180
E2/E3 (250/365 nm)			5.1		4.3		4.2		4.2		
TC (total C)	mg/l	9.4	9.4	25.3	24.0	21.6	20.5	31.5	30.9	22.5	21.4
IC (inorganic C)	mg/l	2.5	2.4	3.9	3.2	1.7	0.9	1.4	1.4	2.1	1.7
TOC (DOC)	mg/l	6.9	7.0	21.4	20.8	19.9	19.6	30.4	29.5	20.4	19.7
COD_{Mn}	mg/l			97	97	92	91	125	123	94	92
Alkalinity	meq/l			0.17	0.11	0.13	0.13	0.10	0.10	0.06	0.05
Total hardness	mg/l			1.2	1.2	1.0	1.0	1.2	1.1	0.6	0.7
N_{tot}	mg/l			1.1	1.0	0.36	0.25	0.5	0.6	0.5	0.4
P_{tot}	mg/l			0.028	0.025	0.17	0.01	0.08	0.05	0.04	0.03
Fe_{tot}	mg/l			1.8	1.5	0.7	0.5	1.3	1.2	0.6	0.5
Al_{tot}	mg/l			0.49	0.40	0.25	0.25	0.40	0.40	0.22	0.24
Mg	mg/l			2.7	2.7	2.2	2.3	3.0	3.0	1.0	1.3
Ca	mg/l			3.8	3.8	3.2	3.3	3.6	3.0	2.4	3.2
Na	mg/l			4.1	4.4	2.9	2.5	1.8	1.7	2.9	2.1
K	mg/l			1.1	0.91	0.57	0.57	1.0	1.0	1.1	1.1
SiO_2	mg/l			8.2	3.8	6.5	5.7	8.2	7.6	7.8	7.8
Cl^-	mg/l			5.5	5.5	1.5	1.3	1.3	1.1	1.3	1.2
SO_4^{2-}	mg/l			21.9	15.3	17.0	12.6	21.4	18.0	12.4	10.4

Isolation and Fractionation Procedure

Organic solutes are generally divided into hydrophobic and hydrophilic (base, acid and neutral) fractions on the basis of their adsorption on nonionic (mostly XAD-8) and ion-exchange resin adsorbents [8]. The XAD-8 resin is specific to hydrophobic organics. In our work attention was mainly paid to the separation of hydrophobic acid and neutral fractions using ultrafiltration and the XAD-8 technique [7] as isolation and fractionation procedures. After prefiltration the original water samples were divided in two parts in the ratio of approx. 5 to 1. The scheme used for the isolation and fractionation of DOM is given in Fig. 1.

A tangential flow membrane filtration [15-17] on a 4 GPM Pellicon Cassette System by Millipore was used to obtain three DOM concentrates, with nominal molecular mass cutoffs, 10^5 (UF.I), 10^4-10^5 (UF.II) and 10^3-10^4 (UF.III), and a filtrate $<10^3$ (UF.IV) from the major part (ca 280 l) of original water samples. The above UF-concentrates (UF.I-UF.III) were divided into three parts. One of them was evaporated to dryness (freeze-drying) without further treatment (method a). Only cations and some organic constituents were removed from DOM of the second fraction on a strong cation-exchange resin (Dowex 50W, X8 200-400 mesh, column 2.7x40.5 cm, flow rate 0.5 l/h) before the fraction was freeze-dried (method b). After acidification to pH 2, the third fraction was treated on a 3.3x60.5 cm Amberlite XAD-8 (20-50 mesh) column with a flow rate 1.6 l/h; humic substances adsorbed onto the resin were eluted at pH 13 directly onto the strong cation-exchange resin (method c).

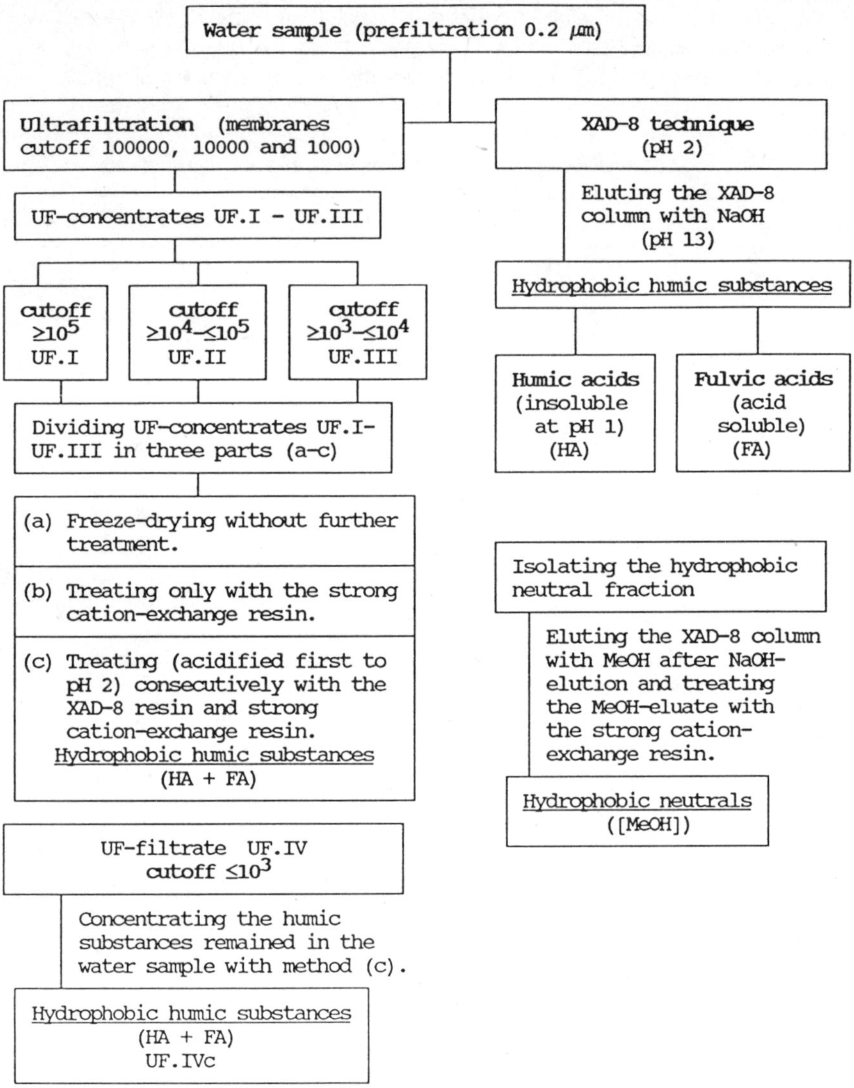

Figure 1 The scheme used for the isolation and fractionation of the dissolved organic matter (DOM).

Humic substances remaining in DOM of the UF-filtrate (UF.IV) were isolated with the method (c). Fractions UF.Ic-UF.IIIc, as well as UF.IVc (total humic substances), were not routinely divided into fulvic and humic acids.

The minor part (ca 55 l) of original water samples was treated with the XAD-8 technique, see method (c) above. Hydrophobic humic substances were adsorbed onto

XAD-8-type acrylic ester polymer-based resin at pH 2; humic substances become nonionic on this resin [1,2,5-7]. It is possible to use a two-step desorption procedure, for instance by eluting with a $NaHCO_3$ (pH 8) buffer to desorb carboxylic acids and then with 0.1 M NaOH (pH 13) to desorb phenolic compounds, respectively [2,18]. In this study the total amount of humic substances adsorbed onto the resin were eluted at pH 13 (0.1 mol/l NaOH). The hydrophobic acids isolated on the XAD-8 resin corresponded to aquatic humic substances, and included mainly fulvic acids (FA, acid soluble) with a trace of humic acids (HA, acid insoluble at pH 1). After separating HA from FA by adjusting the pH to 1 with concentrated HCl, chlorides were removed with XAD-8 resin. The organic material which was not eluted from the adsorption resin by NaOH, i.e. the hydrophobic-neutral fraction (a mixture of hydrocarbons and carbonyl compounds [8]), was eluted by methanol, and the cation-exchange was carried out.

Results

Preparative Isolation and Fractionation of Dissolved Organic Matter

Table 2 lists the amounts and distributions of DOM in the different UF-concentrates (UF.I-UF.IV) fractionated with the methods (a-c) that was used on the water samples S, M1, M2 and M3 as described above. The hydrophobic humic substances isolated with method (c) from the UF-concentrates were divided into HA and FA groups only in the case of sample M3. Table 2 also lists the influence of methods (b) and (c) on the level of DOM in the different UF-concentrates in comparison to the original amount (method a). The total amounts of humic substances (HA+FA) isolated with method (c) from different UF-concentrates are also given in Table 2. Table 3 shows the amounts of the hydrophobic humic and neutral substances isolated with the XAD-8 technique directly from the original water samples. The carbon contents of the effluents (hydrophilic dissolved organic matter) in isolating humic substances with XAD-8 technique from the original water samples are given in Table 4.

By treating UF-concentrates (UF.I-UF.III) with XAD-8 resin (method c), the over-all content of hydrophobic neutrals ([MeOH]) in the M3 sample was 0.72 mg/l.

Chromatography of Dissolved Organic Matter

The void volume and total effective volume of the columns (Sephadex gels and TSK G3000SW) were determined by using Blue Dextran 2000 and acetone, respectively. For the TSK G3000SW column the standard calibration curve was determined using Pharmacia Calibration Kit proteins (ribonuclease A, chymotrypsinogen A, ovalbumin and albumin) and pyridoxal-5'-phosphate, sucrose, sodium deoxycholate, sodium taurocholate, tryptan blue, cyanocobalamin, tannic acid and gammaglobulin. The absorbance was measured at 254 nm. Sephadex gel chromatography gave a similar wavy elution profile - based mainly on three molecular sizes - for the DOM-material of all the studied water samples, as reported previously, e.g. for Lake Mekkojärvi water [11,13]. Polyphenolic DOM-material was observed to adsorb - even if insignificantly - to the dextran gels. The resolution of DOM-materials by HPSEC was good (eight peaks were resolved). The HPSEC-chromatogram of sample S (eight peaks) was quite similar to that given in reference [14]. The HPSEC-peaks were very sharp, and their relative heights were used instead of their areas [14] in calculating relative molecular mass (M_r) distributions of DOM. A detailed discussion on the above chromatographic methods can be found in citation [19] and references given therein.

Table 2 Amounts and distributions of the UF-concentrates fractionated with different methods (a-c). Influence of the methods (b) and (c) on the decreasing (d-%) of DOM.

(S) Savojärvi				(M1) Mekkojärvi				(M2) Mekkojärvi				(M3) Mekkojärvi					
	mg/l	d-%	UF-%		mg/l	d-%	UF-%		mg/l	d-%	UF-%		mg/l	mg/l HA+FA	d-%	HA/FA %	UF-%
UF.I	(nominal molecular mass cutoff of UF-membrane $>10^5$)																
(a)	4.4		7.8	(a)	1.3		2.9	(a)	6.9		9.6	(a)	2.9				10.1
(b)	4.2	4.5		(b)	1.2	7.7		(b)	6.6	4.3		(c)HA	0.92	1.6	44.8	57.5	
(c)	3.1	29.5		(c)	1.0	23.0		(c)	5.5	21.0		(c)FA	0.68			42.5	
UF.II	(nominal molecular mass cutoff of UF-membrane 10^4-10^5)																
(a)	10.7		19.1	(a)	15.2		34.0	(a)	25.2		34.8	(a)	16.9				58.0
(b)	8.7	18.7		(b)	12.8	16.0		(b)	22.1	12.2		(c)HA	4.4	10.4	38.5	42.3	
(c)	7.7	28.0		(c)	11.4	24.9		(c)	19.5	22.6		(c)FA	6.0			57.7	
UF.III	(nominal molecular mass cutoff of UF-membrane 10^3-10^4)																
(a)	40.3		71.8	(a)	28.1		63.1	(a)	40.4		55.7	(a)	9.3				31.9
(b)	22.7	43.7		(b)	22.1	21.5		(b)	32.9	18.6		(c)HA	0.05	4.3	53.8	1.2	
(c)	17.8	55.8		(c)	13.7	51.5		(c)	21.8	45.9		(c)FA	4.2			98.8	
UF.IV	(nominal molecular mass cutoff of UF-membrane $<10^3$)																
(c)	0.72			(c)	2.2			(c)	2.1			(c)HA	0	6.0			
												(c)FA	6.0			~100	

Method (a) = original UF-concentrate, method (b) = treated only with cation-exchange resin and method (c) = treated with XAD-8 resin (XAD-8 technique). Total amounts (mg/l) of hydrophobic humic substances in the UF-concentrates (UF.Ic+UF.IIc+UF.IIIc+UF.IVc; HA+FA): S: 29.3; M1:28.3; M2:48.9; M3: 22.3.

Table 3 Amounts of hydrophobic humic (HA+FA) and neutral ([MeOH]) substances isolated with the XAD-8 technique from the original water samples.

(S) Savojärvi			(M1) Mekkojärvi			(M2) Mekkojärvi			(M3) Mekkojärvi		
	mg/l	HA/FA %		mg/l	HA/FA %		mg/l	HA/FA %		mg/l	HA/FA %
HA	4.3	17.8	HA	4.3	18.2	HA	7.7	21.2	HA	7.2	31.6
FA	19.8	82.2	FA	19.3	81.8	FA	28.7	78.8	FA	15.6	68.4
HA+FA	24.1		HA+FA	23.6		HA+FA	36.4		HA+FA	22.8	
(MeOH	1.6		(MeOH)	1.1		(MeOH)	4.6		(MeOH)	0.79	

Table 4 Carbon contents (mg/l) of the XAD-8 effluents of humic substances isolated from the original water samples at pH 2.

(S) Savojärvi			(M1) Mekkojärvi			(M2) Mekkojärvi			(M3) Mekkojärvi		
TC	IC	DOC	TC	IC	DOC	TC	IC	DOC	TC	IC	DOC
12.6	0.5	12.1	11.9	0.5	11.4	14.1	1.5	12.6	12.6	0.6	12.0

TC = total, IC = inorganic and DOC = dissolved organic carbon.

Table 5 Average molecular masses of the original water samples.

	(S) Savojärvi	(M1) Mekkojärvi	(M2) Mekkojärvi	(M3) Mekkojärvi	Kevojärvi
Mw	12400	10800	11800	15700	6900
Mn	2300	4000	4900	5700	1800
Mw/Mn	5.39	2.70	2.41	2.75	3.83

Because all columns always have different separation efficiencies for the standard compounds as well as for humic substances (chemically different compounds), relative molecular masses (M_r) determined from the calibration curve are not absolute. The M_r-values and their corresponding relative peak heights [19] were used to calculate the average molecular masses (Mw, weight average and Mn, number average) for DOM. Mw-values are always greater than Mn-values, since in polydisperse mixtures the smaller individuals make a larger contribution upon counting the molecules. The quotient Mw/Mn is thus a measure for the inhomogeneity of the mixture. In the present study M_r-distributions were estimated as the average Mw-, Mn- and Mw/Mn-values.

Table 5 gives the molecular mass distributions for the DOM-material of the original water samples S, M1, M2, M3 and Kevojärvi. Table 6 demonstrates the influence of the separation methods (a), (b) and (c) on the molecular mass distributions of the different UF-concentrates. Only in the case of sample M3 were humic substances isolated with method (c) from UF-concentrates and divided into HA and FA fractions, and method (b) was not applied at all. At the end of Table 6 the molecular mass distributions of UF-filtrates are also given. Table 7 shows the molecular mass distributions of the different hydrophobic humic substances separated from the original water samples by use of the XAD-8 technique. The molecular mass distributions of hydrophobic neutrals are also presented in Table 7.

Table 7 Average molecular masses of humic and fulvic acid fractions separated from the original water samples by use of the XAD-8 technique. Average molecular masses of hydrophobic neutrals ([MeOH]).

	(S) Savojärvi	(M1) Mekkojärvi	(M2) Mekkojärvi	(M3) Mekkojärvi	Nordic References
Fraction	Humic acids (HA)				
Mw	26200	25500	29400	34600	26800
Mn	6400	9900	6400	6600	7300
Mw/Mn	4.09	2.58	4.59	5.24	3.67
Fraction	Fulvic acids (FA)				
Mw	8400	11300	12500	10600	8100
Mn	5500	6700	5800	4400	5600
Mw/Mn	1.53	1.69	2.16	2.41	1.45
Fraction	Hydrophobic neutrals ([MeOH)]				
Mw	6400	6600	9000	7500	
Mn	4000	5100	3500	3500	
Mw/Mn	1.60	1.29	2.57	2.14	

Table 6 Influence of the methods (a)-(c) on the average molecular masses of the UF-concentrates. Average molecular masses of UF-filtrates.

	(S) Savojärvi	(M1) Mekkojärvi	(M2) Mekkojärvi	(M3) Mekkojärvi
UF-concentrate	UF.I (nominal molecular mass cutoff of UF-membrane $>10^5$)			
method	(a)	(a)	(a)	(a)
Mw	150000	82700	102000	93500
Mn	150000	66000	99200	82300
Mw/Mn	1.00	1.25	1.03	1.14
UF-concentrate	UF.I (nominal molecular mass cutoff of UF-membrane $>10^5$)			
method	(b)	(b)	(b)	(c) HA
Mw	126900	46400	69500	65800
Mn	100100	12100	20600	16600
Mw/Mn	1.27	3.83	3.37	3.96
UF-concentrate	UF.I (nominal molecular mass cutoff of UF-membrane $>10^5$)			
method (c)	HA+FA	HA+FA	HA+FA	FA
Mw	43900	28500	67200	23000
Mn	16600	11200	18200	6100
Mw/Mn	2.64	2.54	3.69	3.77
UF-concentrate	UF.II (nominal molecular mass cutoff of UF-membrane 10^4-10^5)			
method	(a)	(a)	(a)	(a)
Mw	74400	33000	44400	36800
Mn	42800	15400	30900	13200
Mw/Mn	1.74	2.14	1.44	2.69
UF-concentrate	UF.II (nominal molecular mass cutoff of UF-membrane 10^4-10^5)			
method	(b)	(b)	(b)	(c) HA
Mw	21400	23200	27300	26500
Mn	11100	13100	17000	7900
Mw/Mn	1.93	1.77	1.61	3.35
UF-concentrate	UF.II (nominal molecular mass cutoff of UF-membrane 10^4-10^5)			
method(c)	HA+FA	HA+FA	HA+FA	FA
Mw	20900	22500	25900	19600
Mn	10200	8900	10900	7400
Mw/Mn	2.05	2.53	2.38	2.65
UF-concentrate	UF.III (nominal molecular mass cutoff of UF-membrane 10^3-10^4)			
method	(a)	(a)	(a)	(a)
Mw	5400	7200	9700	9500
Mn	3100	4800	5200	6200
Mw/Mn	1.74	1.50	1.87	1.53
UF-concentrate	UF.III (nominal molecular mass cutoff of UF-membrane 10^3-10^4)			
method	(b)	(b)	(b)	(c) HA
Mw	5700	7200	8900	22600
Mn	3400	5000	5900	6400
Mw/Mn	1.68	1.44	1.51	3.53
UF-concentrate	UF.III (nominal molecular mass cutoff of UF-membrane 10^3-10^4)			
method (c)	HA+FA	HA+FA	HA+FA	FA
Mw	7000	7400	10000	9800
Mn	3700	5100	5900	6300
Mw/Mn	1.89	1.45	1.69	1.56
UF-filtrate	UF.IV (nominal molecular mass cutoff of UF-membrane $<10^3$)			
method(c)	(HA)+FA	(HA)+FA	(HA)+FA	(HA)+FA
Mw	3500	5100	6200	8800
Mn	2000	2900	1800	3400
Mw/Mn	1.75	1.76	3.44	2.59

method (a) = original UF-concentrate, method (b) = treated only with cation exchange resin and method (c) = treated with XAD-8 resin (XAD-8 technique).

Discussion

Colour and DOC values were 2-3 times greater and COD_{Mn} values 7-9 times greater for Lakes Savojärvi and Mekkojärvi than for the averages of literature data for 35 forest lakes and 10 water reservoirs in Finland [19]. Colour value, total (TC) and inorganic carbon (IC) contents, as well as silicate and sulfate concentrations of the water samples decreased most during the prefiltration (0.2 μm). The difference between spring and autumn samples was greatest for colour, COD_{Mn}, TC and TOC values and N_{tot}, Fe_{tot}, Al_{tot}, Mg and K concentrations. The allochtonous origin of humic substances had an obvious effect on their M_r-values. However the absorption at 365 nm was not necessarily strengthened - opposite to empirical conclusions [10] - along with the increasing Mw-values (or Mn-values) when compared to absorption at 250 nm [19].

The results of Sephadex gel chromatography supported previous conclusions (e.g. in [20]) about its inadequacy and limited usefulness. It did not seem to be a recommendable analytical method for characterization of the DOM-material, although it can be applied successfully to some predetermined experiments [19]. With the HPSEC technique it was, however, possible to estimate the M_r-distribution of the DOM-material in the water samples studied with acceptable accuracy on the basis of the peak heights only.

The Mw-values of DOM in the original water samples S (Savojärvi), M1 (Mekkojärvi), M2 (Mekkojärvi), M3 (Mekkojärvi) and Kevojärvi, were between 7000-16000 and Mn-values between 1800-5700 (Table 5). The M_r-distributions (Mw/Mn-values) were fairly high for each water sample. DOM of water samples S and M1 consisted of about 65% of M_r-values 10^3-10^4 and about 26% of 10^4-10^5, whereas samples M2 and M3 consisted of about 42% of M_r-values 10^3-10^4 and about 55% of 10^4-10^5. The HPSEC-chromatograms of samples M1-M3 were, however, fairly similar and the only difference appeared in their M_r-distributions. This is probably caused by the seasonal variation between samples M1-M3. DOM of the Kevojärvi water consisted almost solely (approx. 92%) of M_r-values 10^3-10^4. The HPSEC-chromatograms of the water sample groups S, M and Kevojärvi did not resemble each other.

As shown in Table 2 humic substances isolated with method (c) from water samples S, M1 and M2, consisted of about 51% of the nominal molecular mass cutoff 10^3-10^4 of UF-concentrates and about 35% of 10^4-10^5. In the case of sample M3 the largest fraction of humic substances fell into the nominal molecular mass cutoff 10^4-10^5 (about 47%) of UF-concentrates, and the next largest into the cutoff $<10^3$ and 10^3-10^4 (approx. 27% and 19%, respectively). This is also probably caused by the seasonal variation between samples M1-M3.

As to the UF-concentrates, the decrease of the DOM-material was largest (on the average about 52%) for UF.III (cutoff 10^3-10^4) in the treatment with method (c) (XAD-8 technique). The decrease of the DOM-material was also largest (on the average about 28%) for UF.III in the treatment with method (b) (treated only with cation-exchange resin). When treating the other UF-concentrates (UF.II or UF.I, cut-off 10^4-10^5 and 10^5, respectively) with methods (c) or (b), the decrease of the DOM-material was on the average about 29 and 10%, respectively. This suggests that the greatest changes in the block structure of DOM happens within the nominal molecular mass cutoff 10^3-10^4. The most significant change toward smaller M_r-values (estimated with Mw/Mn-values in Table 6) was found when treating UF.I-concentrates with methods (b) or (c).

The highest HA content for the humic substances of the different UF-concentrates was about 58% (UF.I) and the smallest about 1% (UF.III). Accordingly, the contribution of HA seems to be insignificant within the nominal molecular mass cutoff $<10^3$. The total amounts of humic substances (HA+FA) separated from the UF-concentrates were about 25% higher than those obtained directly from the original water samples (comparison of Tables 2 and 3). The M_r-distributions of HA for the original water samples did not resemble any of those (HA+FA together) obtained for the different UF-concentrates. Some similarity could, however, be found in the M_r-distributions of FA for the original water samples and humic substances (HA+FA) obtained for the UF-concentrates within the nominal molecular cutoff 10^3-10^4. Apparently some splitting of the larger molecular sizes occurred during the acid treatment (as also reported earlier [21]) - particularly in the acid separation (at pH 1) of HA and FA from the XAD-8 techniques. The proportion of HA recovered by the XAD-8 technique directly from the original water samples was about 22% (10-15% in ref.[2]) and that of FA about 78% (85-90% in ref. [2]), which together with hydrophobic neutrals (about 9% of the FA content; ref. [22] gives about 10%) formed about 46% of DOC. Fairly high Mw- and Mn-values for the hydrophobic neutrals (Table 7) can result, e.g. because of a formation of metal-organic complexes [17] or traces of humic substances which are not eluted from the resin with NaOH [23]. The similarity of the M_r-distributions (in Table 7 Mw/Mn-values) of HA and FA fractions separated by the XAD-8 technique from the Savojärvi water and those of the corresponding Nordic References suggests a similar origin (runoff from swamped land); whereas Lake Mekkojärvi has weakly swamped shores, and its drainage basin consists mainly of surface runoff [19].

The DOM-material with M_r-values of 1300-2800 according to HPSEC has been stated to cause most mutagenicity in chlorine disinfection of drinking water [11, 14]. Of the water samples studied in the present study that of Lake Savojärvi exhibited most of the DOM-material (about 24%) within the above M_r-range [19]. About 9% of the DOM-material in the UF-concentrates, about 7% of the humic acids and 8% of the fulvic acids separated by the XAD-8 technique also fell within this M_r-range.

The HPSEC-analyses on WSHS-materials isolated with the two different methods proved that the colloidally dispersed humic acids and the dissolved fulvic acids in humic waters are predominantly present as distinct entities. Therefore a suggestion [24] that the quality and quantity of, e.g. the fulvic acids, is mainly dependent on the isolation method appears to be an overstatement. Humus chemistry is a very demanding and, in its way, exceptional area of research, which more than deserves the phrase "working with humic materials is often frustrating, always laborious and seldom rewarding" [25].

Acknowledgement

The authors wish to thank the Maj and Tor Nessling Foundation and the Academy of Finland, the Research Council for Natural Sciences, for financial support.

References

1. Thurman, E. M. and R. L. Malcolm. In: R. F. Christman and E. T. Gjessing, Eds., Aquatic and Terrestrial Humic Materials, pp. 1-23 (Ann Arbor: Ann Arbor Sci. Press, 1983).

2. Thurman, E. M. Organic Geochemistry of Natural Waters (Dordrecht: Martinus Nijhoff/Dr.W. Junk Publishers, 1985).

3. Steinberg, C. and U. Muenster. In: G. R. Aiken, D. M. McKnight, R. L. Wershaw and P. MacCarthy, Eds., Humic Substances in Soil, Sediment and Water, pp. 105-145 (New York: John Wiley & Sons, 1985).

4. McKnight, D., E. M. Thurman and R. L. Wershaw. Ecology **66**:1339 (1985).

5. Aiken, G. R. In: G. R. Aiken, D. M. McKnight, R. L. Wershaw and P. MacCarthy, Eds., Humic Substances in Soil, Sediment and Water, pp. 363-408 (New York: John Wiley & Sons, 1985).

6. Aiken, G. R., E. M. Thurman and R. L. Malcolm. Anal. Chem. **51**:1799 (1979).

7. Thurman, E. M. and R. L. Malcolm. Environ. Sci. Technol. **15**:463 (1981).

8. Leenheer, J. A. Environ. Sci. Technol. **15**:578 (1981).

9. Pihlaja, K. and J. Peuravuori. - Presented in this publication.

10. De Haan, H. In: R. F. Christman and E. T. Gjessing, Eds., Aquatic and Terrestrial Humic Materials, pp. 165-182 (Ann Arbor: Ann Arbor Sci. Press, 1983).

11. Kronberg, L., B. Holmbom and L. Tikkanen. Sci. Tot. Environ. **47**:343 (1985).

12. Jones, R. I., K. Salonen and. H. De Haan. Freshwater Biol. **19**:357 (1988).

13. De Haan, H., R. I. Jones and K. Salonen. Freshwater Biol. **17**:453 (1987).

14. Vartiainen, T., A. Liimatainen and P. Kauranen. Sci. Tot. Environ. **62**:75 (1987).

15. Michaels, A. S. In: E. S. Perry, Ed., Progress in Separation and Purification, pp. 297-334 (New York: John Wiley & Sons, 1968).

16. Ganzerli Valentini, M. T., L. Maggi., R. Stella and G. Ciceri. Chem. in Ecol. **1**: 279 (1983).

17. Maggi, L., R. Stella and G. Ciceri. Annali di Chimica **74**:257 (1984).

18. MacCarthy, P., M. J. Peterson., R. L. Malcolm and E. M. Thurman. Anal. Chem. **51**:2041 (1979).

19. Peuravuori, J. Chemical and physical properties of humus and their variation in different natural conditions. The significance of a reference humus. Part I. Isolation and fractionation of dissolved organic matter (DOM) from natural waters. Characterization using gel chromatography (in Finnish), (Turku: University of Turku, 1989).

20. Hine, P. T., and D. B. Bursill. Water Res. **18**:1461 (1984).

21. De Haan, H., G. Werlemark and T. De Boer. Plant and Soil **75**:63 (1983).

22. Visser, S. A. Environ. Sci. Technol. **17**:412 (1983).

23. Aiken, G. R. In: F. H. Frimmel and R. F. Christman, Eds., Humic Substances and Their Role in the Environment, pp. 15-28 (Chichester: John Wiley & Sons, 1988).

24. Fuchsman, C. H. Peat. Industrial Chemistry and Technology, p. 160 (New York: Academic, 1980).

25. Schnitzer, M. In: Schnitzer, M. and S. U. Khan, Eds., Soil Organic Matter, p. 57 (Amsterdam: Elsevier, 1978).

Dating of Groundwaters by ^{14}C-analysis of Dissolved Humic Substances

Catharina Pettersson and Bert Allard
Department of Water and Environmental Studies
Linköping University, S-581 83 Linköping (Sweden)

Abstract

Fulvic acids of the DOC of five deep (139-409 m) groundwaters were recovered (adsorption on DEAE-cellulose) and used for age determination (^{14}C determined by accelerator mass spectrometry), and compared with fulvic acids recovered from a shallow groundwater and a surface water. The composition of the seven different fulvic acids was similar, despite variations in hydrochemical conditions and residence times, indicating a high stability of this molecular weight fraction. The ages calculated from the ^{14}C-content of the fulvic acid fraction (600 to 10000 y) are less than the ages indicated from analyses of the dissolved carbonate (data available for three sites). Using a fraction of the DOC with high stability (like the fulvic acid fraction) as a ^{14}C-source when assessing the ages of subsurface waters, appears to be superior to using dissolved carbonate.

Introduction

Groundwater residence times can be assessed from measurements of the decay of radionuclides of atmospheric origin. The nuclides of greatest interest for hydrological purposes are tritium and ^{14}C, although other radionuclides have also been used for dating of subsurface waters (^{85}Kr, ^{39}Ar, ^{32}Si, ^{81}Kr, and ^{36}Cl) [1].

Incorrect estimation of groundwater age may result from using carbonate as a source for ^{14}C-dating, due to uncontrollable processes in the subsurface [1,2]. Chemical and isotopic exchange with CO_2 will, depending on the nature of the system (open or closed), have a direct impact on the resulting ^{14}C-content of the dissolved carbonate fraction in the water. The carbonate equilibrium is sensitive to changes in temperature and chemical conditions, which can lead to either dissolution or precipitation of carbonate ions. Of greater importance are the exchange and substitution reactions between dissolved carbonate and solid carbonate minerals; carbon of organic origin can be converted to carbonate ions and thus enter the water. There might also be a subsurface production of ^{14}C, which will end up in the carbonate species. This additional production will affect the age estimation of water older than approx. 50000 years [3]. Although measurements of the $^{12}C/^{13}C$ ratio can be used for correction of some of these processes, the uncertainties are still large in assessing the original amount of ^{14}C introduced into the system.

Analysis of the ^{14}C-content in humic substances has been used for dating of soils [4-10], but only a few reports have dealt with groundwater age determination using ^{14}C from dissolved organic carbon (DOC) [11,12]. Humic substances, mainly in the form of fulvic acids, are present even in deep and old groundwaters and can be used for age determination (by ^{14}C-analyses). These humic substances, which largely originate from surface soil organic matter, could give information on the residence time of the water in the ground. By using a technique that will not concentrate the low-molecular-weight, dissolved organics, when recovering humic substances, it is possible to neglect the part of the DOC that may originate from, e.g. old hydrocarbons in the aquifer. The inorganic carbonate reactions will generally not influence the DOC content of the water [11].

Groundwater residence times assessed from ^{14}C-analysis of dissolved carbonate, as well as of a DOC fraction (fulvic acids), are compared and discussed in this paper.

Fulvic Acids from Groundwaters

Sampling

Fulvic acids were recovered from five deep groundwaters, one shallow groundwater and one surface water. The five deep groundwaters were sampled from crystalline granitic bedrock: Fjällveden (Fj), approx. 100 km S of Stockholm; Finnsjön (Fi), approx. 100 km N of Stockholm; Gideå (Gi), approx. 500 km N of Stockholm; Lansjärv (L), approx. 1000 km N of Stockholm and Fanay-Augères (F-A), Massif Central, France. The shallow groundwater (T) was collected from a moraine horizon (Tiveden, approx. 250 km SW of Stockholm). A surface water fulvic acid (B) was isolated from a bog area (Bersbo, approx. 200 km S of Stockholm). The chemical composition of the waters are presented in Table 1.

The fulvic acids were isolated in the field by using a weak anion exchange resin, DEAE-cellulose, without any pH-adjustments of the water. After elution the fulvic acids were desalted on an XAD-8 resin, then passed through a cation exchange resin and lyophilized [12].

Chemical Characterization

The fulvic acids were characterized with respect to the following parameters [12]:

- Elemental analysis: C, O, H, N, S
- Molecular weight: Determined with a GPC-technique performed on a HPLC-equipment. Polystyrene sulphonates of known molecular weights were used as reference substances.
- Acid-base properties: Determined by potentiometric titration in aqueous and non-aqueous media [13].
- Age: Measurements of ^{14}C on accelerator mass spectrometer (The Swedberg Laboratory, Uppsala, Sweden).

Results of the characterization of the fulvic acids are presented in Table 2.

Table 1 Hydrochemical data for the sampling sites.

Sampling site	B	T	F-A	Fj	Fi	Gi	L
Depth (m)	0	0	280	409	232	157	139
Age (y)[a]				4235	8090	6450	
Tritium (TU)			16	19	<3	<3	
pH	5.3	4.4	6.2	7.5	7.7	8.8	7.6
Cond. (mS/m)	10.5	6.1		30	531	27.5	9.7
HCO_3^- (mg/l)	5.5	0.4	24	170	260	161	55
SO_4^{2-}	36	12.2	10	0.2	140	0.8	1.6
Cl^-	7.2	3.4	6.2	7	1500	4.4	0.8
Na^+	5.8	3.0	8.5	32	650	49	5.1
K^+	1.8	0.2	1	2.5	8.7	2.2	0.9
Mg^{2+}	8.6	0.6	1	0.4	40	2.6	2.9
Ca^{2+}	8.3	1.1	4	21	320	10	10.4
Fe(tot)	3.1	1.0		6.5	0.87	0.13	0.8
SiO_2				13	16	22	
Ref.	[14]	[15]	[16]	[17]	[18, 19]	[19, 20]	[21]

[a] From the ^{14}C-content of the CO_3-fraction, after correction for the ^{13}C-content.

Ageing Effects

Fulvic acids in surface waters and groundwaters are decomposed by microbial activity and possibly also by chemical processes which lead to a change in elemental composition. In general during decomposition processes, the oxygen and nitrogen content decreases while the carbon content increases [11]. In the present investigation all the fulvic acids studied were quite similar in elemental composition, and in the fulvic acids of greater age (i.e. about 1000 years), there was indeed an increase in carbon content as well as a decrease in oxygen content.

Acid capacity was quite similar for the different fulvic acids. In fact the carboxylic content was almost constant, whereas the content of hydroxylic groups showed some variation. For the fulvic acids from Fi and Gi, the relatively low value of the acid capacity is consistent with the low oxygen content.

Table 2 Elemental composition, molecular weight and acid capacity of fulvic acids from various groundwaters and a surface water.

Sample	B	T	F-A	Fj	Fi	Gi	L
Depth (m)	0	0	280	409	232	157	139
Age (y)[a]			625	1270	4610	5250	9675
C (%)	52.5	50.9	50.5	50.8	53.0	53.7	52.9
H	3.6	3.9	4.3	3.9	3.8	4.4	4.8
N	1.1	0.7	1.4	1.7	0.9	0.5	1.1
O	38.8	42.2	39.8	39.7	36.2	37.5	36.3
S	1.0	0.3	0.6	0.8	1.1	0.5	0.6
Ash	3.0	2.0	3.4	3.1	5.0	3.4	4.3
M_n	1750	1250	750	1250	1750	1150	710
M_w	2650	1850	850	1700	2650	1600	820
M_w/M_n	1.51	1.48	1.13	1.36	1.51	1.39	1.15
Aq. acid cap. (meq/g)	4.65	4.78	5.07	5.14	4.98	5.42	
Non-aq. acid cap. (meq/g) COOH	4.78	5.33	5.40	5.56	4.16	5.33	
OH	1.35	0.51	2.09	2.53	0.83	1.05	
Total	6.13	5.83	7.49	8.09	4.99	6.38	

[a] From the ^{14}C-content of the fulvic acid fraction of the DOC

The progressing decomposition of the fulvic acids was indicated from the slight decrease in molecular weight for the older materials except for the acids from Finnsjön, which were recovered from a saline water. Otherwise, neither the high salinity of the Finnsjön sample nor the high pH of the Gideå water seemed to have any drastic effects on decomposition rates. The lowest molecular weight observed for the fulvic acids from L, is consistent with a longer residence time and a higher degree of decomposition; the F-A fulvic acids, however, deviate from this pattern.

The ratio M_w/M_n expresses the inhomogeneity of the fulvic acids. The closer to unity, the more homogenous the sample, which indicates a more decomposed material. The similarity between the fulvic acids from the various waters suggests that decomposition of fulvic acids in groundwater is a very slow process compared to the decomposition rate in soils [11].

Discussion and Conclusion

Groundwater systems are generally complex mixtures of waters of different origin and ages. Therefore, age estimated by ^{14}C-analysis represents the average residence time of a water in the ground. Two of the deep groundwaters (Fanay-Augères and Fjällveden) contain, in fact, significant fractions of young, shallow groundwaters [15,16], as indicated from chemical data, flow observations and by the presence of tritium. The ^{14}C-data for the fulvic acids (Table 2) confirm that both the F-A and Fj acids may contain a fraction of recent origin. The addition of a fraction of water of less age could explain the similarity in elemental analysis between the fulvic acids from F-A and Fj, and the surface water fulvic acid.

Isotopic data (deuterium, ^{18}O) indicate that the waters from Fjällveden, Finnsjön and Gideå all have a meteoric origin [17-19]. No data are available for the other sites.

The correlation between the inorganic-^{14}C age (originating from the carbonate) and the organic-^{14}C age (originating from the fulvic acid) for three of the deep groundwaters is illustrated in Fig. 1.

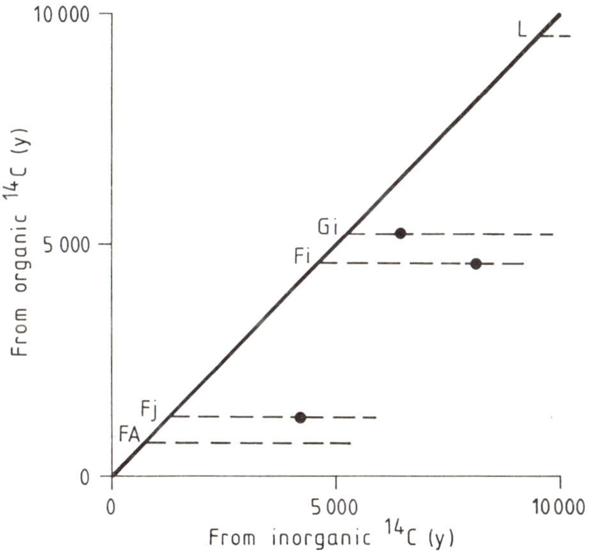

Figure 1 Correlation of age determination of waters based on analyses of organic and inorganic ^{14}C for the various fulvic acids.

The inorganic-^{14}C age indicates an older water than the organic-^{14}C age. The dissolved inorganic carbon (carbonate) content could have be modified by several chemical reactions, as mentioned above, which would have altered the ^{14}C-content and hence influenced the age determination, generally towards a greater age. At the present, it is not known if any side reactions significantly affect the ^{14}C-content of a fulvic acid. The recovering technique used does not concentrate the low-molecular-weight fractions, which means that there would not have been any significant influence on the ^{14}C-content from old hydrocarbons in the aquifer.

The first reports of using humic substances as ^{14}C-sources in dating groundwaters have only recently been published. Thurman [11] discussed age determinations using a fulvic acid. The technique used by this author requires a large amount (2 g) of humic material, and since deep groundwaters often have a very low (< 1 mg/l) concentration of humic substances, it is difficult to collect a large enough quantity to use Thurman's method. Comparatively, however, the accelerator mass spectrometer used in the present study is much more practical to employ since only a few milligrams of humic substance is needed.

Because of the similarity of the various fulvic acids, in terms of molecular weight, chemical composition and acidity, chemical data alone do not provide conclusive information concerning the age of such material. However, ^{14}C-analysis of fulvic acids appears to be a convenient way to estimate theaverage residence time for the water in an aquifer. The absence of obvious side reactions would generally make fulvic acid a better ^{14}C-source for age determination than the dissolved carbonate fraction.

Acknowledgement

The authors are grateful to the Swedish Nuclear Fuel and Waste Management Company for their financial support.

References

1. Davis, S.N. and H.W. Bentley. In: L.A. Currie, Ed., Nuclear and Chemical Dating Techniques: Interpreting the Environmental Record, ACS Symposium series, no 176, pp. 189-222 (Washington D.C.: Am. Chem. Soc., 1982).
2. Wigley, T.M.L., L.N. Plummer and F.J. Pearson Jr. Geochim. Cosmochim. Acta **42**:1117 (1978).
3. Zito, R., D.J. Donahue, S.N. Davis, H.W. Bentley and P. Fritz. Geophysical Letters **7**:235 (1980).
4. Kovalev, R.V., B.M. Klenov and K.A. Arslanov. Izv. Sib. Otd. Akad. Nauk SSSR, Ser. Biol. **3**:6 (1972).
5. Scharpenseel, H.W. Pflanzenernaehr. Bodenk. **133**:241 (1972).
6. Balesdent, J. and B. Guillet. Bull. Assoc. Fr. Etude Sol **2**:93 (1982).
7. Nowaczyk, B and M.F. Pazdur. Rocz. Glebozn. **33**:145 (1982).
8. Tolchelnikov, Y.S. and A.S. Kostarev. Pochvovedenie **3**:15 (1983).
9. Anderson, D.W. and E.A. Paul. Soil. Sci. Soc. Am. J. **48**:298 (1984).
10. Yamada, Y. Nogyo Kankyo Gijutsu Kenkyusho Hokoku **3**:23 (1986).
11. Thurman, E.R. In: G.R. Aiken, D.M. Mc Knight, R.L. Wershaw, Eds, Humic Substances in Soil, Sediment, and Water, pp. 87-104 (New York: John Wiley and Sons, Inc., 1985).
12. Pettersson, C., I. Arsenie, J. Ephraim, H. Borén and B. Allard. Sci. Tot. Environ. **81/82**:287 (1989).
13. Ephraim, J.H., H. Borén, C. Pettersson, I. Arsenie and B. Allard. Environ. Sci. Technol. **23**:356 (1989).

14. Karlsson, S. Influence of Hydrochemical Parameters on the Mobility and Redistribution of Metals from a Mine Waste Deposit (Department of Water and Environmental Studies, Linköping University, Sweden, 1987).

15. Aastrup, M. Pers. comm. (1989).

16. Moulin, V. Pers. comm. (1988).

17. Laurent, S. Analysis of groundwater from deep boreholes in Fjällveden, SKB-TR 83-19 (Stockholm, 1983).

18. Smellie, J., E. Gustavsson and P. Wikberg. Groundwater sampling during and subsequent to air-flush rotary drilling: Hydrochemical investigations at depth in fractured crystalline rock, SKB-AR 87-31 (Stockholm, 1987).

19. Smellie, J., N.-Å. Larsson, P. Wikberg, L. Carlsson. Hydrochemical investigation in crystalline bedrock in relation to existing hydraulic conditions: Experience from the SKB test-sites in Sweden, SKB-TR 85-11 (Stockholm, 1985).

20. Laurent, S. Analysis of groundwater from deep boreholes in Gideå, SKB-TR 83-17 (Stockholm, 1983).

21. Wikberg, P. Pers. comm. (1988).

Characterization of Organic Materials by Means of Electrofocusing

Marco Govi, Claudio Ciavatta, Livia Vittori Antisari and Paolo Sequi
Institute of Agricultural Chemistry, University of Bologna
Viale Berti Pichat, 10-40127 Bologna, Italy

Abstract

Electrofocusing has been applied to characterize the organic matter in pig slurries during incubation. In general, raw pig slurries are resolved in bands that localize mainly in the acid region of the pH gradient, whereas after some months of incubation a much more complex pattern of bands develops, characteristically, in the neutral region of the gradient. Such development of new fractions is markedly evident if incubation is carried out in summer, though in winter the process occurs at a slower rate. Stabilization of pig slurry leads to electrofocusing profiles that are somewhat similar to those of organic matter in the soil. From a qualitative point of view, electrofocusing is suggested to be a fast, reliable, and perhaps a more simple method of assessing stabilization (i.e., maturation or humification) of the organic matter in the slurries.

Introduction

During the last 30 years many procedures have been devised to assess organic-matter quality in soils, fertilizers, sludges and other organic matrices. For a thorough characterization, different parameters (e.g., apparent molecular weight by gel chromatography [1], C/N ratio [2], cation exchange capacity [3]) have been proposed, but either their application is often not generalizable or they can not give enough information for a correct evaluation.

Selective adsorption of humic fractions on different materials, like polyamide, Amberlite XAD-8 or polyvinylpyrrolidone, has been used by many researchers. Several humification indexes have been suggested [4,5] and then utilized by different authors [e.g., 6,7]. Such methods have proven to be useful for evaluating the evolution of organic matter from a quantitative point of view, but are unable to qualitatively distinguish organic matter from different sources.

A technique that easily serves this purpose and, in addition, produces good evidence of the organic matter evolution during humification, is electrofocusing (EF), which can be considered as an extension of the electrophoretic principle. EF is carried out in a polyacrylamide gel tube where an electrophoretic carrier ampholyte has been previously reblended to give a pH gradient from a lower to a higher value. During a run, each macromolecule moves and, under an electric potential, searches out its isoelectric point (IEP). Since 1972, many authors have utilized EF to characterize soil enzymes [e.g., 8], humic substances extracted from soil [e.g., 9], or rivers [e.g., 10] and to characterize other organic materials [e.g., 11]. Although some authors have discussed reliability of EF [e.g., 12], others have considered the EF findings as artefacts. In general, they believed [13] that humic-substance separation by the EF technique was caused by the interaction of the substances with the carrier ampholyte.

Recently, the integrity of humic substances during EF application has been demonstrated, and the possible interaction between humic substances and ampholytes has been assessed as not being responsible for the appearance of some bands as artefacts in the gels [14]. In a previous study [15,16] we followed the evolution of organic matter during incubation of pig slurries, both in summer and in winter, by use of the above-mentioned humification indexes. The objective of this work was the application of the EF technique to the same research material.

Materials and Methods

Pig Slurries

Raw pig slurry was stored in tanks and incubated for 120 days; during incubation, periodic sampling was performed. The incubation was carried out near the livestock at Basilica Goiano, Parma, Italy, both in summer and in winter. More characteristics of livestock effluent have been already given [15,16].

Organic Matter Extraction

Two grams of sample was extracted under N_2 with 100 ml of 0.5 M NaOH for 24 hours. The extracts were centrifuged for 20 minutes at 12,500 rpm, and the supernatants were filtered with 0.45 μm Millipore filters. The filtered fractions were subsequently desalted with the H_3O^+ form of Amberlite IR 120 (Serva, West Germany) to a pH value lower than 7. The solutions so obtained were separated into 1 ml-subsamples and stored in a freezer.

Determination of Organic Carbon

Organic carbon was determined according to a modification of a dichromate oxidation method [17]. An aliquot (1-2 ml) of extracted sample was transferred to a digestion flask together with 5 ml of 2 N $K_2Cr_2O_7$ and 20 ml of conc. H_2SO_4. The solution was immediately heated to 160°C and maintained at the same temperature for 10 minutes; then the solution was cooled, and the excess dichromate titrated with 0.4 N $FeSO_4$.

Electrofocusing (EF)

Gel for the EF separation was cast in glass tubes (150 x 4 mm). Each tube was filled with 12 cm of a gel rod containing 5% polyacrylamide (acrylamide: bisacrylamide 1:26) and 2% preblended Ampholine (pH range 3.5 to 10; from LKB, Sweden) as carrier ampholyte.

Aliquots containing about the same amount of organic carbon (0.2 mg C/tube) were charged on the gel rods; the electrophoretic cell (Bio-Rad, Model 175 Tube Gel-USA) was cooled to 4°C, with a 2177 power supply (LKB). The runs were carried out under constant power (1200 V) for three hours and 30 minutes; then the pH range was immediately verified in each tube gel by using a specific surface pH electrode (Ingold-Switzerland).

Focused bands were coloured with brilliant blue Coomassie G 250 (Merck) and, after 24 hours, decoloured with 0.05 M KOH. The coloured tube gels were scanned at 633 nm with a LKB Ultroscan XL laser densitometer.

Results and Discussion

Fig. 1 shows EF profiles of pig slurries incubated in summer. At the beginning of the incubation period, only a few fractions focalize, mainly in the pH gradient region from 4.0 to 5.5; a few bands are present at higher pH values, one of which is a single characteristic band displayed at about pH 6.3. Practically no fractions focalize in the neutral region of the pH gradient. The EF profile of slurries at the end of incubation is much more complex; some similarities are shown in the acid region, while bands displayed up to pH 6.5 are more strengthened. A marked difference, however, occurs mainly in the neutral region of the pH gradient, where a complex pattern of 7-8 new bands develops.

It is known [11,12] that EF profiles of humic substances extracted from soil are characterized by a great heterogeneity of bands in the neutral region of the pH gradient. So, the evolution of organic matter during incubation of pig slurries is associated with a tendency to develop EF bands more similar to those of humified materials. Our results agree with those reported [18] for sewage sludges, in which, during stabilization, new bands also develop in the neutral region of the pH gradient. EF profiles of samples of pig slurries incubated during the cold season are shown in Fig. 2. Data not reported here show that the EF profiles of pig slurries at the beginning of the incubation period were similar both in summer and in winter, as expected. The first profile in the figure refers to a slurry sampled about 15 days after the beginning of the incubation period. It is possible to see some development of bands in the neutral region of the pH gradient. In the same region, a more complex pattern of bands is shown in the sample of slurries incubated for 120 days, while the complexity is much lower than in the case of a summer incubation.

It seems reasonable to infer that two main factors may control the stabilization of organic matter in pig slurries: time and temperature. In the hot season an easy

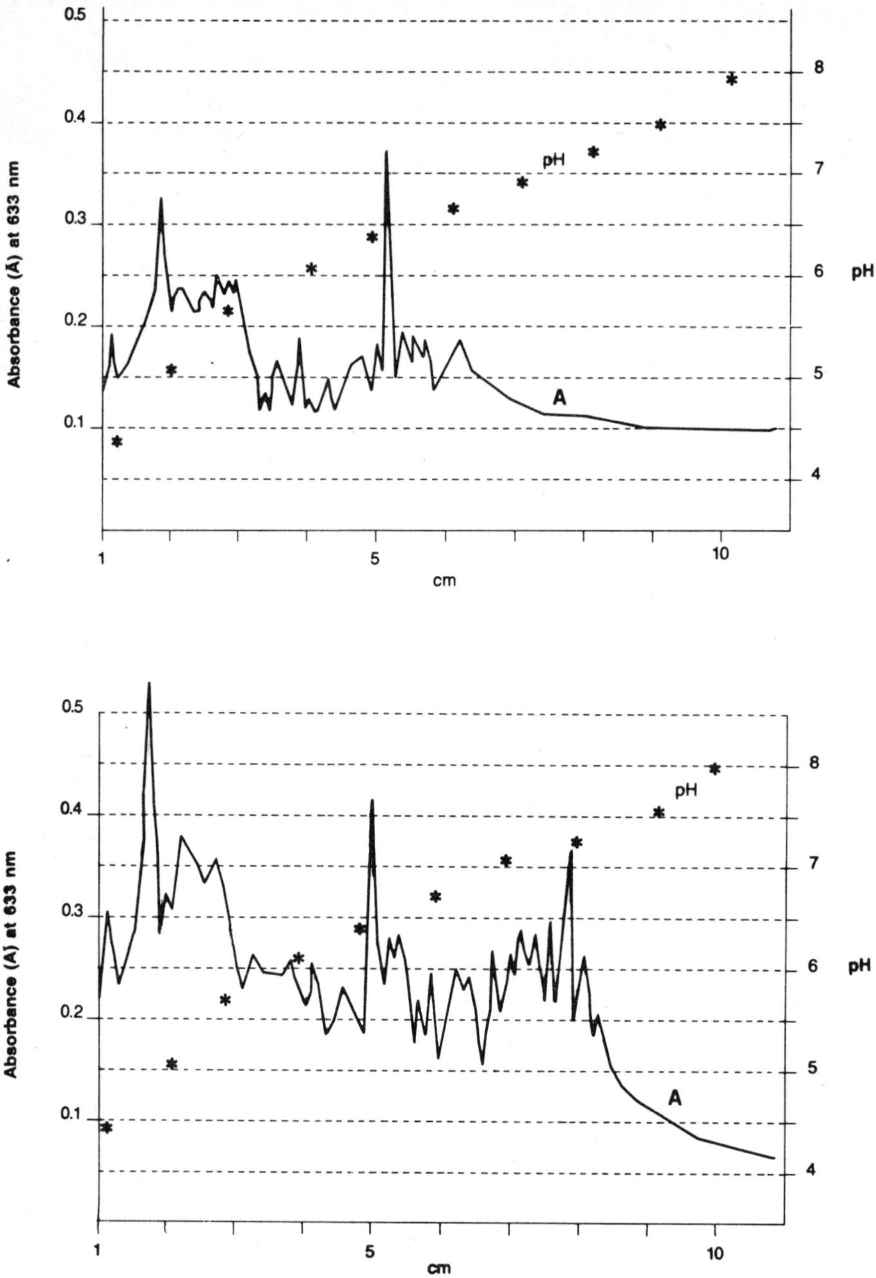

Figure 1 Electrofocusing profile of a 0.5 M NaOH extract of pig slurries incubated for 120 days in summer, at the beginning (above) and at the end (below) of the incubation period.

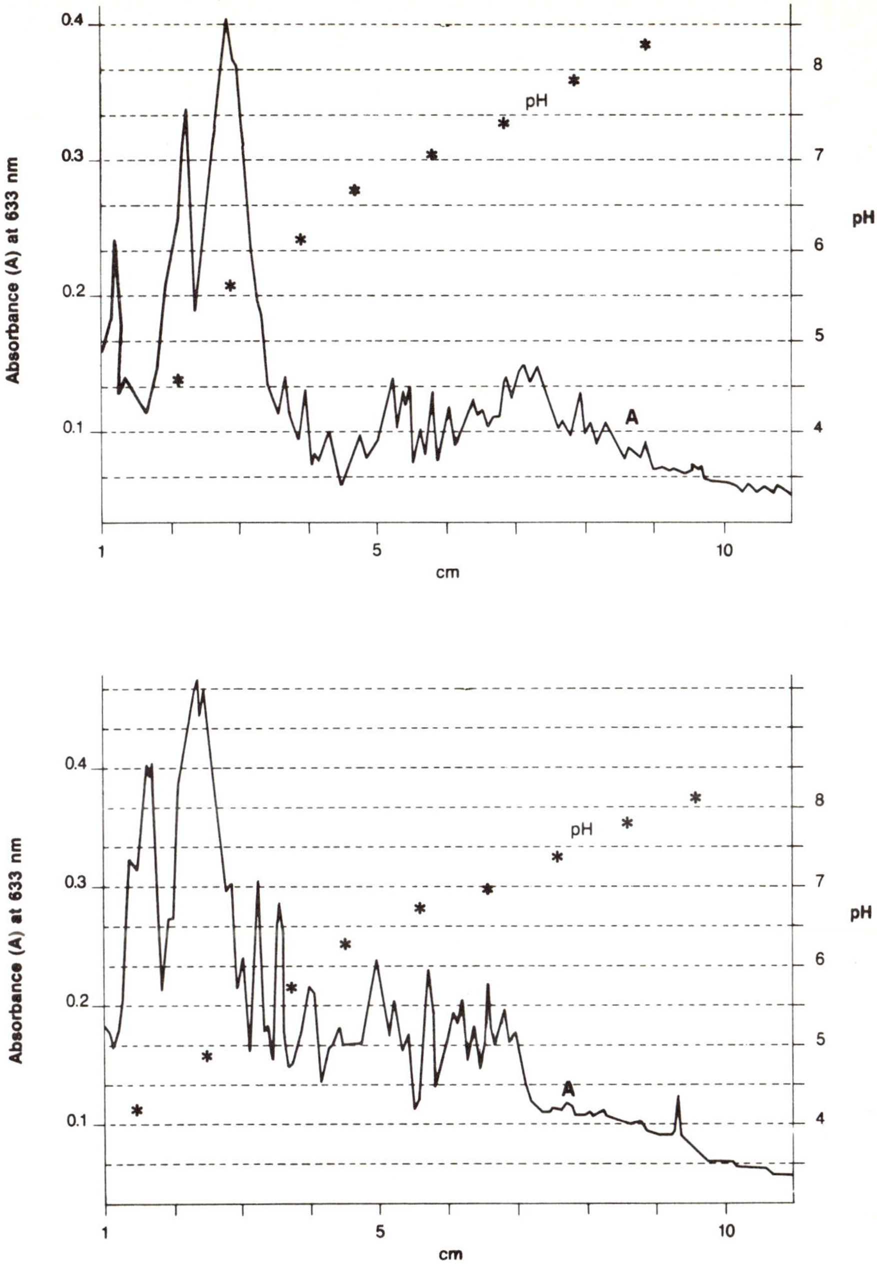

Figure 2 Electrofocusing profile of a 0.5 M NaOH extract of a pig slurries incubated for 120 days in winter, after 15 days (above) and at the end (below) of the incubation period.

stabilization occurs during a period of four months; during the same lenght of time in the cold season, stabilization proves to be much less efficient.

Such results are confirmed by the previous findings that showed significant increases in the humification rate (the ratio between the sum of fulvic and humic fractions and the total organic carbon content in the sample) of the samples of pig slurries incubated in summer, whereas the trend was much less definite in winter [16]. As a matter of fact, humification parameters (degree of humification, DH, and humification ratio, HR) seem to be the only values quantitatively related to organic matter stabilization during the so-called maturation of organic materials. Other characteristics of organic materials do not vary quantitatively in the course of maturation processes. Some authors, for instance, showed variations of some functional groups during maturation of organic materials, especially increases of methoxyls and decreases of phenols; such variations, however, are not constant [19,20] and, when expressed with respect to the organic carbon, may become negligible [21]. Variations of molecular weights generally consist of an increase in the fractions with higher apparent molecular weight. However, the increase proceeds without a direct proportionality [22-24], and in addition, the determination leaves doubt because of the artefacts often occurring with humified materials. In other words, the methods most commonly used for the characterization of humic substances are of little use in evaluating formation of humic substances during maturation of organic materials.

Our findings confirm that electrofocusing is a reliable technique that can be used successfully to monitor differences in the organic matter quality of raw, non-purified materials. Using such materials, interferences are very limited, and resolution appears to be satisfactory.

Another advantage of electrofocusing is the possibility of demonstrating differences between materials, like pig slurries incubated in summer and winter, respectively, that are apparently very similar when tested by other quantitative procedures. In conclusion, as for other organic wastes, EF profiles are suggested to reflect stabilization processes (i.e. humification) of the organic matter on a purely qualitative basis. Although different organic materials are easily recognizable from their characteristic EF profiles, the appearance, in the neutral region of the pH gradient, of new bands typically present in soil organic matter extracts, is a common denomination after complete maturation.

References

1. De Nobili, M., E.T. Gjessing and P. Sequi. In: M.H.B. Hayes, P. MacCarthy, R.L. Malcolm, and R.S. Swift, Eds, Humic Substances II, pp 561-591 (Chichester: Wiley, 1989).

2. Roletto, E., R. Chiono and E. Barberis. Agricultural Wastes **12:**261 (1985).

3. Roig, A., A. Lax, J. Cegarra, F. Costa and M.T. Hernandez. Soil Science **146:**311 (1988).

4. Sequi, P., M. De Nobili, L. Leita and G. Cercignani. Agrochimica **30**:175 (1986).

5. Ciavatta, C. and P. Sequi. Fertilizer Research **19**:7 (1989).

6. Saviozzi, A., R. Levi and R. Riffaldi. BioCycle **198**:54 (1988).

7. De Nobili, M. and F. Petrussi. J. Ferment. Technol. **66**:577 (1988).

8. Cacco, G., A. Maggioni and G. Ferrari. Soil Biol. Biochem. **6**:145 (1974).

9. Cacco, G. and A. Maggioni. Soil Biol. Biochem. **8**:321 (1976).

10. Gjessing, E.T. and T. Gjerdahl. Proc. Int. Meeting on Humic Substances, pp. 43-51 (Wageningen: Pudoc, 1972).

11. De Nobili, M., G. Cercignani and P. Sequi. In: G. Giovannozzi-Sermanni and P. Nannipieri, Eds, Current Perpectives in Environ. Biochem. pp. 85-94 (Rome: CNR-IPRA, 1985).

12. Ceccanti, B., P. Nannipieri and M.T. Bertolucci. Recent Developments in Chromatography and Electrophoresis **10**:75 (1980).

13. Aak, O.V., V.A. Galynkin, A.P. Kaskin and V.I. Jakovlev. Prikladnaja Biochimiai Microbiologija **20**:290 (1984).

14. De Nobili, M. J. Soil Sci. **39**:437 (1988).

15. Govi, M. Maturazione della sostanza organica negli effluenti degli allevamenti suinicoli. M.Sc. Thesis (Bologna: University, 1988).

16. Govi, M., C. Ciavatta, L. Vittori Antisari and P. Sequi. In: A. Frigerio, Ed., Acque Reflue Fanghi, pp. 422-431 (Milano: Centro Scientifico Internazionale, 1989).

17. Ciavatta, C., L. Vittori Antisari and P. Sequi. Communications in Soil Science and Plant Analysis **20**:759 (1989).

18. De Nobili, M., M. Cercignani and P. Sequi. Communications in Soil Science and Plant Analysis **10**:1109 (1986).

19. Riffaldi, R., R. Levi-Minzi, A. Pera and M. De Bertoldi. Waste Management and Research **4**: 387 (1986).

20. Saviozzi, A., R. Riffaldi and R. Levi-Minzi. In: M. De Bertoldi, M.P. Ferranti, P. L'Hermite and F. Zucconi, Eds, Compost: Production Quality and Use pp 359-367 (Udine: CEC, 1986).

21. Riffaldi, R. and R. Levi-Minzi. Agrochimica **XXVII**:271 (1983).

22. Katayama, A., K.C. Ker, M. Hirai, M.Shoda and H. Kubota. In: M. De Bertoldi, M. P. Ferranti, P. L'Hermite and F. Zucconi, Eds, Compost: Production, Quality and Use pp 341-350 (Udine: CEC, 1986).

23. Spaggiari, G.C. and G.L. Spigoni. In: M. De Bertoldi, M. P. Ferranti, P. L'Hermite and F. Zucconi, Eds., Compost: Production, Quality and Use pp 100-107 (Udine: CEC, 1986).

24. Roletto, F. and M. Cerruti. Agricultural Wastes **13**:137 (1985).

Nature of Humic Substances of Mollisol and Luvisol in the Canadian Prairies

A. Singer and P.M. Huang

Department of Soil Science, University of Saskatchewan,
Saskatoon, SK S7N OWO Canada

Abstract

The nature of humic substances (0-25 cm) of a Mollisol (Udic Haploboroll) and a Luvisol (Typic Cryoboralf) in the Canadian Prairies was investigated. The amount of humic acid (HA) was higher in the Mollisol than in the Luvisol. The opposite trend was observed for fulvic acid (FA). There was no great difference in elemental composition of HAs between the Mollisol and Luvisol. The contents of N and S of the FAs were higher in the Mollisol than in the Luvisol. Infrared spectra of the two HAs were essentially similar. The infrared data also indicated that compared with HAs, FAs contained more carboxyl groups and less aromatic C. ^{13}C NMR spectra of the HAs and FAs showed the presence of the pronounced peaks at 20-35 ppm (aliphatic C in alkyl chains), 75 ppm (carbohydrates), 135 ppm (aromatic ring C), and 165-175 ppm (COOH groups). The peaks at 60 ppm (methoxy groups) and 150-155 ppm (O and N substituted aromatic C) were also present in ^{13}C NMR spectra of both HAs. HAs in both Mollisol and Luvisol were substantially more aromatic than FAs. Aliphaticity of HAs was more pronounced in the Luvisol than in the Mollisol. The NMR evidence was in accord with the data obtained from the measurement of the extinction coefficient (E_{280}) and differential thermal analysis. The HA and FA of the Mollisol appeared to contain a higher proportion of more condensed aromatic structures than those of the Luvisol. Both differential thermograms and X-ray diffractograms indicated that HAs and FAs in the Mollisol were better ordered than those in the Luvisol. The data indicated that differences in pedogenesis exerted the influence on the nature of humic substances in grassland and parkland in temperate region.

Introduction

Humic acids (HAs) in some Saskatchewan soils have been described by Anderson *et al.* [1-3]. The samples from the A_h horizon of a Brown, Dark Brown, Black, Dark Gray, and Gray Luvisolic soil sequence were studied. Their study suggested an increase in the proportion of aromatic components in the HAs from the Brown to the Gray Luvisolic soils. Free radical concentrations were lowest for the Brown and Dark

Brown soils, intermediate for the Black and highest for the Gray Luvisolic soils. On the other hand, various studies have shown that HAs and fulvic acids (FAs) extracted from soils formed under widely differing geographic and pedological environments from Eastern, Western and Northern Canada, as well as Japan, West Indies and Argentina have similar analytical characteristics and chemical structures [4]. Chen *et al.* [5] reported that although the general characteristics of HAs and FAs extracted from soils in Israel and Italy were similar to those observed in Canada, there were some differences in the content and composition of inorganics in the FAs, which were reflected in chemical analysis, ESR spectra, DTA curves and SEM micrographs. Therefore, it is not at all clear whether differences in pedogenesis can result in a differentiation in soil humic matter characteristics.

The objective of this study was to investigate the nature of FA and HA of Mollisol and Luvisol in Saskatchewan, Canada. This information is useful in understanding the influence of pedogenesis on the nature of humic substances.

Materials and Methods

Soils

The soil samples studied were the upper horizons (0-25 cm) of an Orthic Black Mollisol (Udic Haploboroll), a member of the Oxbow Association, located at 106° 37'W and 52° 32'N, and of an Orthic Gray Luvisol (Typic Cryoboralf), a member of the Meeting Lake association, located at 107° 20'W, and 52° 52'N. Both soils had been under cultivation for an extended period.

Extraction procedure

The soil was air-dried and then passed through a 1 mm sieve. The extraction and purification of HA and FA was carried out according to the procedure recommended by the International Humic Substances Society [6-9]. Twenty g of soil was placed in a 200 ml centrifuge bottle. An aliquot of 200 ml water, acidified with 1 M HCl to pH 1-2, was added to the soil, and the suspension was shaken for 2 h. The supernatant after centrifugation was saved for FA purification. The residue was neutralized with 1 M NaOH to pH 7 and extracted with 0.1 M NaOH in the presence of N_2. The extract was acidified with 6 M HCl to pH 1. HA was separated from FA by decantation and centrifugation.

The HA fraction was purified by redissolving the HA in 0.1 M KOH, in the presence of 0.3 M KCl, and removal of sediments by centrifugation, and from silicate mineral matter by suspending in a 0.1 M HCl: 0.3 M HF mixture. Finally, the HA, from which Cl^- had been removed by prolonged dialysis in distilled water, was freeze-dried.

The FA fraction was purified by passing repeatedly over columns of XAD-8 resin, and eluting the adsorbed FA back with 0.1 M NaOH. The eluate was immediately acidified with 6 M HCl to pH 1. Silicate mineral matter was removed by treatments with 0.3 M HF. Finally, the FA, from which Na^+ had been removed by passing through a column of H-saturated cation exchange resin (Bio-Rad AG-MP-50), was freeze-dried.

Analytical methods

The analyses of C and H were carried out by the method of Houde and Champy [10]; N, O, S, and P were analyzed by the method of Merz [11], Unterzaucher [12], ASTM [13], and Ma and McKinley [14], respectively. Molecular weights of FAs were determined by vapor pressure osmometry (Huffman Laboratories, INC., Wheat Ridge, Colorado). Visible and UV spectrophotometry was carried out on a PYE SP6-500 UV-VIS spectrophotometer. For the E_4/E_6 determinations, 3 mg of material was dissolved in 10 ml of 0.05 M $NaHCO_3$ buffered at pH 7 [15]. E_{280} was determined on Na-humate solutions containing 17 ppm humate buffered to pH 7 in 0.05 M $NaHCCO_3$ [2]. IR spectra were recorded with KBr pellets on a PerkinElmer model 983 IR spectrometer. Thermal analysis was performed on a Stanton Redcroft model STA 781 combined DTA & TGA apparatus. XRD pattems were recorded on a Philips diffractometer. The ^{13}C CP/MAS - NMR solid state spectra were obtained at 22.6 MHZ (90 MHZ protons) on a magnachem M 100-S spectrometer; a contact time of 1 μsec and a repetition time of 1 sec were employed; 40,000 scans were run on each sample.

Results and discussion

Amounts of HA and FA

HA yields were 10.2 and 7.9 mg/g soil for the Mollisol and Luvisol, respectively. FA yields were 0.62 mg/g of the Mollisol and 0.87 mg/g of the Luvisol. Thus, the Mollisol contained more HA than the Luvisol, while the latter contained more FA than the former. The weight ratios of HA/FA were 16.5 in the Mollisol and 9.1 in the Luvisol. These ratios were considerably higher than those reported for similar uncultivated soils and are probably the consequence of cultivation. FA is far less resistant towards decomposition than HA, and therefore, the former has a much lower mean residence time (MRT) than the latter [16].

Elemental composition

The chemical composition of the HAs of the Mollisol and Luvisol (Table 1) is very similar to that reported for similar Saskatchewan soils by Anderson *et al.* [1,2]. There was no considerable difference in elemental composition between the two soil types. The O/C atomic ratios were close to 0.5, the value common for many soils [17]. The N/C ratios were similar to the average values for neutral soils (op. cit.). Sulfur contents were within the range given for cool-temperate, neutral soils [4]. The HAs of these two soils were similar in P content.

The FAs had a considerably lower C and higher O content than the HAs (Table 1). The O/C ratios of the FAs were well below the averages given for many soils [4]. This difference may reflect lower carbohydrate content [17] in the two fulvic acids. The H contents of the FAs were in the range given by Schnitzer [4] for acid soils and significantly below that for neutral soils. As a result, the H/C ratios were unusually low, possibly suggesting a relatively high degree of aromatic condensation [18]. The N content in both FAs was relatively high, giving rise to relatively high N/C ratios. The N and S contents of the Mollisol FA were considerably higher than those of the Luvisol FA.

Table 1 Some chemical and physical characteristics of humic acids and fulvic acids separated from the two Saskatchewan soils studied.

Analysis	HA-Mollisol	HA-Luvisol	FA-Mollisol	FA-Luvisol
Element		g/kg		
C	54.10	53.38	48.04	48.97
O	32.69	32.35	34.39	39.44
H	4.61	4.64	4.30	4.06
N	4.23	4.43	3.62	2.29
S	0.75	0.62	1.21	0.56
P	0.16	0.14	0.79	0.72
Ash	1.29	1.25	3.50	0.90
Molecular weight	n.d[t]	n.d.	334	1370
E_4/E_6	4.2	5.1	18.3	15.8
E_{280}	0.600	0.544	0.285	0.275

[t]Not determined

Infrared spectra

Both humic acids displayed strong absorption bands around 3400 cm^{-1}, characteristic for hydrogen bonded OH, but no absorption at 3090-3070 cm^{-1} associated with aromatic C-H stretching (Fig. 1). The presence of small but sharp absorption bands at 2925 cm^{-1} indicated aliphatic C-H stretching, a very broad adsorption band near 2500 cm^{-1} was attributed to hydrogen bonded CO_2H [5]. The bands at 1715 cm^{-1}, associated with C=O of COOH and C=O of ketonic carbonyl, were weak, particularly in the Luvisol HA; in contrast, the bands at 1618-1625 cm^{-1}, indicative of aromatic C=C, COO^- and hydrogen bonded C=O were strong and sharp, particularly in the Luvisol HA. The weak absorption at ~1380 cm^{-1} was probably due to O-H bending vibrations of alcohols and carboxylic acids and bending vibration of aliphatic C-H groups. Somewhat stronger absorption at 1230-1243 cm^{-1} was attributed to C-O stretching and OH deformations of COOH groups [19]. Absorbances at the lower frequencies of 1026 and 600 cm^{-1} are most likely due to Si-O-Si and Si-O-C valence and deformation vibrations [5].

Compared with the Luvisol HA, the Mollisol HA appeared to have a somewhat stronger absorption at 1715 cm^{-1} (Fig. 1). However, both HAs differed from those extracted by Chen et al. [5] from Mediterranean soils by weaker 1715 cm^{-1} absorptions, indicating relatively lower proportions of COOH and/or ketonic groups; aromaticity, as indicated by the prominent absorption at 1618-1625 cm^{-1}, appeared to be more strongly developed in the Saskatchewan soils.

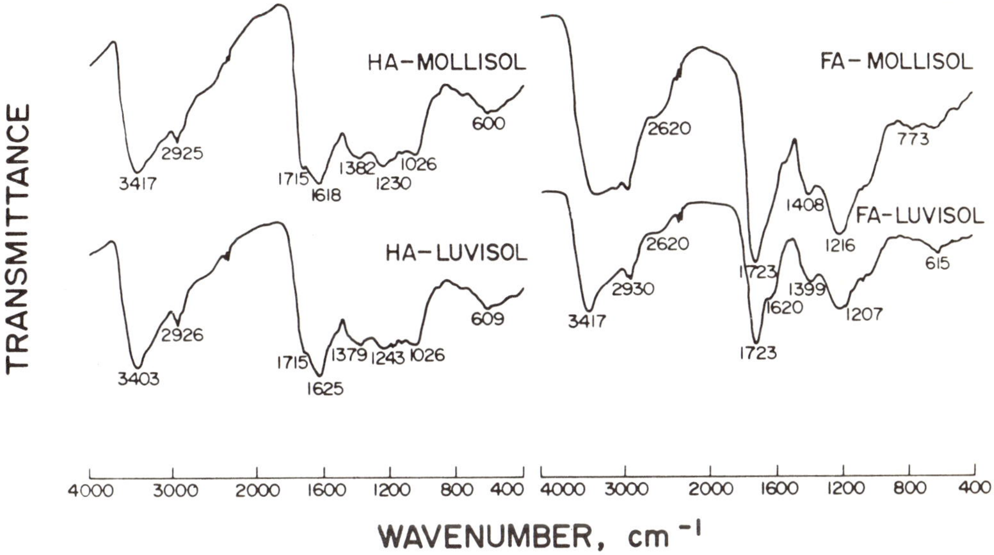

Figure 1 Infrared spectra of humic and fulvic acids of the Mollisol and Luvisol.

The FAs had a strong absorption band (3349-3417 cm^{-1}) in the region of H-bonded OH groups; the frequency of this band in the Mollisol FA (3349 cm^{-1}), however, is considerably lower than in the Luvisol FA (3417 cm^{-1}) (Fig. 1). The 2930-2942 cm-^{1} absorption band, characteristic for aliphatic C-H stretching, and the broad band at 2620 cm^{-1}, possibly associated with hydrogen bonded COOH, were present in the FAs of both Mollisol and Luvisol. The 1723 cm^{-1} absorption bands, characteristic for C=O of COOH, were very strong and sharp in both FAs (Fig. 1). A band at 1620 cm^{-1}, associated with aromatic C=C, COO^- and hydrogen bonded C=O, was evident in the Luvisol FA only. In both fulvic acids a small band at 1399-1408 cm^{-1} was indicative of CH_2 and COO^-; strong and sharp bands at 1207-1216 cm^{-1} were possibly due to C-O stretching and OH deformation of COOH groups [19]. The two FAs differed in that in the Mollisol FA, the C=O stretching vibration (1723 cm^{-1}), due mainly (though not completely) to carboxyl groups, appeared to be more strongly developed, whereas the aromatic C=C "double bonds" conjugated with C=O and/or COO^- (1620 cm^{-1}) was apparently stronger in the Luvisol FA.

NMR spectra

^{13}C NMR spectra of the HAs and FAs of the Mollisol and Luvisol are shown in Fig. 2. Both HAs and FAs showed pronounced peaks in the aliphatic (0-105 ppm), aromatic (135-155 ppm) and COOH groups (165-180 ppm) range.

The prominent 20-35 ppm peaks were due to aliphatic carbons in alkyl chains. The peak at 20 ppm in the Luvisol HA and FA, that may be attributed to terminal CH_3, was not present in the Mollisol HA and FA. The peak at 30 ppm in both HAs and the Luvisol FA was probably due to CH_2 groups in long alkyl chains; possibly the peak at 35 ppm in the Mollisol HA and FA was also associated with these groups. The relative peak areas of these peaks (20-35 ppm) suggest the presence of a larger amount of alkyl chains and aliphatic groups in the Luvisol HA than in the Mollisol HA.

In the 50-105 ppm region, aliphatic carbons substituted by oxygen and nitrogen are commonly observed [20]. The distinct peaks at 60 ppm in both HAs were likely due to OCH_3 groups. The peaks at 70 to 80 ppm in both HAs and FAs of the two soils were attributed to various carbohydrates.

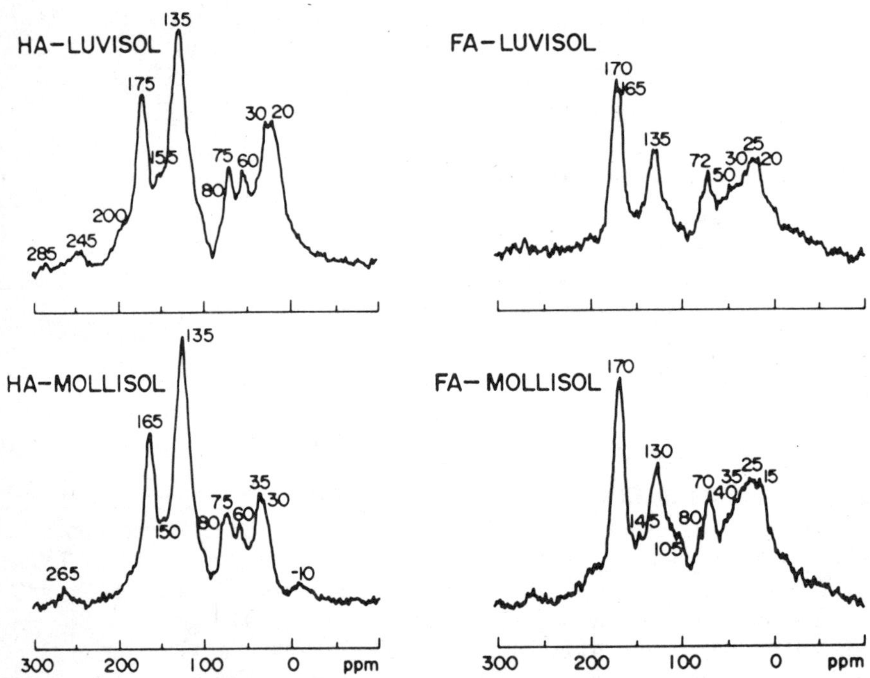

Figure 2 ^{13}C CP/MAS NMR spectra of the humic and fulvic acids of the Mollisol and Luvisol

The sharp and strong peak at 135 ppm, that appeared in both HAs and FAs, was from aromatic ring carbons, in which the ring was not substituted by oxygen or nitrogen. A small peak at 145 to 155 ppm was probably due to O and N substituted aromatic C (phenolic OH or aromatic NH_2). HAs in both Mollisol and Luvisol were substantially more aromatic than FAs. The sharp peak at 165 to 175 ppm was attributed to C in COOH groups.

Ultraviolet and Visible Spetroscopic Data

The E_4/E_6 ratios of the HAs and FAs were within the range reported for soil humic substances by Schnitzer, [4]. The E_4/E_6 ratio of the Luvisol HA was somewhat higher than that of the Mollisol HA, suggesting the Luvisol HA had a smaller particle size (Table 1). A similar trend was observed by Anderson *et al.* [2] in their study of Saskatchewan soils. The E_4/E_6 ratios for the FAs were unusually high, far above the range given for soil FAs by Schnitzer [4]. In their extensive study on humic substances, Chen *et al.* [21] related the E_4/E_6 ratios primarily to particle size or particle/molecular weight. The very low molecular weight of the Mollisol FA thus may be reflected in its high E_4/E_6 ratios (Table 1). The molecular weight of the Luvisol FA was much higher than that of Mollisol FA; the E_4/E_6 ratio followed the same trend (Table 1). However, the E_4/E_6 ratio was still above the common range. Chen *et al.* [21] noted that what they actually measured was particle size. Thus it is possible that both FAs that were examined had very small particle sizes. Chen *et al.* [21] dismissed the notion that E_4/E_6 ratios can be used as a measure of the concentration of condensed aromatic rings in humic materials.

The extinction coefficient (E_{280}) was lower for the Luvisol HA than for the Mollisol HA. Anderson *et al.* [2] postulated that this coefficient increased with the increase in the aromaticity of the humic matter. These data thus suggest that the aromaticity of the Mollisol HA was greater than that of the Luvisol HA. The same trend was observed for the FAs.

X-Ray Diffraction

A broad band at 3.5 Å, appearing in the diffractograms of many HAs, has been attributed to condensed aromatic rings [22]. More recently, Matsui et al. [23] and Shindo *et al.* [24] associated this band with type A humic acid that has a graphite-like structure. Fig. 3 indicated that the band at 3.5 Å was more pronounced in the Mollisol HA than in Luvisol HA. A fairly pronounced reflection at 4.7 Å from the Mollisol FA (Fig. 3) possibly indicated the presence of a broken network of condensed aromatic rings with disoriented aliphatic or alicyclic chains around the edges [25]. The diffractograms (Fig. 3) distinctly suggest that the Mollisol HA and FA contained more ordered structures than the Luvisol HA and FA.

Differential Thermal Analysis

The major reaction in the Mollisol HA was a very strong exotherm at 544°C (Fig. 4), indicating strongly condensed aromatic rings [4]. A much smaller exotherm at 464°C suggested the presence of some less condensed aromatic compounds. In the Luvisol HA, the major exothermic reaction was at 464°C, whereas the high temperature exotherm at 505°C appeared in the form of a shoulder. A low temperature weak exotherm at 288°C was attributable to the decarboxylation of stable COOH groups and aliphatic structures [26]. These results suggest that the Mollisol HA contained principally highly condensed aromatic compounds, whereas in the Luvisol HA the aromatic compounds were of a less condensed nature.

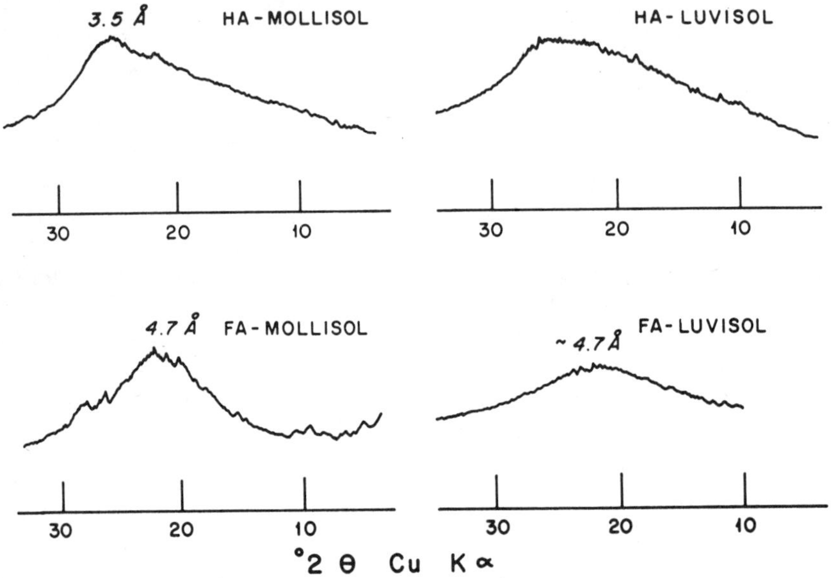

Figure 3 X-ray diffractograms of the humic and fulvic acids of the Mollisol and Luvisol

The slight endotherms at 72-75°C displayed by both FAs are due to dehydration. The prominent exotherms at 272 and 288°C for the Mollisol and Luvisol FAs, respectively, may be attributed, as in the HAs, to the decarboxylation of stable COOH groups and aliphatic structures. At about 300°C, the Mollisol FA exhibited three exotherms and the Luvisol FA had two exotherms. Reactions at these intermediate temperatures have been attributed by Kodama and Schnitzer [26] to the decomposition of the aromatic "nuclei". The exotherms around 400°C probably signified the combustion of very poorly condensed aromatic structures, whereas those at 464 and 472°C (Mollisol) and 480°C (Luvisol) were associated with the pyrolysis of more condensed aromatic structures. According to this interpretation, the Mollisol FA contained a higher proportion of more condensed aromatic structures than the Luvisol FA. The Mollisol FA apparently contained more of the stable aromatic compounds than the Luvisol FA. Neither FA contained the high temperature (~500°C) exotherms reported by Chen et al. [5] for the Mediterranean region FAs which were attributed to silicate impurities. The shape of the curves indicating the exothermic reactions suggested that the humic compounds in the Mollisol were better ordered and better defined than in the Luvisol. This is in accord with X-ray diffraction data (Fig. 3).

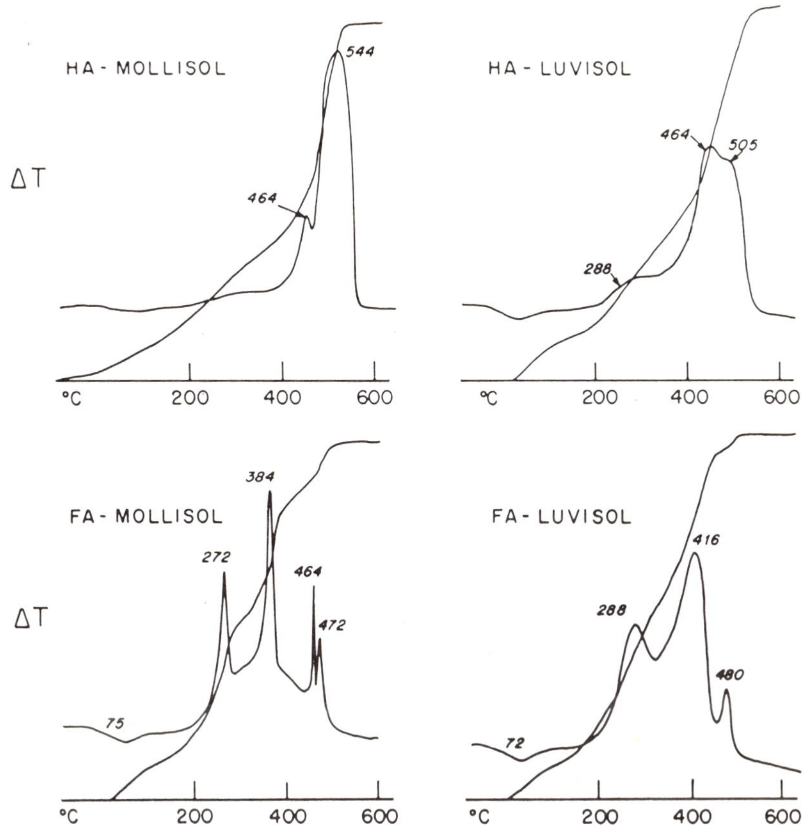

Figure 4 Differential thermal analysis and thermogravimetric analysis traces of the humic and fulvic acids of the Mollisol and Luvisol.

Summary

The amount of HAs was higher in the Mollisol than in the Luvisol. The opposite was true for FAs. The contents of N and S of the FAs were lower in the Luvisol than in the Mollisol. On the other hand, there was no great difference in elemental composition of HAs between the two soils. Aliphatic C in alkyl chains, carbohydrates, aromatic ring C, and COOH groups were present in both HAs and FAs of the soils. Methoxy groups and O and N substituted aromatic C were present in both HAs. Compared with the Mollisol, HAs in the Luvisol were more aliphatic. HAs and FAs in the Mollisol were more well ordered and appeared to contain a higher proportion of more condensed aromatic structures than those of the Luvisol. The nature of humic substances was evidently influenced by differences in soil formation processes in grassland and parkland in the Canadian Prairies region studied.

Acknowledgements

Contribution No. 633, Saskatchewan Institute of Pedology, The University of Saskatchewan, Saskatoon, SK, S7N OWO, Canada. This study was supported by the Natural Sciences and Engineering Research Council of Canada, Grant A 2383-Huang.

References

1. Anderson, D.W., D.B. Russell, R.J. St. Arnaud and E.A. Paul. Can. J. of Soil Sci.**54**:317(1974).

2. Anderson, D.W., E.A. Paul and R.J. St. Arnaud. Can. J. of Soil Sci. **54**:447 (1974).

3. Anderson, D.W. J. of Soil Sci. **30**:77 (1979).

4. Schnitzer, M. In: Soil Organic Matter Studies, Part 2, pp. 117-132. (Vienna: International Atomic Energy Agency, 1977).

5. Chen, Y., N. Senesi and M. Schnitzer. Geoderma **20**:87 (1978).

6. Aikens, G.R. In: G.R. Aiken, D.M. McKnight, R.L. Wershaw and P. MacCarthy, Eds, Humic Substances in Soil, Sediment and Water, pp. 363-385 (New York: John Wiley & Sons, 1985).

7. Hayes, M.H.B. In: G.R. Aiken, D.M. McKnight, R.L. Wershaw and P. MacCarthy, Eds, Humic Substances in Soil, Sediment and Water, pp. 329-362 (New York: John Wiley & Sons, 1985).

8. Leenheer, J.A. In: G.R. Aiken, D.M. McKnight, R.L. Wershaw and P. MacCarthy, Eds, Humic Substances in Soil, Sediment and Water, pp. 409-429 (New York: John Wiley & Sons, 1985).

9. Swift, R.S. In: G.R. Aiken, D.M. McKnight, R.L. Wershaw and P. MacCarthy, Eds, Humic Substances in Soil, Sediment and Water, pp. 387-408 (New York: John Wiley & Sons, 1985).

10. Houde, M. and J. Champy. Microchem. J. **24**:300 (1979).

11. Mertz, W. Zeitschr. Anal. Chem. **237**:272 (1968).

12. Unterzaucher, J. Ber. **73B**:391(1940).

13. ASTM. Method B, High Temperature Option, ASTM D4239-83, pp. 484-495, Section 5, Volume 5.05 (Philadelphia: American Society of Testing & Materials, 1983).

14. Ma, T.S. and J.D. McKinley, Jr. Microchim. Acta **1**:4 (1953).

15. Chen, Y., N. Senesi and M. Schnitzer. Soil Sci. Soc. Am. J. **41**:352 (1977).

16. Vaughan, D. and B.G. Ord. In: D. Vaughan and R.E. Malcolm, Eds, Soil Organic Matter and Biological Activity (Dordrecht: Martinus Nijhoff/Dr. W. Junk Publisher, 1985).

17. Steelink, C. In: G.R. Aiken, D.M. McKnight, R.L. Wershaw and P. MacCarthy, Eds, Humic Substances in Soil, Sediment and Water, pp. 457-476 (New York: John Wiley & Sons, 1985).

18. Kononova, M.M. and I.V. Alexandrova. Geoderma **2**:157 (1973).

19. Stevenson, F.J. Humus Chemistry-Genesis, Composition, Reactions (New York: John Wiley & Sons, 1982).

20. Schnitzer, M. In: P.M. Huang and M. Schnitzer, Eds, Interactions of Soil Minerals with Natural Organics and Microbes. Soil Science Society of America Special Publication 17 (Madison: Soil Science Society of America, 1986).

21. Chen, Y. and M. Schnitzer. Soil Sci. Soc. Am. J. **40**:682 (1976).

22. Pollack, S.S., H. Lentz and W. Ziechmann. Soil Sci. **112**:318 (1971).

23. Matsui, Y., K. Kumada and M. Shiraishi. Soil Sci. Plant Nutr. **30**:13 (1984).

24. Shindo, H., Y. Matsui and T. Higashi. Soil Sci. **141**:84 (1986).

25. Kodama, H. and M. Schnitzer. Fuel **47**:87 (1967).

26. Kodama, H. and M. Schnitzer. Soil Sci. **109**:265 (1970).

Session 2: Biological and Chemical Transformation and Degradation

Influence of Vegetation Changes on Soil Organic Matter

Per Nørnberg

Institute of Geology, University of Aarhus, DK-8000 Aarhus C, Denmark

Abstract

In a heath region at Hjelm Hede in Denmark oak trees are invading a *Calluna/Empetrum* vegetation. In less than a century the oak invasion has caused considerable changes in the soil: what was once an O-horizon under *Calluna* has changed to an A-horizon under oak; the *Calluna* E-horizon has lost its distinct appearance; and the sharp boundary between E and Bh has been obliterated. The directly visible changes are associated with a rise in pH of about one unit in the top horizon under the oaks, an increasing content of organic matter in the E-horizon, a decreasing content of organic matter in the Bh-horizon, and a fall in the C/N ratio.
In order to estimate the total microbiological activity, cotton strips were placed in the upper soil horizons. The loss in tensile strength during two summer months was 10-15% under *Calluna*, but more than 50% under oaks. Initial attempts to find differences in the type and content of organic matter showed that the most abundant low-molecular organic acids extracted from the Of-horizons were 3,4-dihydroxybenzoic acid (protocatechuic acid), 4-hydroxybenzoic acid and 4-hydroxy-3-methoxybenzoic acid (vanillic acid). The extraction was done in 0.1 M sodium pyrophosphate at pH 10.2. The organic compounds were determined by HPLC. The 3,4-dihydroxybenzoic acid was relatively the most important compound under the *Calluna* heath, whereas 4-hydroxy-3-methoxybenzoic acid was most important under oaks. Extractions were performed on water samples from field lysimeter experiments to determine whether the substituted benzoic acids in the soil water arose under transport. These extractions exposed a ppm concentration of 2,4-dichlorobenzoic acid, a compound believed to originate from microbial decomposition of lysimeter material.

Introduction

In the middle of the 19th century, large parts of Jutland, Denmark were covered with heath. A common species in these areas was heather (*Calluna vulgaris*), which played an important role in the ecosystem of heath farming. When the heath farming system was replaced by a rotation crop system, the heath was either ploughed, reclaimed, just left to itself or used for conifer plantations.

If the heath vegetation was not maintained, *Empetrum nigrum*, grasses like *Descampsia flexuosa*, and after that oak (*Quercus robur*), took over. However, aspen

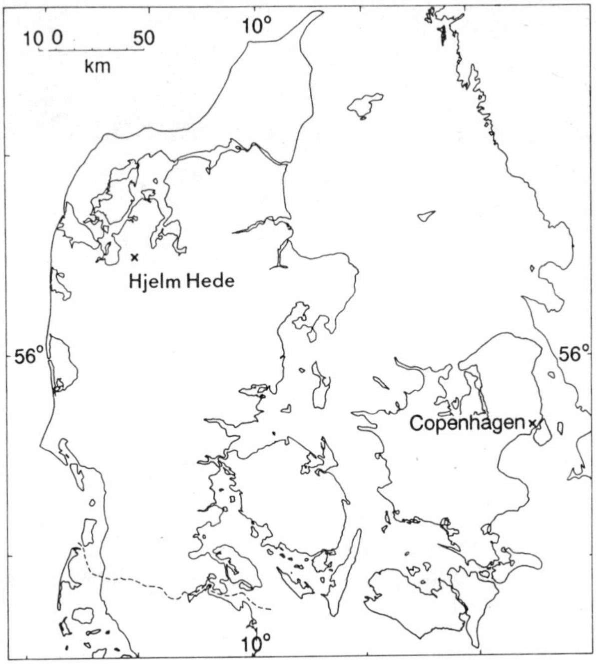

Figure 1 Location of the study area.

(*Popula tremula*) and birch (*Betula pendula*) are also important species in ancient heath regions.

Previous investigations in our study area at Hjelm Hede (Fig. 1) have shown that, during the last century, oak has advanced into the heath area of a rate of one metre per year [1], owing to mouse-spread acorns in the *Calluna* vegetation. Soil changes from mor to mull were observed by Müller [2] and indicated by Tamm [3] and Burrichter [4]. This process, called "depodzolisation" by Dimbleby [5] and Miles and Young [6], is easily recognized under oak trees at Hjelm Hede (Nielsen *et al.* [7,8]). It can be detected in the chemical composition of the soil and clearly seen on soil horizon development after only 60 years of tree growth. This study was aimed at investigating changes in the chemical composition of soil resulting from vegetation changes. Soil samples were extracted for the purpose of determining phenolic acids, and lysimeters were installed to collect soil water.

Materials and Methods

Geology, Climate and Site

The parent material is fluvio-glacial meltwater sands of late Weichselian Age

(10.000 yrs). Mean annual precipitation is 650-700 mm/yr, with an actual evapo-transpiration of about 375 mm/yr. The area is well drained. The mean annual temperature is 7.5°C. The main soil types in this material are spodosols. The soils are described according to USDA Soil Taxonomy [9].

A 9 m long trench was dug from the *Calluna* heath to the stem of a 60-year-old oak tree. Two profiles along this cross section, a and d in Fig. 2, will be discussed.

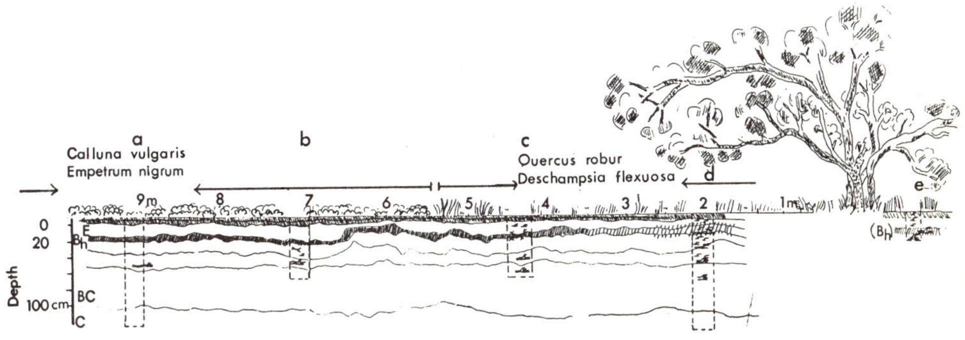

Figure 2 Cross-section of the 9 m long trench with profiles a and d, after Nielsen *et al.* [7].

Lysimeters

Three types of lysimeters were placed in the soil (Fig. 3). Two of them were modifications of the type described by Rasmussen *et al.* [10], in which cores are taken up with a 30 cm wide steel cylinder and placed in a plexiglass cylinder. Of these two, one contained the whole core down to a certain horizon (III), and the other the soil from below the O-horizons and down (II). The third type of lysimeter (I) was a 5 cm high plexiglass cylinder installed from a tunnel and filled with soil material from the depth at which it was placed. The cylinder was kept in contact with the above-lying soil by wedging a PVC cylinder in under the lysimeter. This principle worked only because the experimental area was very sandy and allowed water to drain freely out of the soil.

Laboratory Methods

Samples for soil analysis were sieved in a 2 mm sieve and ground to <125 µm for extractions and carbon and nitrogen determination. Iron and aluminium were extracted with dithionite-citrate-bicarbonate after Mehra and Jackson [11] and with 0.1 M Na-pyrophosphate, pH 10.2, after Bruckert and Metche [12]. Carbon was determined in the extractants by the Walkley-Black method modified by heating to 180°C for 5 min. Total carbon was determined by dry combustion and nitrogen by the Kjeldahl method. pH was determined in a 1:1 soil-liquid mixture. Exchangable bases were leached with 1 M ammonium acetate of pH 7 [13] modified by washing out excess ammonium ions with ethanol and centrifugation instead of filtration. Elements were determined by AAS. For the determination of particle size distribution the samples were peroxidized and dispersed in Na-pyrophosphate. Particles >38 μm were sieved, and particles <38 μm were sedimented in Andreasen pipettes.

Low-molecular-weight phenolic compounds were determined in the pH 10.2, pyrophosphate-extracted fulvic fraction. The extraction procedure is described by Vance *et al.* [14] and shown in the flow diagram (Fig. 4). The extraction from the fulvic fraction was done with diethyl ether three times, and the samples were then evaporated to a volume of 10 ml. The samples were concentrated 5 times before determining the phenolic compounds by reverse-phase HPLC. The HPLC equipment was Perkin Elmer with a series 10 isocratic pump. A 200x4.6 mm column packed with Spherisorb S5, C18 material, and the eluent was acetic acid: 1-butanol: water in the ratio 2:15:983, at pH 3.08.

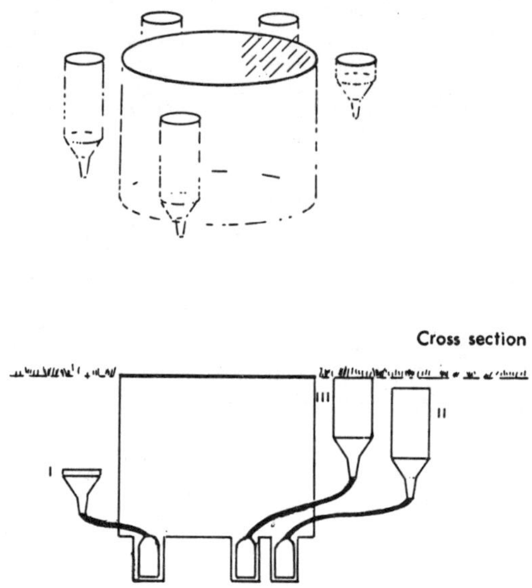

Figure 3 The three types of lysimeters.

Analysis for 2,4-dichlorobenzoic acid was done by reverse-phase HPLC on a Hewlett Packard HP1090 chromatograph equipped with a photodiode array detector. A 100x4.6 mm, 5 μm, C18 column was used. The elution was isocratic with an eluent of distilled water buffered with a phosphate buffer to pH 2.5 and mixed with 40% acetonitrile. The flow rate was 1.5 ml/min., and the column temperature 50°C. Detection was done at 210 nm and 278 nm, with 550 nm as reference.

For control analysis soil samples were centrifuged to collect soil water equivalent to the lysimeter water samples. Stainless steel centrifuge tubes with double bottoms

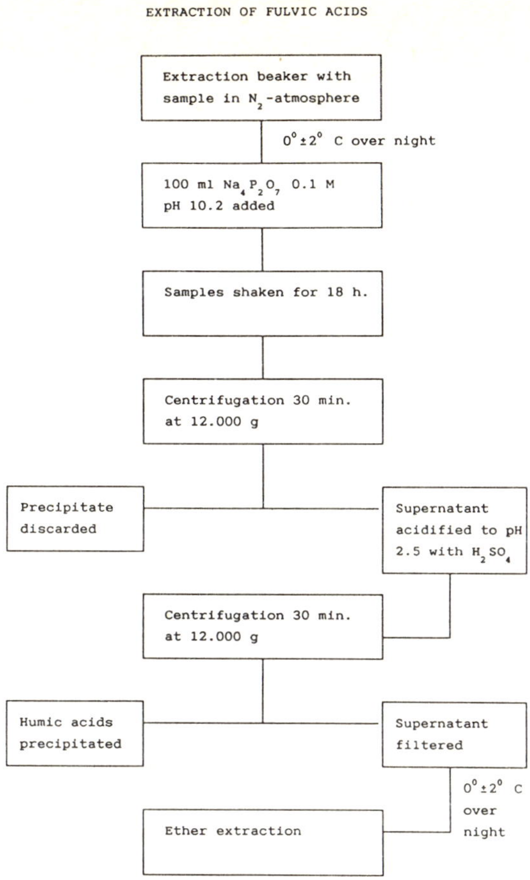

Figure 4 Procedure for extracting fulvic acids.

were used. Centrifugation was done at a minimum of 400 g and a maximum of 700 g, depending upon rotation diameter, for 20 minutes.

Results

Profile Description

Profile a. 910-930 cm from the stem of the oak. Heath vegetation was mainly *Empetrum nigrum* with some senile *Calluna vulgaris*.

Oi, Oe 6-3 cm Black (10YR 2/1) heather peat: *Calluna* remains held together by roots and mosses.

Oa	3-0 cm	Black decomposition products of heather peat more massive than in Oi- and Oe-horizons: held together by numerous roots; few pale mineral grains; abrupt smooth boundary.
E	0-17 cm	Dark brown (7.5 YR 4/2) sand with a few lighter brown spots: friable, structureless, and single grain; few fine roots; clear, wavy boundary.
Bh	17-20 cm	Black (5YR 2.5/1) sand: thickness ranges from 3 to 10 cm; friable, structureless, and single grain; few fine roots; clear, wavy boundary.
Bhs	20-37 cm	Strongly mottled sand, with pockets of strong brown (7.5 YR4/6), looser than the matrix of dark reddish brown (5YR 3/3) sand: weakly coherent, structureless, and single grain; gradual, wavy boundary.
Bs	37-57 cm	Weakly mottled sand with the same colours as above, but paler ones dominant: one coarse horizontal oak root; gravelly layer marks boundary.
BC	57-98 cm	Yellowish brown (10YR 5/8) sand: loose, structureless, and single grain; few dispersed stones; few coherent dark yellowish brown (10YR 4/4) balls of finer texture.
C	98-126 cm	Yellowish brown (10YR 5/4) sand.

Profile d. 190-220 cm from the stem of the oak. Vegetation is *Deschampsia flexuosa* and a few herbs, such as *Trientalis europeae* L. and *Majanthemum bifolium* L. The profile is shaded by the canopy of the oak.

Oi, Oe and Oa	10-0 cm	Black (10 YR 2/1) moder, with upper part more brownish than lower part: many hyphae in the leaves from the oak; lower part held together by roots; many pale sand grains in lower part; clear, smooth boundary.
E	0-8 cm	Dark brown (7.5 YR 3/2) sand: (7.5 YR 4/2) dry, loose, structureless, and single grain; many fine and coarse oak roots; clear, wavy boundary.
Bh	8-15 cm	Dark reddish brown (5YR 2.5/2) sand, thickness varies somewhat, loose, structureless, and single grain; many fine roots, few coarse roots; clear, wavy boundary.

Bhs | 15-31 cm | Weakly mottled sand with spots of strong brown (7.5YR 4/5) in matrix of dark brown (7.5YR 3/4): loose, structureless; more roots than in Bh-horizon; gradual, smooth boundary.

Bs | 31-53 cm | Sand with colours as in Bhs horizon but with fewer mottles: loose, structureless; many oak roots.

BC | 53-94 cm | Yellowish brown (10YR 5/8) sand: loose, structureless, and single grain; few dispersed stones; few coherent dark yellowish brown (10YR 4/4) balls of finer texture.

C | 94-130 cm | Yellowish brown (10YR 5/4) sand.

The main morphological differences between the profiles are seen in the O-horizons, the E- and Bh-horizons and in the boundaries between them. The clear distinction between O-horizons in the *Calluna/Empetrum* profile a is less clear in the oak profile d, where Oa can not be separated from the upper O-horizons. The E-horizon is not as deep and not as light under the oak as under the heath. The Bh-horizon is deeper and less distinct under the oak. The root intensity is higher in the lower part of the oak profile.

Table 1 Total carbon, C/N, pH, pyrophosphate extractable C, Fe and Al, DCB extractable Fe and Al, CEC, and base saturation.

Profile	Horizon	C %	C/N	pH H_2O	pH KCl	C_p %	Cp/Ct	Fe_{DCB} (o/oo)	Al_{DCB} (o/oo)	Fe_p (o/oo)	Al_p (o/oo)	CEC NH_4OAc pH 7	Base Sat. %
a	Oi,Oe	29.3	24	4.25	2.86	1.50	0.05	2.51	0.99	0.79	1.06	67.2	n.d.
	Oa	9.29	37	3.85	2.71	0.81	0.09	0.89	0.51	0.42	0.52	29.4	3.5
	E	0.72	36	4.28	3.10	0.12	0.17	0.57	0.12	0.08	0.07	1.9	4.5
	Bh	1.33	34	4.02	3.28	0.49	0.37	4.68	0.86	3.16	0.93	6.7	1.6
	Bhs	0.93	33	4.50	3.98	0.26	0.28	4.40	1.94	2.87	1.95	4.2	1.6
	Bs	0.38	27	4.67	4.40	0.05		2.91	2.45	0.52	1.63	n.d.	n.d.
	BC-1	0.29		4.61	4.46	tr.		2.01	1.74	0.19	1.22	n.d.	n.d.
	BC-h	0.38		4.60	4.30	-		2.85	2.70	0.44	2.32	n.d.	n.d.
	C	0.10		4.72	4.30	-		1.29	0.60	0.22	0.57	1.2	2.5
d	Oi,Oe	9.40	15	5.02	3.77	0.75	0.08	1.67	0.31	0.42	0.29	15.6	18.6
	Oa/A	4.77	25	4.51	3.12	0.64	0.13	1.39	0.23	0.43	0.24	7.0	9.9
	E	1.27	28	4.30	3.10	0.31	0.24	1.48	0.20	0.45	0.24	3.1	4.6
	Bh	1.01	27	4.15	3.30	0.33	0.33	2.76	0.37	1.69	0.44	4.0	2.6
	Bhs	0.78	28	4.70	4.05	0.24	0.31	3.51	1.44	1.90	1.35	2.8	2.6
	Bs	0.43	26	4.54	4.28	0.10	0.23	3.25	1.96	1.24	1.67	n.d.	n.d.
	BC-1	0.24	34	4.50	4.42	n.d.	n.d.	2.66	1.93	0.27	1.38	n.d.	n.d.
	BC-h	0.35	25	4.50	4.40	-	-	2.64	2.58	0.37	2.12	n.d.	n.d.
	C	0.17		4.69	4.40	-	-	1.12	0.60	0.19	0.57	0.8	3.9

Basic Soil Data

Compared to the *Calluna/Empetrum* profile, in the top horizons under the oak the total carbon content was markedly lower (Table 1), in the E-horizon higher, and, again, in the Bh-horizon lower. The C/N ratio was lower under the oak, and the pH in the top horizons was nearly one unit higher. The pyrophosphate extractable carbon, iron and aluminium were levelled among horizons under the oak as compared to heath. The same holds for iron and aluminium extracted with diethinite-citrate-bicarbonate (DCB). The cation exchange capacity (CEC) follows the content of organic matter and base saturation pH.

Cotton Strips Experiment

Cotton strips were placed vertically in the soil horizons down to Bh. The strips were left in the soil for two summer months, and their loss of tensile strength was taken as a measure of cellulose decomposition (Fig. 5). The original tensile strength

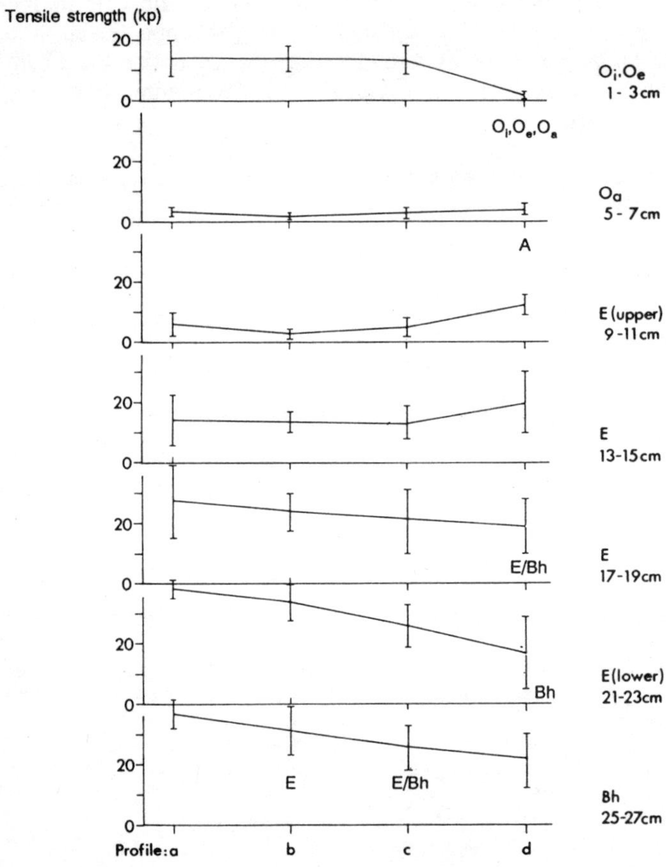

Figure 5 Loss in tensile strength of cotton strips. The bars represent standard deviation of 10 measurements. Vertical scale in kiloponds (after Nielsen *et al.* [8]).

of the strips was 42 kp. In the heath end of the trench (a-c) the upper O-horizons tended to dry out and, for that reason, gave a confusing picture. The strips were too decomposed in Oa to show any difference. From the middle of the E-horizon (17-19 cm), where the soil kept some moisture during the entire study period, the loss in tensile strength was higher under the oak.

Phenolic Compounds

The phenolic compounds that were expected to be found in the extracts of fulvic acids are listed in the standard chromatogram (Fig. 6). In the O-horizons of both

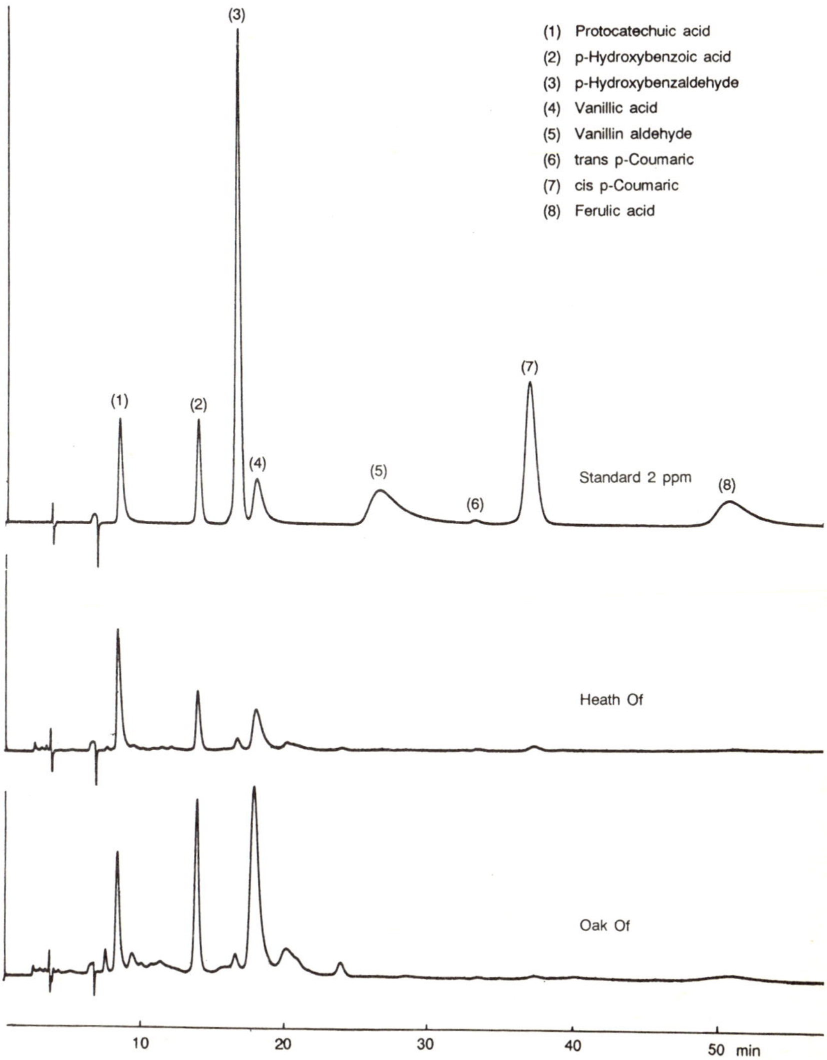

Figure 6 Chromatogram of phenolic compounds in soil extracts from heath Of- and oak Of-horizons.

profiles, 3,4-dihydroxybenzoic acid, 4-hydroxybenzoic acid and 4-hydroxy-3-methoxybenzoic acid were most frequently detected. We found a relatively higher content of 3,4-dihydroxybenzoic acid than of 4-hydroxybenzoic acid and 4-hydroxy-3-methoxybenzoic acid in extracts from the heath Of-horizon, whereas the latter two were more dominant in Of under the oak.

The soil water samples from June 1988 had a high content of 2,4-dichlorobenzoic acid (up to 14 ppm) which decreased during July and August and was not detectable in the samples during the winter. The chlorinated benzoic acid was earlier detected at the same time of the year in samples from 1987 and was still present in the stored samples. In 1989 the compound was again found, starting in late June. Samples of rain water and out-centrifuged water from parallel soil samples were analysed for 2,4-dichlorobenzoic acid during the spring and summer of 1989, and no trace (< 0.1 mg/l) of the compound was detected in these samples.

Discussion

The soil profiles under heath vegetation and oak, respectively, clearly show the difference in soil development. The distinct horizon boundaries under heath vegetation are broken up under oak and material is mixed between O-, E- and Bh-horizons. The roots of the oak trees penetrate deeper into the soil than the roots of heath vegetation. The grasses under the oak probably also play an important role in the soil changes. The carbon content and C/N-ratio (Table 1) clearly show that the mor layer under oak is under decomposition. The distribution of organic matter and the cellulose decomposition (Fig. 5) indicate the increase in biological activity. This is in accordance with the higher level of pH under oak and the higher base saturation status.

It was expected that the polyphenol content of the oak soil would be lower than under *Calluna*, as the polyphenols are known to have a restrictive effect on biological activity. It has also been stated that the 3,4-dihydroxybenzoic acid was responsible for one of the main podzolisation processes, namely the chelation and transport of iron and aluminium, as discussed by Vance *et al.* [15]. The relatively higher content of 3,4-dihydroxybenzoic acid compared to other phenolic acids in the heath Of-horizon might indicate a stronger podzolisation effect than under oak, where the content of 3,4-dihydroxybenzoic acid is relatively lower compared to other compounds. These results are from only one set of extractions, and the work is being continued by collecting soil water instead of performing pyrophosphate extractions. So any conclusive results shall not be drawn.

However, the original purpose of the analysis of soil water collected from lysimeters was changed, since the lysimeter samples contained 2,4-dichlorobenzoic acid in surprisingly high concentrations. First of all the lysimeter materials were suspected to contain the acid and extractions and dissolutions of the materials were carried out to trace the compound or any substance from which it could be derived. No chlorinated compounds were found. The chlorinated benzoic acid could not be detected in rain water or in parallel, centrifuged samples of soil water. It is therefore believed that, when the soil temperature is high enough (in June), the microorganisms destroyed some of the plastic material from which the lysimeters were made. Possibly, the concentration of the acid gets high enough to restrain growth of the microorganisms, which might explain why the concentration decreases during the summer. Further work to trace the source of the acid is in progress.

Acknowledgements

The author is grateful to Lina Asmussen, who did a major part of the laboratory work on extracted phenolic compounds, to Knud Erik Nielsen, in whose project the lysimeter installations were first used, to Dr. Ludvig Haumaier, University of Bayreuth, who first detected chlorinated compounds in the soil water samples, and to laboratory leader Elsa Voldbjerg, Cheminova A/S, who has run a number of quantitative determinations of dichlorobenzoic acid and identified the compounds by UV-spectroscopy and GC-MS. Special thanks are due to Professor Jens Tyge Møller for valuable discussions of the manuscript and to Bodil Bay Nielsen for discussions of technical details and typing of the text.

References

1. Jensen, T.S. and O.F. Nielsen. Oecologia (Berlin) **70**:214 (1986).
2. Müller, P.E. Tidsskrift for Skogbruk **7**:1 (1884) in Swedish.
3. Tamm, O. Proc. 2nd Int. Congr. Soil Sci. **5**:178 (1932).
4. Burrichter, E. Zeitschr. Pflanzenern. Bodenk. **67**:150 (1954).
5. Dimbleby, G.W. J. Ecol. **40**:331 (1952).
6. Miles, J. and W.F. Young. Bull. Ecol. **11**:233 (1980).
7. Nielsen, K.E., K. Dalsgaard and P. Nørnberg. Geoderma **41**:79 (1987).
8. Nielsen, K.E., K. Dalsgaard and P. Nørnberg. Geoderma **41**:97 (1987).
9. U.S.D.A. Soil Taxonomy, Agricultural Handbook No 436 (Washington D.C: U.S. Govn. Printing Office (1975).
10. Rasmussen, L. Bull. Environ. Contam. Toxicol. **36**:563 (1986).
11. Mehra, P.O. and M.L. Jackson. Clay Clay Minn. Monogr. Earth Sci. Ser. **5**:317 (1960).
12. Bruckert, S. and M. Metche. Bull. Ecol. Nat. Supér. Agron. Ind. Alim. **14**:263 (1972).
13. Chapman, H.D. Meth. of Soil Anal., Publ. 9. pp. 891-900 (Madison: A.S.S.A. 1965).
14. Vance, G.F., S.A. Boyd and D.L. Mokma. Soil Sci. **140**:412 (1985).
15. Vance, G.F., D.L. Mokma and S.A. Boyd. Soil Sci. Soc. Am. J. **50**:992 (1986).

Humification Parameters of Organic Materials Applied to Soil

Claudio Ciavatta, Livia Vittori Antisari and Paolo Sequi

Institute of Agricultural Chemistry, University of Bologna,
Viale Berti Pichat, 10-40127 Bologna, Italy

Abstract

A fractionation method to separate humified and non-humified substances based on chromatography on solid polyvinylpyrrolidone (PVP) is described. Also presented is the adoption of three new humification parameters to characterize organic matter in soils, fertilizers and sewage sludges.

Introduction

A knowledge of the ratio between humified and non-humified materials in natural substrates, such as soils, fertilizers, composts, or sludges, can be very important from agronomical and environmental points of view. Addition of humified materials to the soil is equivalent to the addition of stabilized organic carbon. In contrast, due to biological activities, non-humified materials yields humified compounds and metabolic energy. The actual humification degree in soil is depending on the organic materials applied, soil type, and climatic conditions.

A first and often neglected problem is the determination of organic carbon itself. It is difficult to find routine procedures for fractions isolated from very different organic materials, when the most commonly employed method applied to just one of the various kinds of materials of interest, i.e. surface soils, requires a correction factor that ranges from 1.19 to 1.33 to face underestimation arising from the method [1]. A second problem is to find a universal method of extraction and fractionation, reliable for and applicable to a wide range of soils, fertilizers, sludges, etc. A third problem is to express the results obtained. These topics have been the subject of our research for the last few years. The results we have obtained are collected and integrated in the present paper.

Determination of Organic Carbon

All the various methods for the determination of organic carbon in soils or fertilizers involve conversion of organic C to CO_2 by wet or dry combustion and subsequent quantitative determination of the amount of CO_2 formed. Dry combustion methods are precise enough, but require the use of expensive equipment and are most convenient for solid samples that do not contain carbonates. The more widely used wet combustion methods involve a rapid oxidation of the organic carbon in solid or liquid samples by a dichromate-acid mixture, with [2-7] or without [8] external heating. There are many problems related to time and temperature of heating of the mixture. Data obtained without external heating must be corrected. Usually a correction factor of up to 1.33 is used for surface soils [1], because of incomplete oxidation of the organic matter. However, the methods for the determination of organic carbon in soil have only rarely been extended to e.g. organic fertilizers. It would be unwise to propose the same correction factor for fertilizers and surface soils, since the factor will change considerably from one material to another. Work was undertaken [9] to determine if it is possible to find a universal method for both soils and fertilizers that also avoids the necessity for a correction factor. The approach taken was that of using the highest possible heating temperature. The German school of the 1950's had suggested 165°C for 10 minutes [4], whereas more recently, Italian scientists [10] and the Italian Association for Normalization of Analytical Methods in the Chemical Industry, have recommended a lower temperature for this purpose (160 ± 2°C for 10 minutes). In our work the lower temperature and 10 minutes of heating was more appropriate, because at higher temperatures maintaining the initial dichromate concentration was impossible (Fig. 1).

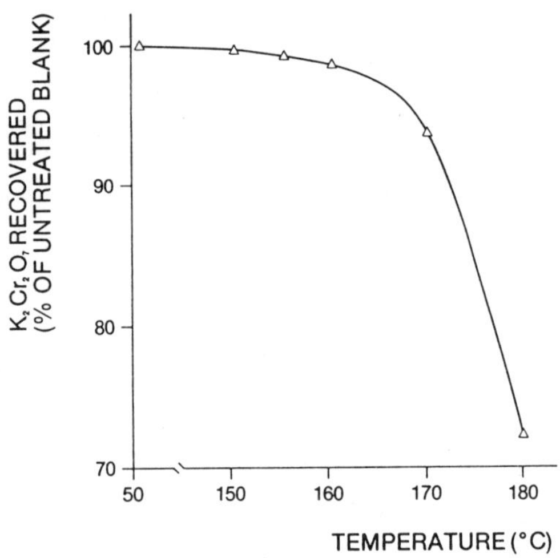

Figure 1 Decomposition of potassium dichromate with increasing temperature (heating time 10 minutes).

Another difficult problem is to maintain a precise temperature (160°C) within such a small range of variation (±2°C) for exactly 10 minutes. The digestion flasks illustrated in Fig. 2 were found to be especially suitable for this purpose. The lower part of the flask is cylindrical, and the intermediate part is a truncated cone. The upper part is again cylindrical, but of smaller diameter, with a glass joint on top which accommodates a silicon stopper. Such a flask acts as a reflux condenser for most of the vapors developed during digestion. The silicon stopper has two small concave hollows on opposite sides to allow excess vapors to escape; a central hole in the same stopper accommodates a thermometer, graduated from 40 to 200°C, with the bulb 0.5-1 cm from the base of the flask. Even heat diffusion from the flame can be assured by the use of a CERAN-type glass-ceramic protection plate instead of a normal glass mantle.

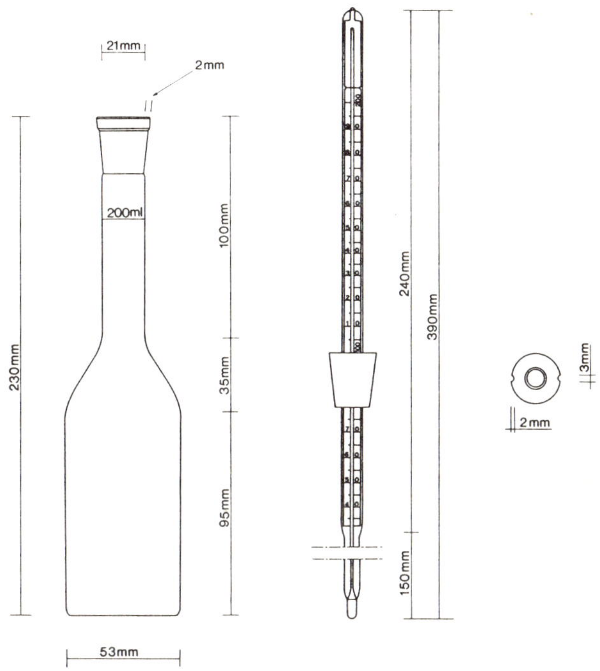

Figure 2 Digestion flask for the determination of organic carbon in soils and fertilizers.

A final important point is that, according to the original procedure [4], the digestion mixture is added with a known excess of $FeSO_4$, then titrated with $K_2Cr_2O_7$. Theoretically, therefore, (i) the amount of potassium dichromate necessary to back-titrate the excess $FeSO_4$ is directly correlated with the amount of organic carbon oxidized during digestion, and (ii) a "blank" is not required.

First of all, we verified that titration with $K_2Cr_2O_7$ by $FeSO_4$, is even more reliable than the reverse titration of $FeSO_4$ with $K_2Cr_2O_7$. Fig. 3 shows that the two reactions develop in very different ways. The end point of $Cr_2O_7^{2-}$ titration with Fe^{2+} is characterized by a sharp change in potential, but the irreversible shift in the voltage arises slowly at the end point of Fe^{2+} titration with $Cr_2O_7^{2-}$. The experiments were carried out potentiometrically, after inserting a combined platinum-calomel electrode in a solution containing known amounts of $K_2Cr_2O_7$ or $FeSO_4$, and titrating with 0.1 M $FeSO_4$ or $K_2Cr_2O_7$, respectively [9].

Since the accuracy of determinations depends on the volume of the reagent needed to achieve a voltage shift, it is apparent that direct titration of dichromate ions with Fe^{2+} is more reliable than the back titration originally suggested. We found this modified method fast, more reliable than other commonly used procedures, and also accurate for organic extracts. The procedure has been adopted in our laboratories for serial determination of carbon in soils and other organic materials and for the aqueous extracts of such materials.

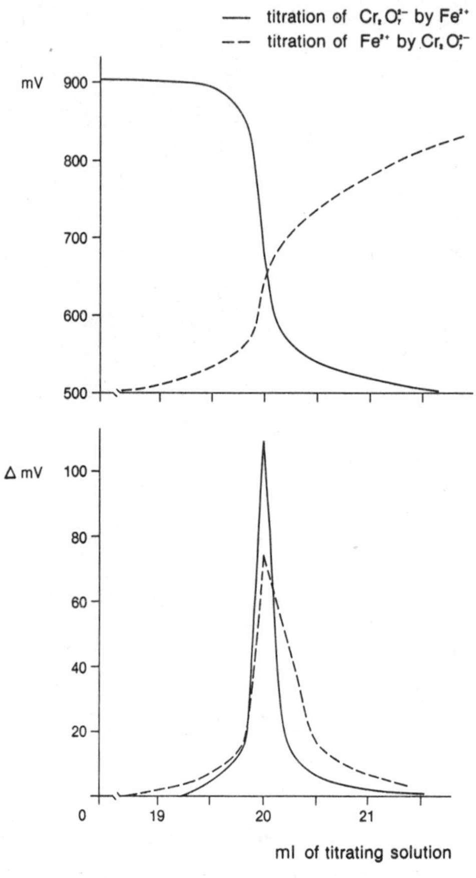

Figure 3 Potentiometric titration of $K_2Cr_2O_7$ with $FeSO_4$, and $FeSO_4$ with $K_2Cr_2O_7$.

Quality Criteria for Organic Matter

The methods used to assess quality of organic matter in soils, manures, sludges or composts are still not well defined. Some authors [11] have attempted the extraction of organic matter with a solution of 0.1 M NaOH plus 0.1 M $Na_4P_2O_7$, but have not applied this procedure to organic materials other than soil. Other scientists have suggested the use of selected spectral properties of humic substances (e.g., the degree of aromaticity [12]) but found that the nominal molecular weight distribution of humic acids decreases as the degree of humification increases. This is opposite to the normally accepted trend. Humification of soil organic matter has been widely studied using the ratio between humic acids (HA) and fulvic acids (FA), but the interpretations of the results are uncertain because they depend on many factors, even including such things as geographic distribution of the soils [13]. Another commonly used spectroscopic variable is the E_4/E_6 ratio (the ratio between the absorbance at λ=465 and λ=665 nm), which is sometimes considered as an index of humification [14]. However, the addition of a small amount of humic substances (e.g. humic acid from leonardite) is sufficient to change the results completely. The cation exchange capacity (CEC) of the organic fractions has also been used to evaluate the degree of humification of organic substances[15], although this criterion is very indirect.

Humic substances have been defined as "amorphous, polymeric, brown-colored compounds, that do not belong to recognizable classes of organic compounds, such as polysaccharides, polypeptides, altered lignins, etc." [16]. In the past, alkaline extracts from soil have been considered as total humic extracts [11], but the suspensions, especially if extracted from organic materials other than soil, contain non-humic substances. Separation of humic acids from fulvic acids is based only on the pH value of the extracts, but the supernatant (FA) also contains many classes of organic materials which are not humic substances (e.g., polysaccharides).

Some authors [17,18] have suggested the use of polyamide columns to retain the colored fractions of FA, whereas other authors have used Amberlite XAD-8 to selectively adsorb humic substances from freshwater [19, 20]. This last procedure is recommended by the International Humic Substances Society (IHSS) for the purification of humic substances from a variety of materials.

Polyvinylpyrrolidone (PVP), a crosslinked adsorbent for the chromatographic separation of aromatic acids, aldehydes and phenols [21,22], has been widely used to separate phenol compounds from organic extracts [23,24]. In acid media, insoluble PVP allows strong adsorptions and, after reelution, good recoveries of humic substances are obtained [24].

Recently, we have suggested the use of selective chromatography on solid PVP to separate humified and non-humified materials in soil, dung, compost and sludge extracts [25]. Presently, the use of solid PVP is required by Italian law for characterization of organic amendments (peat, leonardite and humic extracts).

Fig. 4 shows a fractionation scheme of organic extracts from different matrices (soils, organic fertilizers etc.). The separation of humified (HA+FA) from non-humified (NH) materials is achieved by means of columns packed with insoluble PVP. Non-humified fractions are not retained on PVP. After washing with 0.005 M H_2SO_4, the fulvic fraction is eluted with 0.5 M NaOH and added to the humic acids. This procedure has been applied with good results on organic materials [26,27], and also used to follow the maturation of organic materials in sludges [28] and in piles of compost from urban refuse [29].

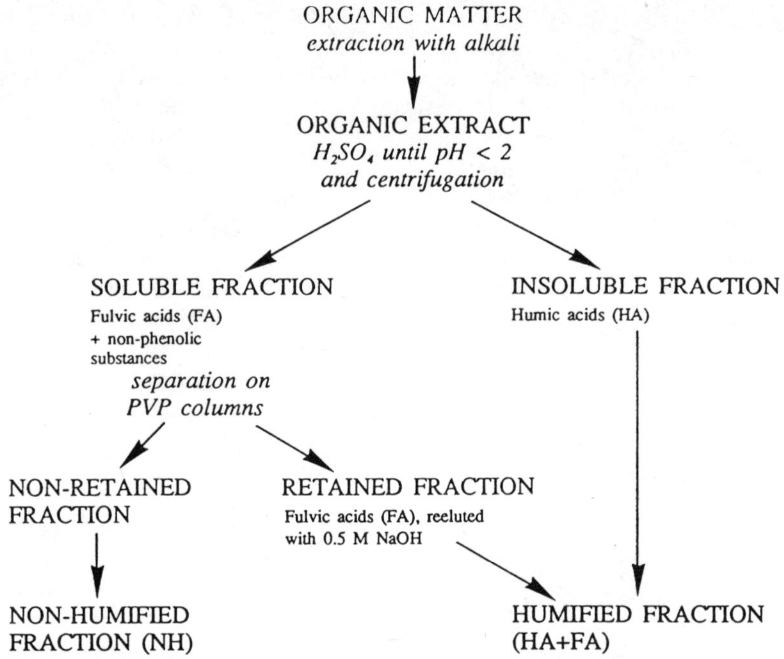

Figure 4 Fractionation scheme for the separation of humified materials from non-humified materials.

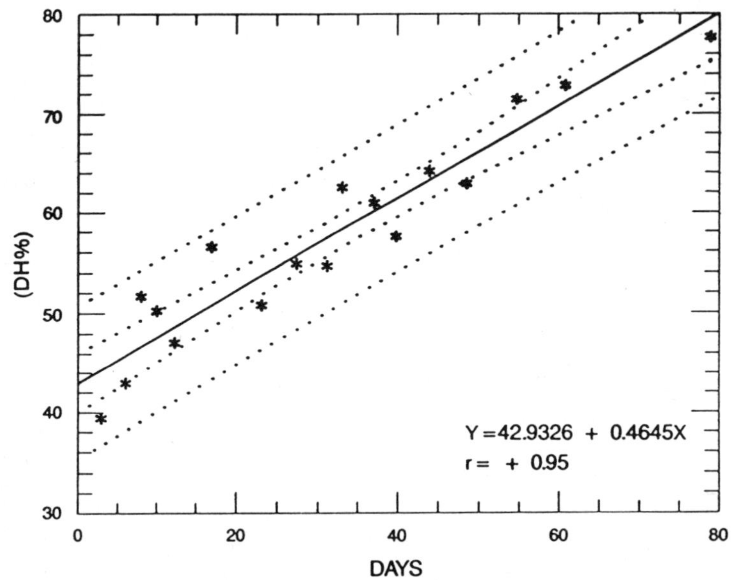

Figure 5 Trend of the degree of humification (DH) during the thermophilic phase of the organic matter stabilization process in a compost from urban refuse. Samples were taken at 15 cm depth. Re-elaborated from [29].

Three new parameters of humification have been proposed:

1 The humification index (HI) [25] HI = NH/(HA+FA) i.e. the ratio between non-humified (NH) and humified compounds (HA+FA).

2 The degree of humification (DH) [30] DH% = [(HA+FA)/TEC] x 100 i.e. the percentage of humified compounds with respect to total extracted carbon (TEC).

3 The humification ratio (HR) [30] HR% = [(HA+FA)/TOC] x 100 i.e. the percentage of humified compounds with respect to total organic carbon (TOC) in the sample.

In general the humification index (HI) is near zero (0-0.5) for humified materials (i.e. soils, organic materials) and much higher than 1 for non-humified materials (i.e. organic fertilizers, raw composts, sewage sludges and swine slurries). This parameter was the first to be proposed, and is the one most often encountered in the literature up to now.

However, the other parameters are very useful in some instances. Fig. 5 shows the trend of the DH during the organic matter stabilization processes in a compost from urban refuse. The value increases continuously, until it asymptotically reaches stable values at the end of the stabilization processes, after 80-120 days of fermentation. Commonly, the degree of humification (DH) is higher than 60% for humified materials (i.e. soils) and is close to 100 % only for leonardites (fossil humic substances). Less humified samples (i.e. organic fertilizers, raw sewage sludges, or swine slurries), show lower values. DH has also been used to monitor the evolution of organic matter from an organic fertilizer (leather meal), even after application to the soil [31], in order to follow a possible chromium release.

Strongly humified materials, such as leonardites, also show a high humification ratio (HR>80%), whereas this parameter appears to be generally lower for soils and organic materials. It must be born in mind that in some materials there is a continuous and immediate transformation of the substances liberated from the organic matrix, so that humic-like substances do not accumulate during the evolution of the organic matter. In this case which can be exemplified by composts from urban refuse, only DH is of practical value, because humified materials are of significant importance in extracts but only slightly relevant with respect to total organic carbon. In other matrices, such as pig slurries, however, HR can be much more significant than DH. Of course, generally speaking, for the majority of organic substrates, both DH and HR change and continuously reflect the evolution of the organic matter.

Conclusions

A simple method of separation of non-humified materials (NH) from apparently humified materials (HA+FA) present in natural substrates such as soils, fertilizers, composts, or sludges has been proposed. The procedure involves a fractionation of the organic extracts by acid precipitation and chromatography on solid polyvinylpyrrolidone. Determination of organic carbon in various organic materials can be made by a reliable wet combustion procedure.

Simple relations among the experimental data yield empirical values of practical concern, defined as "humification parameters". The *humification index,* HI, is the ratio between non-humified (NH) and apparently humified materials (HA+FA); HI was found to be useful mainly to compare diverse organic materials. The *degree of humification,* DH, and the *humification ratio,* HR, are the ratios between the humified materials (HA+FA) and either the total extractable carbon TEC or the total organic carbon TOC present in the sample, respectively. The use of both of the parameters is generally recommended in order to follow the progress of humification processes, though for a particular material only one of them fits in with the actual development of the process. During, for example, the organic matter evolution of composts from urban refuse, only DH shows a progressive increase correlated with the "maturation" of the material, whereas, in general, stabilization processes of slurries from livestock are better described by variations of HR. A better suitability of either DH or HR probably depends on the specific nature or the material considered, i.e., whether the bulk of organic matter in the material is evolved in the maturation process or not. If the humification process is effective only for a small proportion of the material while the bulk remains unaltered, DH can describe the process more properly than HR. Especially in the case of liquid or semiliquid wastes, however, the entire mass of organic matter is simultaneously involved in the stabilization process; under such conditions, only HR can accurately describe the process.

References

1. Nelson, D. W. and L. E. Sommers. In: C. A. Black et al., Eds., Methods of Soil Analysis, Part 2. pp. 539-579 (Madison, Am. Soc. Agronomy, 1982).
2. Schollenberger, C. J. Soil Science **24**:65 (1927).
3. Degtjareff, W. T. Soil Science **29**:239 (1930).
4. Springer, V. and J. Klee. Zeitung für Pflanzenernährung, Dügung und Bodenkunde **64**:1 (1954).
5. Lotti, G. Annali Facolta di Agraria Universita di Pisa **17**:113 (1956).
6. Jackson, M. L. Soil Chemical Analysis (Englewod Cliffs: Prentice-Hall, Inc., 1958).
7. Mebius, L. J. Analytica Chimica Acta **22**:120 (1960).
8. Walkley, A. and I. A. Black. Soil Science **37**:29 (1934)
9. Ciavatta, C., L. Vittori Antisari and P. Sequi. Communications in Soil Science and Plant Analysis **20**:759 (1989).

10. Nigro, C. Personal communication.

11. Schnitzer, M., L. E. Lowe, J. F. Dormaar and V. Martel. Canadian Journal of Soil Science **61**:517 (1981).

12. Tsutsuki, K. and Kuwatsuka. Soil Science and Plant Nutrition **30**:151 (1984).

13. Kononova, M. H. Soviet Soil Science **16**:71 (1984).

14. Chen. Y, N. Senesi and M. Schnitzer. Soil Science Society of America Journal **41**:352 (1977).

15. Roig, A., A. Lax, J. Cegarra, F. Costa and M. T. Hernandez. Soil Science **146**:311 (1988).

16. Hayes, M. H. B. and R. S. Swift. In: D. J. Greenland and M. H. B. Hayes, Eds., The Chemistry of Soil Constituents, pp. 179-320 (Chichester: J. Wiley & Sons, 1983).

17. Sequi, P., G. Guidi and G. Petruzzelli. Agrochimica **16**:224 (1972).

18. Sequi, P., G. Guidi and G. Petruzzelli. Canadian Journal of Soil Science **55**:439 (1975).

19. Leenher, J. A. Environmental Science and Technology **15**:578 (1981).

20. Thurman, E. M. and R. L. Malcolm. Environmental Science and Technology **15**:463 (1981).

21. Olsson, L. and O. Samuelson. Journal of Chromatography **93**:189 (1974).

22. Newton Clifford, M. Journal of Chromatography **94**:261 (1974).

23. Quarmby, C. Journal of Chromatography **34**:52 (1968).

24. Lowe, L. E. Canadian Journal of Soil Science **35**:119 (1975).

25. Sequi, P., M. De Nobili, L. Leita and G. Cercignani. Agrochimica **30**:175 (1986).

26. Petrussi, F., M. De Nobili, M. Viotto and P. Sequi. Plant and Soil **105**:41 (1988).

27. Saviozzi, A., R. Levi-Minzi and R. Riffaldi. BioCycle **198**:54 (1988).

28. De Nobili, M., G. Cercignani, L. Leita and P. Sequi. Communications in Soil Science and Plant Analysis **17**:1109 (1986).

29. De Nobili, M. and F. Petrussi. Journal of Fermentation Technology **66**:577 (1988).

30. Ciavatta, C., L. Vittori Antisari and P. Sequi. Agrochimica **32**:510 (1988).

31. Ciavatta, C. and P. Sequi. Fertilizer Research **19**:7 (1989).

Solid State ^{13}C CP MAS NMR Characterization of the Chemical Structure of Terrestrial Organic Matter from Areas with Differing Vegetational Backgrounds

Marit Krosshavn[1], Jon O. Björgum[2], Timothy E. Southon[2] and Eiliv Steinnes[1]

[1] Department of Chemistry, University of Trondheim, AVH, N-7055 Dragvoll, Norway.
[2] MR-Center, SINTEF, N-7034 Trondheim, Norway.

Abstract

Twelve samples of terrestrial organic matter from areas of different vegetational background were selected for study. The samples were examined with solid state ^{13}C NMR using Cross Polarisation and Magic Angle Spinning (CP MAS). The examination was done on whole soil samples, because isolation and separation of humic substances may alter the chemical structure of the organic matter. The soil samples all appeared to have small amounts of aromatic and larger amounts of aliphatic compounds. This work indicates that vegetational background and degree of decay are important factors influencing the chemical structure of natural terrestrial organic matter.

Introduction

Most methods used in characterization of humus involve isolation and separation of humic substances, which may alter their chemical structure [1]. When using solid state ^{13}C NMR on whole soil this problem is avoided. Solid state ^{13}C CP MAS NMR has been used by other scientists [2, 3] to characterize whole soil.

Solid state ^{13}C NMR was used in the present study to determine the chemical structure [4] and the relative quantities of different functional groups [5] of the soil organic samples. The purpose of this study was also to find out to what extent the type of vegetation [6] and the degree of decay [7] influence the final chemical structure of the organic matter in the soil. This information is important, e. g. in considering the influence that acid rain, heavy metals and other pollutants have on soil.

Materials and Methods

Twelve samples of terrestrial organic matter with different vegetational backgrounds were collected from surface soil layers of various terrestrial ecosystems in the southern part of Norway. A list of the samples is given in Table 1. The samples were taken either in the A_0 layer of podzolic soils or in the upper 10-15 cm of peat bogs. All samples had an organic matter content >85%. The only sample preparation before the NMR experiments was drying and sieving. The samples were investigated by solid state ^{13}C CP MAS NMR. The spectra were acquired on a BRUKER MSL 200 spectrometer. The magnetic field was 4.7 T giving a ^{13}C resonance frequency of 50.3 MHz. The spinning speed was 4 kHz, and a contact time of 2 msec was used.

Results and Discussion

Table 1 shows a list of samples studied in the present work, and Table 2 lists different functional groups together with their respective chemical shift range, as identified in ^{13}C NMR spectra [6].

Fig. 1 shows ^{13}C NMR spectra of soil organic samples from three different pine forests. The samples contain different functional groups, the most important of which are acetals, hydroxyl and carboxylic acid groups and alkyl chains, methyl groups and aromatics. The figure illustrates the fact that samples taken from different sites with the same vegetational background and degree of decay have very similar, but not identical, NMR spectra.

Fig. 2 shows ^{13}C NMR spectra of humus samples in varying degrees of decay from 4 different peat bogs. The degree of decay is an important factor influencing the chemical structure of the organic matter. The NMR spectra show that the signals which arise from hydroxyl (50-90 ppm) and acetals (90-110 ppm) originating from carbohydrates decrease when the organic matter has decayed. The chemical degradation of carbohydrates is due to the metabolism of soil microorganisms [8].

Fig. 3 shows ^{13}C NMR spectra of soil samples taken from sites 50 m apart in the same peat bog. Samples with identical degree of decay and vegetational background have virtually identical spectra.

Fig. 4 shows ^{13}C NMR spectra of soil samples from areas with different vegetation types. This figure illustrates that samples with different vegetational backgrounds give different NMR spectra. An examination of the regions 0-50 ppm and 110-160 ppm shows that samples taken from oak forest, peat bog and pine forest can be distinguished by their NMR spectra. The same sample run by two different operator's one year apart gave two spectra without a residual signal difference.

Table 1 List of samples studied in the present work.

Sample label	Sample site	Type of location	Soil type	Dominant plant species
A	Bygland Aust Agder	Pine forest	Podzol	Pinus silvestris Vaccinium myrtillus Vaccinium vitis-idaea Calluna vulgaris
B	Tonstad Vest Agder	Pine forest	Podzol	Pinus silvestris Vaccinium myrtillus Vaccinium vitis-idaea
C	Svenstjönn Vest Agder	Pine forest	Podzol	Pinus silvestris Betula pubescens Vaccinium myrtillus Juniperus communis Molinia coerulea
D	Elduvatn Rogaland	Oligotrophic bog	Peat	Molinia coerulea
E	Nordbö Vest Agder	Oligotrophic bog	Peat	Sphagnum apiculatum Scirpus caespitosus Molina coerulea Narthecium ossifragum
F	Björnestad Vest Agder	Ombrotrophic bog	Peat	Sphagnum neoreum Calluna vulgaris Scirpus caespitosus
G	Övre Gystöl Rogaland	Oligotrophic bog	Peat	Scirpus caespitosus Sphagnum apiculatum
H	Åtland Vest Agder	Ombrotrophic bog	Peat	Sphagnum neoreum Calluna vulgaris Myrica gale Pinus silvestris
I	Åtland Vest Agder	Ombrotrophic bog	Peat	Sphagnum neoreum Calluna vulgaris Myrica gale Molina coerula
J	Tjellesvik Rogaland	Oak forest	Podzol	Quercus robur
K	Evje Aust Agder	Pine forest	Podzol	Pinus silvestris Vaccinium vitis-idaea Calluna vulgaris Cladina silvatica Dicranum rugosum Pleurozium schreberi

Table 2 Observed chemical shifts of differently bonded carbon atoms in ^{13}C NMR of soil organic matter.

Chemical shift range, ppm	Chemical structure
0 - 120	aliphatic carbon
110 - 165	aromatic carbon
165 - 190	carboxylic acids
0 - 50	alkyl groups from fatty acids
50 - 90	hydroxyl groups from carbohydrates
90 - 110	acetals from carbohydrates
110 - 165	aromates, (alkenes)
165 - 190	carboxylic acids
10 - 25	aromatic terminal methyl groups
20 - 30	aliphatic terminal methyl groups
30 - 50	alkyl chains
50 - 60	amino acids, esters and ethers
60 - 90	ethers and hydroxyl groups
90 - 110	acetals
110 - 120	alkenes
110 - 140	aryl, phenols and furans
140 - 165	phenols, aryl and heteroaromatic carbon
165 - 190	carboxylic acids, carboxylate anions, esters and amides

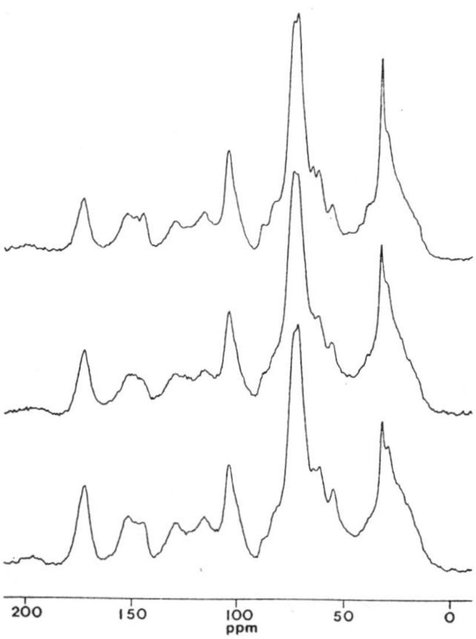

Figure 1 ^{13}C NMR spectra of soil samples from 3 different pine forests.

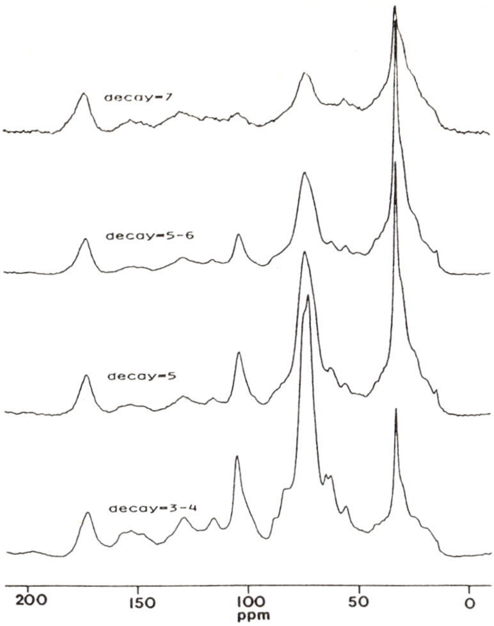

Figure 2 ^{13}C NMR spectra of samples from 4 different peat bogs in different degrees of decay, as estimated using von Post's classification.

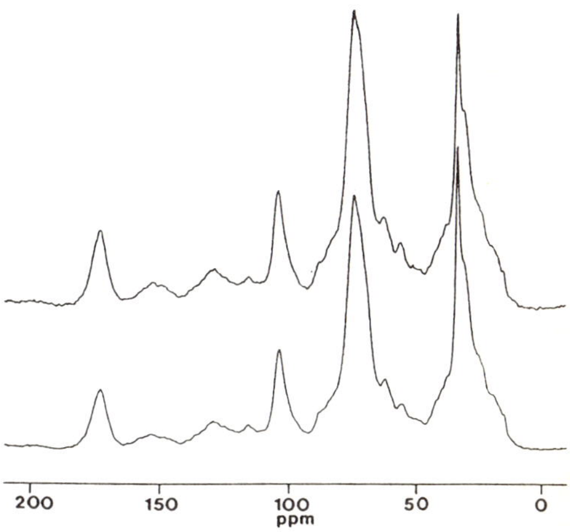

Figure 3 ^{13}C NMR spectra of two samples taken at sites 50 m apart in the same peat bog.

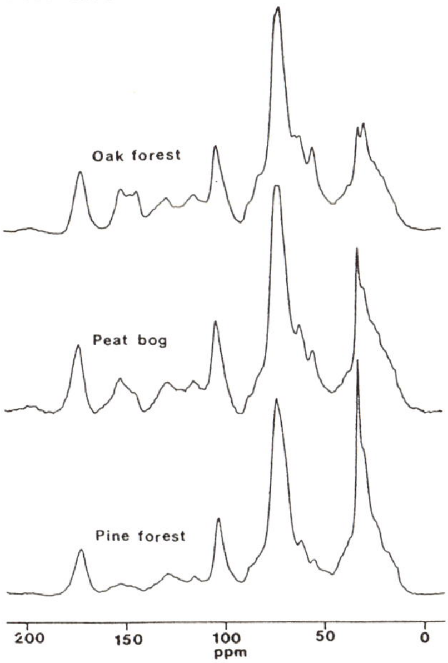

Figure 4 ^{13}C NMR spectra of samples with different vegetational backgrounds, cf. Table 1.

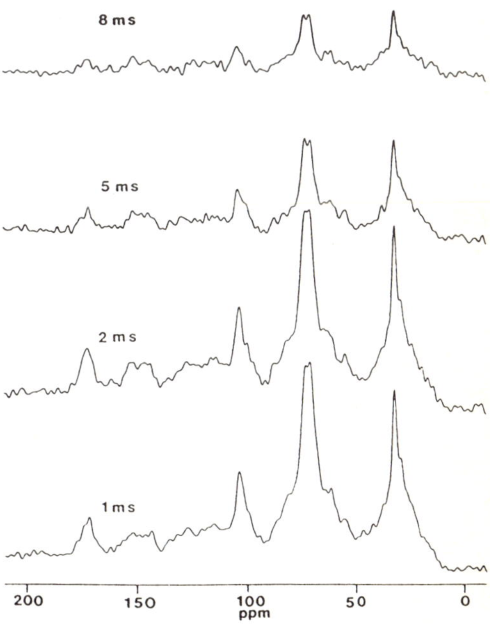

Figure 5 ^{13}C NMR spectra of one sample (K) using different contact times.

^{13}C NMR spectra of the same sample using different contact times are shown in Fig. 5. It appears that the best spectra are achieved with a contact time of 1 or 2 msec.

There is a disagreement concerning the reliability of calculated relative quantities of functional groups identified in solid state ^{13}C NMR spectra (cf. ref 5). To achieve the best quantitative results in solid state ^{13}C NMR, one should consider that

- the contact time must be optimized, so that all carbon atoms in the sample are near their maximum intensity.
- a sufficiently long recycle delay must be chosen ($>5T_1$ for ^1H).
- a well resolved spectrum is required.

The spectra appear to be well resolved. A rather high reliability of relative quantities is therefore expected. The area under the plotted curve was calculated by means of a computer program and the relative signals (%) from the functional groups were estimated. The signal intensity is expected to be proportional to the quantity of the carbon atoms at this frequency. Table 3 shows the vegetation type of soil organic samples together with their respective relative quantity of the most important functional groups. There are considerable differences between the results for individual soil samples, but they all appear to have small amounts of aromatic compounds and larger amounts of aliphatic compounds.

Table 3 The relative abundance of functional groups in different samples of terrestrial organic matter (%).

Vegetation type	Chemical shift range, ppm. 190-165	165-140	140-110	110-90	90-50	50-0	Arom. C 165-110	Aliph. C 110-0
Pine (B)	6.8	6.4	10.8	10.7	42.4	23.0	17.2	76.1
Peat bog (I)	6.6	2.6	6.0	9.3	42.4	33.1	8.6	84.8
Oak (H)	6.7	7.8	10.6	9.6	47.1	18.2	18.4	74.9

Conclusions

Solid state ^{13}C NMR is probably the method which gives the most complete structural information about soil organic matter. It is also possible to calculate relative quantities of functional groups from the NMR spectra, if certain conditions are fulfilled. The samples examined in this work contained small amounts of aromatic compounds [9,10] and larger amounts of aliphatic compounds. Vegetational background and degree of decay seem to be the most important factors influencing the chemical structure of soil organic matter; the present work indicates that vegetational background is the more important of these two.

References

1. Newman, R.H. and K.R. Tate. J. Soil Sci. **35**:47 (1984).

2. Arshad, M.A., J.A. Ripmeester and M. Schnitzer. Can. J. Soil Sci. **68**:593 (1988).

3. Preston, C.M. and J.A. Ripmeester. Can. J. Spectrosc. **27**:99 (1982).

4. Wilson, M.A., A.M. Vasallo, M.E. Perdue and J.H. Reuter. Anal. Chem. **59**:551 (1987).

5. Wilson, M.A. J. Soil Sci. **32**:167 (1981).

6. Verheyen, T.V., R.B. Johns and D.T. Blackburn. Geochim. Cosmochim. Acta **46**:269 (1982).

7. Hempfling, R., F. Ziegler, W. Zech and H.R. Schulten. Zeitschr. Pflanzenernaehrung Bodenk. **150**:179 (1987).

8. Schnitzer, M. and S.U. Khan. Soil Organic Matter (Amsterdam: Elsevier Publishing Company, 1978).

9. Ogner, G. Geoderma **35**:343 (1985).

10. Benzing-Purdie, L.M., M.V. Cheshire, L.W. Berwyn, G.P. Sparling, C.I. Ratcliffe and J.A. Ripmeester. J. Agricult. Food Chem. **34**:170 (1986).

Carbon-13 Nuclear Magnetic Resonance Analysis, Lignin Content and Carbohydrate Composition of Humic Substances from Salt Marsh Estuaries

James J. Alberts[1], Patrick G. Hatcher[2], Mary T. Price[1] and Zdenek Filip[3]

[1] University of Georgia Marine Institute, Sapelo Island, GA USA 31327
[2] Fuel Science Program, Department of Materials Science and Engineering, The Pennsylvania State University, University Park, PA USA 16802
[3] Institut für Wasser-, Boden- und Lufthygiene des BGA, Aussenstelle Langen, Paul-Ehrlich-Strasse 29, D-6070 Langen FRG

Abstract

^{13}C nuclear magnetic resonance spectroscopy, CuO oxidation products of lignin and hydrolyzable carbohydrates were measured for fulvic and humic acids extracted from living and dead *Spartina alterniflora* and salt marsh sediments. With these methods, there was little evidence for early diagenetic alteration of the humic materials. No trends consistent for fulvic and humic acids were observed for either hydrolyzable carbohydrates or lignin derived phenols, and chemical measurements of these fractions did not agree with spectral estimates. Humic acids appear to contain secondary amide linkages typical of proteins and peptides.

Introduction

Previously, we reported on the elemental composition, stable carbon isotope content, spectroscopic characteristics, and metal and organic binding characteristics of a series of humic and fulvic acids isolated from living and dead smooth cordgrass, *Spartina alterniflora* (Loisel) and the sediments on which it grows [1-5]. Here we present the results of further studies on the chemical nature of these humic materials using ^{13}C nuclear magnetic resonance spectroscopy and determinations of lignin derived phenols and neutral sugars by wet chemical analyses.

Materials and Methods

Freshly collected samples of living *S. alterniflora*, standing dead *S. alterniflora* and the sediments on which they grow were collected from the salt marshes of Sapelo Island, GA, USA. These samples were dried, ground, and humic and fulvic acids were extracted and purified as previously described [3]. Elemental analyses for C, H, and N were performed on replicate samples at the University of Georgia Marine Institute with a

Perkin-Elmer Model 2400 Elemental Analyser, and ash content was determined with a Cahn Microbalance.

Solid-state ^{13}C NMR spectra were obtained by the method of cross polarization and magic angle spinning using a Chemagnetics 100S-200L spectrometer as described previously [6]. A contact time of 1 ms and a pulse repetition rate of 1 s were used as conditions which provided the most quantitative data for these particular samples. Relative areas of portions of the NMR spectra were determined using Sigma-Scan and a Jandel Opaque Digitizer. All areas were measured three times and averaged.

Cupric oxide oxidation of the humic and fulvic acids were performed and the phenolic oxidation products determined [7]. The humic and fulvic acids were also hydrolysed and the neutral sugars were analysed [8].

Results and Discussion

Extraction of living and dead plant tissue to produce humic and fulvic acid fractions can also extract other biopolymers, which raise concerns about the integrity of the fractions. However, humic acids have been shown to be almost ubiquitous in living plants [9,10], can be extracted from vascular plant debris [11], and can be released into seawater from *S. alterniflora* by both abiotic and microbially mediated processes [2]. Theories of humic and fulvic acid formation by degradation of biopolymers are widely held [12, 13]. Furthermore, at a recent Dahlem Workshop, a guide to nomenclature for humic substances was proposed which includes extraction, purification and verification steps [14]. We have demonstrated previously that humic and fulvic acids extracted and purified by excepted methods have chemical and spectral characteristics which are very similar to extracts from sediments [1-5]. The findings of this study also verify that these extracts are humic and fulvic acids, whether or not they contain other biopolymers.

Elemental Analyses

The elemental analyses of these materials (Table 1) are very similar to those for earlier extractions of *S. alterniflora* and salt marsh sediments as are the H/C and N/C ratios [3]. Only in the case of the nitrogen content of the humic acids isolated from the live plant material is there a significant difference, with these samples having a lower nitrogen content that is reflected in a lower N/C ratio. High C, and H values and low N values in the fresh humic acids may be caused by the presence of fatty acids which would be in this fraction, as their solubilities are similar to humic acids due to the carboxylic functional groups. In all cases, the samples are very low in ash content.

^{13}C NMR Spectra

The NMR spectra of these materials show many peaks characteristic of fulvic and humic acids, including peaks attributed to alkyl, aromatic, carbohydrate and lignin moieties (Fig. 1). Significant differences exist in the spectra as one compares material extracted from live plants, dead plants, and that extracted from the sediment. The most notable differences are between fulvic acids with the apparent decreases of the peaks at 116 ppm and 148 ppm, which arise from aromatic carbon bound to protons and to oxygen, respectively. In the humic acid spectra, there is an increase in the 56, 148 and 153 ppm

Table 1 Elemental weight percent and atomic ratios of carbon, hydrogen and nitrogen in *S. alterniflora* salt marsh humic matter.

Material	%AFDW[1]	%C[2]	%H	%N	H/C	N/C
Fulvic acids						
Live plant	97.30	53.03 ± 0.16	5.30 ± 0.08	1.24 ± 0.03	1.20	0.020
Dead plant	98.02	50.89 ± 0.45	5.80 ± 0.12	2.62 ± 0.02	1.37	0.044
Sediments	99.03	45.51 ± 0.67	4.98 ± 0.08	3.09 ± 0.05	1.31	0.058
Humic acids						
Live plant	98.92	60.66 ± 0.21	8.25 ± 0.01	4.40 ± 0.01	1.63	0.062
Dead plant	98.95	52.54 ± 0.02	5.15 ± 0.04	4.12 ± 0.01	1.18	0.067
Sediment	94.62	59.96 ± 0.09	6.61 ± 0.05	4.16 ± 0.01	1.32	0.059

[1]Ash free dry weight
[2]Stated error is 1 std dev. of the mean; fulvic acid data from dead plants and sediments (n=7), all others (n=3).

peaks, indicative of an increase in methoxy-carbon units attached to aromatic carbons. As with spectra of other estuarine humic matter isolated from *S. alterniflora* [15], the

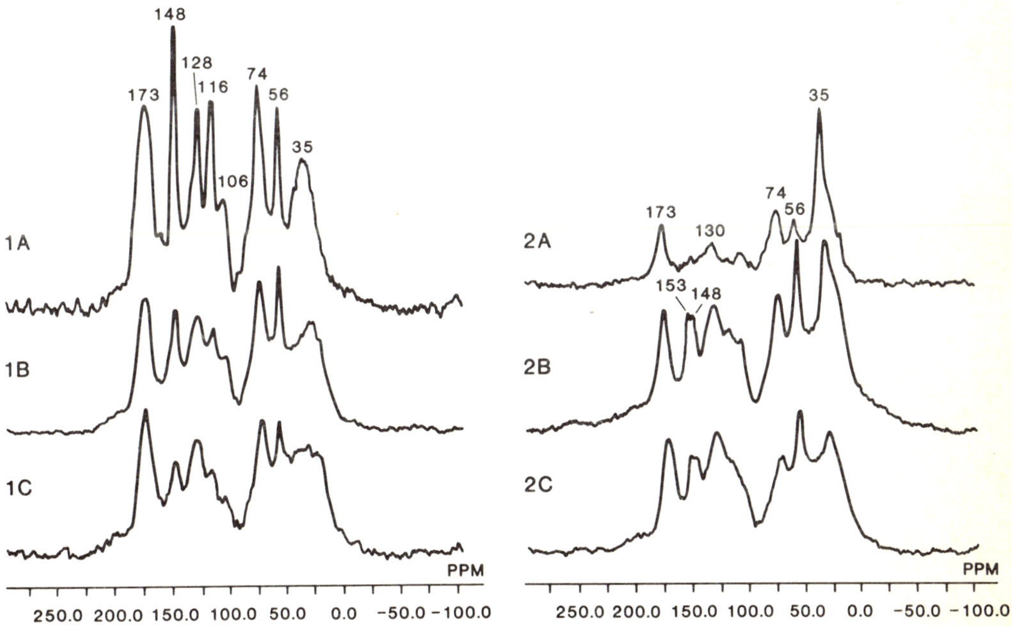

Figure 1 ^{13}C NMR spectra of fulvic and humic acids isolated from a salt marsh estuarine environment. 1A) Fulvic acid from standing live *S. alterniflora*; 1B) Fulvic acid from standing dead *S. alterniflora*; 1C) Fulvic acid from surficial marsh sediments; 2A) Humic acid from standing live *S. alterniflora*; 2B) Humic acid from standing dead *S. alterniflora*; 2C) Humic acid from surficial marsh sediments.

major changes appear to occur between materials isolated from living and dead plants rather than between dead plant and sediment extracts.

For fulvic acids, there is an increase in alkyl carbon and a decrease in aromatic carbon contents from live plant to dead plant to sediment, while O-alkyl and carbonyl carbon remains relatively constant (Table 2). Conversely, for the same order of starting material, humic acids decrease in alkyl carbon content and increase in aromatic carbon. Again, the O-alkyl and carbonyl regions of the spectra stay relatively constant. The trends to greater or lesser aromaticity are not readily apparent in the atomic H/C ratio calculated from the elemental data (Table 1). However, in soil humic acids, the aromatic carbon contents also tend to increase with aging [16]. In that study, however, the O-alkyl carbon increased while there was no consistent trend in the alkyl carbon, which is opposite to what we observed in the estuarine samples.

Table 2 Amount of carbon as a percentage of the total carbon present in the respective chemical-shift range of the ^{13}C NMR spectra.

	220 - 160 (carbonyl)	160 - 110 (aromatic)	110 - 50 (O-alkyl)	50 - 0 (alkyl)
Fulvic acids				
Live plant	18.8	32.2	30.8	19.9
Dead plant	16.3	29.1	34.0	20.9
Sediment	17.8	26.9	31.1	24.7
Humic acids				
Live plant	14.1	15.8	31.1	39.0
Dead plant	14.8	24.6	29.5	31.6
Sediment	14.6	28.3	33.2	23.8

The NMR spectrum of humic acids from live *Spartina* show a large peak for alkyl carbon. We believe that most of this alkyl carbon content arises from the fact that lipids such as fatty acids would probably reside in this fraction. The samples were not pre-extracted with organic solvents to remove lipids; thus, fatty acids would be extracted from fresh plant material with NaOH, and they would then co-precipitate with the acidified humic acids. The high abundance of fatty acids in fresh plant material is well known; however, dead *Spartina* and sediment would contain relatively lesser amounts of these substances. This may partially explain the decreasing trend for alkyl carbon in humic acids from fresh plant to sediment.

Carbohydrate and Lignin Carbon

All the spectra have peaks which are associated with both carbohydrate (74 ppm and 106 ppm) and lignin (56 ppm, 148 ppm - 153 ppm) constituents [15,17]. While the 148 ppm peak appears to decrease in spectra of fulvic acids in the order: living plants, dead plants, sediments, this region increases and two peaks indicative of O-substituted carbons in aromatic rings appear in humic acids from dead plant matter and sediments. The latter two peaks are usually assigned to O-substituted aromatic carbon in guaiacyl (148 ppm) and syringyl (153 ppm) units of the lignin. The syringyl unit should also have a peak at 53 ppm that is equal in intensity to that at 153 ppm, but the signal in the 53 ppm region is usually much greater in intensity than the peak at 153 ppm indicating the presence of other methoxy- or ether units, or possibly N-substituted aliphatic carbons [18].

A detailed discussion of the individual sugars and phenolic oxidation products of lignin which were analysed is beyond the scope of this paper. However, those data allowed the calculation of the amount of carbon present in those compounds, which in turn can be used to compare to estimates based on the NMR spectra.

The total amount of carbohydrate carbon in fulvic and humic acids does not show a trend from living plant to sediment extracts (Table 3). The phenolic oxidation products of lignin decrease in the fulvic acid extracts, but again show no discernible trend in the humic acids. Carbohydrates in soil fulvic and humic acids showed little change in concentration with increasing age, while both had significant decreases in the lignin phenols [16]. The total amount of C per unit of organic matter in carbohydrates and lignin is comparable in the humic acids from estuarine sources and forest soils. However, estuarine fulvic acids appear to have an order of magnitude more carbon in lignin and a tenth as much carbon in carbohydrates as do forest soil fulvic acids [16].

The acid to aldehyde ratio of the vanillin group (Table 3), which has been used to indicate increased diagenetic loss of lignin [19], indicates that there is some diagenetic alteration of the lignin of fulvic acids, but there is no indication of diagenetic change in the humic acids. The relative magnitudes of these ratios for fulvic and humic acids does agree with values for sedimentary and aquatic humic and fulvic materials [19,20].

Table 3 Carbohydrate and lignin contents (mg C/100mg OC) and ratios of vanillic acid to vanillin of humic and fulvic acids isolated from *S. alterniflora* plants and salt marsh sediments

	Carbohydrates	Lignin (S+V+C)	$(Ac/Ad)_V$
Fulvic acids			
Live plant	4.88	24.10	0.63
Dead plant	9.95	9.84	0.77
Sediment	5.63	4.32	1.02
Humic acids			
Live plant	5.52	3.95	0.56
Dead plant	7.74	7.44	0.52
Sediment	2.56	6.81	0.54

Estimates of the relative amounts of carbon in the methoxy-groups of lignin phenols and of carbohydrates may be made from the chemical analyses in Table 3. These values may be compared to the ratio of the areas under regions and peaks of the spectra believed to arise from lignins and carbohydrates (Fig. 1). It was assumed that all the carbon under the regions of the spectra was either lignin methoxy carbon or carbohydrate carbon.

Estimates were made by comparing the relative areas under peaks at 74 ppm to 56 ppm. The comparison of these calculations (Table 4) show that ratios from the chemical estimates are usually greater than the estimates of either NMR based estimation. Thus, the chemical analyses show a significantly higher content of carbohydrate carbon relative to methoxy carbon in fulvic and humic acids.

Table 4 Comparison of the estimates of the ratio of carbohydrate (CHO) carbon to methoxy (OCH_3) carbon in humic and fulvic acids from NMR and chemical analyses.

	CHO/OCH_3	
	NMR	Chemical
Fulvic acids		
Live plant	2.91	1.53
Dead plant	2.21	7.71
Sediment	2.30	7.93
Humic acids		
Live plant	2.88	8.24
Dead plant	1.33	6.09
Sediment	0.84	2.29

The reasons for the discrepancies are unclear. It is possible that differential losses during the chemical oxidations and hydrolyses of the phenolics and sugars are resulting in different ratios [16]. This explanation is particularly true with respect to the lignin oxidation products since the analytical method is specific to the phenols and any minor changes in their structure would result in their not being determined. Another potential source of error is the assumption that all the carbon under a given region is either lignin or carbohydrate derived. The latter is given support by the fact that the peak at 56 ppm should be of equal magnitude to that at 153 ppm if both are derived from syringyl units. However, the peak at 56 ppm often appears of greater intensity, perhaps due to ether linkages or N-substitution on aliphatic carbon. Whatever the reasons, it is apparent that both the chemical and spectral estimation of carbon content of lignin and carbohydrate units, and potential diagenetic trends estimated by these changes, do not agree.

Proteins and Peptides

The strong peak at 173 ppm is often due to carboxyl carbon but is also indicative of secondary amide linkages in peptides and proteins [15,21] and appears in all the spectra. Using samples of humic acids extracted from *S. alterniflora* and humic acids derived from a microbial leaching study of this plant, Filip et al. [15] were able to establish a relationship showing a positive correlation of the ratio of the peak area at 174 ppm to the total spectral area and the atomic N/C ratio. We have recalculated the areas in that study [15] to conform to the areas of this work, allowing us to compare the results.

The three humic samples from this study, while having lower N/C ratios, agree with behavior predicted from the samples in Filip's study (Fig. 2). However, the fulvic acid data from this study have a much higher ratio of the 174 ppm peak area to the total area than would be expected from the N/C ratio of these samples. This higher ratio may be the result of tannins in the fulvic acids which would also contribute to the signal under the 174 ppm peak [22]. The presence of tannins in the fulvic acid but not the humic acids would be expected from the solubility characteristics of the various materials and is supported by the low content of nitrogen and N/C ratios of the fulvic acids. Tannins, especially hydrolyzable tannins, would display NMR peaks at 174, 148, 128, 116 and 74 ppm, all of which are intense in the spectra of fulvic acids.

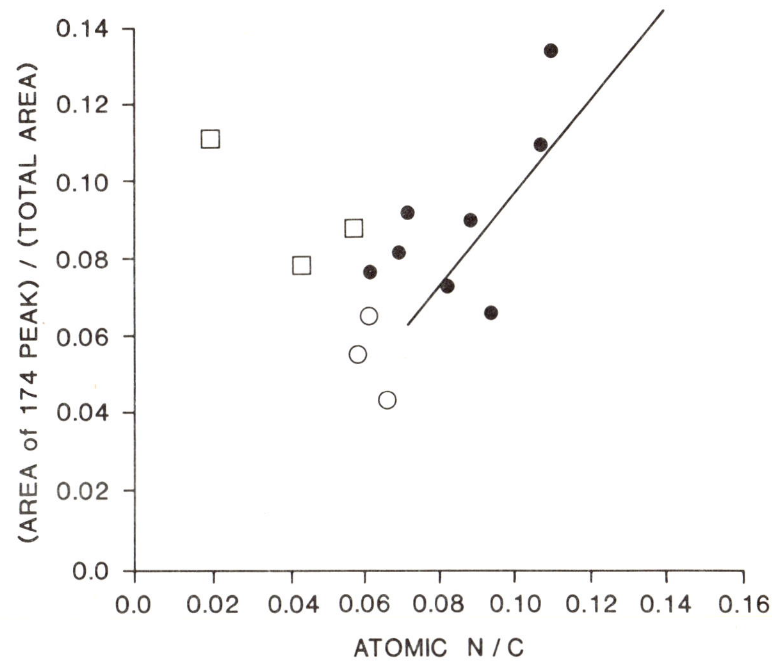

Figure 2 Ratio of the area under the peak at 174 ppm to the area under the entire NMR spectrum compared to the atomic ratios of N/C. Closed circles for humic acids from Filip et al. [15], open circles for humic acids in this study, open squares for fulvic acids from this study. Line is a linear regression using the closed circle points only.

Conclusions

This study of the potential early diagenetic alteration of fulvic and humic acids in a salt marsh show that the greatest changes occur during the time the living plant senesces and dies, but still remains standing in the marsh. The differences in NMR spectra of both humic and fulvic acids from dead plants and the surficial sediments are very similar to each other, but different from the humic and fulvic acids isolated from the living plants. There does appear to be a trend of increasing alkyl character and decreasing aromaticity for fulvic acids during this time period and a decrease in alkyl character and increase in aromaticity in humic acids during the same sequence.

No trends consistent for fulvic and humic acids were observed for either carbohydrates or lignin derived phenols, and chemical measurements of these fractions did not agree well with spectral estimates. Acid to aldehyde ratios of vanillin derivatives did indicate some diagenetic aging of the fulvic acids but none for the humic acids. With the methods employed, this study gave little evidence of diagenetic alteration of the humic matter. However, since the materials extracted from the living plants, dead plants and sediments represent separate and complex assemblages of organic compounds, it may be very difficult to identify subtle, early diagenetic alterations of specific components.

Finally, the humic acids isolated in this study support the suggestion that humic acids contain secondary amide linkages typical of proteins and peptides. However, fulvic acids do not follow the predicted trend, most probably due to tannins associated with the fulvic acids as a result of comparable solubilities under the extraction and purification procedures.

Acknowledgments

We thank Anne L. Bates and Harry E. Lerch, III (U. S. Geological Survey, Reston, VA) for their assistance in obtaining the ^{13}C NMR spectra, and Eileen Hedick and Sarita Marland (UGA Marine Institute) for their help in manuscript preparation. This work was supported in part by the Sapelo Island Research Foundation and National Science Foundation grant INT-8619167. It is contribution no. 639 of the University of Georgia Marine Institute.

References

1. Filip, Z., J. J. Alberts, M. V. Cheshire, B. A. Goodman and J. R. Bacon. Sci. Tot. Environ. **71**:157 (1988).
2. Filip, Z. and J. J. Alberts. Sci. Tot. Environ. **73**:143 (1988).
3. Alberts, J. J., Z. Filip, M. T. Price, D. J. Williams and M. C. Williams. Org. Geochem. **12**:455 (1988).
4. Alberts, J. J. and Z. Filip. Sci. Tot. Environ. **81/82**:353 (1989).
5. Alberts, J. J., Z. Filip and G. J. Leversee. Marine Chem., **28**:77 (1989).
6. Hatcher, P. G. Organic Geochem. **11**:31 (1987).
7. Hedges, J. I. and J. R. Ertel. Anal. Chem. **54**:174 (1982).
8. Cowie, G. L. and J. I. Hedges. Anal. Chem. **56**:497 (1984).
9. Raudnitz, H. Chemistry and Industry (London) 1650 (1957).
10. Raudnitz, H. Science 728 (1958).
11. Ertel, J. R. and J. I. Hedges. Geochim. Cosmochim. Acta **48**:2065 (1985).
12. Hedges, J. I. In: F. H. Frimmel and R. F. Christman, Eds., Humic Substances and Their Role in the Environment, pp. 45-58 (Chichester: Wiley, 1988).
13. Hatcher, P. G. and E. C. Spiker. In: F. H. Frimmel and R. F. Christman, Eds., Humic Substances and Their Role in the Environment, pp. 59-74 (Chichester: Wiley, 1988).

14. Thurman, E. M., G. R. Aiken, M. Ewald, W. R. Fischer, U. Forstner, A. H. Hack, R. F. C. Mantoura, J. W. Parsons, R. Pocklington, F. J. Stevenson, R. S. Swift and B. Szpakowska. In: F. H. Frimmel and R. F. Christman, Eds., Humic Substances and Their Role in the Environment, pp. 31-43 (Chichester: Wiley, 1988).

15. Filip, Z., R. H. Newman and J. J. Alberts. Sci. Tot. Environ., in press (1990).

16. Kogel-Knabner, I., W. Zech and P. G. Hatcher. Z. Pflanzenernahr. Bodenk. **151**:331 (1988).

17. Orem, W. H. and P. G. Hatcher. Org. Geochem. **11**:73 (1987).

18. Gillam, A. H. and M. A. Wilson. Am. Chem. Soc. Symp. Ser. **305**:128 (1986).

19. Ertel, J. R. and J. I. Hedges. Geochim. Cosmochim. Acta **48**:2065 (1984).

20. Ertel, J. R., J. I. Hedges and E. M. Perdue. Science **223**:485 (1984).

21. Piotrowski, E. G., K. M. Valentine and P. E. Pfeffer. Soil Sci. **137**:194 (1984).

22. Wilson, M. A. and P. G. Hatcher. Org. Geochem. **12**:539 (1988).

The Effect of Composting on the Organic Colloidal Fraction from Domestic Sewage Sludge

Gonzalo Almendros[1], Juan A. Leal[2], Francisco Martin[3] and Francisco J. Gonzalez-Vila[3]

[1] Instituto de Edafologia y Biologia Vegetal (C.S.I.C.), Serrano 115 dpdo, 28006-Madrid, Spain
[2] Centro de Investigaciones Biologicas (C.S.I.C.), Velazquez 144, 28006-Madrid, Spain
[3] Instituto de Recursos Naturales y Agrobiologia (C.S.I.C.), Apartado 1052, 41080-Sevilla, Spain

Abstract

Due to the frequent use of domestic sewage sludges as organic additives to soils with low humus content, the chemical characteristics of the extractable polymers from composted and uncomposted sludges were described.

The alkali-soluble, acid-insoluble sludge fraction (humic-like) was isolated after exhaustive lipid extraction, and analyzed by chemical degradation followed by combined gas chromatography mass spectrometry. It was observed that this sludge fraction contained an important amount of lipid compounds (more than 40% by weight). Most of this lipid material can be physically removed, but the residual polymer fractions were also found to be highly aliphatic in nature. The high yields upon degradation of several types of alkanoic acids and the relatively high proportion of polypeptides (30% by weight), as well as carbohydrates, suggested that the humic-like fractions from both the composted and uncomposted sludges consisted of slightly altered microbial and residual biopolymers. The effects of composting sludge mainly concern the selective biodegradation of the less resistant moieties. The great differences between the chemical nature of the humic-like fraction of the composted sludge and that of the soil humic acids are considered to reflect the low proportion of lignified materials in the original sludge.

Introduction

The use of domestic sewage sludge for agricultural purposes requires that the organic composition of the sludge be characterized first, including the analysis of potentially harmful pollutants [1,2]. Sewage sludge seems to be an important source of readily biodegradable organic matter [3], and its favourable effects on the physical properties of soils [4] and its high nitrogen content [5,6] are generally considered to be sufficient to justify the use of sewage sludge on agricultural soil. On the other hand, the often high proportion of heavy metals and phytotoxic substances, as well as the problems derived from certain microbial species, set an upper limit for the direct use of sludge on land; the first aspect is particularly pronounced in acid soils.

The study of the chemical nature of the extractable polymers from sludges is also important in establishing their potential fertility and their degree of "maturity". In general, the humic acid content is often used to evaluate organic fertilizers, although humic substances do not seem to be present, at least *in sensu strictu*, in the products derived from urban wastes [7]. The humic acid fractions are usually defined in terms of the procedures used for their extraction (solubility in alkali and insolubility in acid), although the use of such criteria is inadequate for organic material other than soil [8,9]. For this reason, the organic matter extracted from sludge by the same methods as used for the isolation of humic acids will be referred to in the present paper as the "organic colloidal fraction" (OCF). This sludge fraction is presumably most active in terms of its physico-chemical properties. In addition, during laboratory isolations of soil humus constituents, the sludge OCF's are coextracted together with the humic acids in soils amended with sludge. Consequently, the chemical characterization of such sludge polymer fractions is also useful for predicting the short-term changes in the humic fraction from sludge-amended soils [10]. Differences between the organic matter composition of sludge and humus may be partially due to the reduced period of aerobic evolution of the former; the "sludge humic acids" are frequently compared with the aquatic humic substances [11]. These facts suggest the importance of the oxidative processes in transforming sewage sludge into a material looking more like soil humus. Taking into account the above criteria, the aim of the present study was to examine the presence of humic-type polymer fractions in the composted sludge.

Material and Methods

Sewage Sludge

The sample studied was representative domestic sewage sludge from the treatment station of Camino de la Muñoza, Almeria (Spain). The final product after aerobic and anaerobic treatments was air-dried and ground before analyses.

In the composting experiment the material was incubated for 4 months in a 40-l polyethylene container in a chamber with a constant temperature (30°C). The humidity was kept at 50% of WHC, and the pile was rotated every week to insure even distribution of the material and the temperature and to increase aeration.

Analytical Methods

The ash content was determined in an electric furnace (12 h at 700°C). After treating the sludge with acid (to remove carbonates), the organic carbon was determined with a Carmhograph-12. Nitrogen was determined with a "Technicon" autoanalyzer, and the cation exchange capacity by using barium acetate (pH=8).

The separation of organic fractions from sludge was carried out as follows. The lipid fraction was extracted in a Soxhlet with petroleum ether (40-70°C) during the course of 70 h. The residual sludge was extracted with 0.1 N NaOH under nitrogen, and the soluble extract was acidified to pH 1, thereby obtaining a soluble, colourless fraction and a grayish precipitate (OCF). An attempt was made to concentrate the first fraction in insoluble polyvinylpyrrolidone, but this fraction was not adsorbed on the resin, suggesting a strongly aliphatic character [9]. For that reason, the acid-soluble fraction of this sludge was considered to be highly different from the soil fulvic acids and was not studied. The percentages (as C) of the above fractions were determined in 10-ml decarbonated and desiccated aliquots, using the Carmhograph analyzer.

Since our previous analyses showed an important lipid contribution to the composition the OCF (even after exhaustive lipid extraction of the sludge sample), these freeze-dried, "crude" OCF's were submitted to ultrasonic shaking in petroleum ether, followed by a 12 h continuous extraction in a Blount micro-extractor; the process was repeated during the period of a week, using fresh solvent every day. These treatments were found to be highly effective to remove lipids physically "entrapped" in the OCF, most of them presumably introduced into this fraction by adsorption of the lipids released from sludge during the alkaline extraction.

Characterization of the Sludge OCF

The elementary composition was determined in a CHN analyzer Perkin-Elmer 240 C. The molecular size distribution was studied using Sephadex G-50, recording the densitometric curves at 450 nm. Spectroscopic parameters in the visible range were determined using solutions of 136 mg/l C [12], and the E4/E6 ratio refers to the relation between extinctions at 465 and 665 nm. The analysis of amino acids was carried out in a Biotronik amino acid analyzer LC 7000, after hydrolysis with 6 M HCl. The hydrolyzable sugars were studied by GC as alditol acetates [13], using a Perkin Elmer Sigma-3-10 GC system, and identified by their coincidence with the retention times of authentic compounds.

Oxidative degradation of the OCF's was carried out in two steps, first using a mild reagent (potassium persulphate) [14], then reoxidizing the persulphate-resistant material by alkaline permanganate degradation [15]. In both cases, the oxidation products were extracted with ethyl acetate in a liquid-liquid extractor during the course of 12 h. As in the case of the lipid fraction, these compounds were methylated with ethereal diazomethane, then separated and identified in a GC-MS system Hewlett Packard 5992 B. The identity of chromatographic peaks was estabilished in terms of their mass spectra, but also by comparison with the retention times of authentic compounds. The column used was a 20-m cross-linked capillary OV-101, and the oven temperature was programmed at between 25° and 270°C with a rate of 5°/min.

Results

The whole sludge sample had about 60% organic matter and 2% nitrogen (C/N=17, Table 1). The principal exchange cations were Ca^{2+} and Mg^{2+}. The exchange complex was saturated (absence of exchangeable H^{+}) and the total cation exchange capacity, 115 meq/100 g, was proportional to the high content of organic matter (T/C). The 4-month composting process was responsible for a 22% loss of weight, a value similar to that reported elsewhere [16]. The content of organic matter decreased, but the N content remained constant; a C/N ratio of 12 was observed in the composted sludge. Nevertheless,

Table 1 Effect of composting on the analytical characteristics of sewage sludge.

Sample	%wt loss	% ash	%C org.	%N	C/N	%N-loss	Exchange cations Na+	K+	Ca2+	Mg2+	H+	T	T/C
Original sludge	-	39.8	34.3	1.97	17.4	-	8.3	1.0	88.8	16.9	0.0	115.0	3.3
4-month composted	22.4	51.3	23.9	1.98	12.1	22.0	10.9	1.3	112.0	19.7	0.0	143.9	6.0

T = Total exchange capacity, meq/100g.

when calculating the nitrogen content in terms of the original weight before composting, a 22% loss of N was also observed.

The total exchange capacity increased after composting, as did the T/C ratio, indicating a greater "maturity" of the organic matter in the composted sludge, with a greater proportion of cation-complexing functional groups.

The alkali-extractable fractions of sludge made up to 14% of total C, and the OCF represented about 10% of total C (Table 2). After composting, the percentage of lipids and acid-soluble fractions underwent a small reduction, but the OCF greatly decreased in comparison with uncomposted sludge. Several physico-chemical parameters of the OCF's are given in Table 3. According to the experimental procedure described above, the "crude" OCF was also studied after physical removal of the lipid fraction and degradation with $K_2S_2O_8$, a mild reagent which removes loosely linked organic compounds. These pretreatments were carried out in order to eliminate the "non-humic" matter associated with, or co-extracted with, the OCF, in a attempt to "purify" this fraction and increase its chemical similarity with humic acids.

Table 2 Effect of composting on the yields of extractable fractions from sewage sludge.

Sample		Lipid fraction	Alkali-soluble acid-insoluble (FA-like)	Alkali-soluble acid-insoluble (HA-like)(=OCF)	Alkali-insoluble (humin-like)
Original sludge	1)	1.35	1.50	3.30	28.15
	2)	3.94	4.37	9.62	82.07
4-month composted sludge	1)	0.84	1.32	0.98	20.76
	2)	3.51	5.52	4.11	86.85
	3)	0.65	1.03	0.76	16.13

1) Grams of C from the different fractions in 100 g sludge sample.

2) Grams of C from the different fractions in 100 g sludge organic C.

3) Grams of C from the different fractions after composting 100 g original sludge.

Elementary Composition

The percentages of C, H, O and N of the OCF's were different from those of soil humic acids (Table 3). In particular, the atomic H/C ratios were comparatively high, suggesting a predominance of alkyl compounds [17]. After lipid extraction, the atomic O/C ratio increased, which corresponds to the release of non-polar compounds. The latter parameter further increased after the persulphate treatment of the uncomposted sludge (which eliminated an additional lipid fraction). As expected, the OCF from composted sludge showed an increased O/C atomic ratio. The behaviour of the OCF from the composted sludge tended to be more similar to that of humic acids: lipid extraction induces a relative increase of oxygen-containing groups, but the persulphate treatment induces decarboxylation typical of hydrolytic processes [18]. The changes in the N content were also significant: composting produced a relative increase (lower C/N ratio in the composted OCF's), and lipid extraction and persulphate degradation produced, respectively, an extraction of components without nitrogen, and removal of the easily degradable (carbohydrate and protein) moieties of the OCF. Based on this information the hydrolyzable nitrogen was calculated to 70% (value similar to that of fresh biomass), suggesting a predominant peptidic nature of the N in the OCF. This last parameter decreased as a consequence of composting.

Physical Characteristics

The specific extinction (450 nm) of OCF's was very low in comparison to any soil organic fraction, and the E4/E6 ratio was also relatively low (Table 3). This last parameter was affected by composting, illustrating the changes in the molecular sizes of the sludge OCF's [19]. The MW distribution was studied by gel filtration, which revealed relatively high molecular sizes for this sludge fraction (Table 4). The molecular sizes are affected, in part, by the adsorption of hydrophobic fractions on the OCF. It was found that after the petroleum ether treatment, the separated fractions became more defined, with

Table 3 Analytical characteristics of sludge polymers (HA-like) before and after 4 months of composting.

HA-like	Treatment	Elementary composition					Atomic ratios		% hydro-lyzable N	Spectroscopy/ parameters	
		%C	%H	%O	%N	C/N	H/C	O/C		E4(136 ppm C)	E4/E6
Original	----	55.9	7.9	33.5	2.7	20.7	1.7	0.4	-	0.12	3.6
	Lipid extraction	45.3	5.8	43.0	5.9	7.7	1.5	0.7	68.7	0.14	4.5
	$K_2S_2O_8$ oxidation	37.1	4.4	55.8	2.7	13.7	1.4	1.1	-	n.d.	n.d.
4-month	----	47.3	6.1	41.4	5.2	9.0	1.5	0.6	-	0.20	5.0
	Lipid extraction	32.7	3.9	56.6	6.8	4.8	1.4	1.3	60.8	0.17	5.9
	$K_2S_2O_8$ oxidation	43.0	4.9	48.4	3.7	11.6	1.4	0.8	-	n.d.	n.d.

Table 4 Gel filtration (Sephadex G-50) of organic colloidal fractions from sewage sludges.*

HA-like fraction	K_{av}: 0.0	0.1	0.2	0.3	0.4	0.5	0.6	0.7	0.8	0.9	1.0	1.1
Uncomposted sludge:												
Original	40.0			56.0								4.0
Lipid extracted	59.5			29.8							3.6	7.1
4-month composted sludge:												
Original	28.0			36.0		28.0						8.0
Lipid extraction	31.8					47.6			7.9		12.7	

* Percentages of the total area of the densitometric curve registered at 450 nm.

the development of well-defined peaks of low molecular size (K_{av} 0.3 - 1.1). The significant decrease of the molecular size after composting suggested that the readily biodegradable constituents of the OCF's consisted of the polymers with a high molecular weight.

Infrared Spectroscopy

The IR spectra of lipid-extracted OCF's are shown in Fig. 1. The IR spectra of the uncomposted samples seem to correspond to a complex mixture of fatty, peptidic and polysaccharide substances. Alkyl vibrations were prominent at around 2420 cm^{-1}, and they were also responsible for the bands at 1470 and 720 cm^{-1}.

The sharp vibration near 1720 cm^{-1} may correspond to esters, ketones or acids; the carboxyl contribution was probable, as suggested by the broad band near 2600 cm^{-1}. Both amide bands (1660 and 1550 cm^{-1}) were intense. No definitive evidence for aromatic

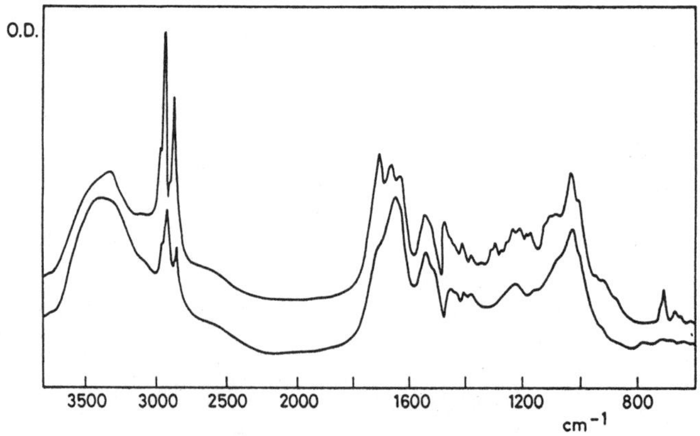

Figure 1 Infrared spectra of the alkali-soluble, acid-insoluble fractions from sewage sludge, after light petroleum, ultrasonic treatment. Top, original sludge; below, 4-month composted sludge.

material was found (the band at 1510 cm^{-1} was only a shoulder and the band at 1610 cm^{-1} was too unspecific). The intensity in the region of 1030 cm^{-1} suggested a certain proportion of polysaccharide. Other bands of the spectrum may be assigned according to methods described in the literature [20, 21], which also suggests the presence of sulfonic groups. After composting, the alkyl bands decreased, but the intense bands of the N-containing components and polysaccharides remained. The C=O vibrations of this composted sample were not strong, as opposed to soil humic fractions.

Degradative Studies

After acid hydrolysis of the OCF most of its constituents split off and could be analysed. The OCF's represented a very high protein content (Table 5). The total amino acid concentration amounted to 30% of this sludge fraction. Glutamic and aspartic acids, leucine and phenylanine were the predominant molecules. After composting, the protein fraction decreased by 50%, phenylalanine and lysine being the less resistant to aerobic transformation. In general, the amino acid patterns and the resistance to degradation of individual molecules are rather similar to those described for soil humic fractions [22].

The hydrolyzable sugars (Table 6) were not predominant (about 5%) in the OCF, but relatively important, if compared with the proportion of soluble polysaccharides in humus. Glucans were the predominant constituents of this fraction, but galactose, man-

Table 5 Effect of composting on the amino acid distribution of the OCFs from sewage sludge. *

Compound	Uncomposted sludge		Composted sludge		Relative losses after composting
Alanine	2.22	(7.38)	1.55	(8.07)	45.8
Arginine	1.76	(5.85)	0.95	(4.95)	58.1
Aspartic acid	3.51	(11.67)	2.56	(13.33)	43.4
Phenylalanine	2.72	(9.04)	1.22	(6.35)	65.2
Glycine	1.80	(5.99)	1.41	(7.34)	39.2
Glutamic acid	3.95	(13.16)	2.86	(14.91)	43.8
Hystidine	1.06	(3.52)	0.76	(3.96)	44.4
Isoleucine	1.20	(3.99)	0.67	(3.49)	56.7
Leucine	2.77	(9.21)	1.55	(8.07)	56.6
Lysine	1.42	(4.72)	0.67	(3.49)	63.4
Proline	1.44	(4.79)	0.99	(5.16)	46.6
Serine	1.72	(5.72)	1.10	(5.73)	50.4
Tyrosine	1.30	(4.32)	0.78	(4.06)	53.4
Threonine	1.64	(5.45)	1.11	(5.78)	47.5
Valine	1.56	(5.19)	1.02	5.31)	49.3
Total aminoacids	30.07	(100.00)	19.20	(100.00)	50.4

* g amino acid in 100 g HA-like fraction (parenthesis: percentage of the total amino acid content).

Table 6 Effect of composting on the sugar distribution of the OCFs from sewage sludge.*

Compound	Uncomposted sludge		Composted sludge		Relative losses after composting
Glucose	2.48	(48.35)	1.30	(32.12)	59.3
Galactose	0.49	(9.55)	0.28	(6.91)	55.6
Mannose	0.49	(9.55)	0.62	(15.30)	1.8
Glucosamine	1.24	(24.17)	1.29	(31.85)	19.2
Galactosamine	0.43	(8.38)	0.56	(13.82)	-1.1
Total sugars	5.13	(100.000)	4.05	(100.00)	38.7

* g sugar in 100 g HA-like fraction (parentheses: percentage of the total hydrolyzable sugars).

nose and aminosugars were also detected. After composting, the polysaccharide fraction decreased to nearly 40% due to the high relative losses of glucose and galactose. On the other hand, mannose and aminosugars were found to be more resistant to composting.

The composition of fatty acids in free lipid fractions was studied for comparison with those incorporated in the OCF's. The proportion of alkanes was relatively small; the most abundant fraction of lipids consisted of n-fatty acids in the C_{12}-C_{28} range. The principal acids were palmitic, stearic and myristic [23]. The C18 unsaturated acids (oleic and linoleic) were also very abundant, whereas palmitoleic acid was not predominant (Table 7).

The distribution pattern of free fatty acids was rather constant after composting. As expected, the total proportion of unsaturated chains decreased, and the total proportion of the chains longer than C_{20} (presumably derived from epicuticular waxes from higher plants), considered highly resistant to degradation [24,25], increased in amount. The lipid fractions separated from the OCF's after ultrasonic treatment in petroleum ether were very abundant, making up 41% and 51% of this fraction from the uncomposted and the composted sample, respectively. The composition of these fractions was rather different to that of free lipid fractions, since they included significant amounts of the branched fatty acids with the configuration iso and anteiso, (C_{14} - C_{19}). These compounds are typical of bacteria and suggest important microbial contribution [25,26].

Successive degradation (persulphate oxidation followed by alkaline permanganate oxidation) yielded compounds similar to those of the soil humic polymers, but in different proportions: most of the products released were alkanoic acids. In Fig. 2 the major chromatographic peaks are labelled and classified in aliphatic series, and in Table 8 the proportions of the different types of products are calculated as percentages of the total chromatographic area, for direct comparison. The degradation yields obtained were as follows: 67% of the uncomposted OCF was degraded by $K_2S_2O_8$ (18% of ethyl acetate-soluble products, for the GC-MS analyses) and 71% of the composted OCF was degraded by $K_2S_2O_8$ (21% soluble in ethyl acetate). The persulphate residues were degraded totally

Table 7 Fatty acid distribution in total sewage sludges and their colloidal organic fractions.*

	C_{12}	C_{13}	C_{14b}	C_{14b}	C_{14}	C_{15b}	C_{15b}	C_{15}	C_{16b}	$C_{16:1}$	C_{16}	C_{17b}	C_{17b}	C_{17}	$C_{18:2}$
F.U	0.5	0.1	0.0	0.1	9.6	0.0	0.0	1.0	0.0	0.3	56.2	0.0	0.0	0.5	9.8
F.C	0.2	0.0	0.0	0.0	6.8	0.0	0.0	5.8	0.0	1.8	54.8	0.0	0.0	0.3	6.5
FHA.U	2.8	0.3	0.2	0.1	15.0	1.1	1.1	1.9	0.0	0.0	36.0	9.8	0.0	1.6	0.0
FHA.C	9.4	0.3	0.7	0.6	11.5	2.5	2.8	1.8	0.0	2.2	32.7	0.0	1.7	1.2	1.4

	$C_{18:1}$	C_{18}	C_{19b}	$C_{19:1}$	C_{19}	C_{20}	C_{21}	C_{22}	C_{23}	C_{24}	C_{25}	C_{26}	C_{27}	C_{28}
F.U	7.0	10.9	0.0	0.0	0.4	0.4	0.2	0.7	0.1	1.6	0.2	0.2	0.1	0.1
F.C	3.3	12.8	0.0	0.0	0.2	0.7	0.2	1.3	0.2	3.7	0.3	0.3	0.5	0.3
FHA.U	10.1	12.9	0.3	2.4	0.3	3.0	0.1	0.2	0.1	0.1	0.2	0.2	0.0	0.0
FHA.C	13.7	11.5	0.8	1.0	1.0	1.9	0.3	0.7	0.1	0.3	0.1	0.1	0.0	0.0

* Percentages of total content of fatty acids.

Abbreviations used: F, free fatty acids of sludge; FHA, fatty acids associated with the organic colloidal fraction (ultrasonic treatment + petroleum ether extraction); U, uncomposted; C, 4-months composted.

by $KMnO_4$, yielding 22% ethyl acetate products (OCF from uncomposted sludge), and 18% in the case of the composted sample.

The products released by persulphate may be interpreted as the degradation products of the more labile polymers present in the OCF's (e.g. glycolipids). The proportion of hydrocarbons was small; the most abundant products were fatty acids. Low proportions of dialkyl phthalates (diethyl, dibutyl, diisobutyl, and dicyclohexyl) were also found; these compounds seem to be frequent in sludges [27].

The lipid fraction released by acid degradation with persulphate was comparatively simple: the C_{14}- C_{18} n-chains were dominant. The content of alkanoic α, ω n-diacids (mainly C_{18}) amounted to around 7% of the total volatile products released by persulphate. No aromatic acid was found in persulphate-removable forms, but hydroxy fatty acids represented 7% or 4% of the volatile products before and after composting, respectively. The products obtained after permanganate degradation confirmed the strongly aliphatic character of the OCF. No alkanes were detected, and the fatty acid fraction was also predominant, including important proportions of iso and anteiso chains. The alkanoic α, ω -diacids (C_6-C_{10}) amounted to nearly 40% of the permanganate degradation products of the original OCF, but that value decreased to 10% after composting. The β-hydroxy fatty acids and a smaller amount of ω-OH acids were also present, and aromatic acids (typical constituents of soil humic fractions) were represented only in minimal proportions, even after the 4-month composting process.

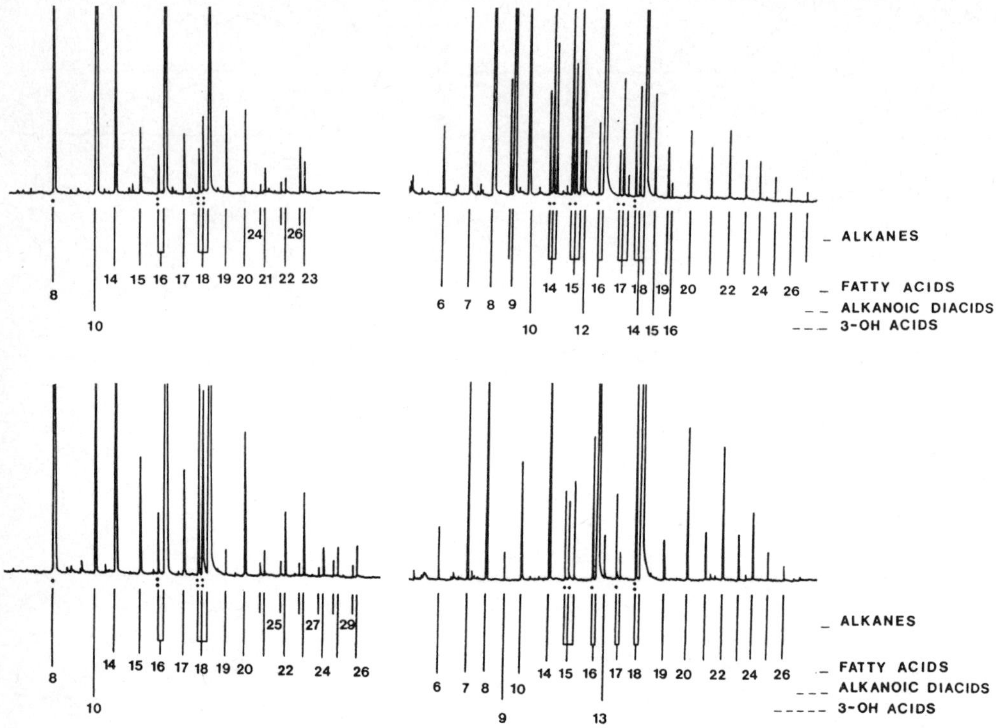

Figure 2 Left: chromatographic separation of the persulphate degradation products from the OCF from uncomposted sludge (A) and from the 4-month composted sludge (B). Right: products released by alkaline permanganate oxidation from the persulphate degradation residue (˙ = branched, : = unsaturated).

Table 8 Effect of composting on the relative yields of degradation products of the OCF from sewage sludge after degradation with $K_2S_2O_8$ followed by alkaline $KMnO_4$ oxidation of the persulphate residue.*

Type of compound	$K_2S_2O_8$ Uncomposted sludge	$K_2S_2O_8$ Composted sludge	$KMnO_4$ Uncomposted sludge	$KMnO_4$ Composted sludge
Alkanes	2.31	2.37	0.00	0.00
Fatty acids: Total	82.29	79.88	57.52	83.55
Branched	0.00	0.00	13.95	9.42
Unsaturated	3.03	5.62	1.61	6.61
Alkanoic α, ω -diacids	5.91	7.10	37.23	10.00
OH fatty acids	6.77	4.29	2.88	2.15
Phenolic acids	0.00	0.00	0.10	0.30
Benzene carboxylic acids	0.00	0.00	1.20	1.80

* Percentages of the total volatile compounds, as methyl esters.

Conclusions

The analytical results suggest that the composition of the OCF from the domestic sewage sludge was very different from soil humic acids. These sludge fractions mainly consisted of protein with a certain proportion of carbohydrates, incorporating a large amount of lipids by physical or chemical bonds. A portion of the OCF was found to be resistant to mildly oxidative procedures - and presumably to biological degradation- [28, 29], but they were also different in composition than any humic fraction. The high proportion of short-chain fatty acids, and the presence of β-hydroxy acids (both frequently showing the iso and anteiso configurations) may be interpretated as the effect of microorganisms responsible for the formation of sludge polymers, such as bacterial lipopolysaccharides. On the other hand, the small amounts of ω-OH acids could be attributed to the presence of altered biopolyesters from higher plant wastes [30].

The very small amount of aromatic constituents was not sufficient to justify a conspicuous contribution of lignin or humic-type polymers. After composting, the physico-chemical properties of the OCF became more similar to those of humic substances, but, from a structural viewpoint, they cannot be considered to contain a predominant fraction of a polymer similar to a terrestrial humic acid.

Nevertheless, even when considering the peculiar composition of the OCF's from composted sludges, it was observed that they coincided with humus fractions in regards to several physico-chemical properties which are adequate to improve soil physical properties (ie., molecular size, cation-complexing functional groups, tendency towards association both with hydrophobic and hydrophilic molecules). The very high proportion of free and linked lipids may be unfavourable to soil properties, because it is well known that these substances may enhance the water-repellent properties of soil. This circumstance is not, however, frequently reported, probably due to the fact that most sludge lipids are of low molecular size and may be considered as easily biodegradable in soil [24].

The advantages and inconveniences of extended composting of sludge must be carefully evaluated. Aerobic conditions do transform sludge into a substratum with increased stability and a certain resemblance to soil humus in terms of several of the parameters examined. But composting also causes dramatic losses of important constituents, such as the total organic matter, protein, and the total content of the OCF. In addition, during the 4-month experiment, the OCF did not turn into a polymer similar to humic substances. These facts suggest that composting time should be calculated in terms of the period necessary for the destruction of harmful microorganisms and phytotoxic substances [31,32] since the composition of the original sludge does not seem to be especially rich in materials of a prehumic nature.

References

1. Truesdale, G.A. and R.A. Wellings. Agrochim. **27**:79 (1983).

2. Hunter, J.V. and H. Heukelekian. Journal WPCF **37**:1142 (1965).

3. Lineres, M., C. Juste, J. Tauzin and A. Gomez. In: P. L'Hermite, Ed., Processing and Use of Organic Sludge and Liquid Agricultural Wastes. pp. 290-303. (Dordrecth: D. Reidel Pub. Co., 1985).

4. Kladivko, E.J. and D.W. Nelson. Journal WPCF **51**:325 (1979).

5. Sommers, L.E., D.W. Nelson, J.E. Yahner and J.V. Mannering. Proc. Indiana Acad. Sci. **82**:424 (1973).

6. Sommers, L.E. J. Environ. Qual. **6**:225 (1977).

7. Almendros, G., E. Dorado and A. Polo. In: Recuperacion de Recursos de los Residuos. Tecnologias. pp 405-416 (Proc. II. Congreso Nacional. Soria, Spain, 1984).

8. Almendros, G., A. Polo and E. Dorado. Agrochim. **27**:439-454 (1983).

9. de Nobili, M., G. Cercignani and L. Leita. In: J. H. Williams, G. Guidi and P. L'Hermite Eds, Long-term Effects of Sewage Sludge and Farm Slurries Applications, pp. 204-209. (London: Elsevier, 1985).

10. Hohla, G.N., R.L. Jones and T.D. Hinesly. J. Environ. Qual. **7**:559 (1978).

11. Boyd, S.A., L.E. Sommers and D.W. Nelson. Soil Sci. Soc. Am. J. **44**:1179 (1980).

12. Kononova, M.M. Soil Organic Matter (London: Pergamon Press, 1961).

13. Laine, R.A., W.J. Esselman and C.C. Sweeley. In: S.P. Colowick and N.O. Kaplan, Eds, Methods in Enzymology, Vol 28, pp. 159-167. (New York/London: Academic Press, 1972).

14. Martin, F., C. Saiz-Jimenez and F. J. Gonzalez-Vila. Soil Sci. **132**:200 (1981).

15. Matsuda, K. and M. Schnitzer. Soil Sci. **114**:185 (1972).

16. Miller, R. H. J. Environ. Qual. **3**:376 (1974).

17. Sposito, G., K. M. Holtzclaw and J. Baham. J. Soil Sci. Soc. Am. J. **40**:691 (1976).

18. Riffaldi, R. and M. Schnitzer. Soil Sci. **115**:349 (1973).

19. Chen, Y., N. Senesi and M. Schnitzer. Soil Sci. Soc. Am. J. **41**:352 (1977).

20. Gerasimowicz, W.V., D.M. Byler and E.G. Piotrowski. Soil Sci. **136**:237 (1983).

21. Sposito, G., G.D. Schaumberg, T.G. Perkins and M. Holtzclaw. Environ. Sci. Technol. **12**:931 (1978).

22. Schnitzer, M. and S.U. Khan. Humic Substances in the Environment (New York: Dekker, 1972).

23. Viswanathan, C.V., B. Meera Bai and S.C. Pillai. Journal WPCF. **34**:189 (1962).

24. Moucawi, J., E. Fustec, P. Jambu, A. Ambles and R. Jacquesy. Soil Biol. Biochem. **13**:335 (1981).

25. Simoneit, B.R.T. and M.A. Mazurek. Atmospheric Environ. **16**:2139 (1982).

26. Kaneda, T. J. Bacteriol. **93**:894 (1967).

27. Strachan, S.D., D.W. Nelson and L.E. Sommers. J. Environ. Qual. **12**:69 (1983).

28. Terry, R.E., D.W. Nelson and L.E. Sommers. Soil Sci. Soc. Am. J. **43**:494 (1979).

29. Terry, R.E., D.W. Nelson and L.E. Sommers. J. Environ. Qual. **8**:342 (1979).

30. Kolattukudy, P.E. and R.E. Purdy. Environ. Sci. Technol. **7**:619 (1973).

31. Vigerust, E. In: J. H. Williams, G. Guidi and P. L'Hermite Eds, Long-term Effects of Sewage Sludge and Farm Slurries Applications, pp. 168-176 (London: Elsevier, 1985).

32. Mori, T., A. Narita, T. Amimoto and M. Chino. Soil Sci. Plant. Nutr. **27**:477 (1981).

Some Effects of Ozonation of Humic Substances in Drinking Water

Dag Hongve, Vidar Lund, Gunvor Åkesson and Georg Becher
National Institute of Public Health, Geitmyrsveien 75,
N-0462 Oslo 4, Norway

Abstract

Ozonation is employed as a method for removal of colour due to humic substances in drinking water. We have examined some effects of ozonation of humic water in the laboratory. Ozonation reduced colour by 80% but had little influence on the DOC concentration and only moderate effect on the UV absorbance at 254 nm. High-performance size-exclusion chromatography (HPSEC) showed that the content of high-molecular-weight substances was reduced while a nearly corresponding amount of low-molecular-weight compounds was produced. The produced substances have acidic properties, are uncoloured and do not absorb UV light at 254 nm. Ozonation also led to higher BOD values. The formed low-molecular-weight compounds were consumed by microorganisms. In the original humic water sample the microbial degradation affected only high-molecular-weight compounds. The higher content of biodegradable organic compounds in ozonated drinking water is probably responsible for accelerated growth of bacteria and production of sludge in the distribution systems of a Norwegian waterwork. The obtained colour reduction seems to be temporary, since the colour of ozonated water increases under the influence of microorganisms.

Introduction

The use of ozone for drinking water disinfection has gained popularity with increasing awareness that chlorination by-products may create possible health hazards. Ozone treatment of drinking water is also employed in order to obtain colour removal, taste and odour control, prevention of growth in basins, oxidation of iron and manganese, and enhancement of coagulation and sedimentation. A negative side effect of ozonation is formation of low-molecular-weight organic substances which are usually more biodegradable than their precursors [1], and thus promote microbial growth in the distribution systems [2]. Modern waterworks employing ozonation must use an appropriate treatment process before the water is distributed.

The Bærum Waterworks (near Oslo, Norway) has used ozone since 1963 for the purpose of colour reduction. Ever since the startup of the ozonation plant, severe problems with formation of sludge in the distribution system have prevailed. The sludge is produced by iron- and manganese-oxidizing bacteria that are nourished by organic substances in the water. In a study in progress the effect of various methods of disinfection on the bacterial growth potential of the treated water is being investigated [3]. The present study deals with some direct effects of ozonation of humic substances in a more coloured water sample, taken from the drainage area of the waterworks.

Several publications have dealt with the reactions between ozone and humic substances. A decrease in molecular weight of the organic substances has been observed [4,5], and linear aldehydes and acids have been identified among the oxidation products of humic substances under various conditions [5,6]. We investigated the change in molecular weight distribution, optical properties, and biodegradability of the organic matter in humic water as a function of relatively low ozone doses. We also looked at the changes in these parameters during microbial degradation of the ozonated and untreated humic substances.

Materials and Methods

Water Quality

We used humic water from a brooklet draining Hellerudmyra, a marsh in the drainage area of the Bærum Waterworks. The sample was highly coloured (179 mg Pt/l), and DOC was 17.6 mg/l. The same source has been used for the isolation of the IHSS Nordic Reference Sample.

Ozonation

Ozone was produced from dried air using an ozone generator. The ozone air mixture was bubbled through 1.5 l water in a graduated cylinder. The potential reaction rate between ozone and dissolved substances was measured to 5.33 mg O_3/l/min by oxidation of dissolved KI and iodometric titration. The duration of the ozone exposure was varied in three subsamples of humic water in order to obtain various degrees of colour removal. The ozonation was stopped when the residual colour was 116, 71, and 39 mg Pt/l, respectively.

Water Analyses

Colour was determined (as mg Pt/l) after filtration (0.45 µm) from the absorbance at 410 nm. UV absorbance was measured at 254 nm. DOC was measured with a Technicon AutoAnalyzer II, Industrial Method 451-76W. BOD was measured in bottles which were incubated in darkness at room temperature (≈23°C). The acidity after the ozonation was neutralized with Na_2CO_3, and the samples were inoculated with 1% tap water containing the naturally occurring bacteria from the distribution system of the Bærum Waterworks. Since the natural water bacteria grow slowly, we chose to use the oxygen consumption during the first two weeks as a measure of BOD. For a prolonged incubation we stored neutralized and inoculated samples at 20°C in darkness for 8 weeks. These samples were aerated twice a week.

HPSEC

High-performance size exclusion chromatography was performed using a 7.5x75 mm precolumn and a 7.5x600 mm column, both packed with TSK-G 3000SW (LKB, Sweden) [7], and using detection of both UV absorbance and DOC [8].

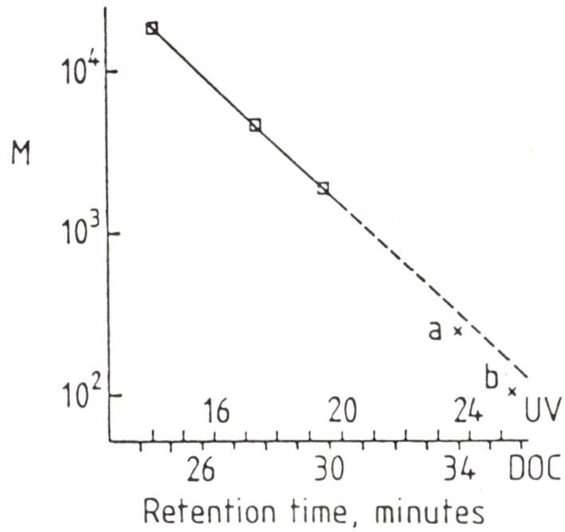

Figure 1 Calibration curve for polystyrene sulfonate standards used to calculate the molecular weights of unknown samples. Results are shown for two low-molecular-weight compounds: potassium phthalate (a) and potassium acetate (b).

For calibration of molecular weights (M) we used standards of sulfonated polystyrene (Polysciences, Inc.) with weight-averaged molecular weights (M_W) of 1800, 4600 and 18000. A linear relationship was obtained between retention time and log M (Fig. 1). The retention time of some substances with low molecular weight were in good agreement with the extrapolated calibration curve.

Results and Discussion

Increasing ozonation had only minor effects on the total DOC concentrations while colour and UV absorbance were more effectively reduced (Fig. 2, Table 1). We also observed increases in the concentrations of strong acid HNO_3 and weak acids causing an increase in the buffer capacity when the samples were titrated between pH 4.0 and 7.0. Based on previous results obtained in a closed system [9], in our samples the amount that had reacted should be less than 1 mmol O_3 per mmol C.

The effects of ozone depend on the vulnerability of various moieties of the humic compounds to oxidation. Some chromophores are easily attacked by ozone as well as by other oxidants, e.g. chlorine [10]. The aromatic moieties causing UV absorbance are less affected. Of the formed aliphatic products, ozone will only react with those having easily oxidizable groups, such as hydroxyl and some carbonyl

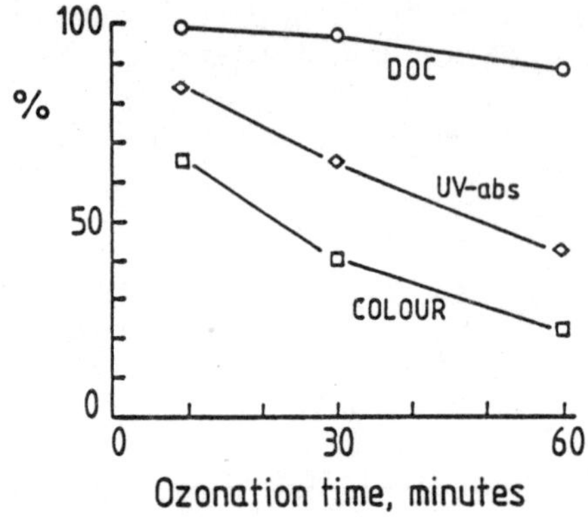

Figure 2 Percentage reduction in DOC, UV absorbance and colour during the ozonation.

Table 1 Changes in chemical parameters during ozonation

Sample no.	1	2	3	4
Ozonation time, minutes	0	9	30	60
Colour, mg Pt/l	179	116	71	39
UV absorbance (254 nm) cm^{-1}	0.904	0.763	0.586	0.382
DOC, mg/l	17.6	17.4	17.1	15.5
NO_3^--N, mg/l	<0.01	0.19	0.65	1.20
pH	4.40	4.26	4.01	3.86
Buffer capac. pH 4.0-7.0, μeq/l	101	122	139	149
M_W	1200	820	690	500
BOD_{14}, mg O_2/l	1.40	1.82	2.70	3.12

groups [2], and aliphatic acids will consequently be the end products of ozonation. A number of low-molecular-weight acids have been identified after ozonation of an isolated fulvic acid [5]. Linear aldehydes occur only as trace substances after ozonation of natural humic water [6].

The size exclusion chromatogram of the untreated humic water shows a continuous distribution of compounds with molecular weights ranging from ≤100 to 15000 (Fig. 3). The mean M by weight (M_W) is 1200. M_W was reduced with increasing ozone dosage. The chromatograms show that the reduction is a result of degradation of substances with high M. The highest ozone dose reduced the maximum M to ≤5000. At the same time, the content of substances with M≤600 increased. Two distinct peaks of degradation products were developed: one had a retention time corresponding to M≈310 and the second M≈170, according to the extrapolated standard curve (Fig. 3A). The areas of the peaks (shaded in Fig. 3A) correspond to concentrations of 2.0 and 1.2 mg DOC/l, respectively.

The degradation products showed no significant absorbance at 254 nm (Fig. 3B), which indicates that they are aliphatic. The M≈170 peak is probably composed of low-molecular weight acids. These acids are eluted as potassium salts, since potassium phosphate buffer is used as eluant. Succinic acid and malonic acid have been identified as the major oxidation products after ozonation of isolated fulvic acids [5]. The M≈310 peak may represent aliphatic C_{15}-C_{17} acids which are produced with low doses of ozone [5].

Increasing ozone dosage led to an increase in the biological degradability (Table 1). In the untreated reference microbial degradation exclusively affected high-M compounds (Fig. 4A). In contrast, the ozonated samples showed decreases of both high- and low-M fractions, and the low-M compounds produced during the ozonation were completely consumed during the incubation (Fig. 4B). The original low-M compounds (>34 minutes retention time) in the humic water seem to be unaffected by ozonation and incubation, as this part of the chromatogram was left almost unchanged by the treatment (Fig. 4A, before incubation, and 4B, after incubation). The latter result seems reasonable, since the low-M compounds present in the humic water which was stored for months before the experiment, should be quite refractive. This is in contrast to the high-M compounds, which were susceptible to both chemical and biological oxidation.

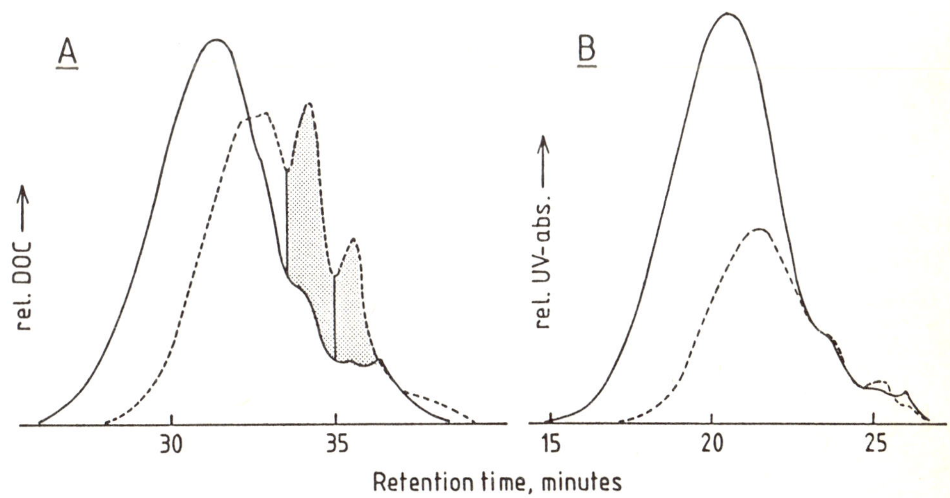

Figure 3 High-performance size exclusion chromatograms of the natural water (——) and after 60 minutes ozonation (---). A, with a DOC analyzer as detector; B, with UV detection (254 nm).

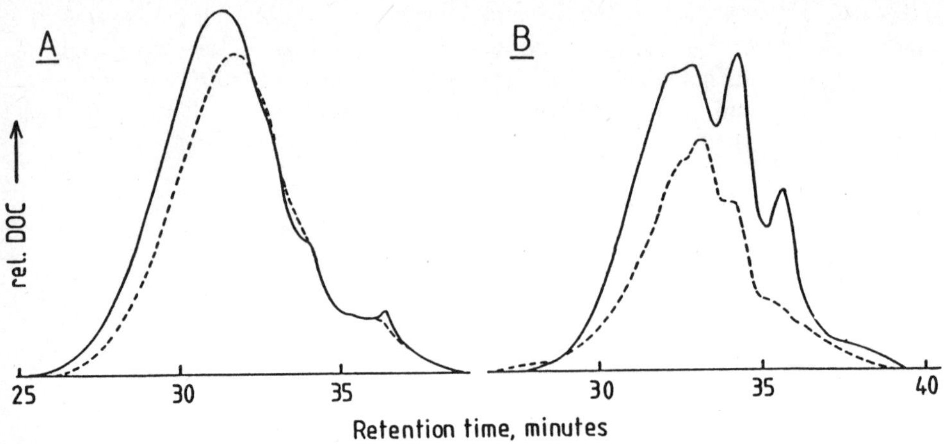

Figure 4 Effects of microbial degradation on HPSEC chromatograms. A, natural humic water; B, sample ozonated for 60 minutes. Before incubation (——) and after 8 weeks incubation (---).

Table 2 Changed water quality parameters after 8 weeks incubation at 20°C.

Sample No.	1	2	3	4
Colour	192	154	115	66
UV absorbance	0.850	0.728	0.583	0.351
DOC, mg/l	15.5	14.5	13.8	10.0

An unexpected result of the incubation was the increase in visible colour of all the samples. This increase varied from 7% in the untreated sample to a quite significant 69% in the most ozonated one (Table 2). Control samples stored at 4°C without inoculation during the time the other samples were incubated, showed no increase in colour.

Conclusion

The study has confirmed that ozone transforms a part of the natural organic matter present in raw water from the Bærum Waterworks into low-molecular-weight compounds which are available as a source of carbon for microorganisms in the distribution system. Two distinct groups of aliphatic compounds were formed: these were probably acids with molecular weights of their potassium salts of about 310 and 170, respectively. Our results indicate that the colour removal which can be obtained with small ozone doses may not be permanent, since the colour of ozonated samples increased with storage in the presence of microorganisms.

References

1. Gilbert, E. Water Res. **22:**123 (1988).

2. Glaze, W.H. Environ. Sci. Technol. **21:**224 (1987).

3. Lund, V. Report (in Norwegian). (Oslo: NTNF. In press).

4. Gilbert, E. Vom Wasser **55:**1 (1980).

5. Anderson, L.J., J.D. Johnson and R.F. Christman. Org. Geochem. **8:**65 (1985).

6. Hoof, F. van, A. Wittocx, E. van Buggenhout and J. Janssens. Anal. Chim. Acta **169:**419 (1985).

7. Becher, G., G.E. Carlberg, E.T. Gjessing, J.K. Hongslo and S. Monarca. Environ. Sci. Technol. **19:**422 (1985).

8. Hongve, D., G. Åkesson and G. Becher. Sci. Total Environ. **81/82:**307 (1989).

9. Anderson, L.J., J.D. Johnson and R.F. Christman. Environ. Sci. Technol. **20:**739 (1986).

10. Oliver B.G. and E.M. Thurman. In: R.L. Jolley, W.A. Brungs, J.A. Cotruvo, R.B. Cummings, J.S. Mattice and V.A. Jacobs, Eds, Water Chlorination. Environmental Impact and Health Effects, Vol. 4, Book 1, pp. 231-241. (Ann Arbor: Ann Arbor Science 1983).

Upgrading the Removal of Humic Substances and Mutagen Precursors in Water Treatment

Ari V.O. Järvinen[1], Markku T. Pelkonen[1] and Terttu Vartiainen[2]

[1] Laboratory of Sanitary and Environmental Engineering, Helsinki University of Technology, SF-02150 Espoo, Finland
[2] The Public Health Institute, SF-70701 Kuopio, Finland

Abstract

This study aimed at investigating different methods of upgrading conventional water treatment plants for improved removal of organic substances. Ozonation, activated carbon filtration and slow sand filtration were tested. Pilot scale experiments were performed at Bodom waterworks in Espoo, Finland. The TOC-value of the influent was 3.2 mg/l (6.7 mg/l COD_{Mn}). The average removal of TOC during activated carbon filtration was 29% (41% removal of COD_{Mn}). Preozonation caused no significant change in treatment efficiency. Mutagenicity (test strain TA100), after chlorination, was lower in ozonated and filtered water than in non-ozonated. The level of mutagenicity achieved was close to that of chlorinated groundwater.

Introduction

The high concentration of humic substances in raw waters is the main problem in water treatment in Finland. In many cases, the most common water treatment, chemical flocculation and sand filtration, cannot remove organics sufficiently. This implies that, during chlorination, halogenated hydrocarbons are formed in considerable amounts. The mutagenicity of Finnish waters has been found to be one of the highest in the world [1]. In the distribution system, residual organics may cause taste and odour problems and corrosion of aluminium.

In this study, different methods of improving the removal of humus were tested in a pilot scale plant at Bodom waterworks in Espoo. Activated carbon filtration, slow sand filtration and ozonation were tested. One target was to achieve biological activity in the filters.

It has been reported that biologically active carbon filters can remove 30-75% of the total organic carbon (TOC). Slow sand filtration may remove 40% of the TOC [2]. Hubele and Topalian [3] have achieved 25-35% TOC removal in an activated carbon filter with preozonation. Without preozonation, the amounts removed were much smaller. Kaastrup [4,5] found that the adsorption capacity with preozonation was almost double the capacity with no preozonation (43 *vs* 25 mg DOC/g GAC (granulated activated carbon)). However, there was no evidence of biological activity influencing treatment efficiency.

In Norrköping, Sweden, it was found that the effect of the GAC filter upon organics was rather short in time, but still a removal of odorous substances was achieved for more than 30 months. Some removal of mutagenic substances by GAC filters was also demonstrated [6]. Loper *et al.* [7] and Kool *et al.* [8] have reported that mutagenicity was reduced efficiently by GAC filtration.

Pilot Test

Pilot scale experiments were performed at Bodom water treatment plant in Espoo in two periods during 1987-89. In this paper only the results of the latter period are referred to. The pilot process raw water was taken after the filtration stage of the full scale process. The pilot process included GAC filtration and slow sand filtration with and without preozonation (Fig. 1).

The filtration rate in the GAC filters was 5 m/h during the first six months (in 1988) and 2.5-3.0 m/h during the last period. The filter depth was 60 cm. The carbons used were Filtrasorb F300 and the so-called "Chinese" BG-09. In the beginning of May -89, the BG-09 filter was replaced by a Filtrasorb F300 filter,

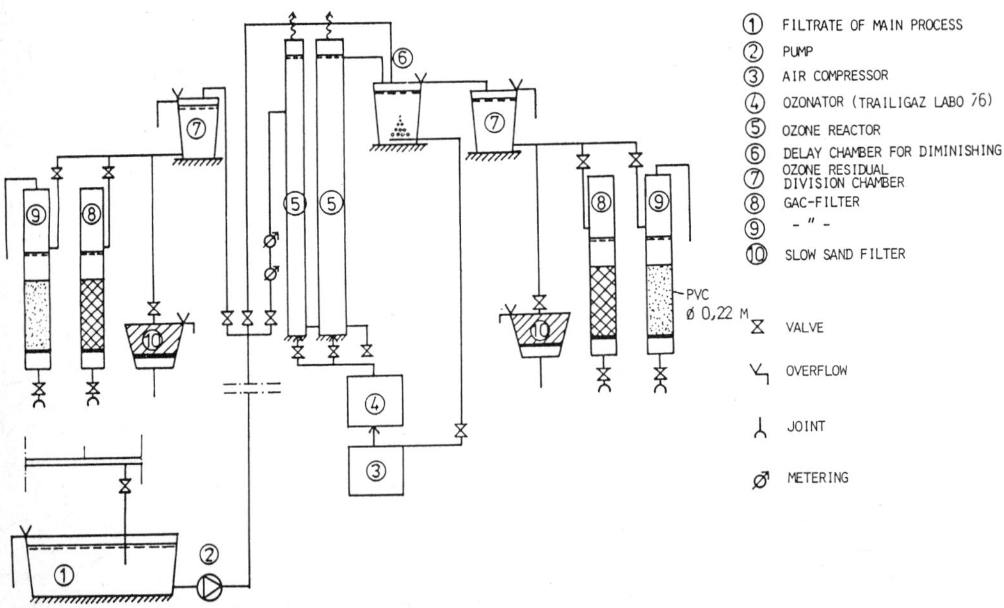

Figure 1 Pilot scale process at Bodom treatment plant, Espoo.

which had already been used for one year in the filter of the main process. The Filtrasorb filter was replaced by a new one of the same type. The filtration rate in the slow sand filters was 0.2 m/h. The filter depth was 60 cm.

The ozone dosage was chosen to yield an ozone residual of 1 mg O_3 /l after the reactor. The contact time was 15 min. According to the yield nomogram of the ozonator, the dosage was kept between 3-5 mg O_3/l. The ozone residual was measured with a comparator (Lovibond).

Samples were taken weekly or every two weeks. The parameters analyzed were TOC, COD_{Mn} , absorbances at wavelengths 254 and 374 nm, turbidity, pH and conductivity. The Ames mutagenicity was tested twice, Jan 17 and May 25, 1989. Two Ames tests from 1987 will also be referred to later.

TOC was measured with an Astro 2001 TOC-analyzer and COD_{Mn} was analyzed according to the Finnish standard SFS 3036 (potassium permanganate as oxidizer). Absorbance was measured with a Perkin-Elmer Lambda 3B spectrofotometer, A_{254} through 100 mm cell and A_{374} through 10 mm cell. Mutagenicities were measured by Ames' tests performed at the Public Health Institute. Test strains were TA100, 98 and 97 without enzymatic activation. Before testing, the samples were chlorinated with a chlorine dose of 0.5 mg/l.

Results

Removal of Organics

The removal of organic substances was measured as removal of TOC, COD_{Mn} and UV-absorbance A_{254} . Fig. 2 shows the results for activated carbon filtration and Fig. 4 the results for slow sand filtration.

In Fig. 2a, b and c, the results for activated carbon filtration are compared with the quality of the raw water and the main process filtration effluent, which was the influent of the pilot. As can be seen, the removal measured as COD_{Mn} was higher than the TOC removal. One explanation for this is the different nature of the organic material represented by TOC and COD_{Mn}, respectively. Conventional chemical flocculation removes more efficiently the COD_{Mn} part of the organics, which are mainly humic acids. TOC also measures the fulvic acid part of the organics, which is less efficiently removed by flocculation and GAC filtration. During the winter season, the main process did not work as efficiently as during the first summer. One reason for this might be the introduction of activated carbon in the sand filters of the main process. This change was carried out in the beginning of June 1988. Equilibrium was apparently achieved in 3-4 months in the main filters. In the pilot filters, it took 4-5 months (Fig. 2b). The removal in the pilot plant filters averaged 10% until new carbon filters were installed in the beginning of May 1989.

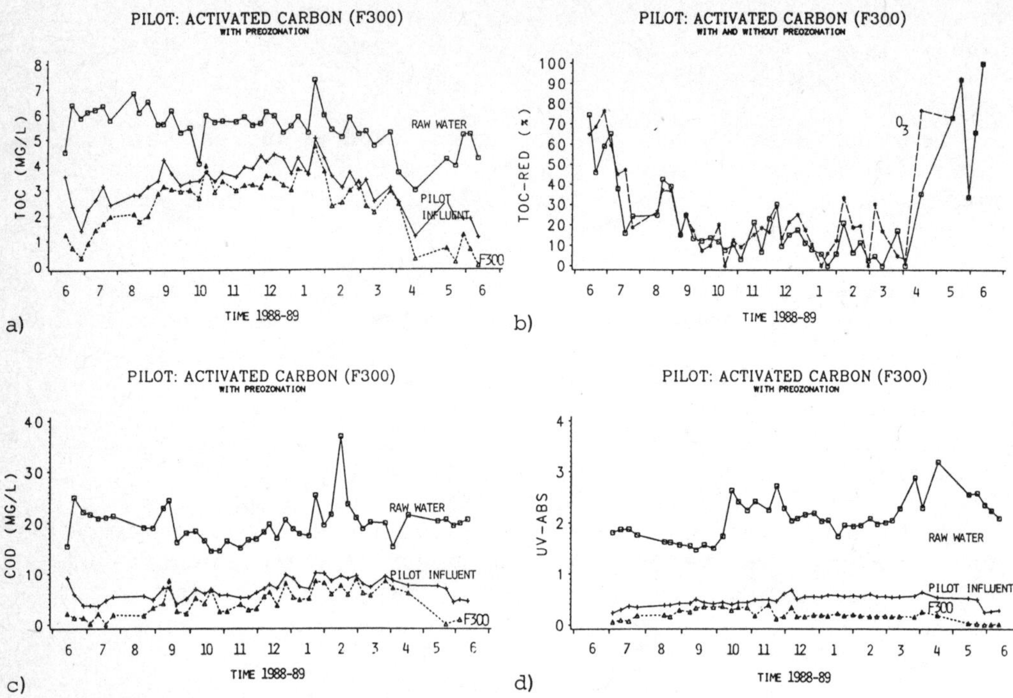

Figure 2 Removal of organics by GAC filter with preozonation. a) TOC removal in mg/l b) TOC removal in % compared with no preozonation c) COD_{Mn} removal in mg/l d) A_{254} removal (Note: carbon changed in the beginning of May 1989)

In Fig. 2b, it can also be seen that preozonation had only a minor effect on the results of the activated carbon filtration. The efficiency started to improve in December 1988, probably because of enhanced biological degradation after 6 months of operation.

To enable on-line measurements of organics, absorbance A_{254} was also determined. In Fig. 2d the removal of UV-absorbance is illustrated, and compared with TOC and COD. In Fig. 3a the correlation of TOC and UV-absorbance A_{254} is shown.

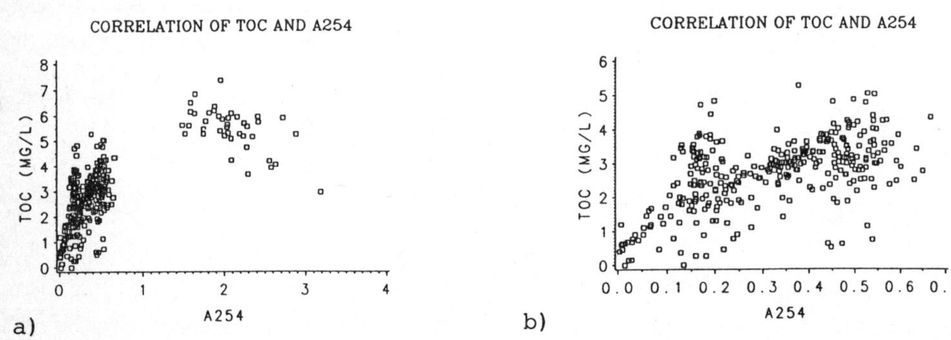

Figure 3 Correlation of TOC and UV-absorbance A_{254}. a) all samples b) without raw water samples.

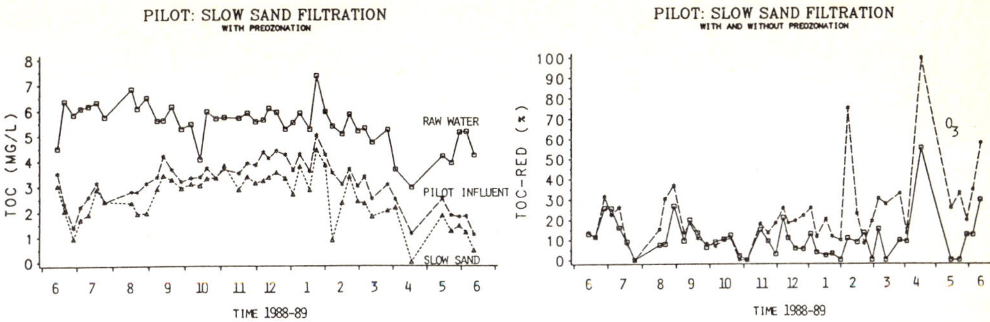

Figure 4 TOC removal during slow sand filtration.

In Fig. 4 the removal of organics during slow sand filtration is shown. TOC removal averaged 10% during the first 6-7 months. Thereafter, a slightly improved removal was observed in the filter with preozonation. The reason for this may probably be attributed to the development of biological activity in the filter. The temperature rise after the winter season may also affect this development. The raw water temperature was 2-5°C from November to April. After this, the water temperature increased to 16°C in June.

The existence of biological activity was measured by observing the balance between dissolved oxygen and carbon dioxide concentrations before and after filtration. In the slow sand filter with preozonation, the oxygen demand was greater than in the non-ozonated one. The CO_2 production was also higher. This was in agreement with the observations of TOC removal. The results from the GAC filters were not reliable because of adsorption of gases in the carbon.

Removal of Mutagenicity

It was possible to test mutagenicity only twice in 1989. As a reference also the results from an earlier experiment in 1987 are referred to here. Mutagenicities (test strain TA100) are shown in Table 1.

Samples were chlorinated in the laboratory like the drinking water in the full scale process (dosage 0.5 mg/l). However, the contact time varied from 1 to 80 hours because of the variability in amount and quality of the organic matter. The chlorine residual after this was 0.2 mg/l. Because the contact time, the dosage and the quality of organics can affect the mutagenicity (see e.g. [7]), the results must be interpreted with some care. According to Table 1 it can be said that the mutagenicity in ozonated and GAC filtrated water was lower than without preozonation. Also slow sand filtration gave good results. But, if looking at the results from January 17, 1989, mutagenicities were not very logical when compared with each others. Some toxicity was found from the sample of the pilot influent, implying that the results remained lower. A raw water in itself is usually not mutagenic. However, in this study it was also chlorinated just for the sake of comparison. Here the possibly insufficient chlorine dosage could have affected the results. No efforts were made to measure the mutagenic potential.

Table 1 Mutagenicity (TA100, net rev/l) of chlorinated waters. Corresponding TOC (mg/l) values in parenthesis.

Sample		Mutagenicity (TOC) 27.2.87	26.10.87	17.1.89	25.5.89
	1)	3 m	9 m	6 m	1 m
Raw vater		1020 (7.2)	740 (6.2)	540 (4.6)	1030 (3.8)
Influent to pilot		610 (3.6)	400 (3.4)	230 (2.9)	100 (1.7)
GAC		540 (2.6)	400 (2.7)	510 (2.7)	100 (<0.5)
GAC + ozone		100 (2.8)	330 (2.6)	570 (3.0)	<100 (1.7)
Slow sand		335 (3.6)	720 (3.1)	540 (2.8)	160 (1.8)
Slow sand + ozone		100 (3.2)	510 (2.9)	- (2.2)	160 (1.1)
Drinking water		- -	790 (3.2)	720 (2.7)	610 (1.7)

1) Duration of run in GAC (months)

Conclusions

The use of preozonation improved only slightly the TOC removal efficiency of the GAC and slow sand filtration. Some biological activity was found in the ozonated processes, but the startup time was long.

The mutagenicity of the GAC or slow sand filtered water was lower with preozonation than without preozonation. Mutagenicities achieved were close to those of chlorinated groundwater.

References

1. Vartiainen T. Mutagenicity of drinking water in Finland (Kuopio: NPHI, 1988).

2. Bouwer E. and P. Growe. J. Am. Water Works Assoc. **90**:82 (1988).

3. Hubele C. and P. Topalian. Aktuelle Probleme der Wasserchemie und Wasseraufbereitung. Engler Bunte-Institut der Universität Karlsruhe **20**: 261 (1982).

4. Kaastrup E. Activated carbon absorption of humic substances and the influence of preozonation on such, Ph. D. Thesis (NTH, 1986).

5. Odegaard H., H. Brattebo, B. Eikebrokk and T. Thorsen. IHSS third international meeting, p. 64 (Oslo, 1986).

6. Tjeder A., M. Lind, A. Kristensson and I. Stening. Vatten **45**:102 (1989).

7. Loper J.C., M.W. Tabor and L. Rosenblum. In: R.J. Jolley, R.J. Bull, W.P. Davis, S. Katz, M.H. Roberts Jr and V.A. Jacobs, Eds, Water Chlorination, Chemistry, Environment Impact and Health Effects, pp.1329-1339 (New York: Lewis Publishers Inc. 1985).

8. Kool H.J. and C.F. van Kreijl. Water Res. **18**:1011 (1984).

9. Kronberg L. Mutagenic compounds in chlorinated humic and drinking water. Ph. D. Thesis (Åbo Akademi, 1987).

Effects of Gamma Irradiation on an Aquatic Fulvic Acid

Irina Arsenie, Hans Borén and Bert Allard
Department of Water and Environmental Studies
Linköping University, S-581 83 Linköping, Sweden

Abstract

An aquatic fulvic acid was irradiated with gamma radiation from a ^{60}Co-source (dose range 0-48 Mrad), as part of a larger study of the transformation and decomposition of humic substances in natural aquatic systems. Experiments were performed at two concentrations (1000 mg/l and 100 mg/l) and at various pH-values (2-10). The fulvic acid transformation was studied by monitoring optical density (UV-spectroscopy), molecular weight distribution (GPC-technique) and total dissolved organic carbon (TOC). A general decrease in TOC with increasing radiation dose was observed: the initial G-value of about 5 decreased with the increasing dose to a minimum value of 0.2-0.3. A simultaneous increase in molecular weight (M_n rose from approximately 2000 to a maximum of about 4000) was observed in the acidic samples (pH 2-4) at a dose below 10 Mrad. Natural background radiation can significantly contribute to the degradation of dissolved humic substances in deep groundwaters, considering the observed G-value for low doses (about 5) and the otherwise high chemical stability of the fulvic acid fraction even after long residence times (10^3-10^4 y) in the ground.

Introduction

Low and intermediate molecular weight humic substances constitute a large, or even dominating, fraction of the dissolved organic carbon (DOC) in many aqueous systems in the environment [1]. Concentrations of humic materials reach the mg/l-level in shallow groundwaters [2]. Even in subsurface waters with long residence times in the ground (e.g. more than 1000 y) the amount of dissolved organics are frequently above 0.1 mg/l, predominantly as fulvic acids (FA) with a molecular weight of less than 2000 [3]. This molecular size fraction appears to be quite stable. A recent study has shown that FA from various localities (surface waters, shallow groundwaters, deep groundwaters) of various ages (up to 10000 y) are similar in terms of elemental composition, molecular weight and functionality (acid capacities from carboxylic and phenolic groups) [4]. However, even if recent and old FA have similar chemical properties, the total concentration, as well as the DOC, generally decrease with increasing residence time [4], indicating a degradation. The mechanisms for this slow degradation (chemical degradation, reduction, hydrolysis, biological degradation and transformation etc) are not known.

The objective of the present work was to study the effects of gamma radiation on a dissolved FA at different pH-values particularly the changes in molecular weight distribution and the complete degradation, as indicated by reduction of the DOC. Only a few similar studies have previously been reported in the literature [5,6].

Experimental

The Fulvic Acid

A FA was recovered from a surface water from Bersbo, Sweden (adsorption on DEAE-cellulose, elution with NaOH, purification on XAD-8 Amberlite resin, treatment with a hydrogen form of a cation-exchange resin and freeze drying) [3]. The FA was characterized with respect to elemental composition, molecular weight (GPC technique; M_n=1750, M_w=2650, where M_n and M_w are the number and the weight average molecular weight, respectively), and functionality (potentiometric titration in aqueous and nonaqueous systems; 4.78 meq/g and 1.35 meq/g capacity of COOH and Ph-OH, respectively) [3,7].

Gamma Irradiation

Solutions of the FA were prepared according to Table 1 and exposed to gamma radiation from a ^{60}Co-source (E =1.17 and 1.33 MeV; dose rate 12 Krad/h). Samples were taken after total doses of 1.8, 3.2, 10, 35 and 48 Mrad and characterized (pH, UV-spectrum, TOC, molecular weight).

Table 1 Irradiated FA systems.

Medium[a]	FA conc. mg/l	$pH_{init.}$	pH_{final}[b]
HCl/NaCl	1000	2.0	1.9
	100	2.1	2.1
NaCl	1000	2.8	4.1
	100	3.8	5.2
Aq.[c]	1000	3.0	3.7
	100	4.1	6.8
Phosphate/NaCl	1000	6.7	6.6
	100	6.7	6.7
Carbonate/NaCl	1000	10.0	9.5
	100	9.5	9.8

[a] I = 0.1 M [b] After 48 Mrad [c] I< 0.01 M

Molecular Weight

Molecular weights (M_n and M_w) were determined with a GPC-technique (HPLC equipment; a Waters 510 pump; a TSK G2000 SW column, 7.5x300 mm plus a precolumn, 7.5x75 mm; a Waters 481 UV-detector, 225 nm). The mobile phase was a 50 mM phosphate buffer at pH 6.8 with a flow rate of 0.5 ml/min. Polystyrene sulphonates of known molecular weights (1600, 4000, 6500 and 16000; Pressure Chemical Co, Pittsburg, PA, USA) were used as standards, [8].

Results

UV-spectra

Colour intensity in the visible region, as well as absorption in the UV region of the spectrum, decreased with increasing radiation dose, as illustrated in Fig. 1. The sharp UV-absorption edge around 400-450 nm for a nonirradiated sample gradually moved towards lower wavelengths with increasing dose.

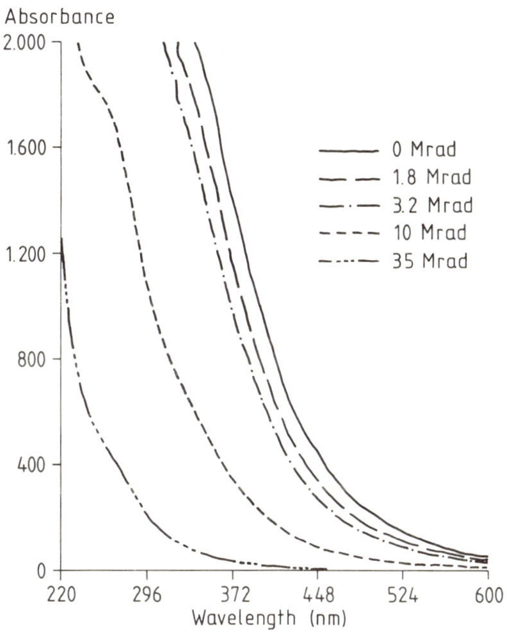

Figure 1 Absorbance *vs* wavelength (Aq.-system, 1000 mg/l).

Molecular Size, TOC

Examples of GPC-measurements are given in Fig. 2, which shows UV-absorbance (at 225 nm) *vs* retention times for samples that had received different doses. There was generally a quantitative correlation between the absorption, defined by the area under the peak (Fig. 2) and the TOC measured in the corresponding solution (Fig. 3). The loss of TOC, as reflected by the absorption reduction, was, however, different in the various systems, as illustrated in Fig. 4.

For the two systems at high pH (phosphate and carbonate) the loss of TOC was higher at a given dose than for the systems at low pH (HCl, NaCl, Aq.). The FA never completely

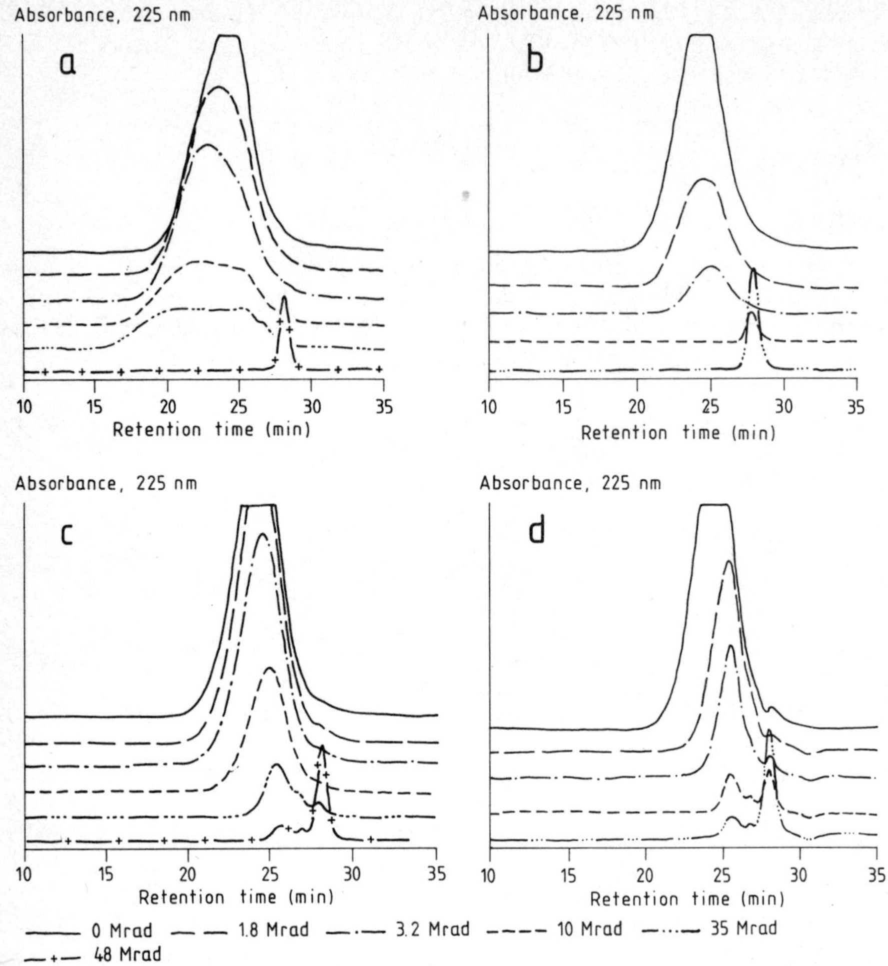

Figure 2 Retention time *vs* dose (GPC): (a) Aq., 1000 mg/l, (b) Aq., 100 mg/l FA, (c) Carb./NaCl, 1000 mg/l FA, (d) Carb./NaCl, 100 mg/l.

dissolved in the HCl system, which led to a lower initial concentration of TOC (and lower absorption), as well as a displacement of the corresponding curve in Fig. 4.

Molecular Weight

Molecular weights calculated from the GPC-measurements are given in Fig. 5a, which shows M_n *vs* dose for various solution systems. In Fig. 5b the same data points are given, although expressed as M_n *vs* the pH that was observed after the various doses. Typically, a general reduction of M_n from an initial value of 1750-2000 down to 500-600 was observed after doses of 10 Mrad (and above) for the samples with low FA concentration (100 mg/l) (Fig. 5a). A simultaneous decrease in the M_w/M_n ratio, from initially 1.2-1.3 to around 1.1, was also noted.

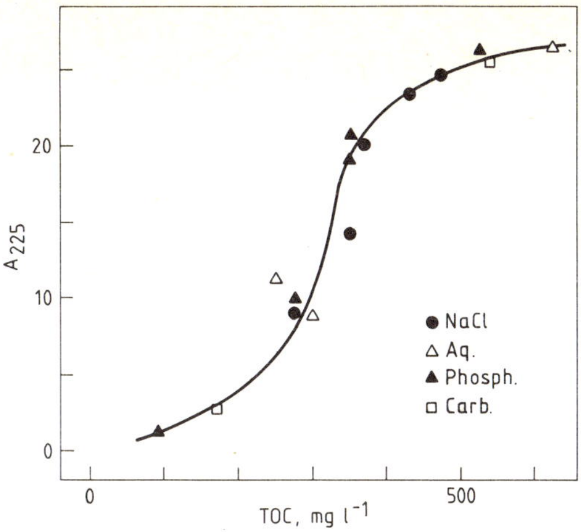

Figure 3 Absorption (A_{225}) *vs* TOC (1000 mg/l FA; doses between 0 and 48 Mrad).

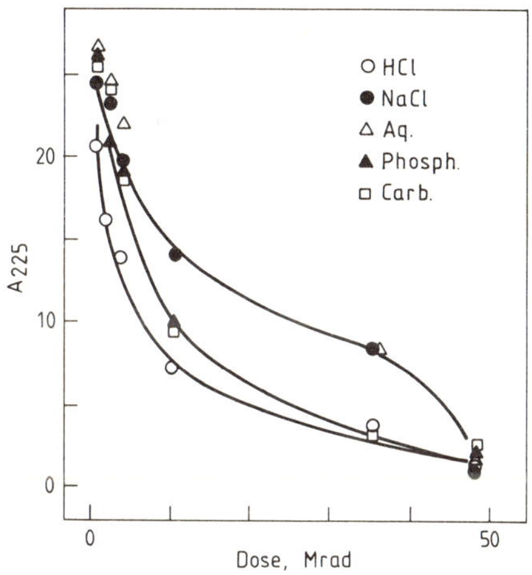

Figure 4 Absorption (A_{225}) *vs* dose (1000 mg/l FA).

At the high FA concentration (1000 mg/l), however, there was a pronounced increase of M_n with increasing dose, particularly in the NaCl- and Aq-systems up to the dose level of 35 Mrad, followed by a sharp decrease of M_n after 48 Mrad. The carbonate system was the only system that did not show any initial increase in M_n with increasing radiation dose. A parallel increase in the M_w/M_n-ratio (up to 2.5 in the NaCl-system after 35 Mrad), followed by a decrease (to approximately 1.1 after 48 Mrad), was observed.

A plot of M_n *vs* pH (Fig. 5b), shows that it was pH rather than the background electrolyte that determined the change in M_n *vs* dose. It is evident that an aggregation took place in the pH range 2-4 at doses up to 35 Mrad; this, was followed by a degradation to species of low molecular weight after 48 Mrad.

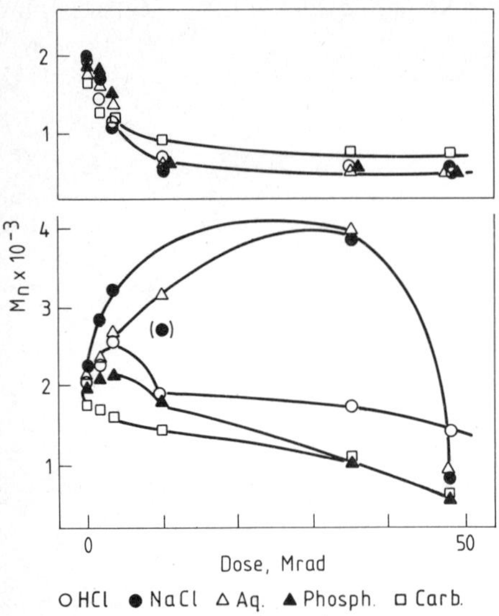

Figure 5a M_n *vs* dose (100 mg/l FA, top, and 1000 mg/l FA, bottom).

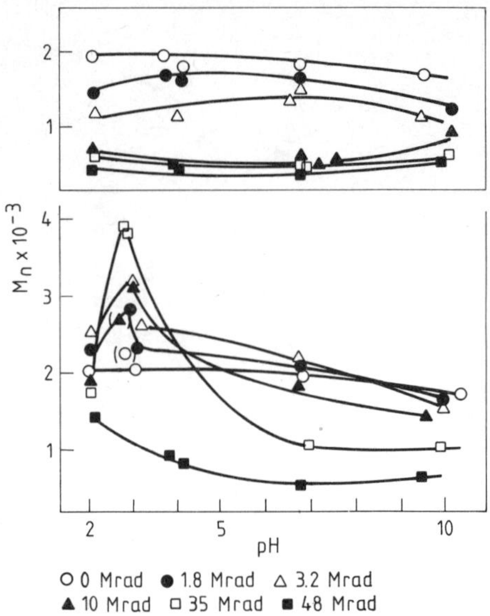

Figure 5b M_n *vs* pH (100 mg/l FA, top, and 1000 mg/l FA, bottom).

Discussion and Conclusions

The radiolysis of water yields H_2, H_2O_2, H^+, OH^-, and as intermediate products H·, HO· and e^-_{aq} in varying amounts. The solvated electron e^-_{aq} as well as H· and H_2 are reducing agents, whereas H_2O_2 and HO· are oxidizing agents. The net effect (reduction or oxidation) is dependant on the other components in the system and particularly on their affinity for the radicals. For organic systems in aqueous solutions, oxidation and hydroxylation predominate, irrespective of their class or molecular complexity [9]. Thus, aromatic compounds undergo oxidative degradation, and e.g. monobasic carboxylic acids tend to be oxidized to diacids and hydroxy acids and finally to CO_2. The presence of oxygen increases the oxidation yield. However, in general, aromatic compounds are known to be more stable towards radiation decomposition than aliphatic compounds, due to resonance stabilization of the benzene ring. Also, aromatic compounds with aliphatic side chains exhibit a stability similar to the purely aromatic substances [10].

The production of low molecular weight acids, like formic and acetic acids, which can account for as much as 5-10% of the loss of TOC, has previously been reported in studies of radiation effects on FA [6]. Also the production of a mixture of CO and CO_2 (with traces of H_2 and CH_4) has been observed [5]. At least at low doses (up to 4 Mrad) the loss of TOC can quantitatively be balanced by the production of CO_2.

In the present study some 30-40% of the dissolved TOC was lost after a dose of 10 Mrad in the systems with a starting concentration of 1000 mg/l FA (Fig. 6). This loss probably corresponds to decarboxylation [9].

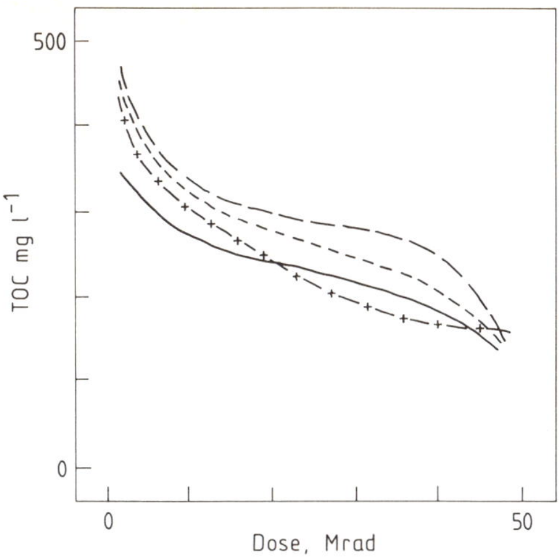

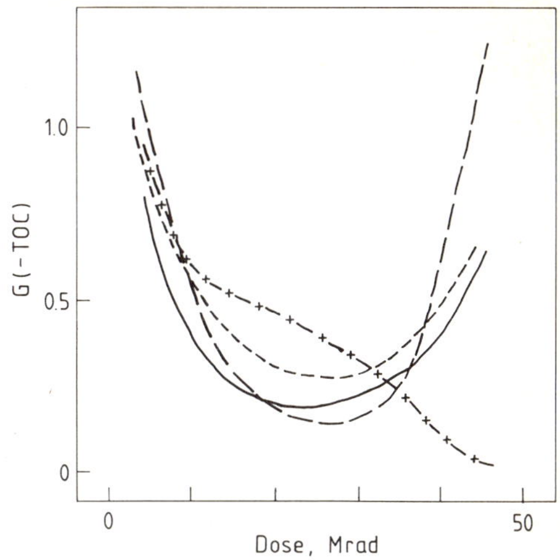

Figure 6 TOC *vs* dose (1000 mg/l FA).

Figure 7 G(-TOC) *vs* dose (1000 mg/l FA).

Increasing doses up to 40 Mrad had relatively minor effects on the TOC, whereas additional doses led to a further degradation. These dose-related degradation processes are further illustrated in Fig. 7, which shows G(-TOC) as a function of the dose. The G(-TOC) is defined as the number of carbon atoms *lost* from the TOC fraction per 100 eV absorbed radiation energy.

The G(-TOC)-values decreased from a maximum of about 5 at low total doses to a minimum value of approximately 0.2-0.3 at intermediate doses (20-30 Mrad), followed by an increase at doses above 30 Mrad. The carbonate system (pH 10) exhibited continuously decreasing G(-TOC) with increasing dose.

It is evident that although TOC decreased in all systems with increasing doses (Fig. 6), the molecular weight distribution (expressed as M_w/M_n or as M_n, Fig. 8) increased at low doses (below 5-10 Mrad) for all systems except the carbonate system. This aggregation had a maximum at pH 2.5-3, when the aggregation progressed with increasing doses up to a total dose of 35-40 Mrad. A release or decomposition of certain parts of the molecule occurs simultaneously with an aggregation, particularly at a low pH. This observation is in qualitative agreement with previously reported data [5]. Evidently this initial aggregation is dependent on the total concentration, since no similar effects were observed at a low concentration level (100 mg/l) (Fig. 8). Further studies of functionality, based on potentiometric titration on irradiated samples are in progress.

Natural gamma radiation levels are in the order of 50-500 mrad/y [11]. Thus, a total dose of about krad/10^4 y would be expected in e.g. a groundwater environment, not considering local higher doses due to contact with fissure minerals containing elements in the U- or Th-series. The chemical changes in dissolved FA under natural conditions are minor after 10000 y [4], and no significant changes in molecular weight distribution

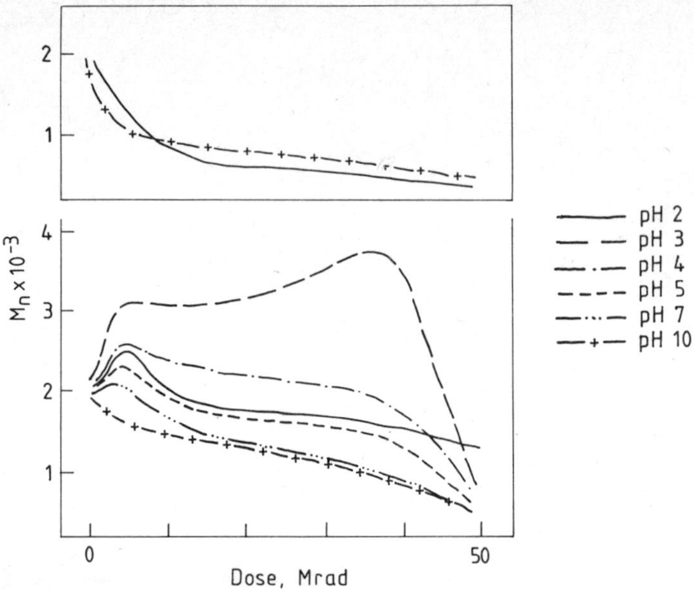

Figure 8 M_n *vs* dose (100 mg/l FA, top, 1000 mg/l FA, bottom).

would be expected, considering the present results. However, a loss of TOC in the order of 0.06 mg/l (equivalent to about 0.1 mg/l FA) would be expected, assuming an energy absorption corresponding to a G value of 5. Since observed FA-concentrations in deep and old groundwaters are approximately 0.001-0.1 mg/l, the abiotic degradation of the FA due to the background radiation can not a priori be disregarded, but it should be pointed out that there is not necessarily a linear dose-response relationship down to natural FA concentration levels. Also, the G-value may be entirely different at the low dose rate expected in natural systems. Formation of larger aggregates with time appears, however, improbable.

Acknowledgements

The gamma irradiation was performed at the Department Nuclear Chemistry at Chalmers University of Technology in Göteborg, Sweden, which is gratefully acknowledged. The authors are also indebted to the Swedish Nuclear Fuel and Waste Management Company for their financial support.

References

1. Mantoura, R.F.C. and E.M.S. Woodward. Geochim. Cosmochim. Acta **47**:1293 (1983).
2. Thurman, E.M. and R.L. Malcom. In: R.F. Christman and E.T. Gjessing, Eds, Aquatic and Terrestrial Humic Materials, pp. 1-23 (Ann Arbor: Ann Arbor Sci., 1983).
3. Petterson, C., I. Arsenie, J. Ephraim, H. Borén and B. Allard. Sci. Tot. Environ. **81/82**:287 (1989).

4. Petterson, C. and B. Allard. In this volume.
5. Senesi, N., Y. Chen and M. Schnitzer. Fuel **56**:171 (1977).
6. Yamasaki, M., T. Sawai and T. Sawai. Radiat. Phys. Chem. **18**:761 (1981).
7. Ephraim, J.H., H. Borén, C. Petterson, I. Arsenie and B. Allard. Environ. Sci. Technol. **23**:356 (1989).
8. Wigilius, B., B. Allard, H. Borén and A. Grimwall. Chemosphere **17/10**:1985 (1988).
9. Haissinsky, M. Nuclear Chemistry and its Applications (Reading: Addison Wesley Publ. Comp., 1964).
10. Friedlander, G., J.W. Kennedy and J.M. Miller. Nuclear and Radiochemistry (New York: John Wiley & Sons, 1964).
11. Choppin, G.R. and J. Rydberg. Nuclear Chemistry. Theory and Applications (Oxford: Pergamon Press, 1980).

Coloured Substances in Swedish Lakes and Rivers - Temporal Variation and Regulating Factors

Tord Andersson, Åke Nilsson and Mats Jansson
Dept of Physical Geography, University of Umeå, S-901 87 Umeå, Sweden.

Abstract

Three large, nationwide sets of data on water chemistry in lakes and rivers, water discharge and groundwater levels were used in a statistical evaluation of the temporal and regional variation of the concentration of coloured substances in Swedish rivers and lakes during the past twenty years. Statistical analysis of collected data revealed remarkable temporal trends. During the study period the concentration of coloured substances increased markedly in both running waters and lakes. Periods of high colour coincided with periods of high precipitation and high runoff. However, even after accounting for changes in runoff by a linear regression model, the concentration of coloured substances in running waters showed an upward trend.

Aim

This work was aimed at examining how and to what extent the concentration and amount of coloured substances in Swedish lakes and rivers have changed during the past two decades. Special emphasis was put on regional variations and the evaluation of possible causes of the observed spatial and temporal patterns.

Materials and Methods

This study utilized water chemistry data from three data bases, here called:

RWDB; Running Waters Data Base (n=18)

LLTV; Lakes - Long Term Variation (n=283)

LSTV; Lakes - Short Term Variation (n=168)

Data on water discharge and precipitation were supplied by the Swedish Meteorological and Hydrological Institute (SMHI). For the evaluation of the importance of precipitation, data from 22 stations, evenly distributed within the selected area (Fig. 1a), were interpolated.

Data were processed using four data programs: LOTUS 1-2-3 and DBASE III for data handling, SURFER for making maps and STATGRAPHICS for statistical treatment. The relative change of absorbance in rivers during the study periods was calculated using simple regression models. Correction for changes in runoff were made by the multiple regression model (1)

$$Abs_t = c_1 + c_2Q_t + c_3Q_{t-1} + c_4 t \qquad (1)$$

where Abs_t and Q_t denote the observed absorbance and the monthly mean discharge during month t. $F > 4$ was used as criterion for parameter acceptance. Model residuals as well as slopes were tested against the water chemistry parameters given in Table 3. Tests on autocorrelation and effect of seasonal variation were made at two stations: River Råne and River Ätran.

Running Waters

The RWDB is based on monthly sampling carried out within the Programme for Environmental Quality Control (PMK) administrated by the National Swedish Environmental Protection Agency (SNV). This monitoring was partly initiated as early as 1965. The chemical parameters are the same as those in Table 3 (except for tot.-Al) and were analysed according to standard methods. The concentration of coloured substances is expressed as absorbance at a wave-length of 420 nm and a path-length of 5 cm. Before analysis, the water samples were filtered (membrane, 0.45 μm).

The analysis of the data material concerning running waters was focused on drainage areas with little industrial activity and negligible river regulation. The geographical distribution of the sampling stations is shown in Fig. 1a. The distribution together with the lenghth of sampling period governed the selection of stations. Monthly sampling had to be carried out for at least 15 years (1972-1986). A few stations were selected to give a longer time perspective (1965-1986).

Lakes

The LLTV data base covers a rather long period of time, from 1972 to 1987, making comparisons with the RWDB possible. The lakes in the LLTV register have a median size of about 2 km^2 . Sampling and analysis is normally carried out by county administrations, and the concentration of coloured substances is measured as colour value (mg Pt/l). Sampling frequency is much lower in LLTV than in RWDB and LSTV. Due to these qualitative shortcomings no detailed studies of regional variations and possible mechanisms were performed on LLTV.

The LSTV data base consists of regularly sampled data (2-4 times/year). However, this program started in 1983 and is consequently not appropriate for studying long-term changes but more to elucidate possible mechanisms and regional variations. The average lake-size in this register is 1.4 km^2 with a median of 0.5 km^2.

The LSTV-lakes are a part of the PMK programme and the same chemical parameters (Table 3) and methods were used as for the RWDB. The geographical distribution of lakes is shown in Fig. 1a (LSTV-register) and Fig. 1b (LLTV-register).

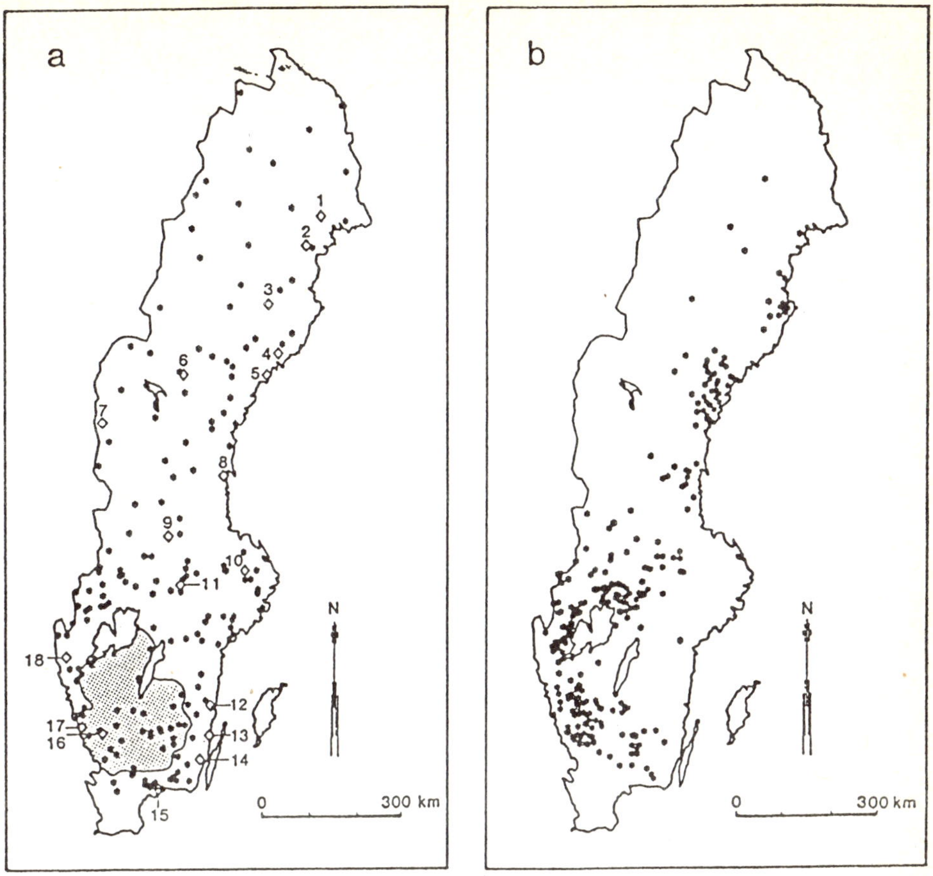

Figure 1 (a) Geographical distribution of the lakes sampled during the years 1984-87 (stars, n=168) and sampling stations for running water (diamonds, n=18). The shaded area denotes lakes included in the evaluation of possible relationships between colour and precipitation. (b) Geographical distribution of the lakes sampled during the years 1972-85 (n=283).

Results and Discussion

Running Waters

Table 1 summarizes the most important results concerning changes in the content of coloured substances in running waters during the past two decades.

It is obvious that the monthly means of absorbance have increased during the period 1972-1986. The relative increase varied between 12% and 150% (Table 1) with the largest increases in the smallest drainage areas. However, it appears as if this increase was a part of long-term variations. This was most obvious at station 12 (Fig. 2a). Similar patterns (Fig. 2b) were found in most other drainage areas. The results emphasized the importance of long-term monitoring when studying temporal variability and trends.

Fig. 2 demonstrates that the colour increase was most marked during the late 1960's, the late 1970's and the mid 1980's. This pattern of variation is strongly related to the runoff (Fig. 3) and consequently also to precipitation. When the colour increase was corrected for temporal variation of monthly mean discharges (dAbs corr, Table 1) according to (1), seven stations showed no significant increase ($p<0.05$). However, there still remained an average increase of 28%, which shows that monthly mean runoff alone cannot explain the colour increase.

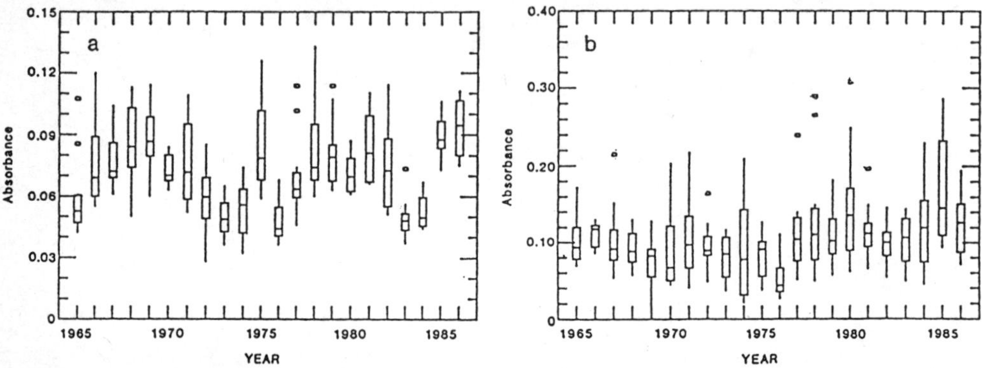

Figure 2 Yearly variation of absorbance 1965-1986 in: (a) River Botorpsström, stn 12; (b) River Ätran, stn 16. Median, lower and upper quartile values, min and max values and outliers are shown.

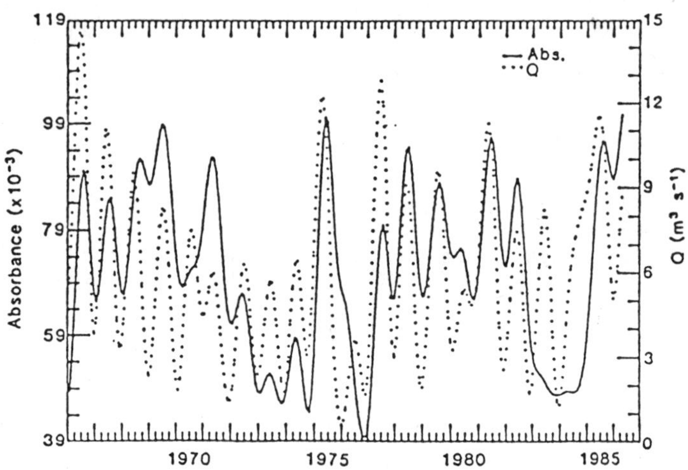

Figure 3 Monthly discharge (Q) and absorbance during the period 1965-1986 in River Botorpsström

Tests on autocorrelations were performed on stations 1 and 16 which both have a large colour increase. Station 1 also has the largest seasonal discharge variation of the 18 studied rivers. The partial autocorrelation coefficients are small (Table 2) showing that seasonal variation of absorbance does not have a constant period length. Adjustment for seasonal variation changed the results (Table 1) for these two stations with less than 5 %.

Table 1 Drainage area, mean runoff and relative increase in the content of coloured substances; absolute (dAbs) and flow corrected (dAbs corr) increase during the two investigated periods.

Stn	River	Area	Q	dAbs 1972-86	dAbs corr 1972-86	dAbs corr 1965-86
		km^2	m^3s^{-1}	%	%	%
1	Råne älv	3768	37.8	48	31	46
2	Pite älv	10797	161.0	23*	16*	
3	Vindelälven	9900	136.7	57	28*	
4	Öre älv	2880	32.3	38	29	
5	Gide älv	3425	36.0	48	27	
6	Ammerån	2463	32.9	37	20	
7	Ljusnan	340	6.5	18*	14*	
8	Delångerån	1992	16.7	93	47	
9	Västerdalälven	8493	146.3	84	74	22
10	Sävjaån	25		150		
11	Arbogaån	88		88		
12	Botorpsström	983	5.7	36	8*	0
13	Emån	4446	26.9	12	25	
14	Ljungbyån	720	4.0	67	34*	33
15	Mörrumsån	3374	23.3	75	8*	18
16	Ätran	2442	31.2	79	45	46
17	Viskan	2160	34.1	30	26	
18	Örekilsälven	1321	23.3	45	23*	
	Mean			57	28	

* Increase not significant at 95% level

Table 2 Partial autocorrelation coefficients for Abs at different "lags" (months) in River Råne (PAC_1) and River Ätran (PAC_{16}).

Lag	PAC_1	PAC_{16}	Lag	PAC_1	PAC_{16}
1	0.57	0.48	13	-0.03	0.00
2	0.18	0.10	14	0.00	-0.09
3	0.09	0.01	15	-0.08	-0.03
4	0.10	-0.07	16	0.00	-0.08
5	0.01	-0.01	17	-0.13	0.02
6	0.07	0.07	18	-0.01	-0.03
7	0.05	0.06	19	0.04	-0.02
8	0.07	-0.03	20	0.09	-0.11
9	0.03	0.03	21	0.02	0.04
10	-0.01	0.16	22	0.10	0.00
11	0.15	0.09	23	0.04	0.08
12	0.20	0.12	24	0.08	0.16

The flow correction model (1) residuals as well as the slope were tested against all other chemical parameters (examplified with pH in Fig. 4) but no significant relationships were found. The geographical distribution did not show any distinct pattern besides a tendency towards lower increase in the eastern parts of southern Sweden.

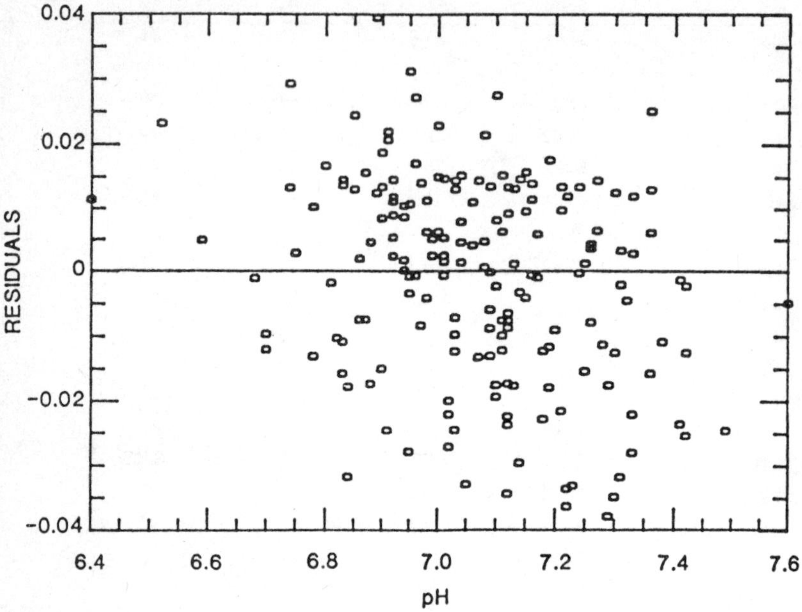

Figure 4 Flow correction residuals for Abs according to (1) *vs* pH in River Botorpsström.

The main reason for the marked increase in absorbances seems to be the increased precipitation and related changes in hydrological conditions between early 1970's and mid-1980's. In most Swedish areas the summers in particular were extremely dry during the initial stage of the study period [1]. This fact is also reflected in low groundwater levels during that time [2].

It is also clear that the variation in the monthly mean discharges alone cannot explain the whole temporal variation of Abs. The results suggest that a more thorough study based on continuous measurements of the hydrological dynamics and detailed knowledge about the drainage area (for example land-use) would supply the information needed.

Long-Term Variations in Lakes

According to the LLTV register (early 1970's to mid-1980's) lake colour has increased with, on average, 20 mg Pt/l. The geographical pattern is shown in Fig. 5. A marked increase (about 100 %) has taken place in large areas of northern Sweden and in southernmost Sweden. In central Sweden there is also an increase, but less pronounced. The average relative increase is approximately the same as in running waters.

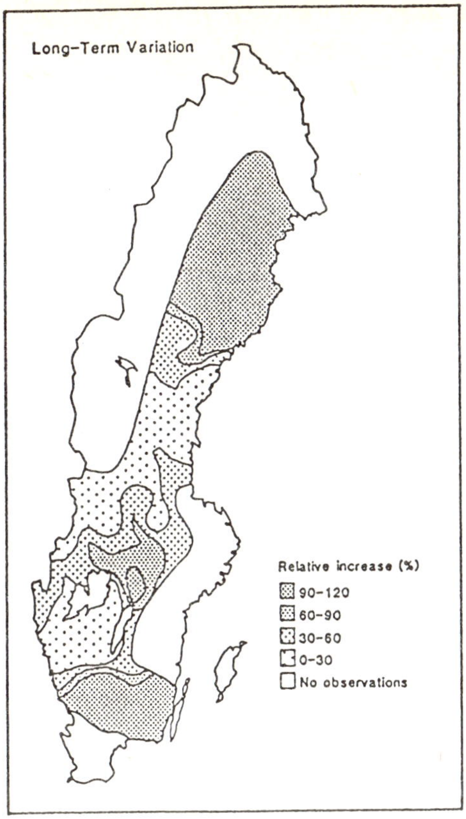

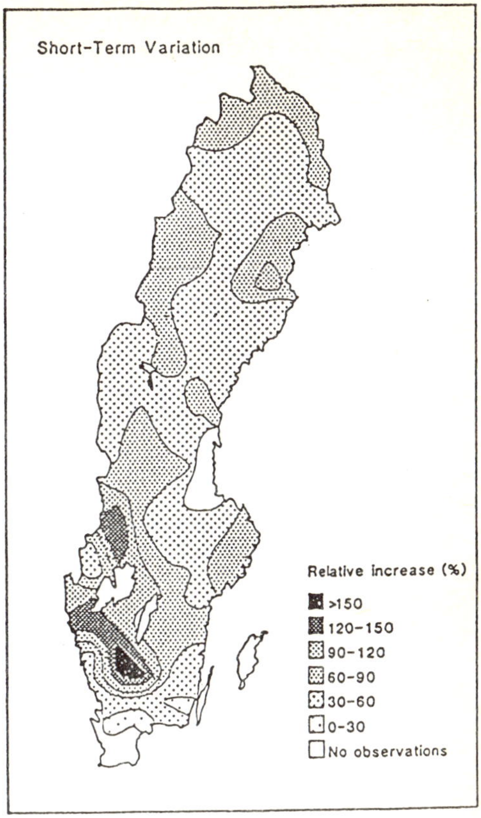

Figure 5 Relative colour change in Swedish lakes between early 1970's and mid-1980's (n=283).

Figure 6 Relative change in absorbance (filtered water, 420 nm) in Swedish lakes during 1984-87 (n=168).

Short-Term Variations in Lakes

Variations of coloured material in lakes during the past few years have been evaluated with the LSTV-register. The chemical analyses of lake water are summarized in Table 3. They are based on annual mean values (2-4 analyses/lake).

The geographical distribution of the change in absorbance can be seen in Fig. 6. The most pronounced increase (>150 %) has taken place in a small area in southern Sweden.

The most significant change (between 1984-87) for any lake water parameter is the increase in absorbance ($p<10^{-11}$). As can be seen from the frequency distribution (Fig. 7) there is a peak between 25-75% increase and the mean increase is 66 %.

Table 3 Lake water characteristics. Mean and standard deviation for 1984 and 1987.

Parameter	Unit	Mean 1984	SD	Mean 1987	SD
pH		6.2	0.7	6.2	0.8
H^+	µeq/l	2.9	6.7	3.5	7.5
NH_4-N	µg/l	36	40	38	58
NO_3-N	µg/l	71	78	73	73
NO_2-N	µg/l	2	1	2	1
Org.-N	µg/l	363	163	402*	186*
Tot.-N	µg/l	472	221	523*	257*
PO_4-P	µg/l	2	2	3	2
Res.-P	µg/l	7	5	9	6
Tot.-P	µg/l	9	7	12	8
Cond.	mS/m	5.7	2.9	5.1	2.6
Ca^{2+}	meq/l	0.22	0.17	0.20	0.15
Mg^{2+}	meq/l	0.10	0.05	0.09	0.04
Na^+	meq/l	0.16	0.10	0.14	0.09
K^+	meq/l	0.016	0.007	0.016	0.007
Alkalinity	meq/l	0.11	0.16	0.10	0.15
$SO_4$2-	meq/l	0.17	0.10	0.15	0.09
Cl^-	meq/l	0.14	0.13	0.12	0.10
Abs	420/5	0.08	0.08	0.13	0.11
$KMnO_4$	mg/l	33	23	38	24
Tot.-Si	mg/l	1.6	1.0	1.7	1.0
Tot.-Fe	mg/l	0.34	0.55	0.48	0.73
Tot.-Mn	mg/l	0.063	0.073	0.057	0.068
Tot.-Al	mg/l	0.12	0.14	0.13	0.12

* value refers to 1985

The correlation between relative change in absorbance and other chemical parameters (1984-87) is rather weak. The strongest correlations ($p<0.001$) are with the parameters expected to be related to colour; namely $KMnO_4$ ($r=0.48$) and Fe ($r=0.46$). There also exists correlations to the change in Org.-N ($r=0.26$) and Tot.-N ($r=0.30$). The study period for these two parameters is 1984-85 due to problems related to a shift of the analytical method of Tot.-N during 1986-87. Among other parameters, Tot.-Si showed the strongest correlation to Abs ($r=0.41$).

Only a very weak positive correlation to the relative change in the pH can be seen ($r=0.17$, $p<0.05$), which is of interest since it has been proposed that acidification may increase the outflow of coloured material [3]. The change in proton concentration was grouped in classes of approximately the same size. Their relation to the change in absorbance is shown in Fig. 8. It is obvious that the median is nearly the same, despite the change in pH. Thus nothing in this material indicates that the observed colour increases are connected to acidification processes. On the other hand this cannot be excluded merely based on correlations; one reason being that the buffer capacity and complex binding capacity of coloured substances may to some extent prevent changes of pH and elements related to acidification.

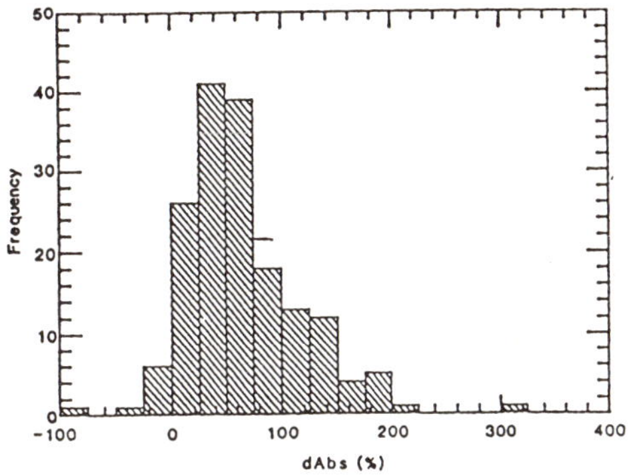

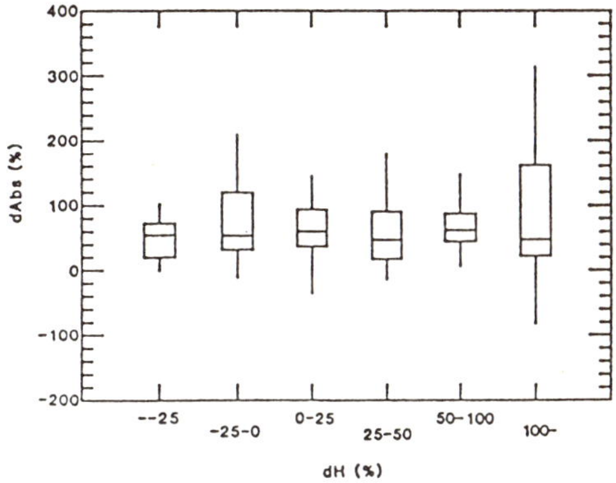

Figure 7 Frequency distribution for relative change in absorbance (filtered water, 420 nm) in Swedish lakes 1984-1987.

Figure 8 Relationship between relative changes in absorbance and the pH. Median, lower and upper quartile values, min. and max. values and outliers are shown.

In running waters, the change in colour was related to changes in runoff. To check if climatic variations are also a governing factor in lakes, an area in SW Sweden where the relative change of colour has a marked gradient (Fig. 1a, Fig. 6) has been used to evaluate the importance of precipitation. Only the years 1984 and 1985 have been used because the most dramatic colour increase during the period of observation took place during these years.

There is a significant correlation between the relative change in Abs (dAbs, 1984-85) and yearly (Sept 84 - Aug 85) precipitation deviation from the normal ($dPrec_{year}$, %):

$$dAbs = -19 + 3.1\ dPrec_{year} \qquad (2)$$

($r^2=0.26$, $p<10^{-2}$, n=30)

It is important to point out that, according to (2), Abs would not increase if the precipitation was normal. It can also be worth mentioning that a one hundred percent increase in precipitation would imply an almost threefold increase in Abs.

If instead the maximum monthly deviation ($dPrec_{max}$, %) during the same period is used, a much more pronounced positive relationship to dAbs is found (Eqn. 3, Fig. 9).

$$dAbs = -71 + 0.81\ dPrec_{max} \qquad (3)$$

($r^2=0.53$, $p<10^{-5}$, n=30)

This indicates that climatical extremes with high precipitation are more effective in enhancing the Abs value than a high yearly mean.

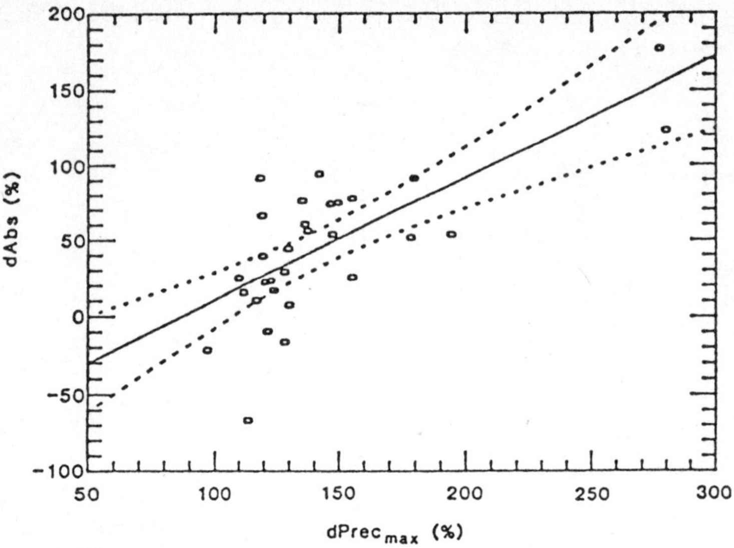

Figure 9 Relationship between relative change in absorbance and maximum monthly deviation in precipitation from the mean during Sept 84 - Aug 85.

The statistical evalutations made in this study have shown dramatic increases in water colour in Swedish surface waters during the past fifteen years. The most obvious findings can be summarized:

- The concentration of coloured substances in running waters has increased with on average more than 50% during the period 1972-1986. In lakes the increase between early 1970's and mid-1980's is between 20 and 120%.

- No clear geographical pattern of variation was found in running waters. Lakes in southernmost Sweden and parts of central and northern Sweden showed a mean increase of more than 100%.

- The main reason for increased colour both in rivers and lakes is an increase in precipitation and runoff. Periods of intensive rain are especially important in this context.

- Even after accounting for changes in discharge, the concentration of coloured substances showed an increasing trend in half of the investigated rivers.

Humic soils in the coniferous forests of Sweden are the dominating source of colour in the investigated rivers and lakes. Considering the great impact of precipitation and runoff fluctuations demonstrated in this study, it is obvious that the water balance and factors that affect the hydrology of such soils must be considered when soil losses of colored substances are discussed. For example, it is possible that measures intensively used in modern Swedish forestry can influence the relationships obtained in this study and to some extent also account for the increase in colour which was not explained by runoff. Examples of such measures are clear-felling, which increase the runoff [4], and ditching, which can increase the leaching of organic material [5].

Acknowledgement

This study was financially supported by the Swedish Environmental Protection Agency. We thank Dr Kjell Johansson for valuable assistance and suggestions throughout the study.

References

1. Eriksson, B. The precipitation and humidity climate of Sweden during the vegetation period, SMHI RMK 46, 73 pp. (1986) (in Swedish).
2. Swedish Geological Research (SGU). The groundwater net. Reports and messages 43. 115 pp. (1985) (in Swedish).
3. Forsberg, C. and S. Löfgren. The water of River Dalälven 1965-1986, water quality and element transportation. The County Board of Kopparberg, 50 pp. (1988) (in Swedish).
4. Grip, H. Changes in runoff after clear-felling. SST, **2**:43 (in Swedish, 1987).
5. Magnusson, T. Effects of forest- and wetland ditching on the acidification of surface waters. Swedish Environmental Protection Agency, PM 1626, 72 pp. (1982) (in Swedish).

Large-Scale Pattern of Mor Layer Degradability in Sweden Measured as Standardized Respiration

Ewa Bringmark and Lage Bringmark
Swedish Environmental Protection Agency, Box 7050, S-75007 Uppsala, Sweden

Abstract

The degradability of forest mor material was described using simple respiration determinations at 20°C under standardized laboratory conditions. A significant geographical pattern was revealed. For mesic coniferous forest sites on podsols, the respiration values were doubled from southern to northern Sweden. This geographical pattern could neither be explained by the temperature response of the respiration nor by the level of pollution. However, there was a clear relationship between standardized respiration values and average temperature parameters at the investigated sites.

Introduction

Within the Swedish Environmental Monitoring Programme (PMK), co-ordinated chemical and biological investigations of, among other things, atmospheric deposition, vegetation, soil, groundwater, and stream water, are performed in small drainage basins in forested and alpine areas. At present, the spatial distribution of the different variables are being analysed. Ultimately, time series of observations will also be used for detecting long-term changes in environments characterized by a considerable natural variability. This paper is focused on standard respiration, which is a parameter that may be regarded as a measure of the biological degradability of the humus material.

Methods

The humus layer (mor) in "homogeneous" soil plots within the drainage basins is sampled in the autumn. Roots and larger particles are removed from the collected cores, and aliquots (20 g) of the remaining fine material are stored at 20°C for 12 days to allow microbiological activity reach a base level. Respiration is then determined by incubation for 20 h in closed 1 l vessels in which NaOH is present in separate small cups. $BaCl_2$ is added to the NaOH, which is titrated with HCl to the endpoint of phenolphthalein. Respiration is calculated as the differences compared to blanks incubated without soil. In 1988, additional respiration measurements were conducted at 10°C for the determination of the Q_{10} factor, *i.e.* the factor by which the respiration is amplified when the temperature is increased by 10°C.

Results and Discussion

Relationship with Latitude

A very significant north-south gradient was found for standard respiration in the mor layers of Swedish coniferous forests (Fig. 1). The close relationship with latitude is valid for sites having mesic pine and spruce forests and, in one case, a beech forest. All these sites are on acidic, podsolised soil. Some northern alpine sites did not conform with the general north-south gradient. Observations from 1988 are shown in Fig. 2. The 1987 monitoring programme produced almost identical results. In the south, forests other than mesic sites with podsols have not been investigated. The absence of a distinct mor layer could make comparisons difficult.

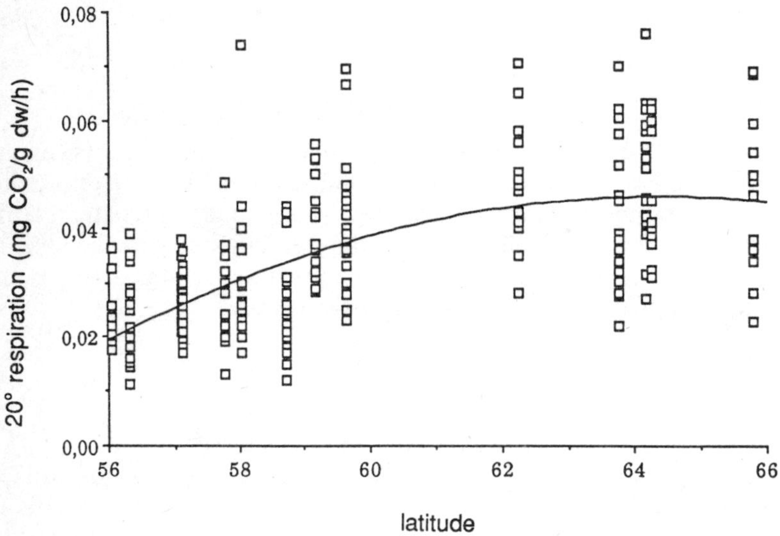

Figure 1 The relationship, in 1987 and 1988, between standard respiration and latitude. Individual values shown (r=0.64***).

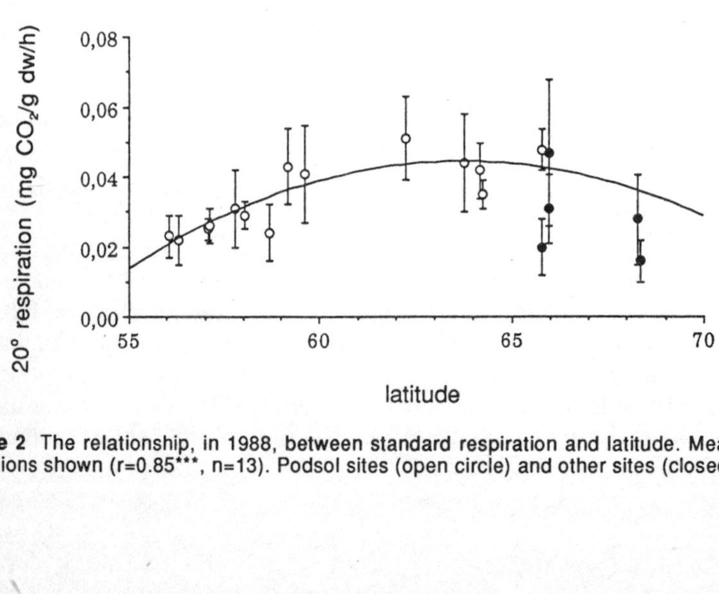

Figure 2 The relationship, in 1988, between standard respiration and latitude. Mean values and standard deviations shown (r=0.85***, n=13). Podsol sites (open circle) and other sites (closed circle)

Temperature Response

The reaction of respiration on instant temperature change was slightly stronger in the south compared to the north. In Fig. 3, this is shown as a somewhat higher Q_{10}-value in low latitudes. The Q_{10}-value for soil respiration is generally between 2 and 3. Higher values have been reported from tundra sites [1], but, in the present study, there were no elevated values for northern forest sites.

The temperature of 20°C for the respiration mesurements is higher than in the usual field situation. However, as Q_{10} does not increase to the north, this cannot explain the observed geographical pattern of standard respiration. In fact, respiration at 10°C measured in 1988 showed a significant north-south gradient, similar to the one at 20°C. This indicates that the observed differences in standardized respiration are not caused by differences in temperature response of the microorganism; they are more likely related to the properties of the organic matter. The absolute value of the respiration at 10°C in northern sites was about the same as at 20°C in southern sites.

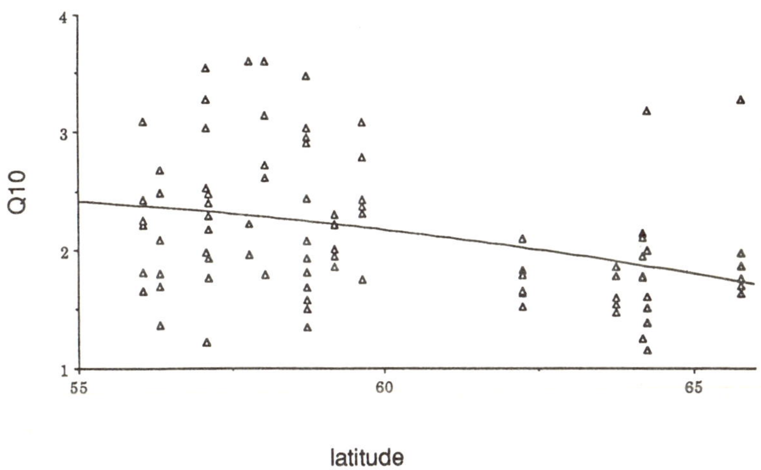

Figure 3 Temperature response (Q_{10}) of respiration in relation to latitude (r=0.35***, n=83). Observations from 1988.

Possible Effects of Environmental Pollution

In the present study, NO_3^- and Pb were selected as indicators of long-range pollution. Since the geographical patterns of different air-borne pollutants in Sweden are often strongly correlated, these parameters may also represent a general level of pollution.

Observed data showed that there was a weak, positive correlation between the standardized respiration and the levels of the selected pollutants (Fig. 4). However, climatic means for temperature show a north-south gradient which is almost identical to that of NO_3^- and Pb. Therefore, the observed correlations may also be explained by climate gradients.

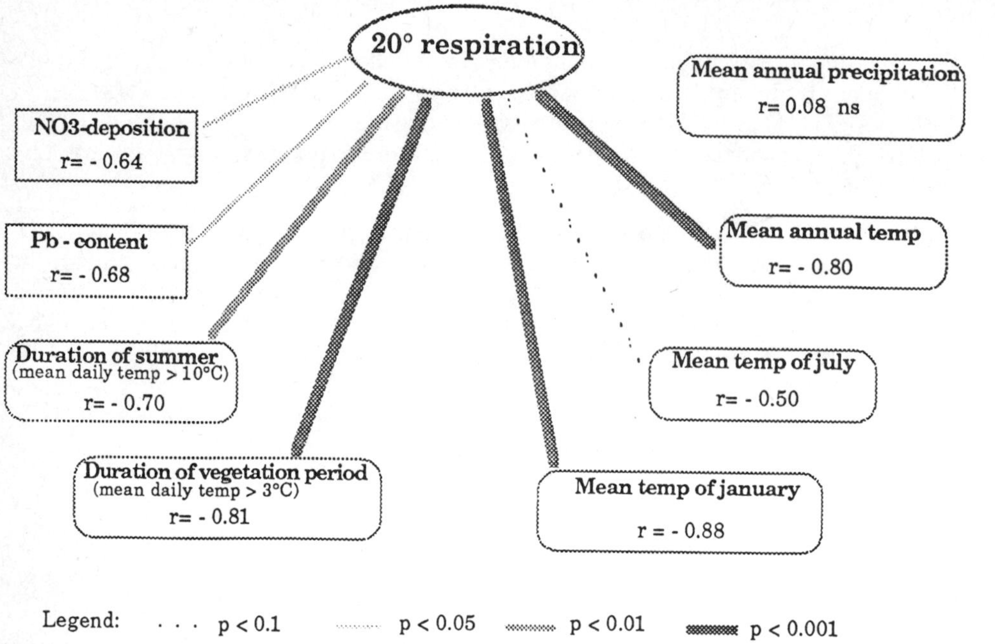

Figure 4 Standard respiration of podsol sites (n=14) in relation to climate and deposition of pollutants. Linear correlation between site means.

Effects of Climate

The site means of standard respiration were combined with climatic means for the period 1901-1930 in a correlation analysis (Fig. 4). The strongest correlations were observed for temperature characteristics of the climate mean, whereas precipitation seems to be less significant. Observed correlations also indicate that differences in winter temperature have a greater impact on standardized respiration than differences in summer temperature.

The mor layers are generally thicker in the south. The organic matter that has developed under different climatic conditions may differ in the content of readily available carbohydrates and the resistence of older fractions [2 - 4]. It is suggested that, although the input of fresh plant litter is larger in the south, and the initial decomposition rate there is higher, a more resistent humus with a higher degree of condensation is developed in the milder climate.

Possible Effects of Nitrogen

There is growing evidence that nitrogen incorporated in the lignin fraction plays an important role in creating resistant organic matter [5,6]. There is generally a larger nitrogen content in southern ecosystems, the reasons being both natural and anthropogenic. However, the nitrogen parameters that may be related to standardized respiration have not yet been identified.

References

1. Bunnel, F.L., D.E.N. Tait, P.W. Flanagan and K. van Cleve. Soil Biol. Biochem. **9**:33 (1977).
2. Berg, B. Scand. J. Forest Res. **1**:359 (1986).
3. Bunnel, F.L., D.E.N. Tait and P.W. Flanagan. Soil Biol. Biochem. **9**:41 (1977).
4. Schnitzer, M. and E. Vendette. Can. J. Soil Sci. **55**:93 (1975).
5. Berg, B., B. Wessén and G. Ekbohm. Oikos **38**:291 (1982)
6. Berg, B., H.Staaf and B. Wessén. Scand. J. Forest Res. **2**:399 (1987).

Session 3:
Complex Formation and Interactions with Solids

Ion Binding by Humic Substances: Considerations Based on the Solution Chemistry and Heterogeneity of Humic Substances

James H. Ephraim[1], Michael M. Reddy[2] and Jacob A. Marinsky[3]

[1]Department of Water and Environmental Studies, Linköping University, S-581 83 Linköping, Sweden.
[2]U.S. Geological Survey, Denver Federal Center, Lakewood, USA.
[3]Department of Chemistry, State University of New York, Buffalo, N.Y. 14214, USA.

Abstract

The development and refinement of a two-phase model for describing the potentiometric properties of crosslinked weak acid polyelectrolyte gels and their linear analogues are discussed. Application of this approach has been facilitated by identifying functional group heterogeneity and separate phase effects as the major complications in the description of ion binding by humic substances. By their separation and quantification, equilibrium properties of these systems have been anticipated. This approach may be described as a "discrete ligand" model with solution chemistry effects taken into consideration. Examples of its applications are enumerated and an attempt is made to compare the approach with others.

Introduction

The study of humic substances grew considerably once it was recognized that these organic acids significantly affect the transport of trace metal ions in the environment. Humic substances are currently viewed as an extremely heterogeneous mixture of molecules with a wide range of weak acid functionalities and with molecular weight distributions ranging from as low as several hundred to as high as about 300,000 [1].

The ion binding behavior of humic substances continues to puzzle environmental scientists even though considerable advances towards a better understanding have been made in the past decade. One barrier to a comprehension of the nature of ionic interaction with dissolved organic matter (DOM), (including humic and fulvic acid) is inadequate knowledge of the solution chemistry of the DOM. Another is its functional heterogeneity. Attempts to remedy this situation by a number of researchers have only considered one or the other of these in their description of metal-humate interactions [2-4]. Recently, we introduced an approach which identified both complicating factors. We were able to correct for their separate contributions to the observed properties once the solution properties were treated as a separate phase effect. This permitted a more realistic description of metal-humate interactions [5,6].

In this paper, the development and refinement of graphical approaches employed for describing polyelectrolyte behaviour in solution and their potential for application to humic substances in solution are discussed. The merit of these graphical approaches are carefully assessed.

Theoretical Background: Philosophy of Two-Phase Model

Analysis of potentiometric behavior of a number of polyelectrolytes as a function of counterion concentration using the Gibbs-Donnan model [7-10] has shown that the linear analogs of crosslinked weak acid polyelectrolyte gels exhibit a counterion concentrating region next to their surface that simulates the separate phase properties of the gels [8]. In both systems, the distribution of diffusible components, e.g. H_2O, NaCl, HCl at equilibrium is controlled by their respective electrochemical potentials which are identical at every point.

For example, at equilibrium, the electrochemical potential, μ_i, for each diffusible species in and out of the polymer subphase may be equated. By assigning the same standard state to the components in the two phases so that they cancel, the following important relationship between the activity ratios of exchangeable counterions, e.g. Na^+ and H^+, in the two phases is obtained as long as the osmotic pressure term is small enough to neglect [7].

$$pH - pNa = pH_p - pNa_p \qquad (1)$$

In the above equation, the p's represent species in the polymer subphase. Incorporation of the above equation into the Hendersen-Hasselbalch equation yields the following:

$$pK_{app} - pK_{int} = (pNa - pNa_p) \qquad (2a)$$

where pK_{app} is the negative logarithm of the apparent dissociation constant of the weak acid repeated in the polyelectrolyte and pK_{int} corresponds to its intrinsic (thermodynamic) constant. For a gel, $\{Na\}_p$, the activity of the Na^+ ion, may be equated to $[(V_bM_b + hV_s)\gamma_{\pm,p}]/V_p$ where V_b is the volume of base, M_b its corresponding molarity, h is the hydrogen ion concentration, V_s is the volume of the solution, $\gamma_{\pm,p}$ is the activity coefficient of Na^+ in the gel phase and V_p is the solvent content of the gel in grams. In the case of a linear polyelectrolyte, the relationship between the counterion concentrations in the solution and in the counterion concentrating zone next to the polyelectrolyte molecule surface may be expressed by the Boltzmann Distribution function:

$$Na_p = Na^+(\exp - \varepsilon\Psi_s/kT) \text{ and } H_p = H^+(\exp - \varepsilon\Psi_s/kT)$$

where ε and Ψ_s are the unit charge and the polymer surface potential, respectively. The difference between polyelectrolyte gels and their linear analogs is due to a difference in charge distribution. The separate well-defined phases of the gel-salt systems are electroneutral, the counterion enriched domain and the solution external to it in their linear analogs are not. The net negative charge due to the fractional release of counterions from the polyelectrolyte domain is balanced by the equal positive charge they produce in the solution phase [7,11]. We discuss below how these concepts can be applied to the characterization of humic substances.

Graphical Methods

pK_{app} versus Degree of Neutralization, α

It is common practice to plot pK_{app} *vs* α, the degree of neutralization, when characterising weak acid polyelectrolyte gels and their linear analogs. For titrations performed at different bulk electrolyte concentrations (i.e. ionic strengths) with gels, such plots yield curves which are essentially parallel to each other when the gel is rigid and the change in its solvent content with a is not strongly affected by changes in counterion concentration levels [7]. The separation between the curves at the different ionic strengths (water activities in this instance) is essentially equal to the difference in the logarithm of the bulk electrolyte concentrations, as predicted by eqn 2. For a flexible gel whose solvent uptake pattern is strongly dependent on ionic strength, curves will no longer parallel each other. Instead they can be expected to diverge with the maximum vertical displacement at elevated α values sizeably smaller than the difference in the logarithm of the bulk electrolyte concentration levels. Since the potentiometric response of linear, weak acid polyelectrolytes to counterion concentration levels mimics that of their crosslinked analogs [8,12], flexibility or rigidity estimates are equally accessible from exactly the same kind of pattern assessment.

pK_{app} versus pH + pX

In an earlier attempt to characterise polyelectrolytes, uniqueness of pK_{app} *vs* pH+pX plots as α function of ionic strengths was attributed to rigidity of the polymer molecule [13]. However, re-examination of the approach following comments by Morel and Cabaniss [14] has led to the following conclusions [15]:

1. For a rigid polymer, pK_{app} *vs* pH+pX is only unique for different ionic strengths as long as α is less than 0.6. For a flexible polymer, no uniqueness is obtained at all.

2. Total uniqueness of pK_{app} *vs* pH+pX over the whole range of α signifies a contraction of the polymer volume, V_p, by as much as a factor of three. These reconsiderations have led to the conclusion that plots of pK_{app} *vs* pH+pX have only limited usefulness in describing the rigidity of polymers.

pK_{app} versus pH

Earlier claims that uniqueness of pK_{app} *vs* pH at different ionic strengths signified that the polyelectrolyte was hydrophobic (impermeable) has been correctly criticised by Morel and Cabaniss [14]. Further consideration [15] has revealed that insensitivity of pK_{app} *vs* pH to ionic strength occurs under these conditions:

1. The molecule is essentially monomeric (as pointed out by Morel and Cabaniss [14]).

2. The slope of the pK_{app} *vs* pH curve for the polymer approaches unity.

On the basis of these further considerations, assignment of hydrophobicity (impermeability) based only on the uniqueness of the pK_{app} *vs* pH curve is totally unacceptable.

α versus (pH-pX) or (pH-pM)

Careful re-examination of the model has shown that greater insight with respect to the conformational properties of polyelectrolyte gels and their linear analogs may be obtained from plots of α *vs* pH-pX for polyanionic or pH+pX for cationic polymers. From such plots, absolute rigidity, conformational stability and variable conformational response at different water activities may be obtained [15].

Plots of α *vs* pH-pM will be unique at different ionic strengths for as long as the polymer remains rigid. Any separation between these curves at different ionic strengths is indicative of changes in the volume of the polymer. If these plots at different ionic strengths are parallel and remain parallel, then it signifies the fact that ionic strength have effect only on the initial volume of the polymer. A detailed description is given elsewhere [15].

Application of Graphical Approaches to Fulvic Acid Systems

Plots of pK_{app} *vs* α for the fulvic acids studied by us [6,16,17] indicate that these molecules behave like polyelectrolytes in solution. These plots essentially parallel each other when $\alpha \geq 0.5$ but tend to converge when α approaches zero. Examples of such plots are shown for the soil-extracted Armadale fulvic acid and the surface water extracted Bersbo fulvic acid in Figs. 1 and 2 respectively.

Plots of α *vs* pH-pNa at different ionic strengths for Finnsjön FA (Fig. 3) show that the shape of the curves (which are parallel to each other) is similar to those obtained for known synthetic gels and their linear polyelectrolyte analogs [7,8]. These observations are, however, difficult to rationalize since the molecular weight of fulvic acids are very much lower than those of typical polyelectrolytes; compare the low number average molecular weight of ~650 measured for the Armadale fulvic acid in vapor pressure osmometry studies [18]. The fact that the magnitude of the pK_{app} is susceptible to counterion concentration levels even though its molecular weight is so low cannot be dismissed. The most reasonable rationalization of this property of the small fulvic acid molecule has to reside in its hydrophobicity. It has been determined that the presence of fulvic acid in solution markedly enhances the solubility of organic compounds. Such significant solubility enhancement of relatively water-insoluble organic substances has been described in terms of a partition-like interaction [19] with the fulvic acid molecule present as a separate discrete entity. Such hydrophobicity has been demonstrated as well by its effective adsorption from solution by methylmethacrylate resin [20]. Advantage has been taken of this property of the fulvic acid molecule in procedures developed for its separation from natural waters.

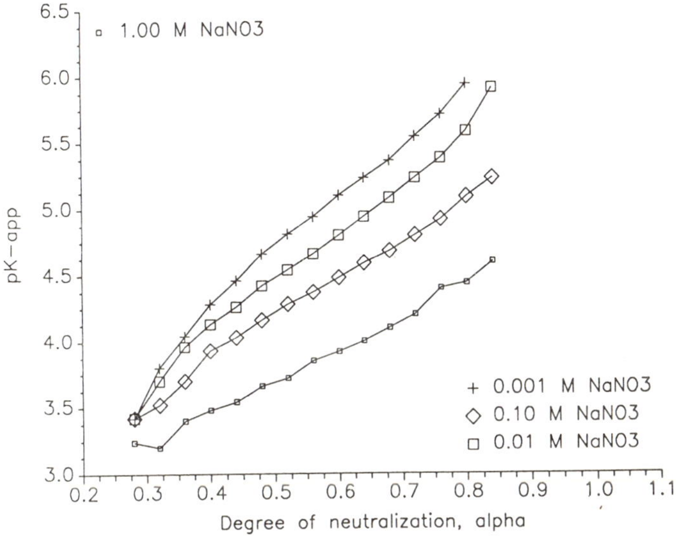

Figure 1 Plots of the apparent acid dissociation constant for Armadale Horizon Bh fulvic acid as a function of the degree of neutralization at 25°C at different background electrolyte ($NaNO_3$) concentrations.

One can envision the formation of a microphase to be facilitated by the hydrophobic property of the FA molecule. In this situation, the pH response to counterion concentration levels will be a reflection of the accessibility of its charged surface potential, ψ_s to the pair of ions, M^+ and H^+, in solution. The greater accessibility of the less solvated counterion to the charged surface of the fulvic acid molecule can be presumed to account for the fact that the vertical displacement of the pK_{app} *vs* α plots, where they parallel each other, is much smaller than the counterion concentration ratio of the fulvic acid solution yielding these results. One would expect the hydrophobic molecule to keep solvent uptake at a small, reasonably constant value so that the alternate explanation of the buffered pK_{app} response to a decrease in the counterion concentration level of the FA solution, by the accompaniement of a sizeable increase in the water content of the macrophase, is highly unlikely. If, however, the source of the observed response of pK_{app} to changes in counterion concentration levels in the $\alpha \geq 0.5$ region is a consequence of the rapid change in the solvent uptake by the counterion concentrating region next to the surface of the FA molecule, advantage can be taken of eqn 2, the Gibbs-Donnan based expression presented earlier, to estimate the magnitude of this quantity.

$$pK_{app} - pK_{int} = (pNa_s - pNa_p) = \Delta pK \qquad (2b)$$

To take advantage of this alternate view, ΔpK at each ionic strength is obtainable from the difference between pK_{app} at the particular ionic strength and the pK_{app} at the highest ionic strength, usually at I=1.0 M where the sensitivity of pK_{app} to changes in salt concentration becomes negligibly small for the fulvic acid samples. The computation of the sodium ion concentration inside the polymer subphase, pNa, becomes feasible and this can facilitate the computation of the solvent content of the polymer domain as ionization proceeds and at the different ionic strengths. The $[Na^+]_p$ content of the polymer domain is equal to the product of the total quantity dissociated, ΣA^- and $(1-f_{Na})$ where f_{Na} is the fraction of Na^+ ion that escapes from the polymer domain at each experimental α. It is calculable by dividing the measured concentration of Na^+ ion in the salt-free fulvic acid samples as a function of α, $C_{exp,\alpha}$, by the stoichiometrically-based estimate of its concentration at that α, $C_{comp,\alpha}$. For this purpose one need only divide the quantity of base added by the volume of the resultant solution.

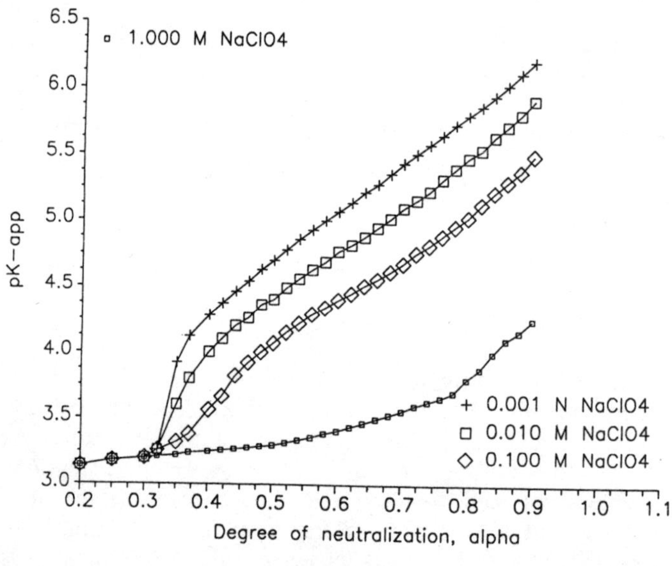

Figure 2 Plots of apparent acid dissociation constant for Bersbo FA versus the degree of neutralization at 25°C as a function of ionic strength ($NaClO_4$) FA=314 ppm. (A combination of Figs 1 and 3 of ref. 16).

With this approach, the volume of the polymer domain is calculable with eqn 3:

$$\log\{V_p/\gamma_p\} = \log\{A^{\cdot}V^2(1-f_{Na})/(m_{Na}V_i\gamma_s)\} \qquad (3)$$

where V_p is the volume of the polymer domain, γ_p is the activity coefficient of the counterion in the polymer domain, V the total volume of the solution, V_i the initial volume of solution, γ_s activity coefficient of the counterion in the bulk electrolyte and m_{Na} is the molarity of sodium ion in the bulk solution.

The most suitable values for V_p and γ_p are then obtainable by an iterative procedure which is initiated by arbitrarily selecting a γ_p value [7]. The ionic strength value that evolves $\{(\Sigma A^{\cdot}[2-f_{Na}])/(2V_p)\}$ is used to define the ionic strength of the polymer domain. Eqn 4, given below, is then used to estimate γ_p at that ionic strength.

$$\gamma_p = (\gamma_{\pm,NaCl})_{Ip}^2/(\gamma_{\pm,KCl})_{Is} \qquad (4)$$

Application in this way of the mean molal activity coefficient reported in the literature for NaCl and KCl at I_p presumes that $\gamma_{K^+} \approx \gamma_{Cl^-} = \gamma_{KCl}$ at the polymer domain molalities encountered. Their transport numbers are nearly equal over a considerable concentration range [21] implying that their hydrated ionic radii are practically the same as well. This lends some support to the extra thermodynamic assumption inherent in eqn 4. The additional assumption that $\gamma_p = \gamma_s$ when $I_p = I_s$ that is also needed does not appear to be unreasonable as well.

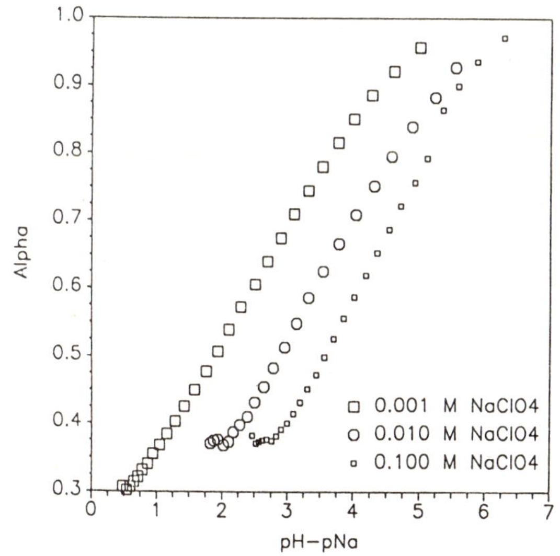

Figure 3 Plots of degree on neutralization versus pH-pNa for Finnsjon FA as a function of ionic strength. FA = 222ppm; bulk electrolyte = $NaClO_4$; Temperature = 25°C.

If the calculated γ_p value is different from the arbitrarily selected value a second γ_p value, based on the one resolved in the first set of operations, is chosen to repeat the cycle. When the two values converge, V_p has been resolved.

An important advantage of the two phase model derives from the capability it provides for separating the perturbations due to "separate phase" (ionic strength effects) and functional group heterogeneity. Fig. 4 shows the various stages of the program used in the application of this approach to metal-humate systems.

The separate phase effect is identified and quantified by performing acid-base titrations as a function of ionic strength from I=0.001 M to as high as I=5.0 M. As the ionic strength increases, the dependence of the resolved pK on the ionic strength is such that the values eventually converge when an ionic strength of 1.00 M is reached. The rise in pK_{app} with α that characterises this limiting curve ($I \geq 1.000$ M) is due exclusively to the functional heterogeneity of the fulvic acid. Its shape has been used to examine site assignment possibilities (number, intrinsic pK and abundances) and to facilitate analysis of the effect of counterion concentrating domain of the fulvic acid molecule on pK perturbations observed at the lower salt concentration levels. The vertical displacement of the lower ionic strength-based pK_{app} *vs* α curves from the limiting curve, ΔpK, when plotted *vs* α provides an accurate measure of this disturbance.

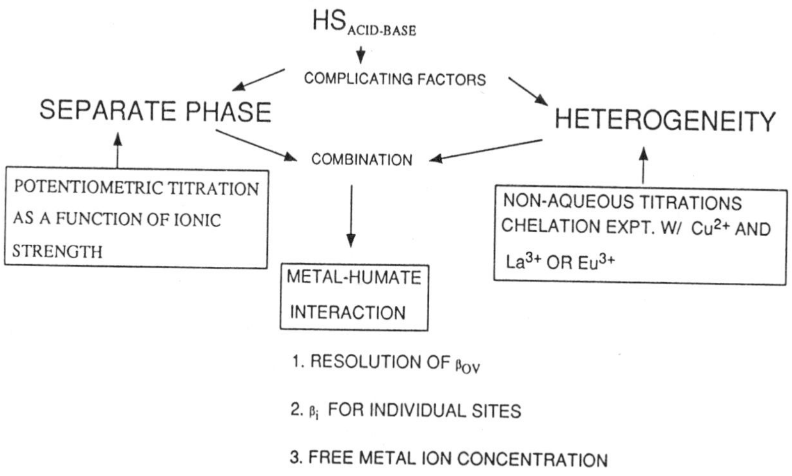

Figure 4 An outline of the stages in the application of the two phase model to humic substances systems.

Functional group heterogeneity of the humic substance is estimated by performing titrations in nonaqueous media with the use of an internal reference compound, para-hydroxybenzoic acid, and by performing titrations in which heavy metal ions, e.g.

Cu(II), and La(III) or Eu(III) are added in varying amounts until present in excess [6,16,17]. Nonaqueous titration data allows estimates of carboxylic acid and -OH group (alcohols and phenols) content. Typical results obtained in nonaqueous titrations of four different fulvic acid samples are shown in Table 1 [6,16,17]. The four FA samples are from different origins; Armadale FA is soil extracted, Suwanee River FA is an aquatic FA, Bersbo FA is from a bog surface water and Finnsjon FA is from a deep groundwater in a granitic bedrock. Small differences are observed in the acid capacities of these fulvic acid samples. For example, the -COOH content for all four samples ranged from 4.78 to 4.15 meq/g FA, while the -OH content ranged from 0.83 to 1.84 meq/g FA. These results suggest that the carboxylic content of fulvic acids are similar and are around 4 to 5 meq/g FA while their corresponding -OH content range from 1 to 2 meq/g FA. This observation may be the consequence of similar methods of extraction [4,20].

Typical results of "chelation" experiments are shown in Table 2. From the FA-Cu(II) titrations, chelating moities with a salicylic acid-like arrangement and a catechol and/or an acetylacetone-like arrangement are indicated. Results of the La(III)/Eu(III)-FA experiments appear to provide confirmation of the participation of catechol and/or acetylacetone-like arrangements. The bog surface water extracted FA, Bersbo, appears to have the highest fraction of chelating moieties while the running water sample, Suwanee FA, has the least.

Table 1 Aqueous and nonaqueous titratable acids for a number of FAs.

	Acid capacity, meq/g FA			
	Aqueous	Nonaqueous		
FA Sample		-COOH	-OH	Total
Armadale	5.60±0.15	4.15±0.05	1.84±0.06	5.99±0.03
Suwanee	5.65±0.03	4.24±0.02	1.41±0.01	5.66±0.03
Bersbo	4.65±0.15	4.78±0.03	1.35±0.03	6.13±0.07
Finnsjön	4.98±0.03	4.16±0.04	0.83±0.03	4.99±0.04

Table 2 Extra acidity obtained from acid-base titrations with excess Cu(II) and La(III).

	Extra acidity expressed as a percentage of total aqueous capacity		
FA sample	Carboxylic	Enol+Phenolic	Total
Armadale	≈25	≈25	≈50
Suwanee	0	≈25	≈25
Bersbo	35-45	25-30	60-70
Finnsjön	5-12	38-40	45-50

Nonaqueous titrations data and titration data obtained in presence of varying quantities of heavy metals [6,16,17] facilitated selection of 4 to 5 acid sites and their abundances assignments which could be used to reproduce the limiting ionic strength pK_{app} *vs* α plot. Results of such an exercise with the four FA samples are presented in Table 3. For all the FA samples except Bersbo FA, the use of four sites reproduced the experimental potentiometric behaviour. In the case of Bersbo FA, five sites were necessary to obtain a good fit. The pK values for all the FA samples ranged from 1.7 to 7.0 [6,16,17]. The functional heterogeneity of a typical fulvic acid sample is, of course, not reproduced from molecule to molecule. These assignments are believed to be averages for each of the four or five weak acid groups encountered in the typical fulvic acid molecule.

After such assignment of intrinsic acid dissociation constants for the most predominant acid sites, metal-humate interaction may be estimated in a number of ways. The first method of estimation uses the overall complex formation function, β_{ov}:

$$\beta_{ov} = \Sigma M_{bi}/\{M_f \gamma_{\pm} 10^{-z\Delta pK}\ \Sigma A^-\} \qquad (5)$$

Table 3 Resolution of predominant acid sites for a number of FAs.

Sample	Site 1 pK_1	Ab_1	Site 2 pK_2	Ab_2	Site 3 pK_3	Ab_3	Site 4 pK_4	Ab_4	Site 5 pK_5	Ab_5
Armadale	1.8	24.5	3.4	30.4	4.2	22.4	5.7	22.7		
Suwanee	2.2	28.0	3.1	25.0	4.3	22.0	5.6	25.0		
Bersbo	1.7	20.0	3.3	25.0	5.0	30.0	6.5	20.0	7.0	5.0
Finnsjön	2.1	20.0	3.0	30.0	4.0	40.0	5.5	10.0		

[a]Ab_i=abundance of the ith site

where ΣM_{bi} is the sum of the metal bound to all the active sites, ΣM_f is the free metal ion concentration, $\gamma_{\pm}$ is the single ion activity coefficient of the metal ion under study, z is the charge on the metal ion, ΔpK is the separate phase effect determined from the metal-free fulvic acid potentiometric titrations and A^-_i is the sum of the dissociated portions of the four or five acid sites at a given pH; it is obtainable from the following relationship:

$$\Sigma A^- = (\alpha_{ov} HA_T - \Sigma M_{bi}) \qquad (6)$$

In the above expression, α_{ov} is the overall degree of neutralization, HA_T is the total fulvic acid in appropriate concentration units. The overall degree of neutralization, α_{ov} be obtained in two ways, i.e. either by using the pK_{app} *vs* pH curve to resolve pK_{app} for use in the Henderson-Hasselbalch equation, pK_{app} = pH-log$\{\alpha_{ov}/(1-\alpha_{ov}\}$ or by computing the degree of neutralization for each site, α_i, ($\alpha_i=1/\{1+10^{(pK+\Delta pK - pH)}\}$) and using the relationship:

$$\alpha_{ov} = \Sigma\alpha_i Ab_i \qquad (7)$$

where Ab_i is the abundance of the ith site. The two methods for computing α_{ov} yield equivalent results. The inclusion of the metal ion concentrating term, ΔpK, in the computation of β_{ov} makes our approach different from the others. It has to be emphasised that the overall complex formation function, β_{ov} is a composite parameter just like the limiting $pK_{(HA)v}$ resolved at $I \geq 1.00$ M. It is not a thermodynamic parameter describing the various binding patterns to the various complexing sites. However, it can be used to compare on a macroscopic level, results obtained from different experimental techniques [22]. Additionally, the overall complex formation function can be used in the three component metal ion-inorganic oxide-humic substances system to quantify the influence of humic substances on the adsorption of metal ions onto these inorganic oxides [23].

The second method for describing metal-humate systems employs calculation of the complex formation constant for each acid site. In a metal-humate system where the pH, ΣM_{bi}, M_f and HA_T are known, the following expression relating the intrinsic formation constant to each complexing site may be written by assuming a one-to-one interaction between the acid sites and the metal ion:

$$\Sigma M_{bi}/\{M_f\gamma_{\pm}10^{-z\Delta pK}HA_T\} = \Sigma\beta_i(\alpha_i Ab_i) \qquad (8)$$

where β_i is the one-to-one complex formation constant to each ith site, α_i is the degree of neutralization of the ith site, Ab_i is the abundance of the ith site. A manipulation of the above expression yields the expression:

$$\beta_{ov} = \Sigma\beta_i(\alpha_i Ab_i)/\Sigma(\alpha_i Ab_i) \qquad (9)$$

Initial estimates of the various β_i's (guided by the results in the nonaqueous and chelation titrations and the literature), in equation 8 may be set up as a matrix. For most of the FA sources a 4x4 matrix would suffice for the iterations which are effected by changing the β_i's until the calculated $\Sigma M_b/M_f$ approaches the experimentally determined $\Sigma M_b/M_f$ [22]. This method of computing site complex formation constants allows association of the complexing functional groups with the assigned acid sites and thus can guide the design of other analytical techniques to facilitate identification of such functional groups [21].

In an earlier approach, use of literature-based stability constants for the Cu^{2+} ion complexed species formed with the different sites deduced from the design of the experimental program led to a capability for predicting the binding of macro quantities of Cu^{2+} to Armadale Horizon Bh fulvic acid as a function of degree of neutralization, α, ionic strength, and metal ion and fulvic acid concentrations [24]. A combination of this procedure with the trial and error iteration may lead to a better assignment of the functional groups constituting the particular fulvic acid under investigation.

A fourth approach provides the capability for predicting the free metal ion concentration of a metal-humate system whose pH, total metal ion and total quantity of humic substance are known [11]. With literature-based assignments of stability constants for the complexes considered to form with the ligands available, trial concentrations of free metal ion, presumed to be in equilibrium with the FA are made and iterations are performed until the molality of the free metal ion selected yields a material balance ($M^{2+}+\Sigma M_{bi}=\Sigma M$). This approach has been used successfully to describe Ca^{2+} and Cu^{2+} uptake by Armadale FA [11,25].

Comparison with Other Approaches

In broad terms, models for describing metal-humate interactions may be categorized as the discrete ligand and the continuous distribution models [26]. Our approach may be categorized as a discrete ligand approach but emphasises the need to incorporate contributions due to ionic strength effects and functional site heterogeneity. Recently, Tipping *et al.* [27] has employed a model also incorporating binding-site heterogeneity and macroionic effects. Philosophically, the approach employed by Tipping *et al.* [27] is identical to ours, i.e. the recognition of "counterion effects" and heterogeneity factors as the perturbations to the acid-base properties of humic substances. However, the number of predominant sites and the method of computing the counter ion correction term are different.

Summary

The protonation and metal ion binding properties of humic substances (fulvic acids) over a range of solution ionic strength has led to a two-phase model. This model asserts that the ionic equilibria of fulvic acid may be described by considering the fulvic acid as constituting a separate microphase in solution. This hypothesis is existent in the literature [19].

The success of this two phase hypothesis mandates the inclusion of ionic strength effects with the functional group heterogeneity of the fulvic acid molecule in modelling metal-humate interactions. The approach outlined in this paper facilitates such an exercise.

Acknowledgement

Financial support from the Swedish Nuclear Fuel and Waste Management Company is gratefully acknowledged.

References

1. Schnitzer, M. and S.V. Khan. Humic Substances in the Environment (New York: Marcel Dekker, 1972).

2. Posner, A.M. J. Soil Sci. **17**:65 (1966).

3. Gilmour, J.T. and N.T. Coleman. Soil Sci. Amer. Proc. **25**:710 (1971)

4. Paxeus, N. Ph.D. Thesis, University of Gothenborg, (1985).

5. Marinsky, J.A. and J. Ephraim. Environ. Sci. Technol. **20**:349 (1986).

6. Ephraim, J., S. Alegret, A. Mathuthu, M. Bicking, R.L. Malcolm and J.A. Marinsky. Environ. Sci. Technol. **20**:364 (1986).

7. Marinsky, J.A., T. Miyajima, E Högfeldt and M. Muhammed. Reactive Polymers, **11**:279 (1989).

8. Marinsky, J.A. and N. Imai. Macromolecules, **13**:271 (1980).

9. Alegret, S., J.A. Marinsky and M.T. Escaleas. Talanta, **31**:199 (1984).

10. Marinsky, J.A. In: W. Stumm, Ed., Aquatic Surface Chemistry, p. 49 (New York: Wiley, 1987).

11. Marinsky, J.A., M.M. Reddy, J. Ephraim, and A. Mathuthu. "Ion binding by humic substances. A Computational procedure Based on functional site heterogeneity and separate Phase behaviour," SKB Technical Report **88-04**, 1988.

12. Nagasawa, M., T. Murose and K. Kondo. J. Phys. Chem. **69**:4005 (1965).

13. Marinsky, J.A. J. Phys. Chem. **89**:5294 (1985).

14. Cabaniss, S. and M. M. Morel. Environ. Sci. Technol. **23**:746 (1989).

15. Marinsky, J.A. Environ. Sci. Technol. **23**:746 (1989).

16. Ephraim, J.H., H. Borén, C. Pettersson, I. Arsenie and B. Allard. Environ. Sci. Technol. **23**:356 (1989).

17. Ephraim, J.H., H. Borén, I. Arsenie, C. Pettersson and B. Allard. Sci. Total Environ. **81/82**:615 (1989).

18. J.A. Marinsky and M.M. Reddy. Anal. Chim. Acta. **232**:123 (1990).

19. Chou C.T., R.L. Malcolm, T.I. Brinton and D.E. Kile. Environ. Sci. Technol. **20**:502 (1986).

20. Thurman E.M. and R.L. Malcolm. Environ. Sci. Technol. **15**:463 (1981).

21. Robinson, R.A. and R.H. Stokes. Electrolyte Solutions, 2nd ed. Appendix 8, (London: Butterworths, 1959).

22. Ephraim, J.H. and H. Xu. Sci. Total Environ. **81/82**:625 (1989).

23. Xu H., J. Ephraim, A. Ledin and B. Allard. Sci. Total Environ. **81/82**:653 (1989).

24. Ephraim, J. and J.A. Marinsky. Environ. Sci. Technol. **20**:367 (1986).

25. Ephraim J.H. and J.A. Marinsky. Anal. Chim. Acta. **232**:171 (1990).

26. Dzombak D.A., W. Fish and F.M.M. Morel. Environ. Sci. Technol. **20**:669 (1986).

27. Tipping E., C.A. Backes and M.A. Hurley. Water Res. **22**:597 (1988).

The Incorporation of Natural Organic Matter - Cation Interaction into the Speciation Code PHREEQE

W. Eberhard Falck

Fluid Processes Research Group
British Geological Survey, Keyworth, Nottingham NG12 5GG, U.K.

Abstract

There is a growing need to consider the influence of organic macromolecules on the speciation of ions in natural waters. It is recognised that a simple discrete ligand approach to binding of protons/cations to organic macromolecules is not appropriate to represent heterogeneities of binding site distributions. A more realistic approach has been incorporated into the speciation code PHREEQE. Because of the concept of the speciation code the discrete ligand approach was retained but the binding intensities are modified using an electrostatic (surface complexation) model assuming the organic molecules form rigid, impenetrable spheres. The theoretical concept is discussed and the relevant changes to PHREEQE are outlined. Examples where the model is tested against literature data are shown and practical problems are discussed.

Introduction

A recent review of models which describe the interaction between natural organic matter and cations/protons [1] has shown that most models are conceptually incompatible with current speciation codes like PHREEQE. The major problem faced by all models is the need to describe adequately the heterogeneity in the distribution of binding sites. Various types of continuous distribution models have been suggested to represent binding site distribution curves [2-5]. In contrast speciation codes require discrete binding sites with discrete properties. The discrete ligand approach does not allow *a priori* for variation in environmental conditions like ionic strength and pH, since stability constants for the complexes used are conditional rather than truely thermodynamic, and non-specific binding (e.g. electrostatic interaction) and site interaction are not taken into account. These effects may strongly influence the binding behaviour of organic poly-electrolytes, see [6] for a comprehensive treatment of this subject.

Theoretical Considerations

The effect of unspecific electrostatic interaction on the binding between organic macromolecules and cations has been treated comprehensively in [7] where the Debye-Hückel theory is applied to rigid, ion impenetrable macromolecules. The free energy of formation ΔG^o for a complex can be separated into two components, the intrinsic standard free energy of formation ΔG^o_{int} and an arbitrary function Φ which describes the change in free energy of formation due to the varying charge Z on the macromolecule, i.e. the electrostatic interaction factor. (For the sake of convenience all associations between protons/cations and macromolecules will be called complexes in the following, regardless whether they are strict complexes):

$$\Delta G^\circ = \Delta G^\circ_{int} + RT \cdot \phi(\theta) \qquad (1)$$

where θ is the degree of site occupation/dissociation, R the gas constant and T the absolute temperature. This can also be written in terms of association constants K:

$$K = K_{int} \cdot e^{-\phi(\theta)} \qquad (2)$$

Tanford [7] relates Φ to Z in the following way:

$$\phi(Z) = 2 \cdot z \cdot Z \cdot w \qquad (3)$$

where z is the charge on the complexing cation and

$$w = W_{el} \cdot N \,/\, RT \cdot Z^2 \qquad (4)$$

where N is Avogadro's number. The free energy of electrostatic interaction W_{el} is a function of the (conformational) model chosen to represent the macromolecule. In the case of a spherical, rigid and impenetrable molecule it can be written as:

$$W_{el} = \frac{Z^2 \cdot e^2}{2 \cdot \varepsilon \cdot \varepsilon_0} \left[\frac{1}{b} - \frac{\kappa}{1 + \kappa a} \right] \qquad (5)$$

where e=electronic charge; ε=relative permittivity; ε_0=permittivity of the vacuum; a=radius of 'gyration' and b=radius of 'closest approach' (see [7] and Fig. 1). κ is the Debye-Hückel parameter or in its inverse form $1/\kappa$ the thickness of the double layer around spherical (macro-)ions and is given by:

$$\kappa = \sqrt{\frac{8 \cdot \pi \cdot N^2 \cdot e^2 \cdot I}{\varepsilon \cdot \varepsilon_0 \cdot R\,T}} = 1.17 \times 10^{10} \cdot I^{1/2} \ [\mathrm{m}^{-1}], \text{ at } 25°\mathrm{C} \qquad (6)$$

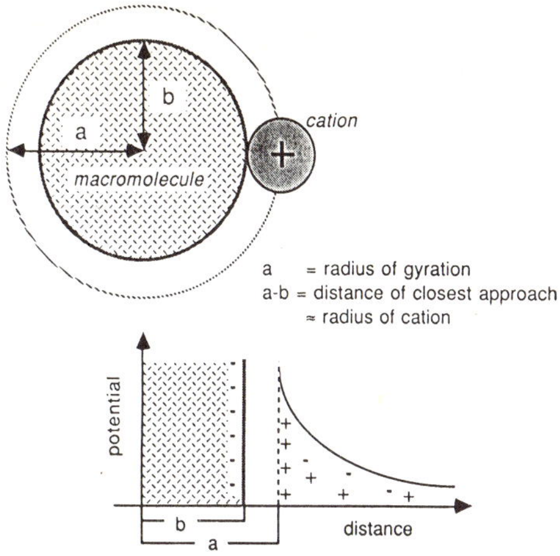

Figure 1 Rigid sphere model with diffuse layer for organic macromolecule (after [6] and [7]).

From equations (5) and (6) it can be seen that the free electrostatic energy and hence its contribution to the binding energy is a function of the ionic strength I and will diminish with increasing I. For instance, in the case of seawater the electric double layer is reduced almost to zero.

The phenomenological effect of electrostatic interaction is that as cations/protons dissociate the negative charge on the macromolecule increases and the remaining cations bind more strongly, i.e. the apparent stability constant K increases. It should be noted that in the case of amphoteric molecules/surfaces this is counteracted by anions associating, decreasing the net charge on the molecule.

Implementation of Theory

In a comprehensive treatment of natural systems it is necessary to consider competition between different cations/protons for specific site types, and the competition between different sites for cations/protons. This is usually done by evaluating multi-component Stern-Langmuir equations [7,8] which yield distribution ratios between bound and free ions. Since negatively charged organic ligands compete for cations with other small anions, e.g. sulphate, phosphate, hydroxide, it seems reasonable to develop one of the speciation codes (e.g. PHREEQE) for this task. Speciation codes solve simultaneously mass action equations for each species and mass balance equations for each chemical element considered. Species are defined as being made up of 'elements' which do not necessarily have to be elements in a chemical sense [9]. For instance, the 'element' representing an organic ligand would be its fully discharged form. This concept has been used widely to model complexation with low molecular weight organic ligands, like EDTA, and to represent heterogeneity in binding sites a range of ligands has been used [10].

Heterogeneity is introduced into the model considered here in two ways: (a) by assuming that more than one type of ligand, each with different intrinsic stability constants are present, and (b) by the introduction of electrostatic interaction, which effectively 'smears out' the stability constants of the discrete ligand model.

In order to calculate W_{el} from eqn (5) one has to know the charge Z on each molecule, but because of the heterogeneity of natural organic matter the charge on individual molecules cannot be determined. Only the total charge (Q) per mass of organic material, i.e. the titratable acidity, can be measured by experiment. However, Z can be calculated from the relationship between Q and an average molecular weight M_w: $Z=M_w \times Q$. Independently determined weight-average molecular weights or those estimated from size distributions can be used as first estimates to constrain M_w. M_w can be used subsequently, together with the related radius of gyration, as a fitting parameter.

A first estimate for the radius of gyration can be obtained from the average number of charges and the surface site density [11]. Since with increasing radius the volume of a sphere increases faster than its surface area, the ratio between 'back-bone' structure and functional groups exposed to the surface has to increase with increasing molecular weight. Otherwise the average density within the sphere would decrease and at the same time the surface site density would increase. This has three alternative consequences: (a) providing the spherical model is valid, not all functional groups can take part in binding, (b) a configuration with lower volume to surface ratio has to be assumed, or (c) the structure has to be penetrable for small ions. It seems likely from these considerations that fulvic acids, having comparatively low molecular weight, will behave more like impenetrable spheres, while high molecular weight humic substances have to be treated as either 'flat' surfaces or penetrable gels. The distance of closest approach (a-b) has been shown not to vary considerably from molecule to molecule and a constant value of 10^{-10} m has been assumed [12].

The intrinsic stability constants for each type of site are other unknowns in the model. It must be emphasised that the model presented here does not attempt to evaluate these constants from experimental data but rather aims to be a predictive tool. To address the former problem some sort of fitting algorithm would be needed to perform this task economically. A simple trial-and-error approach is possible, although time-consuming. The concentration of binding sites has to be determined experimentally, i.e. the concentration of carboxylic acids, dissociable OH-groups and the total acidity has to be measured. Further experimental evidence can be used to estimate the likely nature of carboxylic acids and other dissociable groups to narrow the range of values for stability constants.

Once concentration and stability constants have been estimated it is a straightforward procedure using the speciation code to determine the actual charge Z on the organic macromolecule and consequently its degree of dissociation θ under given sample conditions. It is simply a book-keeping exercise, adding up all organic species and their charge.

Incorporation of a routine to calculate apparent stability constants, however, presents some numerical difficulties. The concept used within PHREEQE assumes all species to be made-up from 'elements' which combine according to the law of mass action to form species. This requires the organic ligands to be introduced as fully discharged, e.g. L^-, species. In nature, of course, introduction of additional metals or protons into a system results in exchange reactions and a rearrangement of equilibria, while during model simulations all components compete simultaneously. This gives rise to very large electrostatic interaction terms in the initial iteration step which in the second iteration step cause equilibria to swing to complete neutralisation of macromolecules. The macromolecule being neutralized, in the following iteration step electrostatic interaction will be zero and the degree of dissociation θ is determined by intrinsic stability constants only. This may result again in a probably excessive electrostatic interaction term. Ideally the iteration should converge to a stable value of θ, the apparent stability constant K being a function of θ which is in turn is a function of K. However, rounding errors on the computer and the singularity at $\theta=1$ can cause oscillations. This was found a problem particularly in weakly buffered solutions, as in pH titrations of humic acid in inert electrolytes (described below). Where strong pH-buffering species are present, the problem of non-convergence seems less likely to occur.

The iteration procedure resembles the problem of calculating activity coefficients for small ions which are a function of the ionic strength, which in turn is a function of the activities of the dissolved species. Therefore it is appropriate to place the algorithm for the electrostatic model in the subroutine GAMMA of PHREEQE. As in the case of solids activity coefficients of macromolecular species are assumed to be unity. The appropriate changes to the original code [9] are reported in detail elsewhere [13]. The latter also outlines an alternative approach using a flat surface model.

Results

The model has been tested against published results for pH-titration experiments with humic/fulvic acids [8,14]. NaCl had to be substituted in the calculations where $NaNO_3$ had been used as supporting electrolyte in experiments [8]. Depending on the redox potential chosen, thermodynamic equilibrium models predict that substantial amounts of total nitrogen will be present as N_2(aq) and the system NO_3^--NH_4^+ acts as pH-buffer. This has repercussions on calculated ionic strength and pH. In nature, however, these redox-reactions are slow and probably far from equilibrium.

Variables within the problem are: the total acidity and the acidity associated with the various functional groups. Fitting parameters are the respective stability constants, the average molecular weight M_w, if not determined independently, and the radius of gyration a. M_w is used together with the total acidity Q to evaluate the average charge Z on organic molecules.

For examples with data from [8], the pK values calculated by these authors have been used which were derived by non-linear least square fitting using the electrostatic model of equations 2 and 3. In [8] an empirical expression for w with two adjustable parameters has been adopted. The parameters in [14] were estimated with a continuous distribution model; however, doubt has to be cast onto the lowest pK value reported, which cannot be determined unambiguously from this experiment. To obtain a reasonable fit to data from [9] slightly different pK values and respective acidities to those reported had to be assumed (Fig. 2).

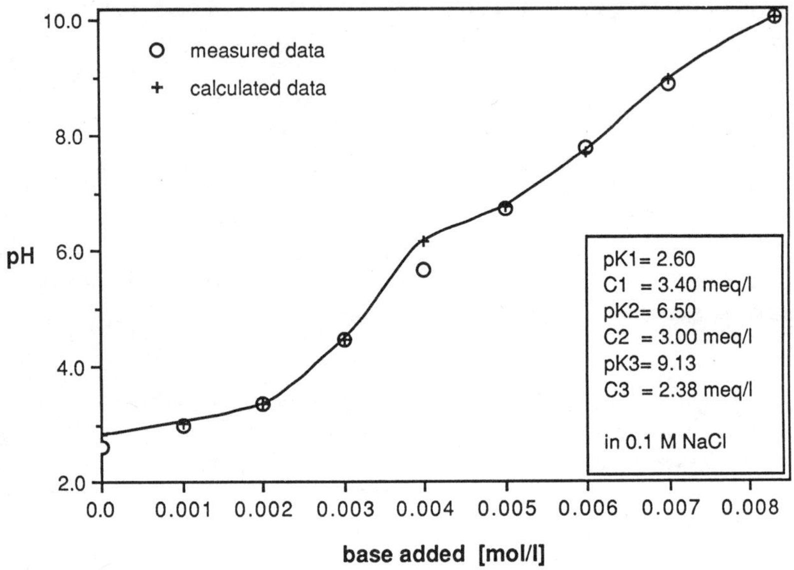

Figure 2 Titration curve of lake sediment humic acid from [14] and calculated curve.

Figs 2 to 4 compare calculated results with measured data. In most cases a three pK-value model has been chosen. pK values in the range of 2.2 to 2.6 and 4.3 to 4.6 account for two dissociation steps, analogous to small organic acids containing carboxylic functional groups. The difference between carboxylic and total acidity of a sample is attributed to a type of functional group containing (phenolic) OH-groups with pK values >10. In fact, to reproduce titration curves one type of pK-value is necessary for every 2 to 3 pH-units covered, as previously noted [3].

Comparing the fitted values for the radius of gyration a (Figs 3 and 4) at different ionic strengths shows an increase in a with decreasing ionic strength which seems reasonable. For the sample in Fig. 4 a three pK-value model was necessary in order to ensure convergence of PHREEQE.

The degree of dissociation θ and its dependence on the amount of base added, and ionic strength of the solution is illustrated in Fig. 5. It should be noted that θ of the different sites would not be a function of ionic strength if no electrostatic

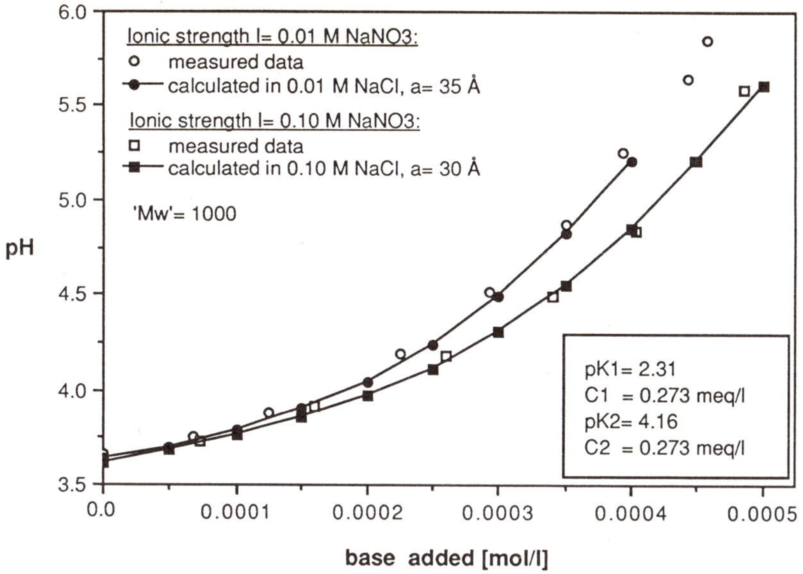

Figure 3 Titration of lake sediment humic acid (sample LFHS from [8]) and calculated curve.

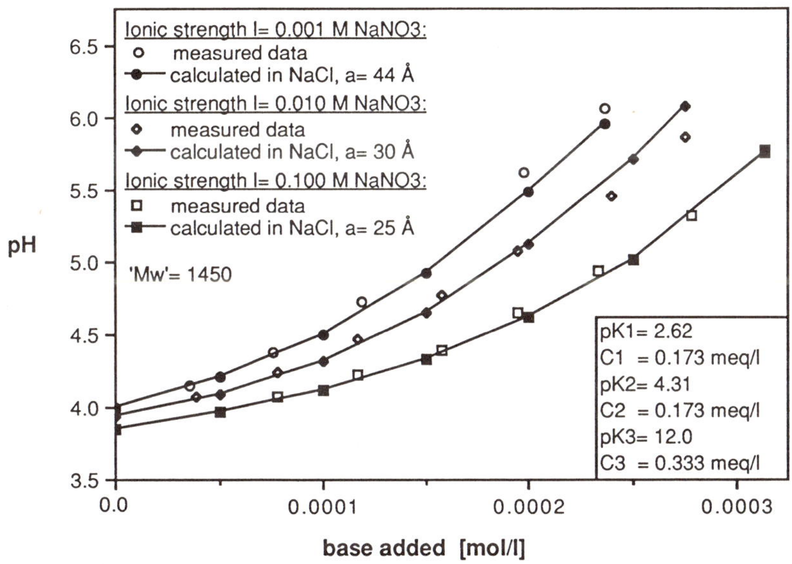

Figure 4 Titration of lake sediment humic acid (sample PRHS-A from [8]) and calculated curve.

interaction were considered. However, θ will vary slightly in actual calculations as a result of changes in activity coefficients of other constituents with varying ionic strength.

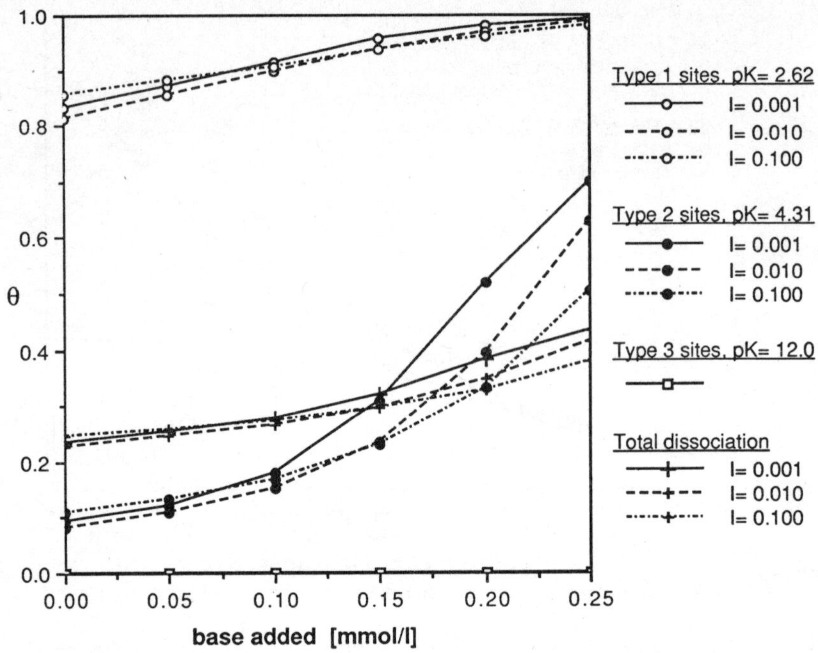

Figure 5 Degree of dissociation θ as function of base added and ionic strength (sample PRHS-A from [8]).

Conclusions

A model has been developed which includes proton/cation interaction with organic macromolecules in inorganic speciation models. It accounts for variations in environmental conditions with physically meaningful fitting parameters. The model has been shown to describe satisfactorily pH-titration experiments carried out at different ionic strength values, which cannot be described using simple discrete ligand models. The potential of the model to handle complex problems of cation - organic interaction in competition with inorganic anions still needs to be demonstrated.

Acknowledgements

This work has been funded jointly by the Natural Environment Research Council (NERC) and the Commission of the European Communities (CEC). The author wishes to express his thanks to E. Tipping, Institute of Freshwater Ecology (IFE), Windermere, G. Williams, J. Higgo and D. Noy, BGS, for helpful discussions. This paper is published by permission of the Director of the British Geological Survey (NERC).

References

1. Falck, W.E. Rep. Fluid Processes Res. Group, Br. Geol. Surv. WE/88/49, Rep. EUR 12531, 76p. (1988).
2. Perdue, E.M. and C.R. Lytle. Environ. Sci. Technol. **17**:654 (1983).
3. Dzombak, D.A., W. Fish and F.M.M. Morel. Environ. Sci. Technol. **20**:669 (1986).
4. Marinsky, J.A. and J. Ephraim. Environ. Sci. Technol. **20**:349 (1986).
5. Altmann, R.S. and J. Buffle. Geochim. Cosmochim. Acta **52**:1505 (1988).
6. Buffle, J., Complexation Reactions in Aquatic Systems: An Analytical Approach (Chichester: Ellis Horwood Ltd., 1988).
7. Tanford, C. Physical Chemistry of Macromolecules 710 p. (New York: John Wiley, 1961).
8. Tipping, E., C.A. Backes and M.A. Hurley. Water Res. **22**:597 (1988).
9. Parkhurst, D.L., D.C. Thorstenson and L.N. Plummer. U.S. Geol. Surv., Water-Resources Investigations Rep. 80-96, 210 p. (1980).
10. Sposito, G. and S.V. Mattigod. GEOCHEM: A Computer Program for the Calculation of Chemical Equilibria in Soil Solutions and Other Natural Water Systems: (The Kearney Foundation of Soil Science, Univ. of California, 1979).
11. Tanford, C. J. Amer. Chem. Soc. **79**:5340 (1957).
12. Tanford, C. J. Amer. Chem. Soc. **79**:5348 (1957).
13. Falck, W.E. Computers & Geosciences (in press).
14. Rhea, J.R. and T.C. Young. Environ. Geol. Water Sci. **10**:169 (1987).

Calcium Binding to an Aquatic Fulvic Acid

Nicklas Paxéus and Margareta Wedborg*

Department of Analytical and Marine Chemistry, University of Göteborg and Chalmers University of Technology, S-412 96 Göteborg, Sweden

Abstract

The degree of binding of calcium to aquatic fulvic acid from the Göta River was estimated from potentiometric titrations. A pH-glass electrode and a calcium-selective electrode were used to monitor the free concentrations of the competing, central ions. The ionic strength and the temperature were maintained constant at 0.1 M and 25°C. The total concentration of fulvic acid was maintained at approximately 1 g l^{-1}, while the total calcium concentration was varied within the range 0-10^{-3} M. Two types of titrations were carried out: (1) back titration with hydrochloric acid from basic solution, roughly within the pH range 10.5-2.5; (2) titration with calcium chloride at a constant total hydrogen ion concentration. The model applied for the calcium binding was an extension of our previous model for the acid-base behaviour.

Introduction

Typical hard cations, such as calcium, prefer oxygen as a donor atom. It can therefore be expected (a) that calcium will bind to the oxygen-containing carboxylic and phenolic groups in humic substances, and (b) that the complexes formed will be of a predominantly electrostatic nature. Calcium is one of the major cations in natural waters and is, consequently, of importance for the state of aquatic fulvic and humic acids.

The aim of the present study was to estimate the strength of calcium binding to an aquatic fulvic acid (FA), and how this binding is affected by pH and total calcium concentration. We have also investigated the possibility of accounting for the competition between protons and metal ions for the complexing sites by applying a model consisting of a limited number of complexing groups. In this model it is assumed that each of the groups is characterised by a concentration, a stability constant for the proton binding and a stability constant fro the metal binding. Application of such a model would result in a set of metal binding constants rather than a conditional, pH-dependent, over-all constant.

---

* Corresponding author

Experimental

The Fulvic Acid

The FA was concentrated from Göta River water with XAD-7 subsequent to an anion exchange preconcentration step (procedure A in [1]). Purification of the FA was performed as previously described [1]. The last step in the purification procedure included dissolution of the FA in ethanol and precipitation in absolute diethyl ether. Since the use of ethanol may result in a partial ethoxylation of the carboxylic groups [2] , a small amount of the FA obtained was subjected to de-ethoxylation by alkaline hydrolysis. This procedure has been described previously [2].

Reagents

All chemicals used were of analytical grade. Solutions of 0.5 M sodium hydroxide and hydrochloric acid were prepared from Merck Titrisol ampoules. The 0.1 M calcium chloride stock solutions were standardized by titration with EGTA-zinc-zincon.

Apparatus

The digital voltmeter (Solartron 7055) had a resolution of 0.01 mV, and was equipped with an eight channel scanner and a voltage follower. Radiometer G202C glass and K4040 calomel electrodes and an Orion 93-20 calcium electrode were used to monitor the free concentrations. Titrant was added from a Metrohm Dosimat 655 burette (resolution 1 µl). The additions from the burette and the voltmeter readings were computer controlled, and the system was thermostated at 25±0.01°C.

Titration Procedures

Performance of the calcium electrode was checked by titrating 0.1 M KCl with 0.1 M $CaCl_2$. Since the useful pH range of this electrode is limited by the protonation of the ion exchanger in the membrane, an additional test was performed, in which the pH was decreased at a constant ionic strength and at a constant total calcium concentration of about 10^{-4} M. No effect on the calcium electrode signal could be observed for pH≥4.

The FA was weighed and dissolved in triple distilled water. This solution was subjected to ultrasonic treatment for about 15 min. Potassium chloride and calcium chloride were added, and the pH was adjusted to about 10.5 by the addition of sodium hydroxide. The back titration with hydrochloric acid was started after two hours. Each titration point was taken as the arithmetic mean of 100 readings, and the time elapsed between successive additions of titrant was 60-120 s. In all titrations, the ionic strength was maintained at 0.1 M, and the total concentration of calcium was 0, 10^{-4} M, 5×10^{-4} M or 10^{-3} M. The concentration of FA was about 1 g l^{-1}.

The total hydrogen ion concentration was adjusted to the desired pH by addition of HCl/NaOH. Calcium chloride was added from the burette at the titration points taken as described above. The pH change during a titration was monitored and found to be 0.1 pH-unit under acidic conditions and 0.3-0.4 under neutral to basic conditions.

Theoretical Model and Calculations

The model used for the calcium complexation is an extension of our previously described model for the acid base properties [3]. This implies modelling the FA as a sum of six monobasic acids, HA_i, where HA_1 is the strongest and HA_6 the weakest acid. The corresponding stability constant for HA_i ($K_i = [HA_i] / ([H]x[A_i])$) and the concentration of each model group are denoted K_i and C_i, respectively. It is assumed that each group, i, can also form a complex with calcium, MA_i, and thus that $C_i = [A_i] + [HA_i] + [MA_i]$.

For the non-linear, curve-fitting calculations, the Levenberg-Marquardt method was utilised [3,4]. These calculations were performed in two steps:

(1) E_H, the glass electrode potential readings, and the volume HCl added were selected as the dependent and independent variables. The six pairs (K_i, C_i) and the intercept in the Nernstian equation for the glass electrode, E_H^0, were selected as parameters to be fitted, as previously described [3].

(2) E_M, the potential reading from the calcium electrode, was selected as the dependent variable, and E_H and the volume HCl as the independent variables, while the stability constants, $K_{Mi} = [MA_i] /([M][A_i])$ and the constant term in the Nernstian equation for the calcium electrode, E_m^0, were the parameters to be estimated. Analogous to step (1), the fitted equation

$$[M]_{tot} = [M]_{calc} \left(1 + \sum_{i=k}^{n} \left(C_i/(1 + K_i[H] + K_{Mi}[M]_{calc})\right)\right)$$

was solved for $[M]_{calc}$ by combining of the bisection and secant methods, the criterion being that $|[M]_{tot,calc} - [M]_{tot,exp}|$ should be smaller than a given tolerance. In this calcu- lation, the results for K_i, C_i, and E_H^0 from step (1) were inserted as constants.

The ratio $[M]_{free}/[M]_{tot}$ was calculated from $\exp((E_M-E_M^0)nF/RT)/M]_{tot}$.

Results and Discussion

Titration Curves

The experimental titration curves are shown in Fig 1. (glass electrode) and Fig. 2 (calcium electrode). The pH range in these figures has been limited to 10.2 - 4, which was the range used for the curve-fitting calculations with E_M as the dependent variable.

Curve-fitting Treatment

In the acid-base model, titration points within the pH-range 10.2 - 2.5 were utilised. The low final pH was necessary to estimate the stability constant and the concentration of the most acidic type of groups. The results for duplicate titrations of solutions with calcium concentrations within the range 0 - 10^{-3} M are shown in Table 1:

The results for the K_i's are reproducible to within 0.1 log unit and for the C_i's to within about 0.1 meq l^{-1}, irrespective of the calcium concentration.

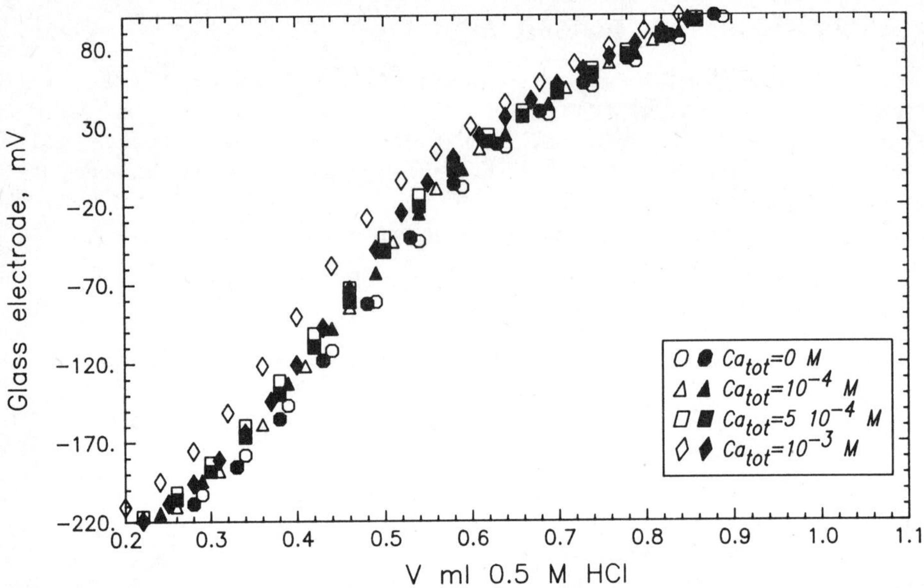

Figure 1 Emf for the glass electrode vs. v ml 0.5 M HCl. Back titration of FA from basic to acidic solution. Duplicate titrations were made for each total concentration of calcium. The v range 0.2 - 1.1 ml corresponds to a pH range of 10.2 - 4.

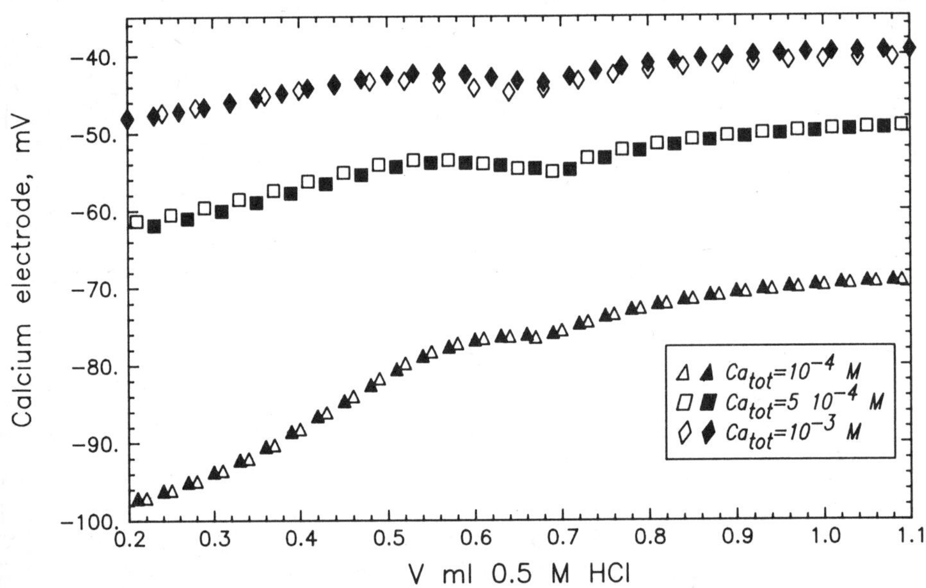

Figure 2 Emf for the calcium electrode vs. v ml 0.5 M HCl. The v range 0.2 - 1.1 ml corresponds to a pH range of 10.2 - 4.

In the calcium complexation model, titration points taken below a pH of approx. 4 were excluded, since the signal from the calcium electrode becomes pH dependent. In

Table 1 Results for K_i and C_i from curve-fitting with E_H as the dependent variable. C_i is given in meq/g and $[M]_{tot}$ in mol/l. The complexing groups are numbered 1 - 6 according to the model.

		1	2	3	4	5	6
$[M]_{tot}=0$		1	2	3	4	5	6
	log K_i	2.60	4.09	5.18	6.34	8.03	9.57
		2.59	4.07	5.16	6.29	8.00	9.62
	C_i	2.07	1.35	1.00	0.58	0.55	0.63
		1.98	1.37	0.99	0.58	0.66	0.65
$[M]_{tot}=1.02x10^{-4}$		1	2	3	4	5	6
	log K_i	2.61	4.12	5.21	6.37	8.06	9.62
		2.60	4.08	5.15	6.30	8.02	9.65
	C_i	2.10	1.41	0.98	0.60	0.53	0.66
		2.02	1.39	0.99	0.65	0.61	0.67
$[M]_{tot}=5.08x10^{-4}$		1	2	3	4	5	6
	log K_i	2.66	4.12	5.16	6.33	8.00	9.60
		2.65	4.16	5.22	6.37	8.01	9.60
	C_i	2.10	1.38	1.02	0.64	0.55	0.69
		2.09	1.43	0.98	0.59	0.55	0.70
$[M]_{tot}=1.01x10^{-3}$		1	2	3	4	5	6
	log K_i	2.58	4.07	5.15	6.35	8.00	9.58
		2.63	4.15	5.25	6.41	8.00	9.55
	C_i	2.10	1.49	1.08	0.65	0.49	0.74
		2.12	1.49	1.03	0.51	0.49	0.70

Table 2 the results are summarised for duplicate titrations at three different total concentrations of calcium, and for a stepwise inclusion of the complex forming groups, starting with the most basic type of ligand. The mean error is

$$\sum_{i=1}^{n} | E_{M,calc} - E_{M,exp} | / n$$

As could be expected (see Fig. 2), the conditions become less favourable as the total calcium concentration is increased, the main reason being that the fraction of calcium, which is complexed, becomes smaller, and thus the change in the calcium electrode emf decreases. The absolute mean errors are of the same order of magnitude as the absolute mean errors from fitting the acid-base model.

The sum of the squares of the errors decreases significantly as groups five, four and three are included in the model, whereas the statistical improvement of the fit is very small when group two is included. An attempt to include the most acidic group resulted in a vanishingly small value for K_{M1} at $[M]_{tot}= 10^{-4}$ M. For the two higher calcium concentrations, the curve-fitting procedure with all six K_{Mi} as parameters did not converge properly, and the variances and co-variances for the parameters became very large. The conclusion to be drawn from this is that the most acidic group does not influence the state of calcium enough to be of statistical significance in the present experiments. Fig. 3 shows the residual vectors for $[M]_{tot}= 10^{-4}$ M. The maximum deviation of the model from the experimental curve occurs in the region v = 0.6 - 0.7 ml (pH 5.5 - 6) where the curve bends. This bend was reproducible in all titrations with hydrochloric acid.

Effect of pH and Total Calcium Concentration

Fig. 4 and Fig. 5 show the fraction of non-complexed calcium as a function of pH and total calcium concentration, from titration with hydrochloric acid and calcium chloride, respectively. The bend in the curves at pH 5.5 - 6 (Fig. 4) becomes more pronounced as the total concentration of calcium is increased. For the two highest calcium concentrations, $[M]_{free}/[M]_{tot}$ actually decreases with decreasing pH within this range, an effect which cannot be explained with an equilibrium model. Whatever causes this uptake of calcium, e.g. a change in structural conformation, the effect is reproducible. In the equilibrium model, this effect is interpreted as a low value for K_{M4} (see Table 2).

Table 2 Results for K_{Mi} from curve-fitting with E_M as the dependent variable. The complexing groups are numbered 2-6 according to the model.

	log K_{Mi}					E_M^0	Error square sum $(mV)^2$	Mean error mV
$[M]_{tot}=1.02\times10^{-4}$	6	5	4	3	2			
	4.2					43.8	1575	3.6
	4.3					43.9	1693	3.8
	3.5	3.9				45.4	527	2.1
	3.6	3.9				45.6	513	2.0
	4.0	3.4	3.5			47.3	127	0.9
	4.0	3.4	3.4			47.7	106	0.8
	4.0	3.8	2.7	3.0		49.5	7	0.2
	4.0	3.7	2.8	3.0		49.8	5	0.2
	4.0	3.8	2.8	3.0	1.8	49.9	7	0.2
	4.0	3.7	2.8	3.0	0.5	49.9	5	0.2
$[M]_{tot}=5.08\times10^{-4}$	6	5	4	3	2			
	3.7					45.2	349	1.8
	3.7					44.9	355	1.8
	2.9	3.4				45.8	230	1.3
	3.0	3.4				45.5	217	1.3
	3.3	1.2	3.0			46.7	138	0.9
	3.4	1.1	3.2			46.5	126	0.9
	3.3	2.7	0.7	2.9		48.5	28	0.4
	3.4	2.8	-1.9	2.9		48.3	24	0.4
	3.4	2.8	0.7	2.8	2.3	49.8	25	0.4
	3.4	2.9	1.1	2.8	2.5	50.0	19	0.3
$[M]_{tot}=1.01\times10^{-3}$	6	5	4	3	2			
	3.5					46.1	178	1.2
	3.7					47.2	151	1.1
	2.6	3.3				46.5	142	1.0
	2.8	3.4				47.6	112	0.9
	3.0	0.5	2.9			47.1	97	0.8
	3.1	0.8	3.2			48.3	69	0.7
	3.1	2.2	1.1	2.7		48.5	26	0.4
	3.2	2.2	1.3	2.7		49.3	19	0.3
	3.2	2.3	0.5	2.5	2.4	50.1	22	0.4
	3.2	2.5	0.7	2.5	2.3	50.8	15	0.3

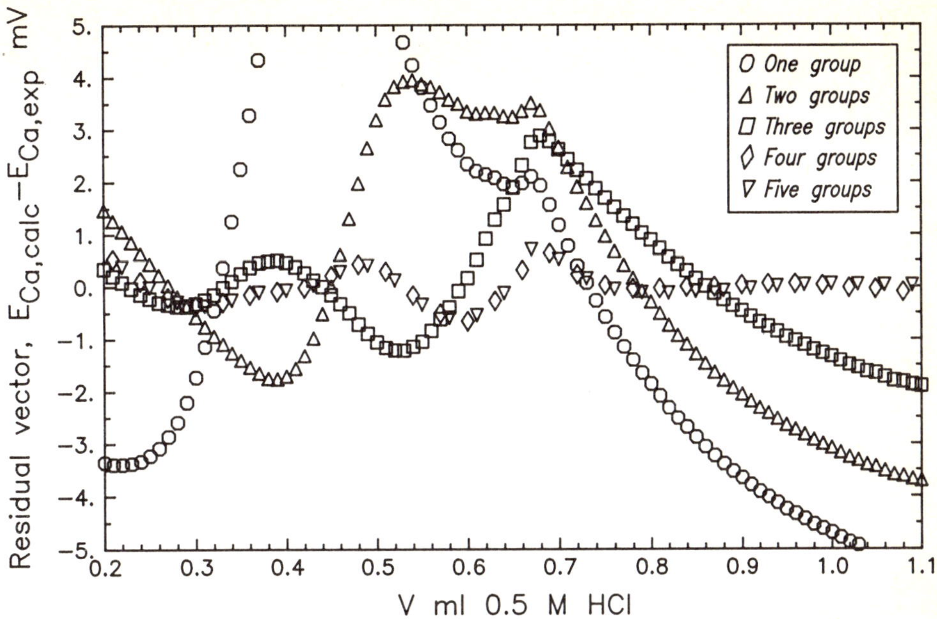

Figure 3 Residual vectors from non-linear curve-fitting of the model for metal binding, assuming that up to five of the six acidic model groups bind to calcium.

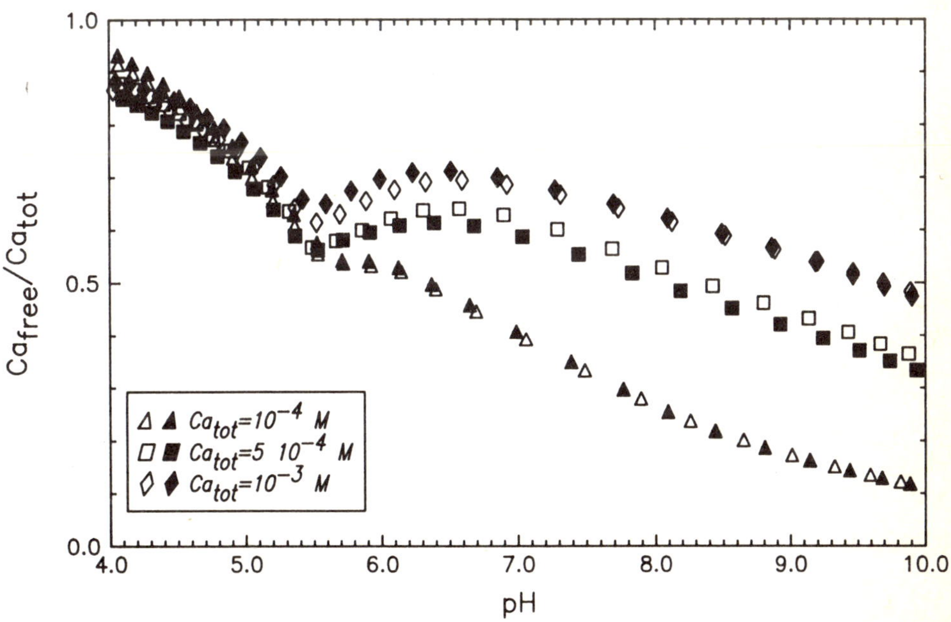

Figure 4 Dependence of $[M]_{free}/[M]_{tot}$ on pH.

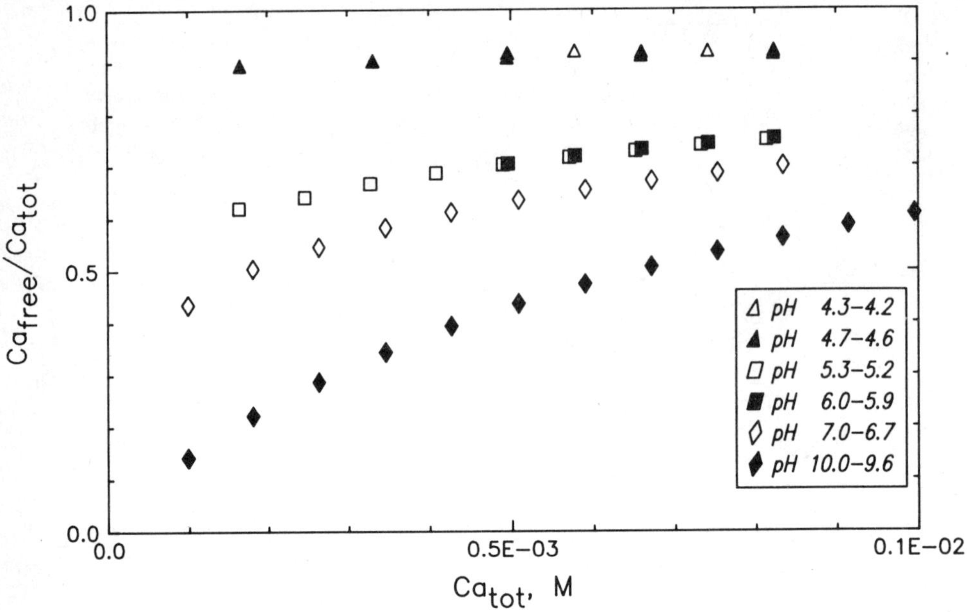

Figure 5 Dependence of $[M]_{free}/[M]_{tot}$ on the total concentration of calcium.

The characteristics of the $[M]_{free}/[M]_{tot}$ vs. pH curves as described above did not change when the membrane of the calcium electrode was replaced or when a Radiometer F2110 Ca electrode was substituted for the Orion electrode. Thus, if this effect was caused by the electrode, it would more likely be associated with the working principle than with the malfunctioning of an individual electrode. Interference from H^+ does not seem to be a possible cause, since the pH range was restricted, and since such interference is expected to increase with decreasing $[M]_{tot}$, which is opposite to the observed effect.

By comparing at similar $[M]_{tot}$ and pH values in Figs 4 and 5, it can be concluded that $[M]_{free}/[M]_{tot}$ is reasonably reproducible, irrespective of the type of titration.

Comparison between FA Batches

The results presented were obtained using the original FA (see experimental section), but care was taken to check this material against the hydrolysed batch. Such comparison revealed very little difference, in general within the experimental reproducibility. The major difference found was for C_1, C_2, and C_3, where the hydrolysed batch gave results which were slightly larger, but still less than 0.1 meq/g above the upper limit of the experimental range.

Comparison with Literature Results

Some results previously reported for calcium complexation with humic substances are summarised in Table 3.

Table 3 Calcium complexation with humic substances. Literature values.

Type of humic material	Method	Ionic strength mol/l	pH	log constant	Reference
Soil FA	UV-VIS (650 nm) ion exchange	0.1	3, 5	2.6, 3.4 2.7, 3.3	[5]
Humic compounds (peat, river, lake, seawater, marine sedim.)	GPC with AAS and UV-VIS (250-402 nm) detection	0.02	8	3.27-4.65	[6]
Aquatic FA	Copper ion selective electrode	0.1	5-6.5	2.44-2.39	[7]
Aldrich technical grade HA	Solvent extraction ^{45}Ca liquid scintill.	0.1	3.9-5	2.25-3.32	[8]
Lake FA	Calcium ion selective electrode	0.1	5 7 9	1.8 -2.4 2.0 -2.4 2.0 -2.6	[9]

Due to the great varability of investigations in regard to the origin of experimental material, the isolation and measurement methods and the total concentrations of calcium and humic substances, straightforward comparison is hardly possible. In all cases but one, an increase in pH resulted in a stronger calcium binding, which is in agreement with our results. The very small decrease of the calcium binding with increasing pH reported in [7], is probably not significant. Our result that an increase in total calcium causes a decrease in the calcium binding is in qualitative agreement with [9]. All of the experiments summarised in Table 3 have been carried out at one or several constant pH's, and, therefore, the bend in the $[M]_{free}/[M]_{tot}$ *vs* pH curve (Fig. 4) has not been previously reported.

The complexation properties of humic substances cannot be well-predicted from models based on known monomeric acids. Nevertheless, it may be of some interest to compare our results in Table 1 and 2 with literature values for the complexation constants of some monomeric acids. Table 4 below includes examples of monomeric acids which contain structural units of relevance for such comparison. In general, the calcium complexation constant of a group of certain acidity in Table 4 is smaller than our experimental estimates for a model group of similar acidity. Exceptions from this are the acids with two carboxyl groups and an additional oxygen (malic acid, diglycolic acid), citric acid with three carboxyl and one hydroxy group, and acetyl acetone, which exhibit similar or stronger complex formation.

Table 4 Literature values of stability constants for some monomeric acids of interest for comparison with functional groups present in humic substances (from Martell and Smith [10]). When available, values valid for 0.1 M ionic strength and 25°C were selected.

Ligand	Stability constant[1] (logarithmic value)					
	K_1	K_{12}	K_{13}	K_{ML}	K_{MHL}	K_{MH2L}
acetic acid	4.6			0.5		
benzoic acid	4.0			0.2		
2-hydroxy-2-methyl-propanoic acid	3.8			0.9		
3-hydroxy butanoic acid	4.3			0.6		
malonic acid	5.3	2.7		1.5	0.5	
succinic acid	5.2	4.0		1.2	0.5	
glutaric acid	5.0	4.1		1.1	0.5	
diethyl malonic acid	7.0	2.0		1.2[2]	0.4[2]	
phtalic acid	4.9	2.7		1.1		
malic acid	4.7	3.2		2.0	1.1	
diglycolic acid	3.9	2.8		3.4	1.2	
citric acid	5.7	4.3	2.9	3.5	2.1	1.1
salicylic acid	13.4	2.8			0.1	
salicylic aldehyde	8.1			1.1		
acetylacetone	8.8			2.1[3]		

[1] K_1=[HL]/[H][L] ; K_{12}=[H_2L]/[HL][H] ; K_{13}=[H_3L]/[H_2L][H]
K_{ML}=[ML]/[M][L] ; K_{MHL}=[MHL]/[HL][M] ; K_{MH2L}=[MH_2L]/[H_2L][M]

[2] Estimate based on the relative strengths of the Ca and Zn complexes with malonic acid

[3] Estimate based on the relative strengths of the Ca and Zn complexes with 4-(methyl ethyl) tropolone

References

1. Paxéus, N., B. Josefsson, D. Dyrssen, and K. Lundquist. In: R.F. Christman and E.T. Gjessing, Eds, Aquatic and Terrestrial Humic Materials, pp. 387-405 (Ann Arbor: Ann Arbor Science, 1983).
2. Lundquist, K., N. Paxéus, M. Bardet and D. R. Robert. Chemica Scripta **25**:373 (1985).
3. Paxéus, N. and M. Wedborg. Anal. Chim. Acta **169**:87 (1985).
4. Nash, J.C. Compact Numerical Methods for Computers (Bristol: Adam Hilger, 1979)
5. Schnitzer, M. and E.H. Hansen. Soil Sci. **109**:333 (1970).
6. Mantoura, R.F.C., A. Dickson and J.P. Riley. Estuar. Coast. Mar. Sci. **6**:387 (1978).
7. Buffle, J., P. Deladoey, F.L. Greter and W. Haerdi. Anal. Chim. Acta, **116**:255 (1980).
8. Choppin, G.R. and P.M. Shanbhag. J. Inorg. Nucl. Chem., **43**:921 (1981).
9. Dempsey, B.A. and C.R. O'Melia. In: R.F. Christman and E.T. Gjessing, Eds. Aquatic and Terrestrial Humic Materials, pp. 239-273 (Ann Arbor: Ann Arbor Science, 1983).
10. Martell, A.E. and R.M. Smith. Critical Stability Constants, Vol. 3 (New York and London: Plenum Press, 1977).

Interaction of Strontium and Europium with an Aquatic Fulvic Acid Studied by Ultrafiltration and Ion Exchange Techniques

Maria Nordén, James Ephraim and Bert Allard
Department of Water and Environmental Studies, Linköping University
S-581 83 Linköping, Sweden

Abstract

The complexation of an aquatic fulvic acid, FA, with Sr^{2+} and Eu^{3+} was studied using an ultrafiltration technique and an ion exchange distribution method. The total amount of bound metal (Sr^{2+} and Eu^{3+}) was measured as a function of pH at low metal concentrations (trace levels) and constant FA concentration. In the Sr-FA system the bound metal fraction increased slightly with pH, and the values obtained from the two experimental techniques were comparable. For Eu-FA, according to the ultrafiltration data, the fraction of bound metal ion was relatively insensitive to pH changes, whereas values from the ion exchange measurements showed a strong and positive dependence on pH. The results are discussed in the light of possible intrinsic problems of the two methods.

Introduction

Metal ion binding by humic and fulvic acids is of significant importance in natural water systems. Unfortunately, the solution chemistry of soluble humic substances is not well understood thus making a comprehensive description of their participation in the transport of metal ions in the environment difficult. Humic and fulvic acid molecules exist as a mixture of different functionalities which vary in size and molecular weight. As a consequence of the inherent heterogeneity coupled with the hydrophobicity of these molecules, it is difficult to interpret ion-interactions in natural water systems containing high concentrations of humic and fulvic acids. Only recently has an approach been developed which recognises the need to separately determine perturbations due to ionic strength and functional group heterogeneity, in order to achieve a comprehensive description of metal-humate interactions [1,2].

To effect a comprehensive modeling of metal-humate interactions, the acquisition of reliable experimental data is an obvious prerequisite. The selection of an appropriate experimental technique for studying a particular metal-humate system is therefore of utmost importance. Ideally, different methods employed in the study of the same metal-humate system must yield similar results if the various methods are measuring the same kind of reaction. Studies of metal-humate systems have been dictated by the complexity of the aqueous chemistry of the metal ion and the availability of techniques for the speciation assessment of the metal ion. One approach used seeks a quantitative determination of the free metal ion in the metal-humate system. In another approach the estimate of metal-bound ligand has been the objective. In the first approach, which includes techniques like ion-specific-electrodes (ISE) and anodic stripping voltametry, the free metal ion is directly measured in situ or via an indirect method. An example of the second approach is Laser Induced Fluorescence (LIF), which determines the quenching of the ligand fluorescence in the presence of metal ions.

In Sweden the present suggestion by the Nuclear Fuel Safety Project (SKB), for final storage of high-level radioactive waste is deposition in granitic bedrock at a depth of 500 m. In the distant future (estimated to about 100 000 y from now), or in the case of accidental leakage, the radioactive waste could penetrate the man-made barriers and reach the only natural barrier, the granitic bedrock. At depths of 500 m the concentration of organic matter is considerable. Between 12-33 % of the organic material consists of humic substances which can form relatively strong complexes with metal ions. The elements Sr and Eu, were chosen as model elements for the present study, as they are fission products in spent uranium fuel and also have properties similar to other longlived radionuclides, e.g. Ra(II), Pu(III) and Am(III). In this paper, results of the interaction of Sr(II) and Eu(III) with a well-characterised fulvic acid sample, studied with two different methods, are compared and interpreted considering inherent problems of the various methods.

Materials and Methods

Chemicals and Equipment

A well-characterised aquatic fulvic acid isolated from surface water in a bog area (Bersbo, about 200 km south of Stockholm) was used [3,4]. In the ion exchange distribution studies, the sodium form of a Dowex 50 WX8 (mesh 50-100) was employed as the exchanger. The isotopes, ^{85}Sr (in aqueous solution) and ^{152}Eu (in 0.10 M HCl), were purchased from Amersham. Milli-Q water was used in preparing all solutions. Analytical grade chemicals were employed. An Amicon ultrafiltration cell, 8050, was used in conjunction with a YM2 diaflo membrane. Radioactivity measurements were carried out using an LKB (Wallac) 1282 Compugamma counter. A Radiometer pHM 85 precision pH-meter and a pH-combination glass electrode, GK 2402B, were employed for pH determinations. All experiments were performed at a constant ionic strength (0.100 M $NaClO_4$), constant total concentration of FA (120 mg/l, corresponding to a total capacity of 5.6×10^{-4} eq/l), and constant total metal ion concentrations of $[Sr]=1.008 \times 10^{-9}$ M and $[Eu]=1.20 \times 10^{-11}$ M. The pH-adjustments were effected with $HClO_4$ or NaOH without the use of buffering agents.

Procedure

The ion exchange distribution method has previously been developed for the determination of stability constants of metal-ligand systems [5]. A modification made to the approach was the determination of the distribution coefficient in absence of the ligand, D_o, as a function of pH. This allowed an experimental correction for the possible perturbation of D_o by the formation of hydroxy complexes of the metal ion under study. A description of the experimental details are presented elsewhere [6].

The ultrafiltration experiments were performed following the procedure outlined previously [7]. The performance of the membrane, YM2, was checked by ascertaining that the rejection coefficient of the fulvic acid was unity, whereas that of the metal ion was zero [8]. A detailed description of the ultrafiltration method may be found elsewhere [7,8].

Results

Ion Exchange Distribution Studies

Results from the ion exchange distribution studies are given in Fig. 1. The quantity M_b/M_f was obtained from

$$M_b/M_f = (D_o\text{-}D)/D \qquad (1)$$

where D and D_o are the respective distribution coefficients in the presence and absence of FA in the system. The quantity D_o was determined at pH values sufficiently low to ensure the absence of hydroxy complexes of the metal ion, Fig. 2 [5,6]. At high pH values the correction for formation of hydroxy species was made by using the literature value of the corresponding stability constants. In the present study, D_o was determined as a function of pH for both Sr^{2+} and Eu^{3+}, Fig. 2. It is observed that at pH values lower than 4, a constant D_o is obtained for the Eu^{3+} system. At these pH values, the formation of the hydroxy complexes of Eu^{3+} (ie. $Eu(OH)^{2+}$ and $Eu(OH)_2^{+}$) is insignificant. However, due to increased formation of the hydroxy complexes of Eu^{3+}, lower D_o are obtained as the pH increases beyond 4. The real lowering of the distribution coefficient due to the fulvic acid complexation is thus obtained from the difference between the D_o and the D (in the presence of fulvic acid), at the given pH. Results from the Sr^{2+} system showed an insensitivity of D_o with pH changes. This suggests that the formation of hydroxy complexes of Sr^{2+} is insignificant, even at high pH values.

Ultrafiltration

The results of the ultrafiltration experiments are presented as plots of $\log(M_b/M_f)$ versus pH in Fig. 1. The quantity M_b/M_f was obtained from

$$M_b/M_f = (C_o\text{-}C)/C \qquad (2)$$

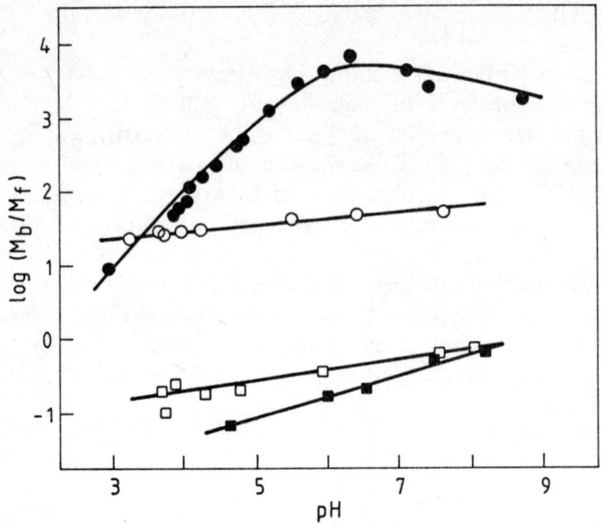

Figure 1 Formation of complexes with FA (expressed as log M_b/M_f) *vs* pH (I=0.100 M $NaClO_4$, FA=120 mg/l, [Sr]=1.008×10^{-9} M, [Eu]=1.20×10^{-11} M). □ Sr-UF, ■ Sr-IEX, ○ Eu-UF, ● Eu-IEX. (UF = ultrafiltration, IEX = ion exchange).

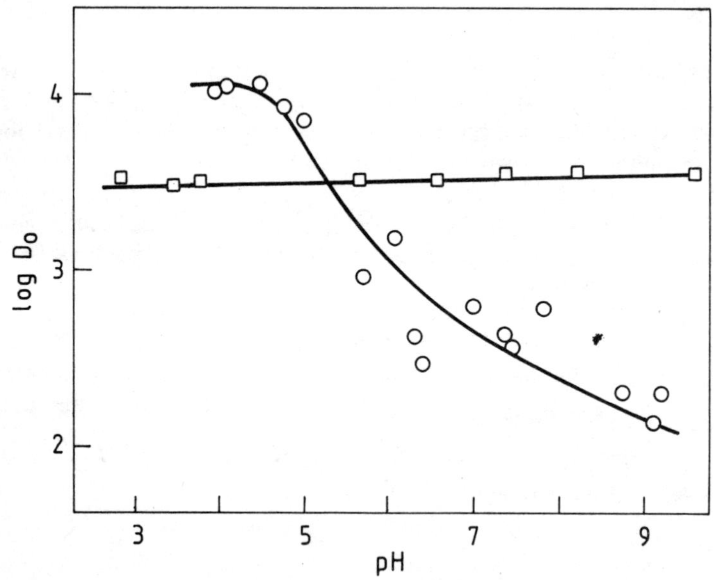

Figure 2 The relationship between D_o, the distribution coefficient in the absence of fulvic acid, and pH for the two metal ions (I=0.100 M $NaClO_4$). □ Sr^{2+} and ○ Eu^{3+}.

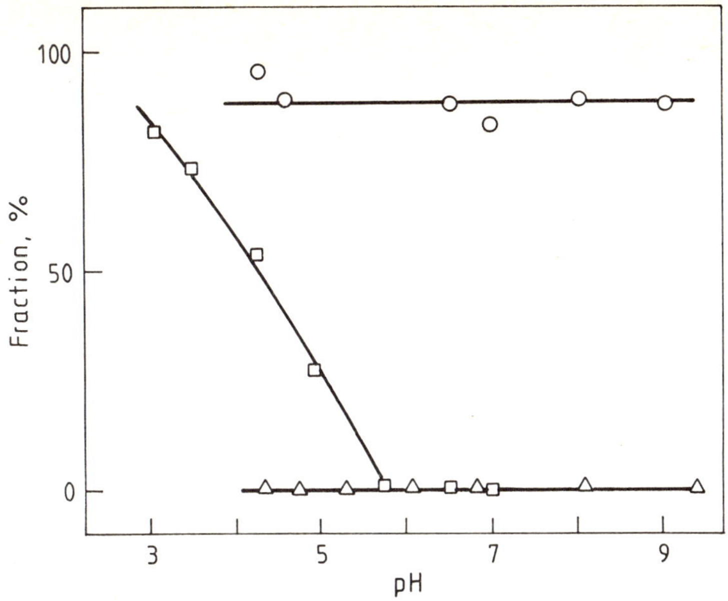

Figure 3 Metal fraction that filtered through the YM2 membrane (cut off 1000). ○ Sr, no FA □ Eu, no FA △ FA only.

where C_o is the initial activity and C the activity after filtration. In this system, M_b is the total bound metal, and M_f is the free metal ion concentrations. A correction for the fraction of the metal that was not complexed by the FA but was retained by the filter, could be made from measurements in systems without FA (Fig. 3), as described previously [8].

Discussion

In the Sr^{2+}-FA system, an apparent sensitivity of $\log(M_b/M_f)$ to pH changes was observed for both methods. At higher pH values, the two methods yielded similar results but at low pHs, the ultrafiltration yielded $\log(M_b/M_f)$ values higher than the ion exchange distribution studies by ~0.5 logarithmic units (pH=4.5). For the Eu^{3+}-FA system, the two different methods yielded different results. The $\log(M_b/M_f)$ at the same pH for the ion exchange distribution studies was higher than their corresponding values in the ultrafiltration at pH values greater than 4. Whereas a pH dependence of $\log(M_b/M_f)$ was observed in the ion exchange distribution studies, an apparent insensitivity of the $\log(M_b/M_f)$ to pH changes was observed in the ultrafiltration studies (Fig. 1).

Calculated formation constants would give the same qualitative pH-sensitivity, as well as differences between the two methods. However, because the objective of this paper is to compare experimentally determined quantities obtained from the two methods, without influences due to choice of model in the computations, only the quantity M_b/M_f is discussed.

The variations in the results obtained from the two techniques may be attributed to different methods of determining the ratio M_b/M_f. In the ion exchange distribution studies, this value is obtained indirectly from Eqn (1), whereas in the ultrafiltration, M_b is obtained by determining the difference in the radioactivity measurements. An intrinsic source of error in the ion exchange distribution method is the accuracy of the determination of D_o. Because of possible ion-dipole interactions between the ion exchange resin and the fulvic acid molecule, and errors in the determination of D_o, higher M_b/M_f values may be obtained in the ion exchange distribution studies [6]. It is conceivable that "clogging" effects could lead to concentration-related errors in the filtration and also to changes in the equilibrium due to the high concentration of the FA before the filtering. These effects would, however, give rise to high M_b/M_f values, but the observed effect was the opposite. In the ultrafiltration technique, the M_f determination is highly susceptible to error [8].

The results in the Sr^{2+}-FA binding studies using the ultrafiltration and the ion exchange distribution methods are similar with some divergency at lower pH which is possibly attributable to systematic errors in the two methods. A similar dependence of M_b/M_f on pH has previously been found for the Cd-FA system using the two different methods [7].

Diverging results for the Eu^{3+}-FA system when using the two methods have been reported by two independent researchers [9,10]. No rationalization of these differences has been given, but it has been hypothesized elsewhere [11] that the possible fragmentation of the fulvic acid molecule by the Eu^{3+} might be responsible for the observed behaviour. These results are corroborated by Al^{3+} binding by humic substances [12], where permeation of humic material through a dialysis membrane was found to increase with increased complexation of Al^{3+}. Further investigation into this hypothesis is still in progress.

The strikingly different results obtained for the Eu^{3+}-FA system suggest that the seemingly complex aqueous chemistry of the Eu^{3+} leads to a characteristically unique behaviour of the Eu^{3+}-FA system.

Conclusions

The two different methods employed in our study of Sr^{2+} and Eu^{3+} binding by a well-characterised aquatic fulvic acid yielded results which confirmed the fact that Eu^{3+} binds more strongly to FA than Sr^{2+}. In the Eu^{3+}-FA system, the ion exchange distribution studies yielded results which were higher than those obtained in the ultrafiltration technique while in the Sr^{2+}-FA system results from the two methods were similar, even though the ultrafiltration results were slightly higher at lower pH values.

To further explain the binding pattern in the metal-humate systems, resolution of binding functions for the various heterogeneous sites in the humic material must be effected [8,11]. It is contended that the extent of discrepancies in results obtained from different methods may be used as a measure of the complexity of the metal-humate interaction. Further studies of possible systematic errors related to the molecules, as well as evaluation of binding constants relevant for specific sites, are in progress.

Acknowledgements

Financial support from the Swedish Nuclear Fuel and Waste Management Company and graphical assistance from Ms Lisbeth Samuelsson are gratefully acknowledged. Technical assistance from Mr Lars Wigforss is additionally acknowledged.

References

1. Marinsky, J.A. and J. Ephraim. Environ. Sci. Technol. **20**:349 (1986).

2. Ephraim, J., S. Alegret, A. Mathuthu, M. Bicking, R.L. Malcolm and J.A. Marinsky. Environ. Sci. Technol. **20**:354 (1986).

3. Ephraim, J.H., H. Borén, C. Pettersson, I. Arsenie and B. Allard. Environ. Sci. Technol. **23**:356 (1989).

4. Pettersson, C., I. Arsenie, J.H. Ephraim, H. Borén and B. Allard. Sci. Tot. Environ. **81/82**:287 (1989).

5. Schubert, J. J. Phys. Colloid. Chem. **52**:340 (1948).

6. Ephraim, J.H., J.A. Marinsky and S.J. Cramer. Talanta **36**:437 (1989).

7. Ephraim, J.H. and H. Xu. Sci. Tot. Environ. **81/82**: 625 (1989).

8. Ephraim, J.H. and J.A. Marinsky. Anal. Chim. Acta **232**:171 (1990).

9. Torres, R.A. and G.R. Choppin. Radiochim. Acta **35**:143 (1984).

10. Caceci, M.S. Radiochim. Acta **39**:51 (1985).

11. Ephraim, J.H. Submitted to Sci. Tot. Environ.

12. Berggren, D. Intern. J. Environ. Anal. Chem. **35**:1 (1989).

Complexation Behaviour of Humic Substances from Granitic Groundwater Towards Am(III)

Valérie M. Moulin, Marco S. Caceci and Michel J. Theyssier
Commissariat à l'Energie Atomique
CEA-IRDI/DERDCA/DRDD/SESD/SCPCS
BP 6 92265 Fontenay-aux-Roses Cedex, France

Abstract

The role of humic substances in the transport of radioelements in natural aquifers is widely recognized. Complexation with humic materials is of particular interest when assessing the speciation of actinides in natural waters for the safety evaluation of nuclear waste storage policies.

The interaction between Am(III), a trivalent actinide, and humic/fulvic acids of aquatic origin (granitic groundwater) was investigated by two different techniques: spectrophotometry and size exclusion chromatography. Stability constants were calculated assuming the formation of complexes of 1:1 stoichiometry. The unit systems of these constants are discussed. Results when using these two techniques on Am(III) and Eu(III) are compared.

Introduction

Humic substances (mainly humic and fulvic acids) constitute the major fraction of the organic carbon of ground waters [1,2]. They have the ability to form rather stable complexes with cations [1-3] and can enhance the mobility of trace elements in natural systems, notably radioelements [4-7]. Americium is one of the most toxic radioelements in nuclear wastes. Like similar trivalent elements, it is expected to form strong complexes with humic substances [5,8].

The interaction between trivalent ions (americium, europium) and natural organic ligands has been investigated either by separative techniques, such as ion-exchange [9], size exclusion chromatography [10], dialysis [11], ultrafiltration [12] and solvent-extraction [13], or by non-separative techniques, such as spectrophotometry [14]. In the present investigation, we studied the use of spectrophotometry, a non-separative method, and size exclusion chromatography, a separative one, to study and characterize the behaviour of aquatic humic substances in the presence of Am(III).

Materials and Methods

Spectrophotometry

Absorbance spectra of samples containing varying concentrations of humic materials, 2.7×10^{-5} M Am, 0.1 M $NaClO_4$, 10^{-4} M acetate buffer; adjusted to pH 4.65 were recorded between 490 and 530 nm with a Cary 17 D spectrophotometer, against a reference of identical humic or fulvic content, pH and ionic strength. All work with americium was performed in a glove box (20°C, in the presence of air).

Conditional stability constants (β) were computed by non-linear best square fit of absorbance data assuming formation of a 1:1 complex, as already described [14], and were expressed in l/eq. The equivalent capacity for Am(III) (EC, in meq/g, defining the amount of complexing sites under these experimental conditions) used for the calculation of the humic/fulvic concentration was obtained from plots of the molar extinction coefficient (ε) at the wavelength corresponding to the complex formation *vs* humic/fulvic concentration.

Size Exclusion Chromatography

The chromatographic system consisting of a Spectra-Physics pump, a Rheodyne model 7125 syringe-loading sample injector (200 μl injection loop), a column (25 x 0.35 cm) packed with Sephadex G15 gel, and an on-line detector for α emitters [15], was located in a glove box; the electronics of the pump and of the detector were outside the box. Detection limit was determined to be less than 1 ng americium.

Size exclusion chromatography has been previously applied [10,16,17] to interaction studies between humic acids and Cu, Eu, Th and U. The present application to Am(III) (choice of experimental conditions: buffer, gel column packing, etc.) was based on results previously obtained with Eu, which is a good analogue of trivalent actinides.

The experimental procedure can be briefly summarized as follows. The column was preequilibrated with a buffered eluent (citrate 10^{-3} to 5×10^{-3} M, pH 5) containing a known concentration of americium (10^{-7} to 10^{-6} M) and electrolyte ($NaClO_4$ 0.1 M). Humic materials were dissolved in the mobile phase (with or without americium), injected on the column, and eluted. The resulting elution profile giving the americium concentration as a function of volume is characteristic of the method (Fig. 1). Detailed descriptions of the interpretation of the chromatograms can be found elsewhere [16,17]. From the area of the positive peak, corresponding to the quantity of bound americium, the conditional stability constant for a 1:1 complex (β, expressed in l/g) can be obtained:

$$\beta = Q_b \, / \, [Am] \, Q_{lig} \qquad (1)$$

where Q_b and [Am] are the concentrations of bound and unbound Am, respectively (considering citrate complexes) and Q_{lig} is the quantity of ligand injected.

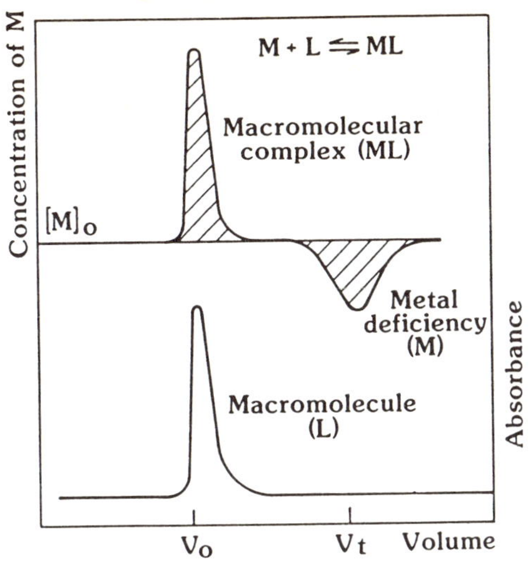

Figure 1 Typical size exclusion chromatogram: a column packed with a porous gel is equilibrated with a solution containing the cation M at the concentration $[M]_o$ (pH and ionic strength fixed); the ligand L is injected in the same medium and eluted. The area of the positive peak (dashed zone) corresponds to the quantity of bound cation M; it allows calculation of the conditional stability constant β of the equilibrium M + L ⇌ ML.

For both techniques it has been verified that the presence of the buffer (acetate or citrate) did not affect the results. No mixed complexes are assumed to exist under those conditions.

Humic Samples

Humic and fulvic acids used in the study were isolated from a granitic groundwater (Fanay-Augeres in the Massif-Central, France). Their recovery and characterization have been described elsewhere [18]. Aldrich humic acid (purified form) was also used.

Results and Discussion

Representative absorbance spectra of Am in the presence of Fanay-Augeres humic and fulvic acids are shown in Fig. 2. The characteristic absorbance peak of americium (III) at 503.1 nm is shifted by increasing concentrations of humic or fulvic acids to 504.5-505 nm. The assumption that only one complex is formed in the system is supported by the presence of an isobestic point. Experimental data (Fig. 3) in the form of molar extinction coefficients (ε) *vs* humic/fulvic concentration produce the values of β (in l/eq) presented in Table 1 together with the measured equivalent capacities (EC) of the ligands.

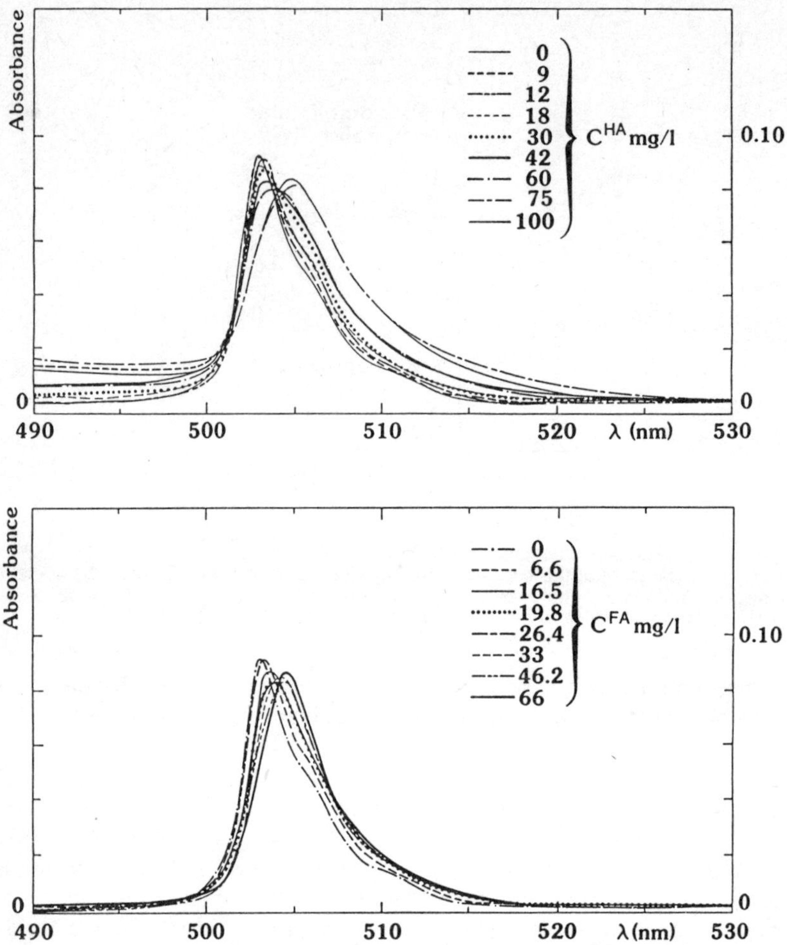

Figure 2 Absorbance spectra of Am(III)-humic (top) and fulvic (bottom) acids (from Fanay-Augeres groundwater) systems: pH 4.65 and I = 0.1 M $NaClO_4$.

Table 1 Conditional formation constants β determined by spectrophotometry (pH 4.65, I=0.1 M) and chromatography (pH 5, I=0.1 M) and expressed in different systems of units.

Sample	EC (meq/g)	Spectrophotometry logβ (l/eq)	logβ (l/g)	logβ (l/eq H^+)	Chromatography logβ (l/g)	logβ (l/eq H^+)
HA Fanay	0.30	7.0±0.2	3.5	6.0	4.6±0.3	7.1
FA Fanay	0.45	6.5±0.2	3.2	5.4	4.2±0.3	6.4
HA Aldrich	0.96	7.0±0.2	4.0	6.2	4.8±0.3	7.0

In typical chromatograms obtained by the size exclusion chromatography technique (Fig. 4), the presence of two peaks (positive and negative) indicates binding between Am and the organic ligand. The separation of the two peaks is attained only in part, mainly because of charge exclusion effects [10,16]. The conditional formation constants (β) (expressed in l/g) obtained at pH 5 and I = 0.1 M $NaClO_4$ are presented in Table 1.

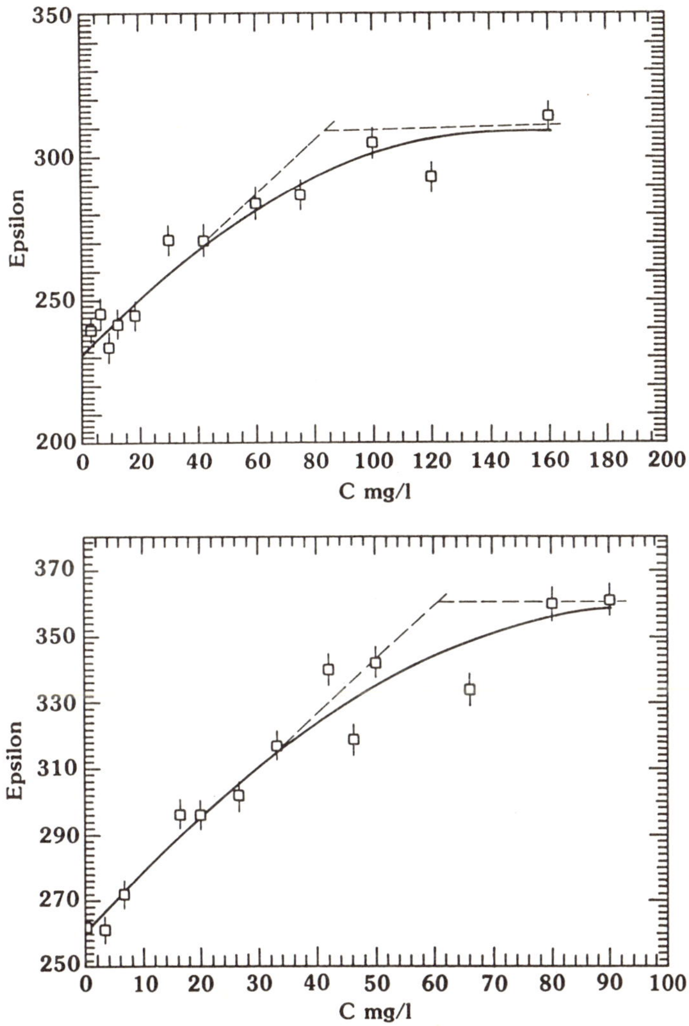

Figure 3 Molar extinction coefficient (ε) as a function of humic (top) and fulvic (bottom) concentration (C in mg/l) at the wavelength corresponding to the complex formation; pH 4.65 and I = 0.1 M $NaClO_4$.

Some literature data of conditional formation constants for Am(III) complexation with humic substances are summarized in Table 2. Assuming formation of 1:1 complexes between Am^{3+} and hypothetical humic or fulvic molecules, the equilibrium

$$Am^{3+} + HA^{x-} \rightleftarrows AmHA^{(3-x)+}$$

is defined by the stability constant β:

$$\beta=[AmHA^{(3-x)+}]/[Am^{3+}][HA^{x-}]$$

Different authors have used different units to express [HA]: g/l, equivalent of protonated sites H^+/l, equivalent of carboxylic groups COOH/l, and mol/l ("author's unit"). This leads to somewhat incomparable results (Table 2), which were normalized here to allow data comparison. Interaction constants were recalculated in common units of l/g (Tables 1-2) to permit a more convenient comparison.

Spectrophotometry yields conditional formation constants (values valid at the pH studied) in fair agreement with previous results (Tables 1,2), despite the relatively large difference in equivalent capacities between different aquatic humic substances. Conditional formation constants obtained by chromatography are in fairly good agreement with spectrophotometric results, although a certain enhancement is observed with chromatography. This may be due to the difference in ligand/metal ratio used in both techniques: at high ligand/metal ratios (as in chromatography)

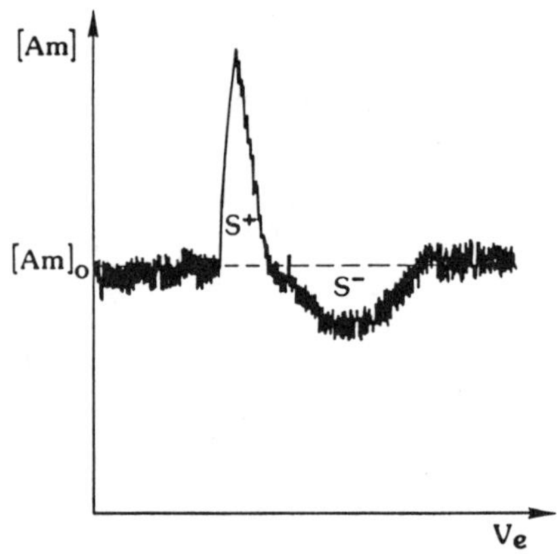

Figure 4 Chromatogram obtained with humic acids and Am(III) on Sephadex G15, in citrate buffer $2.5x10^{-3}$ M, at pH 5 and I=0.I M, [Am] $5x10^{-7}$ M, 500 mg/l HA (500 µl).

strong sites are involved in the complexation, whereas at low ligand/metal ratios (as in spectrophotometry) weaker sites participate. In any cases, formation constants obtained for Am are on the same order of magnitude as the ones obtained for Eu.

Conclusion

The complexation strength of humic substances from a granitic groundwater is similar to that of other aquatic humic materials, although the complexing capacities

Table 2 Literature data for the interaction constants of the systems Am/Eu-humic substances.
SP spectrophotometry — IE ion-exchange
SE solvent extraction — SEC size-exclusion chromatography
UF ultrafiltration — D dialysis

Sample	Method	pH	I	EC (meq/g)	logβ (author unit)	logβ (l/g)	Ref.
Am(III)							
FA-Groundwater	SP	4.65	0.1	0.88	6.4 (l/eq)	3.1	14
HA-Surface water	SP	"	"	1.20	7.0 (l/eq)	4.1	14
FA-Surface water	SP	"	"	1.22	6.0 (l/eq)	3.1	14
HA-Soil	IE	6.5	0.1	-	6.4 (l/moles)	3.1	19
HA-Lake	SP	4.65	0.1	1.03	7.0 (l/eq)	4.0	14
"	IE	4.5	0.1	-	6.83 (l/eq H)	4.4	9
"	SE	4.65	0.1	-	9.3 (l/eq COOH)	6.6	13
Eu(III)							
HA-Aldrich	SEC	5	0.02	-	4.3 (l/g)	4.3	16
"	UF	5	0.1	0.28	4.5 (l/g)	4.5	12
"	D	4.5	0.05	0.22	6.2 (l/eq)	2.5	11
Ha-Lake	IE	4.65	0.1	-	7.4 (l/eq H)	4.9	9
"	SE	"	"	-	8.6 (l/eq COOH)	5.7	13
FA-Sediment	IE	4.5	0.1	-	6.5 (l/eq H)	4.2	20
HA-Soil	IE	9	0.1	-	13.7 (l/eq H)	11.2	21

of the former are lower. Spectrophotometry and size exclusion chromatography are in relatively fair agreement with one another. Nevertheless, both techniques present a limit: they are based on a model where the humic or fulvic acids act as a pure ligand giving 1:1 complexes, this assumption being chosen as a first approach. It appears that for the formation constants, a consistent system of units is necessary to compare data.

Acknowledgements

Aldrich humic acids (as purified samples) were kindly provided by TUM-Garching, München (J.I. Kim and G. Buckau). Messrs. Anav and Robert (CEA/DGR-FAR) are gratefully acknowledged for their support with the electronics. Part of this work was performed in the shared-cost program of the Commission of the European Communities under contract FI1W/0068.

References

1. Thurman, E.M. Organic Geochemistry of Natural Waters (Dordrecht: Nijhoff and Junk Publishers, 1985).

2. Aiken, G.R., D.M. McKnight, R.L. Wershaw and P. MacCarthy. Humic Substances in Soil, Sediment and Water (New York: J. Wiley, 1985).

3. Christman, R.F. and E.T. Gjessing. Aquatic and Terrestrial Humic Materials (Ann Arbor: Ann Arbor Sci., 1983).

4. Choppin, G.R. Radiochim. Acta **44/45**:23 (1988).

5. Choppin, G.R. and B. Allard. In: A.J. Freeman and C. Keller, Eds, Handbook on the Physics and Chemistry of the Actinides, pp. 407-429 (Amsterdam: Elsevier, 1985).

6. Kim, J.I. In: A.J. Freeman and C. Keller, Eds, Handbook on the Physics and Chemistry of the Actinides, pp. 413-455 (Amsterdam: Elsevier, 1986).

7. Ramsay, J.D.F. Radiochim. Acta **44/45**:165 (1988).

8. Moulin, V., P. Robouch, P. Vitorge and B. Allard. Radiochim. Acta **44/45**:33 (1988).

9. Choppin, G.R. J. Inorg. Nucl. Chem. **40**:655 (1978).

10. Lesourd-Moulin, V. Humic Acids and their Interactions with Cations Cu, Eu, Th, U, Report CEA-R-5354 (Saclay: Service of Documentation, 1986).

11. Carlsen, L., P. Bo and G. Larsen. In: G.S. Barney, J.D. Navratil and W.W. Schultz, Eds, Geochemical Behaviour of Disposed Radioactive Waste, Chapter 10 (Washington, DC: ACS-Symp. Series 246, American Chem. Soc. 1984).

12. Caceci, M. Radiochim. Acta **39**:51 (1985).

13. Torres, R.A. and G.R. Choppin. Radiochim. Acta **35**:143 (1984).

14. Moulin, V., P. Robouch, P. Vitorge and B. Allard. Inorg. Chim. Acta **140**:303 (1987).

15. Robert, A., C. Sella and R. Heindl. Nucl. Instr. Methods in Physics Res. **225**:179 (1984).

16. Moulin, V. In: J.I. Kim and E. Warnecke, Eds, Vortrage des 66- PTB Seminars PTB-SE-14 (Braunschweig: ISSN, 1986).

17. Moulin, V., A. Billon, C. Poitrenaud and M. Porthault. To be submitted.

18. Dellis, T. and V. Moulin. In: D.L. Miles Ed., Proceedings of WRI-6, pp. 197-201 (Rotterdam: Balkema, 1989).

19. Yamamoto, M. and M. Sakanoue. J. Radiat. Res. **23**:261 (1982).

20. Marinsky, J.A. Report KBS AR84-21 (Stockholm: Swedish Nuclear Fuel and Waste Management Co, 1982).

21. Maes, A., J. De Brabandere and A. Cremers. Radiochim. Acta **44/45**:51 (1988).

Natural Organic Acids in Acidic Surface Water. Acid-Base Properties, Complex Formation with Aluminum, and their Contribution to Acidification

Ying-Hua Lee
Swedish Environmental Research Institute, P.O. Box 47086,
S-402 58 Göteborg, Sweden

Abstract

In order to assess the natural organic acidity of surface waters a simplified chemical equilibrium model was tested. The application of a potentiometric titration system, allowed the determination of both the concentration and the average proton dissociation constant (pK_a) of the moderately weak acidic carboxylic group $(COOH)_W$ in streamwaters. The observed weak acidity in the pH range 3.8 to 5.0, could be described by assuming that the predominant acidic sites are comprised of $(COOH)_W$ having pK_a=4.46±0.36 (n=84). The concentration of $(COOH)_W$ in the studied streamwaters ranged from 31 to 105 μM. The $(COOH)_W$ content of the DOC in these streamwaters was 7.9±2.3 meq/g C (n=20) during autumn and 6.3±1.8 meq/g C (n=10) during summer. The statistical analysis indicated that the total concentration of $(COOH)_W$ and of DOC were the principle factors controlling the concentration of non-labile organic Al (Alorg) in these streamwaters. The concentrations of monomeric Al and hydrogen ions were less important. Approximately 8 to 18% of $(COOH)_W$, together with the strong acidic carboxylic groups, may be involved in the formation of Alorg of phthalic-like complexes rather than salicylic-like complexes. The acidic contribution from the dissociation of $(COOH)_W$ and Al-organic complexes was estimated.

Introduction

The influence of natural organic acids on the acidification of surface water has been documented [1]. It has been shown, however, that in a large number of lakes, the acidic deposition derived from anthropogenic sources of SO_2 is the major causal factor in recent lake acidification [2-6]. In recent years it has been necessary to include organic acidity in surface water acidification models in order to predict the future trends in surface water quality and the rates of acidification and recovery.

As the natural organic acids are a complex mixture of weak acid polyelectrolytes, consisting mostly of humic and fulvic acids with a number of carboxylic and phenolic groups [7], there are no standard methods available for measuring the organic acidity in surface water. A number of different approaches and models have been used in order to interpret the acid-base characteristics of natural organic acids [8-15]. Recently, it has been suggested that the statistical continuum models may over-estimate the complexity of the acid-base properties [8,9]. The organic acids may exhibit a limited number of sites, and the most dominant acidic sites can be described by the acidic functional groups with proton dissociation constants from $pK_a = 1.7$ to 9.5 [8-10,14,15].

In surface waters with $pH<5.5$, the acidic functional groups with $pK_a<5$ have the most influence on the surface water acidification, because they are able to partially or completely release protons and acidify waters when they enter the watercourse. The strongly acidic groups with $pK_a<3$ (carboxylic acid groups, $(COOH)_S$) are believed to form chelate ligands together with adjacent phenolic groups (i.e. salicylic-like) or to other moderately weak carboxylic, acid groups, $(COOH)_W$ (i.e. phthalic-like), and are involved in the formation of complexes with Al and Fe, [16-18]. The abundance of the $(COOH)_S$ may have a decisive effect on the formation of organic complexes with Al.

The present study focused mainly on characterizing the acid-base properties of organic acids from natural acidic surface waters ($pH<5.5$) without preconcentration treatment. A potentiometric titration system, of high precision and accuracy, and a simplified chemical equilibrium approach were applied to determine the average proton dissociation constant and the concentration of $(COOH)_W$ in actual water samples. (The similar determination of $(COOH)_S$ was difficult, mainly because the uncertainty of the EMF measurement became very great at ambient concentrations of natural organic acids at $pH<3$ [19]). For this study more than fifty streamwater samples were collected in relation to the SWAP (Surface Water Acidification Programme) field treatment project in forest catchments within the Lake Gårdsjön area in southwestern Sweden [20].

Materials and Methods

Sampling and Analyses

The samples used were streamwaters collected from the forested minicatchments F5, L1 and F1 adjoining Lake Gårdsjön in the southwest of Sweden (Fig. 1). DOC was analyzed according to the procedure of Menzel and Vaccaro [21]. The concentration of monomeric Al (Al_m) was determined using the method of Barnes [22]. The concentration of non-labile organic Al (Alorg) was determined by the method of Driscoll [23].

Potentiometric Titration

To prevent side reactions, such as hydrolysis of metal ions prior to the titration, the sample was purified. By removing Al^{3+} and other inorganic Al complexes, as well as the cations of heavy metals, using a strongly acidic cation exchanger (Amberlite

Figure 1 Site location of minicatchments F5, F1, L1 and Lake Gårdsjön.

120). This method [23] does not remove organic acids and their Al complex fractions.

The titrations were performed in a constant ion medium (0.10 M $NaClO_4$) to prevent the variation of the activity coefficient. Different ionic media with different concentrations (0.02 to 0.2 M $NaClO_4$) were also used but only for examining the electrostatic polymeric effect and the influence of the ion media concentration on the acidic property. The pH of the sample was adjusted to 3 and the concentration of the ion medium to 0.1 M by addition of known concentrations of $HClO_4$ and $NaClO_4$. The CO_2 was removed by N_2 purging for 30 min., and 50 ml CO_2-free sample solution was then transferred to the titration vessel. The titration was carried out in a N_2 atmosphere at 25°C and at pH of about 3 to 5, adjusted by current generated hydroxide in an automatic system (current of about 1 mA).

A detailed description of the EMF measurements (equipment, procedures, corrections for liquid junctions etc.) are given elsewhere [24,25]. The total amount of strong acid added and originally present in the sample water was determined using the Gran plot method [26] from the titration point at pH<3.3.

Treatment of Data

Acid-Base

The titration data with pH from 3.05 to 3.3 were used to determine the strong acid concentration. It was assumed that both the dissociation of $(COOH)_S$ with $pK_a<3$ and that of $(COOH)_W$ with $pK_a>4$ had small influences on the titration data because of the low concentration of the organic acids. The assumption was justified, since the difference between the value determined in the sample water and that determined in the separate titration of the strong acid was less than 0.3 mV for most cases. For evaluating the average proton dissociation constant and concentration of $(COOH)_W$, the titration data with pHs from about 3.8 to 5.0 were used. Within the pH range studied it was also assumed that the dissociation of the weak acidic functional groups with $pK_a>5$ had negligible effects, as did dissociation of the organically complexed aluminum. Therefore, as a first approximation, the titration data from pHs of about 3.8 to 5.0 can be treated as a titration of a mixture of strong acid and mono-weak acid $(COOH)_W$ with strong base, which can be described as follows:

$$C_B - C_S = C_{COOH}(1 + K_a^{-1}[H^+])^{-1} + [OH^-] - [H^+] \qquad (1)$$

where C_B is the strong base concentration added during the titration, C_S is the sum of strong acid concentration added and originally present in the sample water determined using the Gran plot [26], C_{COOH} is the total concentration of $(COOH)_W$, and K_a is the average proton dissociation constant of $(COOH)_W$.

For pH<5, the concentration of OH^- can be omitted and eq. (1) can be converted to a linear relation between C_{COOH} and K_a^{-1} as follows:

$$(1 + K_a^{-1}[H^+])([H^+] + C_B - C_S) - C_{COOH} = 0 \qquad (2)$$

"Best-fit" values of C_{COOH} and K_a^{-1} can be determined using the method of least squares. This method has been examined using known concentrations of benzoic acid and phthalic acid.

Multiple Regression

In order to study which chemical parameters are significantly correlated to the concentration of Alorg, and how they are related to each other in stream water, we statistically analyzed the parameters T_{COOH} (= $C_{COOH} + C_{Alorg}$), $[H^+]$, the average $(COOH)_W$ content of the DOC (= $T_{COOH}DOC^{-1}$), Al_m and C_{Alorg}.

Results and Discussion

Dissociation Constant and Concentration of $(COOH)_W$

The results of the titrations of benzoic acid and phthalic acid, as shown in Table 1 and Fig. 2, were examined to see if the dissociation of $(COOH)_S$ and $(COOH)_W$

could give considerable influence on the determination of C_S and the evaluation of C_{COOH} and pK_a. Fig. 2 shows the excellent fit of the titration data of benzoic acid and phthalic acid to the theoretical curve calculated with the best value of C_{COOH} and pK_a using eq. (2). The results suggested that the random errors in the evaluation of C_{COOH} and pK_a were low because of the high precision of the potentiometric titration system. The relative errors of strong acidity and C_{COOH}, shown in Table 1 (less than 8% and 11%, respectively), were mainly caused by the systematical errors of the determinations of E_o for the glass electrode and C_S. It was therefore expected that the phthalic acid having $(COOH)_S$ with pK_a = 2.76 would give rise to greater error of C_S and C_{COOH} than benzoic acid.

Table 1 The results of titrations of benzoic acid and phthalic acid.

Weak acid	$\Delta C_S (C_S)_{add}^{-1}$ %	$\Delta C_{COOH} (C_{COOH})_{add}^{-1}$ %	pK_{exp}	pK_{lit}
Benzoic acid				
C_{add}=98.6 μM	1.4	0.3	4.43	4.01
C_{add}=49.3 μM	0.1	2.4	4.36	
Phthalic acid				
C_{add}=5.0 μM	3.6	7.6	4.79	4.92
C_{add}=100 μM	8.4	1.4	4.81	

$\Delta C_S = C_S - (C_S)_{add}$
$\Delta C_{COOH} = C_{COOH} - (C_{COOH})_{add}$

In natural waters, however, a fraction of $(COOH)_S$ is bound to metal ions (Al and Fe) and is probably slightly dissociated within the pH range studied. Consequently, the amount of $(COOH)_S$ unbound may therefore be low in the natural sample water studied. Comparing the pK_a value obtained from this study with that from the results of other work (27), the difference in pK_a was less than 0.4 pK_a units, which was not surprising since different methods and media were used in the determination.

Fig. 3 shows that the titration data of the same streamwater in two different ionic media (0.02 M and 0.05 M $NaClO_4$) fitted the theoretical curve very well. The average relative error for C_B between regression and added, $((C_B)_{reg} - (C_B)_{add})(C_B)_{add}^{-1}$, from the titration data at the pH range from about 3.8 to 5 for most of the streamwater samples was usually <0.0005 mM. The average dissociation constant pK_a obtained for 84 water samples was 4.46±0.36.

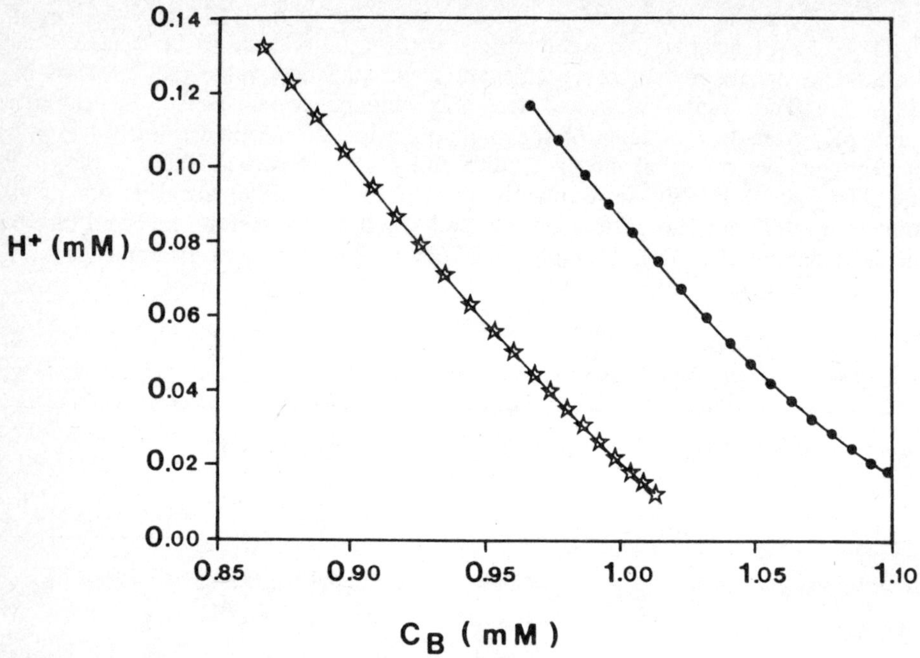

Figure 2 The experimental data C_B and H^+ of benzoic acid (50 μM) and phthalic acid (100 μM). The theoretical curves were calculated with the "best-fit" values of C_{COOH} and K_a^{-1} using eq. (2). $pK_a = 4.36$, C_{COOH} = 50.5 μM for benzoic acid (★) and $pK_a = 4.66$, C_{COOH} = 88.7 μM for phtalic acid (●).

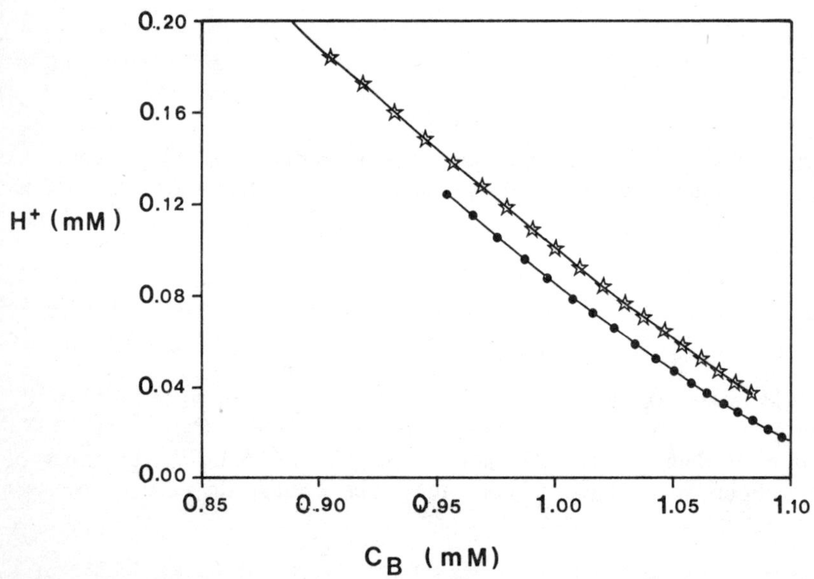

Figure 3 The experimental data C_B and H^+ of the streamwater (F1) in two different ionic media 0.02 M and 0.05 M $NaClO_4$. The theoretical curves were calculated with the "best-fit" values of C_{COOH} and K_a^{-1} using eq. (2). $pK_a = 4.36$, C_{COOH} = 82.3 μM (★), and $pK_a = 4.24$, C_{COOH} = 82.8 μM (●).

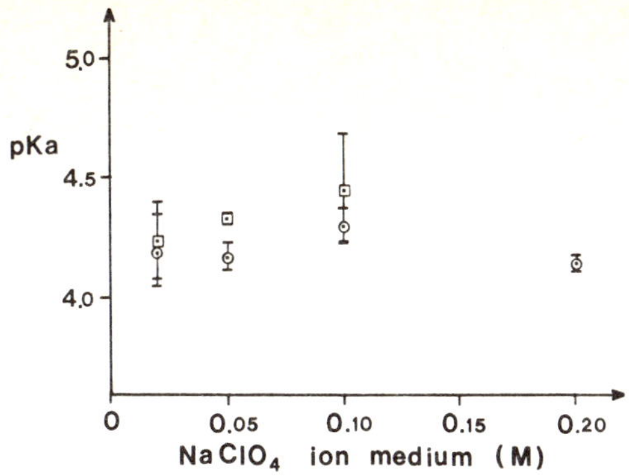

Figure 4 Variation of proton dissociation constant pK_a with different concentrations of $NaClO_4$ ion medium in two streamwater (F1) samples. The error bars correspond to standard deviation of estimated pK_a.

Fig. 4 shows that the dependence of pK_a on different concentrations of the ionic medium was small. A lack of sensitivity of potentiometric behavior to ionic strength implied that the natural organic acids behave as low molecular weight polyelectrolytes. Such an insensitive response has also been observed by other researchers [8,14]. Ephraim *et al.* [8] concluded that it resulted from the salt impermeability of the polyelectrolyte molecule. As an approximation, the average pK_a value determined in this study at ionic medium >0.02 M may also be valid for diluted acidic natural waters.

The agreement between the pK_a value obtained from this study with that from others (pK_a from 4.2 to 4.5) [8,10,14,15] was very good, despite the different methods used for determination. This may imply that these moderately weak carboxylic groups are the most predominant acidic sites, and the method used in this study is appropriate for determining both the concentration and the average proton dissociation constant of $(COOH)_W$ in acidic natural water.

Carboxylic Acid Group Content

The concentration ranges of DOC and T_{COOH} in streamwater are 3.5-13 mg/l and 30 to 95 μM, respectively. The concentrations of DOC and C_{COOH} in the streamwater were controlled by the hydrological flow paths and the degree of microbial mineralization in the upper soil profile. Comparing the seasonal variation of DOC/T_{COOH} observed in streamwater/upper soilwater reported from other papers [28,29], a typical pattern of the seasonal variation of DOC and T_{COOH} in streamwater from the Lake Gårdsjön area was also observed (Fig. 5). The concentration of dissolved organic substances (as DOC and T_{COOH}) increased from summer to early autumn, dropped in December, and rose again during high flow periods (early spring). During dry summers the streamwater mostly consisted of deep soil water and/or groundwater in which the concentrations of dissolved organic substances were low. High flow rates, resulted in high DOC in the streamwater because of the increased contribution of organic horizon leachates to stream charge.

Even though the streamwaters studied were from catchments having different field treatments [20], the results of the linear regression indicated that the correlation between DOC and T_{COOH}/or C_{COOH} was highly significant.

$$DOC(mg/l) = 0.128\ T_{COOH}\ (\mu M) + 0.70 \quad n=51 \quad p=0.00 \quad r^2=0.64$$
$$DOC(mg/l) = 0.117\ C_{COOH}\ (\mu M) + 2.11 \quad n=47 \quad p=0.00 \quad r^2=0.60$$

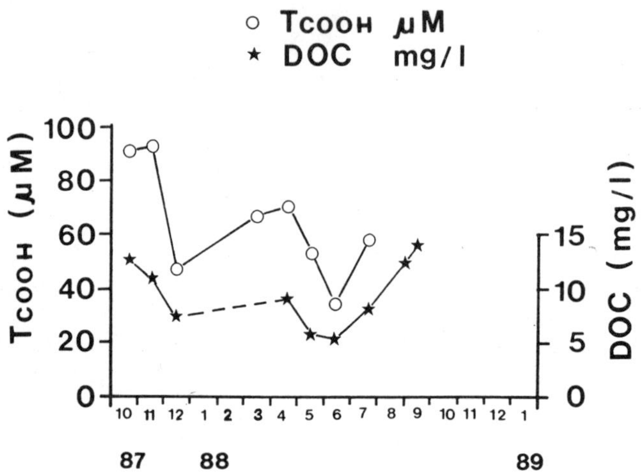

Figure 5 The seasonal variation of DOC and T_{COOH} in F1 streamwaters.

It was also found, that during autumn in these streamwaters the average number of carboxylic groups (including Alorg) per gram carbon (= 7.9±2.3 meq/g C, n=20) was greater than that during summer (6.3±1.8 meq/g C, n=8). The differences in the carboxylic group contents of DOC revealed that, during dry summers, DOC may contain more fractions of neutral species, bases and hydrophilic acids than during autumn [28].

Comparing the present data on the carboxylic group contents of DOC with other earlier studies in which the acid-base titration procedure was used, only small variations are observed, despite the diversity of the water samples and different techniques used. Cronan and Aiken [28] reported carboxyl contents of 6-7 meq/g carbon for O/A horizon soil water. Oliver *et al.* [13] found that the carboxyl content of aquatic humic substances was close to 10 meq/g C, using aquatic humic substances isolated from different locations throughout the United States and Canada.

Aluminum Complexation and the Abundance of $(COOH)_s$

The concentration ranges in μM of T_{COOH}, $T_{COOH}DOC^{-1}$ meq/g C, Al_m, $[H^+]$ and C_{Alorg} in streamwaters are 30.7 to 104.7, 4.4 to 11.8, 13.9 to 62.8, 43.7 to 134.9 and 0.6 to 15.5, respectively. The results of the statistical analysis indicated that the best-fit for C_{Alorg} was obtained when all four independent variables, T_{COOH},

$T_{COOH}DOC^{-1}$, $[H^+]$ and Al_m, were included in the regression, and the data were not transformed to logarithms. The multiple regression equation was

$$C_{Alorg} = 0.13\ T_{COOH} + 0.59\ T_{COOH}DOC^{-1} + 0.09\ Al_m - 0.05\ [H^+] - 2.8 \qquad (3)$$

$(n=40,\ p=0.00,\ r^2=0.604)$

The statistical evidence strongly confirmed that the total concentration of $(COOH)_W$ was the main factor influencing C_{Alorg} (p=0.000, correlation coefficient=0.72). This was consistent with the hypothesis that Al organic complexation is involved in the reaction with phthalic-like chelate ligands in which one of the functional groups is $(COOH)_W$. The term $T_{COOH}DOC^{-1}$ was positively related to C_{Alorg}, as shown by statistical analyses (p=0.03, r=0.47). The interesting implication of this was that the $(COOH)_W$ content of DOC was also important for determining complexation capability with Al. Compared with T_{COOH} and $T_{COOH}DOC^{-1}$, both Al_m and $[H^+]$ were not related to, or only weakly related to, C_{Alorg} (for both variables, p=0.06). The weak correlation between Al_m or $[H^+]$ and C_{Alorg} was unexpected in view of the general conception of simple metal complex formation of phthalic and salicylic acids, and the significance of T_{COOH} and $T_{COOH}DOC^{-1}$ in controlling Al organic complex formation in the acidic streamwaters is unclear. A possible explanation is that only very few salicylic-like moities are available in fulvic/humic acids in comparison to the abundance of Al and the phthalic-like chelate ligand sites. In addition, due to the low pH range (3.8 to 4.3) of the studied streamwaters, the concentrations of Al and H^+ were often positively correlated to each other, which may result in a compensative effect on Al complex formation. Consequently, Al_m and $[H^+]$ will play a less important role in control of the formation of non-labile organic Al.

The results of the multiple regression allow us to estimate the binding capacity of the moderately weak carboxylic group involved in Al complex formation. The value of the coefficient for T_{COOH} in the multiple regression (0.125±0.047; 95% significance) gives a measure of 8 to 18% of T_{COOH} in the Al complex formation. As discussed above, if the phthalic-like complexes dominated the complexation model [18] at low pH, the abundance of $(COOH)_S$ adjacent to the $(COOH)_W$ may be close to that of $(COOH)_W$ involved in Al complexation. The average molecular weight of fulvic acid and humic acid is normally between 1000 to 2000, and about 50% is carbon [30]. The mobile fulvic acid fraction in soil and streamwaters has a molecular weight in the range of 810 to 930 [31]. This means that the fulvic acid must contain 4 to 8 $(COOH)_W$ (if $T_{COOH}DOC^{-1} = 8$), and probably only one $(COOH)_W$ together with one $(COOH)_S$ will be involved in Al phthalic-like complex formation. Therefore high values of T_{COOH} and $T_{COOH}DOC^{-1}$ will favour the formation of Alorg.

Acidity Contribution

In acidic surface water there are two major processes which contribute to the production of organic acidity - the dissociation of the carboxylic group with $pK_a <$ about 5 and the Al (Fe) organic complex formation.

To estimate, as a first approximation, the proportion of acidity contributed by the dissociation of organic acids, we only need to consider the $(COOH)_W$ as a main acidic group releasing protons by dissociation. Assuming that all the strongly acidic carboxylic groups were probably already involved in complex formation, we can quantify the total organic acidity by measuring C_{COOH}, pK_a, pH, and C_{Alorg}, and using the following equation:

$$\text{organic acidity} = [COO^-] + 2C_{Alorg} = K_a^{-1}C_{COOH}([H^+] + K_a^{-1})^{-1} + 2C_{Alorg} \qquad (4)$$

If we know the average carboxyl content of DOC ($= C_{COOH}DOC^{-1} = F$) we can use the equation:

$$\text{organic acidity} = K_a^{-1}(FDOC - C_{Alorg})([H^+] + K_a^{-1})^{-1} + 2C_{Alorg} \qquad (5)$$

Fig. 6 shows that the acid contribution from dissociation of $(COOH)_W$ increases with increasing DOC and pH.

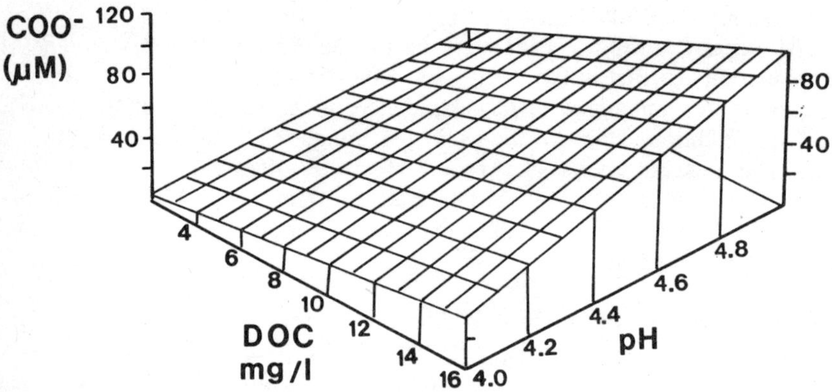

Figure 6 The acid contribution from dissociation of $(COOH)_W$ as a function of DOC and pH. Calculations were based on values of pK_a = 4.46, and an average carboxyl content of DOC = 7.9.

Eqs. 4 and 5 provide the estimation of the total acidity contributed by the organic acids. A part of the organic acidity may be already released in the watershed's soil through weathering and ion exchange reactions.

Fig. 7 shows the total contribution from microcatchment F1 by the dissociation of organic acids and aluminum complexation with organic acids to free acidity in streamwater. During high flow periods the acidity contribution from strong acid sites dominated (about 70%), whereas during autumn more than 70% of free acidity may be attributed to the organic acidity.

Since the reactions between Al and organic acids become much more complicated at pH >5 (such as DOC adsorbed to the particulate Al, etc.), and dissociation from other weak acidic functional groups must also be considered, the simplified approach suggested above would need to be modified.

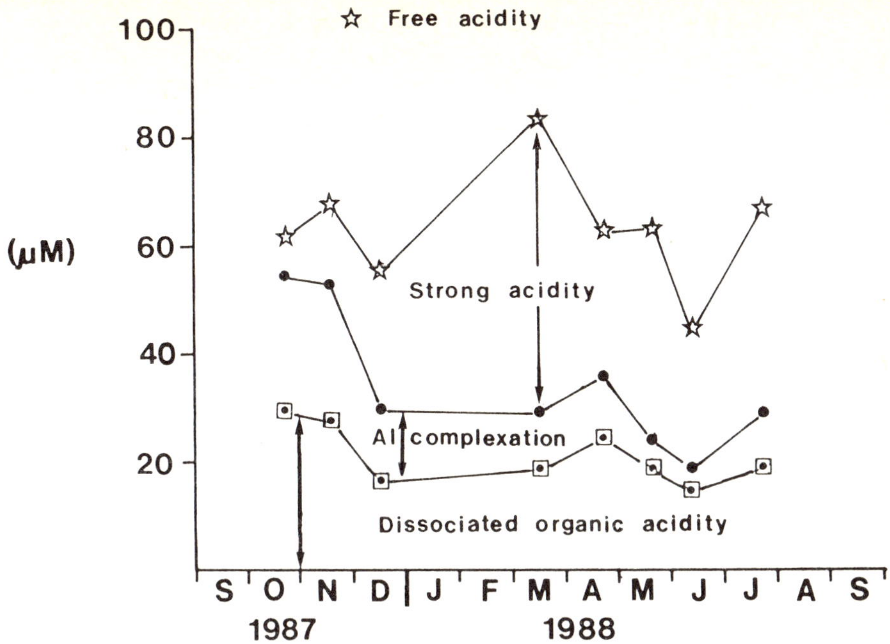

Figure 7 The seasonal variation of the relative contribution of strong acidity and the potential natural organic acidity to free acidity (H^+) in streamwater from microcatchment F1 within the Lake Gårdsjön area.

Acknowledgements

This study was conducted under the support of the board of the Scandinavian Water Acidification Programme (SWAP). Thanks are due to H. Hultberg for useful comments on the manuscript. Thanks also to O. Broberg for supply of DOC data, to I. Torbrink, P. Carlsson and Siming Li for performing the remaining analyses, as well as to S. Larsson for sampling, to R. Gould and D. Cooper for checking the language of this paper and to E. Knudsen for her assistance in the preparation of the manuscript.

References

1. Jones, M.L., D.R. Marmorek, B.S. Reuber, P.J. McNamee and L.P. Rattie, Eds, Brown waters. LRTAP Workshop No. 5, (Toronto: Environmental and Social Systems Analysts Ltd. (1986).

2. Lazerte, B.D. and P.J. Dillon. Can. J. Fish. Aquat. Sci. **41**:1664 (1984).

3. Brakke, D.F., A. Henriksen and S.A. Norton. Nature **329**:432 (1987).

4. Gorham, E., J.K. Underwood, F.B. Martin and J.G. Ogden, III. Nature **324**:451 (1986).

5. Nilsson, S.I. In: F. Andersson, and B. Olsson, B., Eds, Lake Gårdsjön - An Acid Forest Lake and its Catchment, pp. 311-318 (Stockholm: Publ. House of the Swedish Research Councils, 1985).

6. Hultberg, H. and P. Grennfelt. Water Air Soil Poll. **30**:31 (1986).

7. Stevenson, F.J. Humus Chemistry. (New York: Wiley Interscience, 1982).

8. Ephraim, J., S. Alegret, A. Mathuthu, M. Bieking, R.L. Malcolm and J.A. Marinsky. Environ. Sci. Technol. **20**:354 (1986).

9. Ephraim, J.H., H. Borén, C. Pettersson, I. Arsenie and B. Allard. Environ. Sci. Technol. **23**:356 (1989).

10. Paxeus, N. and M. Wedborg. Anal. Chim. Acta **169**:87 (1985).

11. Gramble, D.S. Can. J. Chem. **50**:2680 (1972).

12. Perdue, E.M. and C.R. Lyttle. Environ. Sci. Technol. **17**:654 (1983).

13. Oliver, B.G., E.M. Thurman and R.L. Malcolm. Geochim. Cosmochim. Acta **47**:2031 (1983).

14. Lövgren, L., T. Hedlund, L.-O. Öhman and S. Sjöberg. Water Res. **21**:1401 (1987).

15. Tipping, E., C.A. Backes and M.A. Hurley. Water Res. **22**:597 (1988).

16. Schnitzer, M. and S.U. Khan. Humic Substances in the Environment (New York: Marcel Dekker, 1972)

17. Gramble, D.S., A.W. Underdown and C.H. Langford. Anal. Chem. **52**:1901 (1980).

18. Bloom, P.R. In: Chemistry in the Soil Environment, Spec. Publ. No. 40, pp. 129-150 (Madison: American Soc. of Agronomy, 1981)

19. Dempsey, B.A. and C.R. Melia. In: F.R. Christman and E.T. Gjessing, Eds, Aquatic and Terrestrial Humic Materials, pp. 239-273 (Ann Arbor: Ann Arbor Sci., 1983).

20. Lee, Y.H. and H Hultberg. In: The Final Conference of the Surface Water Acidification Programme, London, March 19-23, 1990. (Phil. Trans. R. Soc. Lond. A, 1990, 10pp).

21. Menzel, D.W. and R.F. Vaccaro. Limnol. Oceanogr. **9**:138 (1964).

22. Barnes, R.B. Chem. Geol. **15**:177 (1975).

23. Driscoll, C.T. Intern. J. Environ. Anal. Chem. **16**:267 (1984).

24. Lee, Y.H. In: F. Andersson and B. Olsson, Eds, Lake Gårdsjön - An Acid Forest Lake and its Catchment, pp. 109-119. (Stockholm: Publ House of the Swedish Research Councils, 1985).

25. Rossotti, F.J.C. and H. Rossotti. The Determination of Stability Constants, (London: McGraw, 1961).

26. Gran, G. Analyst **17**:661 (1952).

27. Yasuda, M., K. Yamasaki and H. Ohtaki. Bull. Chem. Soc. Japan **33**:1067 (1960).

28. Cronan, C.S. and C.R. Aiken. Geochim. Cosmochim. Acta **49**:1697 (1985).

29. Nilsson, S.I. and B. Bergkvist. Water Air Soil Poll. **20**:311 (1983).

30. Plechanov, N., B. Josefsson, D. Dyrssen and K. Lundquist. In: F.R. Christman and E.T. Gjessing. Eds, Aquatic and Terrestrial Humic Materials, pp. 387-405 (Ann Arbor:Ann Arbor Sci, 1983).

31. Dawson, H.J., B.F. Hrutfiord, R.J. Zasaski and F.C. Ugolini. Soil Sci. **132**:191 (1981).

Investigations of the Interaction of Transuranic Radionuclides with Humic and Fulvic Acids Chemically Immobilized on Silica Gel

Robert A. Bulman[1] and Gyula Szabo[2]

[1] National Radiological Protection Board, Chilton, Didcot, OX11 0RQ, England

[2] Present address: Frederic Joliot-Curie National Research Institute for Radiobiology and Radiohygiene, PO Box 101, H-1775 Budapest, Hungary

Abstract

The first preparation of humic and fulvic acids chemically immobilized on silica is reported. Two different procedures have been used for immobilization of humic acid. These chemically immobilized humic substances have been evaluated for their potential in modelling the uptake of Pu(IV) and Am(III) by humic and fulvic acids; their adsorptive properties are quite distinct from the parent silica. These chemically immobilized forms of humic and fulvic acids could be of value for modelling the behaviour of transuranics in soils.

Introduction

Extensive research over the years has established that humic substances play an important role in the geochemical cycling of metal ions and organic pollutants. In recent years several reports have indicated that humic substances might influence the speciation of plutonium and americium. These papers have been briefly reviewed [1]. More recently non-aqueous extraction procedures have shown an association of plutonium with humic substances in estuarine sediments and a salt marsh which are only a few kilometres from Britain's nuclear fuel reprocessing facility [2,3].

Improvements in our understanding of the interception of pollutants by humic substances still call for further research into the chemistry of humic substances. Unfortunately, the solubility characteristics of the humic substances limit investigations of their interaction with some pollutants. In an endeavour to overcome such limitations, for instance humic acids are semi-solid gels at pH 1-2, which are also the conditions which suppress the hydrolysis of Pu(IV) and Am(III), we have selected highly definable heterogeneous phase reaction conditions so that we can study the interactions of Pu(IV) and Am(III) with humic substances. Suitable conditions are generated by chemical immobilization on an inert support, a procedure now used to study a variety of chemical and biochemical reactions, of humic and fulvic acids on silica [4-6]. In this account we report the preparation of these materials and evaluate their interaction with Pu(IV) and Am(III), species which undergo ready hydrolysis above pH 2-3 unless stabilized against hydrolysis by complexation. In addition, the use of an inert support for the humic substances might give a better approximation of the environmental reaction conditions of humic substances.

Materials and Methods

Chemicals

Sodium dithionite, silica gel (Kieselgel-100, 0.063-0.2 mm) and p-nitrobenzoylchloride were purchased from BDH. Humic acid and 3-aminopropyltriethoxysilane, glutaraldehyde and trimethylchlorosilane were purchased from Aldrich Chemical Co. Fulvic acid was isolated from a freshly composted garden soil by the method of Schnitzer and Skinner [7].

Chemical Immobilization Procedures

An outline of the reaction procedure used to immobilize humic and fulvic acids on silica by diazotization is given in Fig. 1 and the method used to immobilize humic acid via a glutaraldehyde spacer arm is given in Fig. 2. In both cases established synthetic procedures [8-11] were used to derivatize the silica gel. Moisture-free silica gel (100 g of Kieselgel 100, 0.063-0.2 mm) and 3-aminopropyltriethoxysilane were refluxed under anhydrous conditions in toluene for 8 h to yield aminopropylsilica (*I*). In the diazotization procedure, *I* was heated at 80°C for several hours, after extensive washing with toluene, methanol and acetone. When cool, the product was reacted in a sealed flask at 50°C with a large excess of p-nitrobenzoyl chloride in chloroform, to which was added sufficient triethylamine to consume liberated hydrogen chloride, to yield *II*. Any free silanolic groups remaining on the silica were masked by reaction of *II* with 10% trimethylchlorosilane (TMCS) in toluene under reflux for 4 h [12]. Excess TMCS was removed by extensive washing with toluene and the purified N-p-nitrobenzoyl-(3-aminopropyl)-silica (*II*) converted to N-p-aminobenzoyl-(3-aminopropyl)-silica (*III*) by reduction with a solution of sodium dithionite (5%) at 100°C for 60 min. This product was washed with a large excess of distilled water to remove excess sodium dithionite whereas a small amount of sulphur was extracted with ethanol. Diazotization of *III* with 2% sodium nitrite in 1 M HCl at 0-5°C for 60 min yielded *IV*. Reaction of the diazotized material with a large excess of an aqueous solution of either 1% humic acid or fulvic acid at pH 7.5 yielded, respectively, chemically bound humic acid (CBHA) and chemically bound fulvic acid (CBFA). Both substances were washed with a large excess of distilled water, to remove unreacted humic or fulvic acids. The products were air-dried after washing with methanol and acetone. In the glutaraldehyde procedure derivatization of dried aminopropylsilica (*I*) proceeded by washing it with distilled water and 0.5 M potassium phosphate buffer, pH 7.5. Activation of the wet gel with 10 volumes of

5% aqueous glutaraldehyde for 5 h and a wash with 15 volumes of distilled water yielded *II*. The wet activated gel was stirred at ambient temperature for 8 h with 1% humic acid solution (1000 ml), pH 7.5. The product was washed with 10 volumes

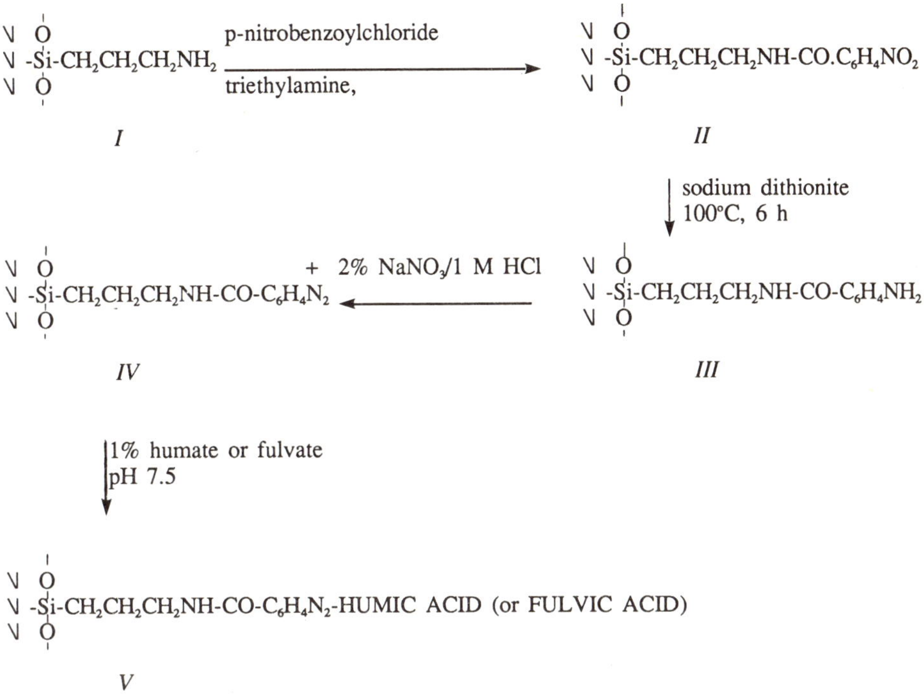

Figure 1 Reaction sequence for the preparation of CBHA and CBFA by chemical immobilization of humic and fulvic acids on silica by a diazotization procedure.

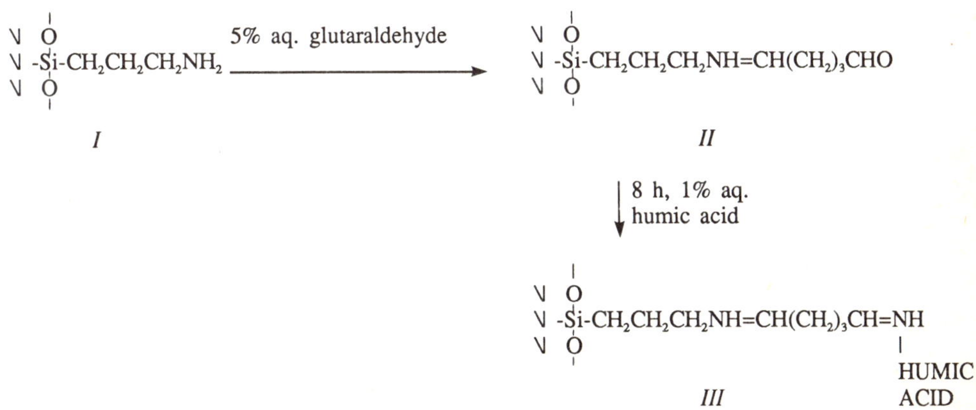

Figure 2 Reaction sequence for the preparation of CBHAglu by chemical immobilization of humic acid on silica using glutaraldehyde.

of 0.5 M phosphate buffer and a similar volume of distilled water to yield *III*. Any free aldehyde groups remaining on *III* were masked by reaction with a buffered solution of ethanolamine, pH 7.5, for 3 h. The chemically bound humic acid gel (CBHAglu) was then washed with 15 volumes of distilled water and dried at 60°C for several hours.

Characterization of CBHA, CBFA and CBHAglu

The organic content of the gels was determined by thermogravimetric analysis from 200 to 1000°C [13].

Determination of Metal-Binding Capacity

The metal-binding capacity of the immobilized biogeopolymers was determined by shaking 200 mg of the gels with 10^{-4} M solutions of calcium chloride, copper(II) nitrate and iron(III) nitrate for 2 h in an Erlenmeyer flask. The gels, which were collected by centrifugation, were washed with distilled water which was discarded. The extent of uptake of metal ions was then determined by washing the gels with 0.1 M HCl, and distilled water and then measuring the metal ion content of the two solutions, when combined, by atomic absorption spectrometry.

Radiochemical Procedures

Citrates of ^{239}Pu(IV) and ^{241}Am(III) were prepared from stock solutions of the nitrates in 4 M HNO_3. An aliquot of the stock solution was evaporated to dryness and the residue dissolved in 0.01 M HNO_3 to which an equal volume of 2% trisodium citrate was added. The pH of this solution was about 6.5. The solution was passed through a membrane of porosity 0.025 µm (Millipore (UK) Ltd, Wembley) to minimize on the presence of polymeric transuranic species [14]. The radionuclides were determined by liquid scintillation counting with a commercial scintillant cocktail (Beckman Ready Value™) in a Beckman liquid scintillation counter.

Determination of Adsorption Isotherms

The uptake of ^{239}Pu and ^{241}Am on to silica and the immobilized substrates was determined by suspending the derivatized silica (500 mg) in solutions (20 ml) containing either Pu(IV) citrate (25 Bq; 1.09×10^{-8} g; 4.5×10^{-5} µmol) or Am(III) citrate (25 Bq; 1.97×10^{-10} g; 8.147×10^{-7} µmol). The solutions used were either 0.1 M phosphate (pH 1-8) or 0.1 M citrate (pH 1-8), where necessary the pH was adjusted with 2 M HCl or 2 M NaOH. Aliquots (1 ml) were removed at 1, 10, 30, 60 and 120 min from the solutions for liquid scintillation counting.

Results

Chemical immobilization of humic acid and fulvic acid on silica gel yielded the coloured products listed in Table 1. Titration of CBHAglu with dilute acid showed that the material was a weak acid with an acid content of 275-325 µmol/g of CBHAglu. The characteristics of CBHA, CBHAglu and CBFA are summarized briefly in Table 1.

The uptake from citrate solution, at pH 4, of Pu(IV) by silica and of Pu(IV) and Am(III) by CBHA and CBHAglu is presented in Fig. 3. A comparison of the adsorption isotherms for the uptake of Pu(IV) and Am(III) by CBHA, CBHAglu and silica from pH 1-8 is presented in Fig. 4. In Fig. 5 the adsorption isotherms for the uptake of Pu(IV) and Am(III) from citrate and phosphate solutions by CBHA and CBFA are presented.

Table 1 Characteristics of gels bearing chemically immobilized humic and fulvic acids.

	CBHA	CBFA	CBHAglu
Particle size[a], µm	63-200	63-200	63-200
Surface area[a], m^2/g	320	320	320
Pore diameter[a] nm	10	10	10
Humic substance content, mg/g gel	18	19.2	16.2
Divalent cation uptake capacity, µmol/g	171	ND[b]	179
Fe (III) uptake capacity, µmol/g	236	ND[b]	245
Humic acid content of the gel, mg/g gel	26	25	27
Colour	brown	dark orange	dark brown

[a] Manufacturer's values for the starting material
[b] ND=not determined

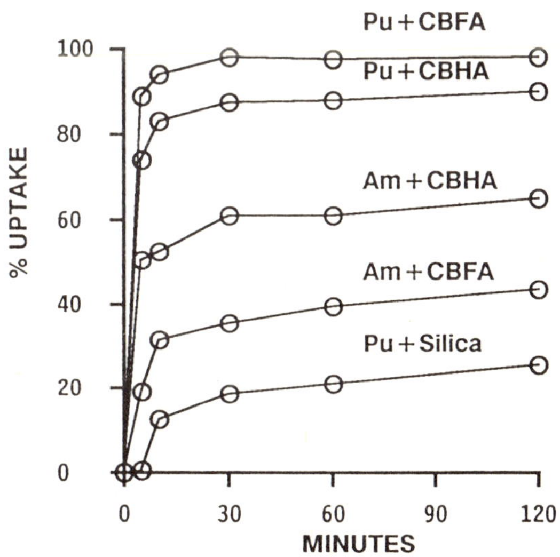

Figure 3 Uptake of Pu(IV) and Am(III) on to CBHA, CBFA and silica at pH 4 in the presence of citrate.

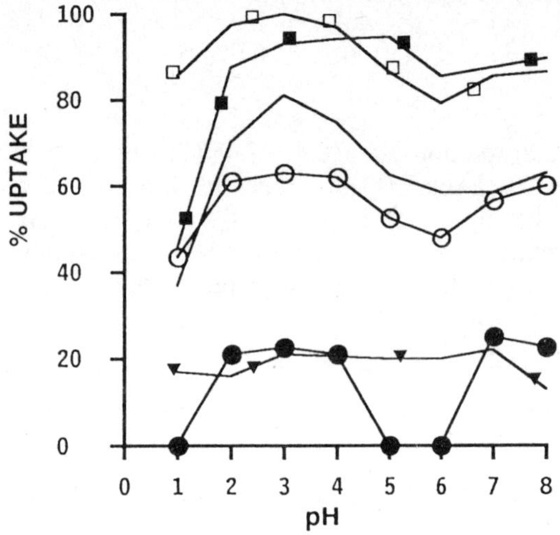

Figure 4 Adsorption, at 60 min, from citrate of Pu(IV) by: CBHAglu (□), CBHA (■) and silica (●) and Am(III) by: CBHAglu (-), CBHA (o) and silica (▼).

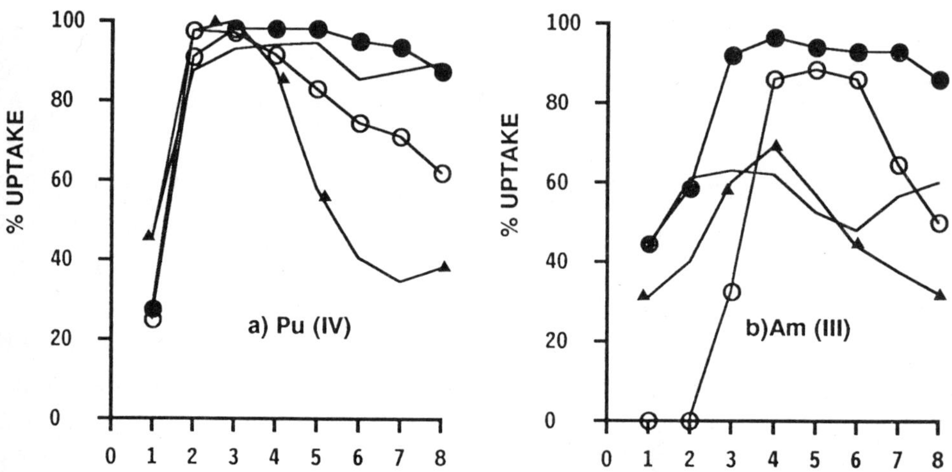

Figure 5 Adsorption, at 60 min, of Pu(IV) and Am(III) by CBHA from phosphate (●) and citrate (-), and by CBFA from phosphate (o) and citrate (▲).

Discussion

From a quite extensive search of *Chemical Abstracts* it appears that this is the first reported chemical immobilization of humic substances upon inert supports.

The evidence for chemical bonding of humic and fulvic acids to silica must be drawn by inference. The reaction of "diazoniumsilica" (*IV*) with humic and fulvic acids is likely to proceed by the well established reaction sequence, namely interaction with the ortho- or para- directed positions of phenols. Whereas the reaction of diazoniumsilica yielded the coloured products listed in Table 1, the products obtained on physical adsorption of fulvic and humic acids on to aminopropylsilica were off-white to light brown. Any interaction of humic and fulvic acids with the naked surface of silica, that is those sites free of the propylamino moiety, should have been minimized by the trimethylsilylation of silanolic groups after the preparation of *II*. The interaction of SiO_2, as opposed to ->Si-OH, with humates and fulvates might still occur but the adsorption will be weak and uptake of humates and fulvates through covalent bond formation will predominate.

The free flowing properties of CBHA, CBHAglu and CBFA at acidic pHs, the conditions which produce gelatinous precipitates of humic acid, facilitate a rapid reaction of Pu(IV) and Am(III) with the ligands of the humic acids at acidic pHs. By using immobilized humic acid it is possible to obtain a more uniform reaction of Pu(IV) and Am(III) with the ligand sites of humic acid.

In the investigation of the uptake of Pu it has been assumed that it is present as Pu(IV) although it might be reduced to Pu(III), an oxidation state which can not be excluded in view of the observations of Bondietti *et al* [15] who showed that up to 15% of Pu(IV) could be reduced to Pu(III) at pH 4. Because of their similarities Pu(III) and Am(III) should give the same basic adsorption isotherms. An examination of Fig. 5 shows a maximum uptake of Pu at pH 3 whereas the maximum uptake of Am(III) occurs at pH 4. Perhaps the maximum uptake of Pu at low pH represents the uptake of Pu(IV).

From an examination of Fig. 3 it is apparent that the uptake of Pu(IV) and Am(III) by CBHA and CBFA is similar but that the affinity for Pu(IV) is greater than that for Am(III). It is also evident from Fig. 3 that the uptake of Pu on to silica is much less. As the adsorption isotherm for the uptake of Am(III) onto silica was similar it is not presented.

Comparison of the adsorption isotherms for the uptake of Pu(IV) and Am(III) by CBHA and CBHAglu (Fig. 4) indicates that the mechanism of immobilization does not appear to have a significant influence upon the uptake of the radionuclides. The only differences are the slightly greater uptakes of the radioelements by CBHAglu.

Examination of Fig. 5 shows that the adsorption isotherms of Pu(IV) and Am(III) by CBHA and CBFA, at 60 min, in the presence of 0.1 M citrate and 0.1 M phosphate exhibit significant differences. The principal features of adsorption are summarized in Table 2. The differences in adsorption point to the regulation of adsorption by some specific chemical phenomena and are obviously not due to general adsorption by silica. In fact, adsorption by the silanolic groups of silica can be excluded as the treatment of *II* with TMCS should block the silanolic groups which might have acted as binding sites for the cations. Further support for our contention that the immobilized humic substances are acting as binding sites for these cations comes from the observations of Kerndorff and Schnitzer [16] who obtained similar shaped graphic representations for the interaction of Al(III) with humic acids.

Table 2 Principal features of adsorption of ^{239}Pu(IV) and ^{241}Am(III) by CBHA and CBFA.

	Citrate				Phosphate			
		CBHA		CBFA		CBHA		CBFA
	pH	% uptake	pH	% uptake	pH	% uptake	pH	% uptake
^{239}Pu(IV)	3-5	95	3	100	3-5	97	2-3	97
^{241}Am(III)	3	60	4	67	4	97	5	97

Obviously such differences in uptake are plainly the manifestations of specific chemical reactions between complexed forms of Pu^{4+}, or Am^{3+}, with ligands in the binding sites. In general the maximum adsorption of the radionuclides onto CBHA and CBFA occurs at around pH 2 to 4. Presumably this is the region in which the ligands in the metal-binding sites ionize. At present it is not possible to make detailed comment upon the nature of the interaction of a low molecular weight chelate with a macromolecule bearing chelating moieties. The diversity of the shapes of the adsorption isotherms obtained in these studies implies that several reaction processes must be occurring. For example, from Fig. 5 it is apparent that there is a second metal-binding ligand on CBHA which becomes available for interaction with Am^{3+} at pH 7-8. The reduction in adsorption by CBHA at pH 6 of Am(III) and Pu(IV), in the presence of citrate, followed by an increased uptake at pH 7-8 might indicate at least two common binding sites for these cations. Similar observations have been made by Nash and Choppin [17] who, on the basis of the interactions of Th(IV) with humic acids, proposed the existence of two types of thorium-binding sites in the polymer containing one or two carboxylate groups. However, in a later publication four Th-binding sites were proposed [18].

It is evident that the uptake of Pu(IV) by CBHA from citrate and phosphate media is similar. In contrast, there is a divergence in the uptake of Pu(IV) by CBFA in these media above pH 4. The divergence in uptake characteristics for the two substrates is even greater for both cations in the two media. A general feature of the adsorption is the greater degree of uptake of the transuranics from phosphate media. It is not clear if this increased uptake arises from the participation of phosphate in the binding sites of CBHA and CBFA or is because of the overall greater weakness of the mixed ligand phosphato-citrato complexes of Pu(IV) and Am(III).

Evidence to indicate an association of Pu(IV) and Am(III) with CBHA and CBFA as complexed cations, and not as hydrolyzed species, comes from the demonstration that chelating agents such as ethylenediaminotetraacetic acid (EDTA) and similar chelating agents mobilize a portion of the radioactivity from CBHA and CBFA. Plainly, the stability constant for the complexed forms of these radiocations with the chemically immobilized humic substances must lie between 3.5×10^{15}, the stability constant for Pu(IV) citrate, and 1×10^{27}, the stability constant suggested by Smith and Martell [19] for Pu(IV) EDTA. Torres has reported a value of 9×10^{18} for Pu(IV)-humate [20]. Evidently, immobilization of humic acid on silica has not dramatically reduced the affinity for cationic species such as Pu^{4+}.

In summary, this study has demonstrated that chemical immobilization of humic substances could be a valuable aid for investigations of the interception of metal ion pollutants by humic substances. Our studies with high performance liquid chromatography columns packed with silica bearing chemically immobilized humic acid indicate that such materials could also be of use for modelling the environmental fate of polycyclic aromatics [21] and perhaps other organic pollutants.

Acknowledgements

This work was supported in part by research grant BIG-B-048-UK from the Commission of the European Communities. The award of a research fellowship by the International Atomic Energy Agency to Gy. Szabo is gratefully acknowledged.

References

1. Bulman, R.A. In: P.J. Coughtrey, Ed., Ecological Aspects of Radionuclide Release, pp. 105-113 (Oxford: Blackwell, 1983).

2. Bulman, R.A., T.E. Johnson and A.L. Reed. Sci. Tot. Environ. **35**:239 (1984).

3. Szabo, Gy., R.A. Bulman and A.J. Wedgwood. J. Environ. Radioactivity. In press.

4. Hodge, J.P. and D.C. Sherwood, Eds, Polymer Supported Reactions (London: John Wiley, 1980).

5. Leznoff, C.C. Acc. Chem. Res. **11**:327 (1978).

6. Stark G.R. Ed, Biochemical Aspects of Reactions on Solid Supports (New York: Academic Press, 1971).

7. Schnitzer, M. and S.I.M. Skinner. Soil Sci. **105**:392 (1968).

8. Hill, J.M. J. Chromatogr. **76**:455, (1973).

9. Weetall, H.H. Biochim. Biophys. Acta **212**:1 (1970).

10. Sugawara, K.F., H.H. Weetall and G.D. Schucker. Anal. Chem. **46**:489 (1974).

11. Szabo, Gy., I. Guczi and D. Stur. J. Radioanal. Chem. **111**:441 (1987).

12. Unger, K.K. Porous Silica, its Properties and Use as Support in Liquid Column Chromatography (Amsterdam, Elsevier 1979).

13. Fulcher, C., M.A. Crowell, R. Bayliss, K.B. Holland and J.R. Jezorek. Anal. Chim. Acta **129**:29 (1981).

14. Smith, H., G.N. Stradling, R.A. Bulman and G.J. Ham. Health Phys. **30**:318 (1976)

15. Bondietti, E.A., S.A. Reynolds and M.H. Shanks, In: Transuranium Nuclides in the Environment, STI/PUB/410, pp. 273-287 (Vienna: International Atomic Energy Agency, 1975).

16. Kerndorff, H. and M. Schnitzer. Geochim. Cosmochim. Acta **28**:1701 (1980).

17. Nash, K.L. and G.R. Choppin. J. Inorg. Nucl. Chem. **42**:1045 (1980).

18. Choppin, G.R. and K.L. Nash. J. Inorg. Nucl. Chem. **43**:357 (1981).

19. Smith, R.M. and A.E. Martell. Sci. Tot. Environ. **64**:125 (1987).

20. Torres, R.A. Ph.D. Thesis, Florida State University, 1982.

21. Szabo, Gy., L.S. Prosser and R.A. Bulman. These proceedings and submitted to Chemosphere.

Complexing Properties of Humic Substances Isolated from Sea Water; the Contribution of these Substances to Complexing Capacities of Water from the Baltic Sea and Geochemical Implications of this Phenomenon

Janusz Pempkowiak
Institute of Oceanology
Polish Academy of Sciences, P.O. Box 68, Sopot, Poland

Abstract

Some 20% of the cadmium, 60% of the lead and 70% of the copper dissolved in the Baltic Sea occur in the form of organic complexes, as revealed by differential pulse anodic stripping voltammetry (DPASV) analyses. Complexing capacities in the ranges 0.8-1.7 μg/l (Pb), 1.4-3.2 μg/l (Cu), and 0.3 μg/l (Cd) were also found. Complexing capacities of humic substances account for 70-90% of those of sea water, indicating that these substances are responsible for the major part of the complexing properties of the Baltic water. Aqueous solutions of humic substances exhibit decreasing complexing capacities with increasing salinity. This suggests major changes in proportion of dissolved species in the mixing zone of fresh and saline water. Some 60% of the complexed lead and 40% of the complexed copper were released from organic complexes in water from the Vistula River when salinity was increased in laboratory experiments.

Introduction

The view that speciation of trace metals in water is essential for their geochemistry (migration and distribution between phases) and bioaccumulation has found increasing support [1-4]. Investigation of physicochemical forms of metal ions in water is difficult due to sensitivity of the equilibria between species [2,5-7] and a risk of contamination in the course of separation [8,9]. Therefore, a number of methods have been tried in speciation studies, including ultrafiltration, fluorescence, dialysis, the use of ion selective electrodes, and polarography [5,10,11]. From among the methods utilized for speciation in sea water, anodic stripping voltammetry (ASV) is often employed, because determinations are carried out without separation of different forms (no contamination), and if properly executed, the method does not influence the equilibria. The main disadvantage of the method is excessive selectivity. Moreover, several species are determined at the same time, giving the so-called electroactive fraction [2,12,13]. Therefore, the speciation of determinable forms is unprecise. Adsorption of organic ligands on the mercury surface [2,6], local supersaturation of ligands with trace metals during the dissolution step [2], and a heterogeneous nature of organic ligands [2,14] is also suspected. Nevertheless, the general operational simplicity and sensitivity of ASV have encouraged its fairly widespread adaptation [6].

Most often, ASV measurements are performed on filtered water samples before and after UV oxidation of organic matter; the difference found is attributed to release of metal ions bound to organic ligands. In addition, complexing capacities (concentrations of various metals converted into non-electroactive forms due to complexation with organic ligands originally present in water samples) utilized in toxicological studies can be easily evaluated by applying ASV to samples titrated with a given metal [6,15]. Recently, Nürnberg [16] proposed a more elaborate speciation procedure, which includes ASV determination of a sample acidified to pH 2. This makes it possible to evaluate the amounts of metal ions bound in weak organic complexes as well. The procedure has been successfully employed in speciation studies of cadmium, lead, copper and bismuth, in samples collected in the North Sea off the Belgian coast [17].

Organically bound metal ions are common in humus rich waters. Humic substances, due to their huge concentrations and their chemical properties, greatly influence complexing capacities of sea water [4,7,18,19]. This is particularly true in the Baltic Sea, because the concentration of humic substances is an order of magnitude higher than in ocean water, and half of the substances present are abundant in functional groups of substances of terrestrial provenience [4].

The present paper presents data indicating that concentrations of organically bound cadmium, lead and copper, as well as complexing capacities of the Baltic water, are caused by dissolved humic substances. Complexing properties of the substances depend to a great extent on both the pH and salinity of an aqueous solution of isolated substances, which suggests changing proportions of dissolved metal species in the mixing zone of river and sea water.

Materials and Methods

Water samples were collected at sampling stations indicated on the map presented in Fig. 1. Samples were collected with a plastic bathometer suspended on a plastic lined wire. Filtration and other operations were carried out in a laminar flow bench. Before analyses, samples were stored at 4°C. Determinations were carried out with a PPO4 polarograph and a rotating glassy carbon electrode using differential pulse anodic stripping voltammetry. The following operating parameters were used: deposition potential, −1000 mV; deposition time, from 400 to 700 s; potential ramp 7 mV/s, pulse height 50 mV; rotation rate of the electrode, 1500 rpm. All surfaces in contact with the samples were washed with 1:1 nitric acid, distilled water (Elgastat - England) and actual water sample. All chemicals used were Suprapure (Merck). Using the described procedure, the following precision of determinations (mean of five determinations ± σ) of cadmium, lead and copper in a water sample collected from a quay in Sopot were recorded: Cd, 0.11±0.03; Pb, 0.21±0.04; Cu, 1.83±0.12 (μg/l). Humic substances were isolated from water samples collected in polyethylene containers from the open sea and the Vistula River. After filtering, water was acidified to pH 2 and passed through glass columns filled with Amberlite XAD-2 resin (10 bed volumes per hour). After about 4000 bed volumes of water had passed through, columns were washed with distilled water and humic substances were desorbed with 1 M ammonium hydroxide. The eluate was concentrated in a rotary evaporator at 45°C until a concentration factor of about 4000 was reached. Concentration of humic substances in the concentrate was determined gravimetrically. Titration with a solution of an appropriate ion was used to evaluate complexing capacity of both the water samples and aqueous solutions of humic substances.

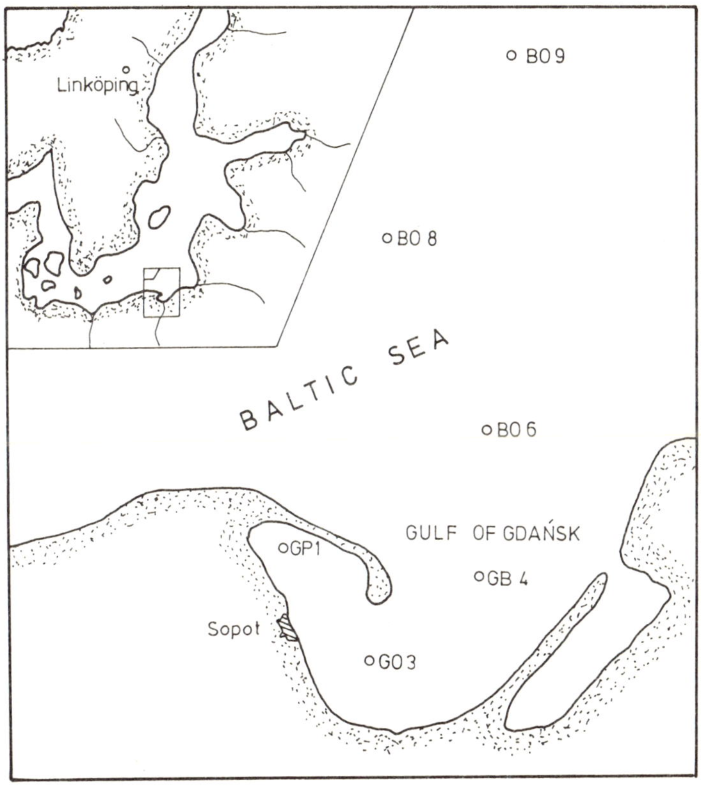

Figure 1 Map of the Southern Baltic Sea showing surface (10 m deep) water sampling stations.

Results and Discussion

Concentrations of various cadmium, lead and copper species in water samples collected at the stations shown in Fig. 1 are presented in Table 1.

Table 1 Concentration and speciation of cadmium, lead and copper in the Baltic water.

	Concentration of various species (µg/l)								
Sampling station	Cd			Pb			Cu		
	A	B	C	A	B	C	A	B	C
GPI	0.12	0.14	0.16	0.15	0.21	0.38	0.52	0.79	1.65
GO3	0.13	0.14	0.15	0.17	0.20	0.34	0.71	1.15	1.85
GB4	0.05	0.04	0.04	0.08	0.11	0.23	0.40	0.84	1.30
BO6	0.03	0.03	0.04	0.14	0.21	0.28	0.38	0.62	1.42
BO8	0.03	0.04	0.03	0.23	0.25	0.25	0.33	0.47	1.16
BO9	0.03	0.03	0.04	0.19	0.23	0.27	0.25	0.52	1.28

A - fraction determined at natural pH ("free" ions, hydrated ions, inorganic complexes, some labile organic complexes)
B - fraction determined at pH 2.0 (metals bound in weak organic complexes plus fraction A)
C - metal concentration after UV-irradiation (metals bound in strong organic complexes, plus weak organic complexes and fraction A)

The total concentrations of the metals are characteristic of coastal and open areas of the Baltic Sea [12,20,21]. This indicates that sampling and analytical procedures used prevented contamination of the investigated samples. The results of speciation presented in Table 1 indicate a large and relatively stable proportion of organically bound copper, also a very large but rather variable proportion of lead, and a small fraction of organically bound cadmium. A small fraction of organically complexed cadmium was reported earlier in water from the Baltic [12,20] and North Seas [17]. As for copper, both small proportions [18,22] and high proportions [20,21,23] were reported; the former results refer to copper bound with hydrophobic organic compounds only, thus they are not representative. Variable concentrations of both total and organically complexed lead can be noticed. This is most likely caused by a large amount being transported through the atmosphere, creating patches of very high lead concentration [12]. Complexation of trace metals in the Atlantic Ocean has also been reported, however to a much smaller extent [7]. This is caused by the presence of smaller concentrations of organic substances in the ocean than in the Baltic Sea [4,7]. Another reason could be the low salinity of the Baltic water. Using mathematical modelling Mantoura *et al.* [3] established that the fraction of organically bound trace metals decreases rapidly with increasing salinity.

Complexing capacities of the Baltic Sea water towards cadmium, lead and copper are presented in Table 2. The complexing capacity towards copper (1.4-3.4 µg/l) is about ten times that towards cadmium (0.3 µg/l); the complexing capacity towards lead is in between (0.8-1.7 µg/l). Such a sequence has been reported for other bodies of water [13,15,17,24]. Strength of the complexes has been conveniently explained on the basis of stability constants and free energy of complex formation [25,26]. Differences in complexing capacity established in this study indicate the possibility of copper being bound by donor centers inactive for other metals [23]. Complexing properties of humic substances isolated from sea water are also presented in Table 2. The substances were isolated using a procedure of sorption from acidified water on XAD-2 resin and a subsequent desorption with ammonium hydroxide [4,23].

Table 2 Complexing capacities (c.c.) of humic substances towards cadmium, lead and copper.

	Water c.c.			Humic acids			
Sampling station	Cd	Pb	Cu	conc.	c.c. (μg/mg)		
		μg/l		mg/l	Cd	Pb	Cu
GP1	-[a]	1.6	3.4	2.73	0.08	0.61	1.24
GO3	0.3	1.7	3.2	2.13	0.09	0.76	1.59
BO8	-	0.8	1.4	1.93	0.05	0.34	0.62

[a] - below detection limit (~0.1 μg/l)

Isolated humic substances contained small amounts of metals, which changed little upon passing through an ion exchange column. The aqueous solutions of the substances were then titrated with metals, and the complexing capacities were calculated from the titration curves. The small complexing capacities of humic substances towards cadmium draw attention. This is due to the property of the cadmium ion which as a soft acid, forms stable complexes with soft bases (according to Brönsted classification). From among organic electron donors, the sulfhydryl functional group is particularly well-suited for forming complexes with cadmium [4]. Elemental analysis of humic substances isolated from sea water indicated very little, if any, sulphur [4]. Comparing complexing capacities of sea water and humic substances one comes to a conclusion that the substances account for 70-90% of complexing capacities of the water. Thus, humic substances, which constitute 30-40% of dissolved organic matter in the Baltic Sea water [4], are almost entirely responsible for the observed complexing properties. Apart from differing complexing capacities towards cadmium, lead and copper, humic substances exhibit different strength of complexes. Earlier investigations of humic substances indicate that the most stable complexes are formed with copper and the least with cadmium [4,19]. The stability of humic acid-lead complexes is comparable with that of cadmium, although complexing capacities are much larger. The results are based on the shape of titration curves of humic substances with metal ions [4].

The influence of the pH of an aqueous solution of a humic substance on its complexing capacity is presented in Fig. 2. In the case of cadmium and lead, the dependence was investigated at pH above 4.2 only, since a distortion of the polarographic curves occurred in acidic solutions. This was caused by adsorption of humic substances on the mercury surface [2,6]. The complexing capacity towards copper decreases rapidly when pH decreased from 4 to 3. This is caused by protonation of carboxylic and phenolic groups capable of acting as electron donors only when dissociated [7,18]. In the pH range between 7 and 8, characteristic of estuaries, complexing capacities do not change appreciably due to this phenomenon.

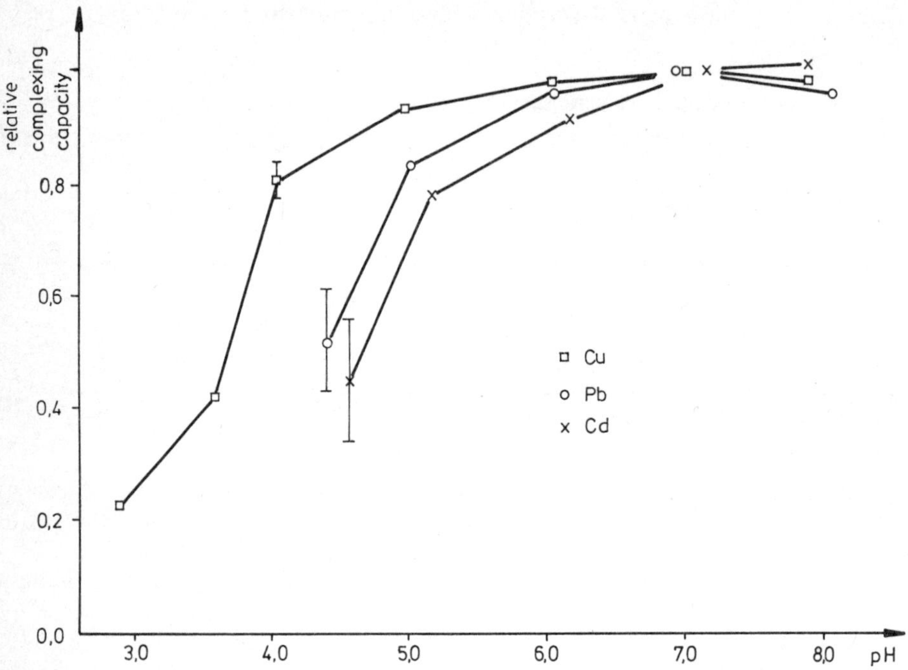

Figure 2 The dependence of relative complexing capacities (c.c.) of humic substances on the pH of a solution (15 mg/l of humic acids isolated from the BO8 water sample dissolved in 0.1 M NaCl; c.c. at pH 7 is equal to 1.0. For numerical values of c.c., see Table 2.). Bars indicate standard deviation of 5 determinations.

Changes of complexing capacities of humic substances isolated from the Baltic Sea water occurring in the course of salinity (ionic strength) increase are presented in Fig. 3. The decrease of complexing capacities is caused by competing action of Ca^{2+} and Mg^{2+} ions present in sea water at concentrations exceeding those in river water some fifty-fold. The decreases of both lead and cadmium complexing capacities are more pronounced than the decrease for copper, which agrees well with the lower apparent stability constants of humic-lead and humic-cadmium complexes [4]. Complexing capacity towards copper decreases slowly; furthermore there seems to be about 0.3 μg/mg of residual complexing capacity present even at large salinities, which again suggests copper to be bound to complexing sites specific for this element, possibly for steric reasons. Increasing salinity (ionic strength) causes conformational changes within molecules of humic substances [26]. This influences stability constants to some extent [25]. It remains to be seen whether the changes could be due to something other than the competition of Ca^{2+} and Mg^{2+}.

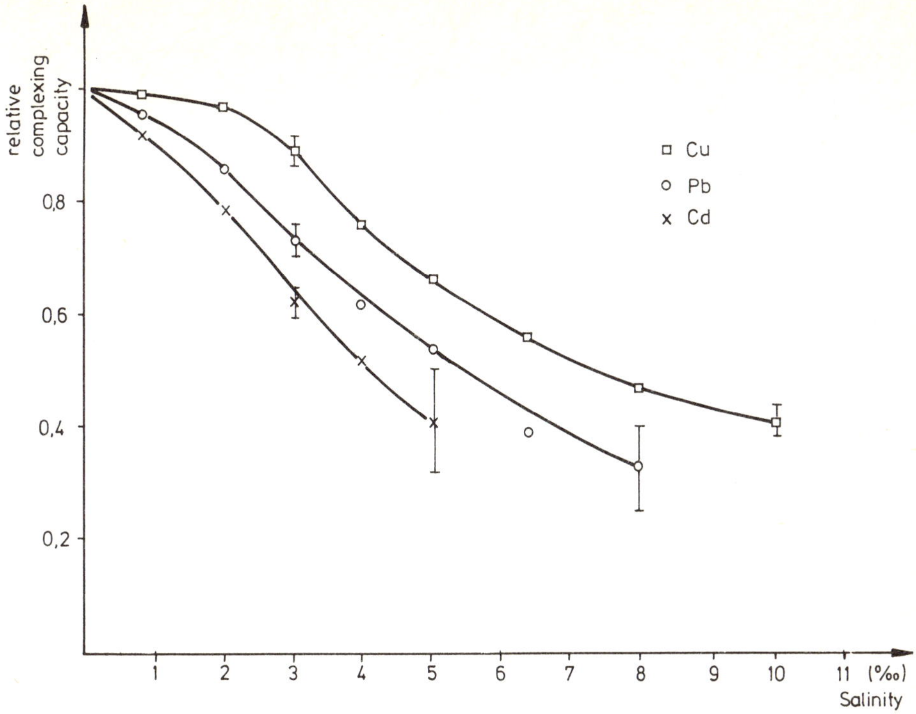

Figure 3 The dependence of relative complexing capacities (c.c.) of humic substances on the salinity of an aqueous solution (15 mg/l isolated from the BO8 water sample dissolved in a mixture of distilled and UV-irradiated ocean water; c.c. at 0‰ salinity is equal to 1.0. For numerical values of c.c., see Table 2.). Bars indicate standard deviation of 5 determinations.

The decrease in complexing capacities towards ions of transition metals may be important for understanding the phenomena occurring in the estuaries. Both trace metals and humic substances concentrations in river water exceed several times those in sea water. For instance, in the Vistula River the average concentration of humic substances is equal to 7.8 mg/l, whereas for the Baltic Sea water this value is 2.1 mg/l [4]. The increase of salinity in a mixing zone of fresh and saline water should, therefore, release ions from both dissolved and flocculating humic substances. The overall concentration of dissolved species of metals can remain unchanged, whereas the proportion of various species changes, influencing, for instance bioavailability and geochemical behaviour of metals. Results of an experiment presented in Fig. 4 support this conclusion. In the experiment, ocean water was added to the Vistula water so that a final salinity in the range of 1.0 to 10.0‰ was reached. It was found that increasing salinity caused increasing concentrations of electroactive (not complexed with humic substances) forms of lead and copper. The increase found is equal to some 65% of the lead and 40% of the copper originally bound in organic complexes in river water. In the case of cadmium no increase was found, however the precision of determinations was rather poor.

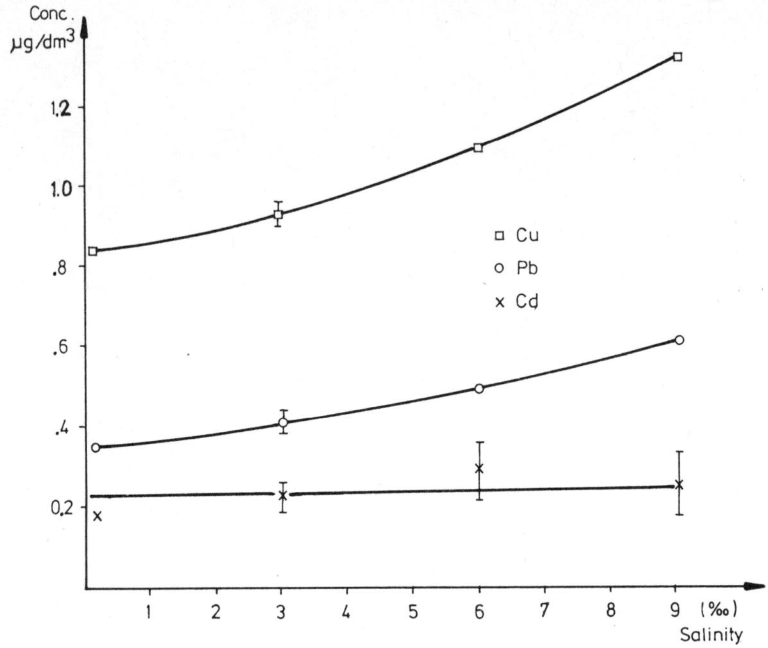

Figure 4 The dependence on salinity of electroactive species concentrations of cadmium, lead and copper in Vistula River water mixed with UV-irradiated ocean water.

Conclusions

The interaction with humics should influence the geochemistry of trace metals, which, upon entering (with fresh water) the marine environment, encounter a geochemical barrier caused by increasing pH and salinity (ionic strength). Presented results indicate that increasing salinity exerts great influence on the amount of major dissolved species of lead and copper and, to a smaller extent cadmium, causing a decrease in the organically bound fraction of the metals. Such important geochemical characteristics as distribution between various phases (including biota), residence time, concentrations, migration, and sedimentation, depend on the organically bound fraction. Quantitative evaluation of the influence of the phenomenon calls for elaborated studies. It has already been established, however, that in the presence of humic substances, accumulation of trace metals by algae decreases (three- to five-fold, depending on metal concentrations), as does sorption of metal ions on suspended matter [4]. The latter phenomenon was attributed to both modified properties of clay surface caused by humus sorption and the modified speciation within the dissolved fraction.

Acknowledgement

This work was financially supported by the Polish Academy of Sciences grant C.P.B.P. 03.10.2.2.

References

1. Anikiev, W. Short-term geochemical processes and the ocean pollution (Moscov: Nauka, 1987).

2. Buffle, J., J. Vuilleumier, M. Tercier and N. Parthasarathy. Sci. Tot. Environ. **60**:75 (1987).

3. Mantoura, R., A. Dickson and J. Riley. Est. Coast. Mar. Sci. **6**:387 (1978).

4. Pempkowiak, J. Concentrations, properties and origin of humic substances in the Baltic Sea (Wroclaw: Ossolineum, 1989).

5. Florance, T. and G. Batley. Talanta **24**:151 (1977).

6. Gorman, W., R. Skogerboe and P. Davis. Anal. Chim. Acta **187**:325 (1986).

7. Piotrowicz, S., G. Harvey, D. Boron, C. Weisel and M. Springer-Young. Mar. Chem. **14**:333 (1984).

8. Kramer, C., Y. Guo-hui and J. Duinker. Fresenius Z. Anal. Chem. **317**:383 (1984).

9. Mart, L. Talanta **29**:1035 (1982).

10. Florance, T. Talanta **29**:345 (1982).

11. Tuschall, J. and P. Brezonik. In: C. Kramer and J. Duniker, Eds, Complexation of Trace Metals in Natural Waters, pp. 83-94 (The Haque: M. Nijdorf/W. Junk Publ., 1984).

12. Brügmann, L., T. Gian and H. Berge. Acta Hydrochim. Hybrobiol. **16**:457 (1988).

13. Hasle, J. and M. Abdullah. Mar. Chem. **10**:475 (1981).

14. Turner, D., M. Varney, M. Whitfiels, R. Mantoura and J. Riley. Sci. Tot. Environ. **60**:17 (1987).

15. Tao, S., J. Chen and F. Tand. Environ. Tech. Lett. **8**:433 (1987).

16. Nürnberg, H. In G. Leppard, Ed., Trace Element Speciation in Surface Waters and its Ecological Implications, pp. 211-230 (New York: Plenum Publ.Corp., 1983).

17. Gillain, G. and C. Brihaye. Oceanol. Acta **8**:231 (1985).

18. Osterroht, C., W. Wenck and K. Kremling. Mar. Ecol. Prog. Ser. **22**:273 (1985).

19. Pempkowiak, J. and L. Brügmann. Complexing capacities of humic substances isolated from the Baltic environment towards cadmium, lead and copper. Proc. XIV Conf. Baltic Oceanogr. (Gdynia: Comm. Sea Expl, 1984).

20. Bordin, G., M. Parttilä and H. Scheinin. Mar. Pollut. Bull. **19**:325 (1988).

21. Brügmann, L. and P.-V. Rapp. Reun. Cons. Int. Esplor Mer. **186**:329 (1986).

22. Osterroht, C., K. Kremling and W. Wenck. Mar.Chem. **23**:153 (1988).

23. Pempkowiak, J. Oceanol. **16**:167 (1983).

24. Szpakowska, B., J. Pempkowiak and Z. Baloniak. Arch. Hydrobiol. **108**:259 (1986).

25. Saar, R. and J. Weber. Environ. Sci. Technol. **16**:510 (1982).

26. Plavsic, M., H. Bilinski and M. Branica. Mar. Chem. **21**:151 (1987).

An Evaluation of Chemically Immobilized Humic Acid: A New Stationary Phase for RP-HPLC Prediction of Uptake of Organic Pollutants by Soils and Sediments

Gyula Szabo[2], S. Lesley Prosser[1] and Robert A. Bulman[1]

[1] National Radiological Protection Board, Chilton, Didcot, OX11 0RQ, England.

[2] Present address: Frederic Joliot-Curie National Research Institute for Radiobiology and Radiohygiene, PO Box 101, H-1775 Budapest, Hungary.

Abstract

Humic acid has been chemically immobilized on HPLC grade silica and used as the stationary phase for RP-HPLC. This new phase has been evaluated by determination of the soil-organic carbon/water partition coefficent (K_{oc}) for a series of aromatics. Log K_{oc} has been determined for acenaphthene, acenaphthylene, fluorene and fluoranthene; the respective values are 3.79, 3.83, 4.15 and 4.74.

Introduction

To predict the environmental fate of anthropogenic substances requires an understanding of their transport processes through soils and sediments. An accurate prediction of these transport processes requires an understanding of their interactions with humic substances which are now well recognized as interceptors of many organic pollutants [1-4].

An important parameter in predicting the environmental fate of organics is the partition coefficient (K_p) for sediment/water or soil/water. As direct determination of K_p is difficult, its value has been determined by using a variety of relationships. For the sake of brevity these relationships and their definitions are summarized in Table 1. In recent years reverse phase - high performance liquid chromatography (RP-HPLC) has been evaluated for determining the values of some of the parameters which are involved in these relationships [5,6]. In order to mimic the adsorptive properties of the organic phase of soils investigators have used octadecyl- and cyanopropylsilica as the stationary phases. Obviously, the closest prediction of the uptake of anthropogenic compounds by soils will be obtained from stationary phases which bear a strong resemblance to soil or sediment. Here we report our development of a new stationary phase, SiO_2-CBHA, formed from silica bearing chemically bound humic acid (CBHA). Its potential has been monitored by its prediction of the soil adsorption coefficient (K_{oc}) for a series of aromatics.

Table 1 Relationships and terms used in determination of uptake of pollutants by solid phases from aqueous phases.

Term	Definition	Relationship
K_p	Solid phase/wate partition coefficient	$\log K_p = \log K_{ow} + \log f_{oc} + b$ f_{oc} is the fractional organic content of particulates and a and b are constants
K_{ow}	n-octanol/water partition coefficient	
K_{oc}	$\frac{\mu\text{g adsorbed chemical/g organic carbon (from soil or sediment)}}{\text{concentration in aqueous solution } \mu\text{g/ml}}$	Above equation can be simplied to $\log K_{oc} = a \log K_{oc} + b$
k'	HPLC capacity factor	
k'_w	Theoretical HPLC capacity factor; represents the capacity factor of the solute with pure water as the mobile phase	Obtained by extrapolation from a binary solvent phase to 100%; $\log k' = \log k'_w + S\Phi$ where Φ is the volume fraction of organic solvent in the water-organic mixture and S is a constant for a given solute - eluent combination. $\log K_{oc} = \log k'_w + b$

Materials and Methods

The aromatic hydrocarbons used in this study were obtained from Aldrich Chemical Co Ltd and were used without further purification.

Preparation of SiO_2-CBHA

Humic acid (Aldrich Chemical Co Ltd) was immobilized on Hypersil WP 300 5 µm (Shandon, Cheshire) by the reaction sequence described elsewhere [7,8]. The essential steps are outlined in Fig. 1. Basically, 3-aminopropyltriethoxysilane (Fluka Chemicals Ltd) was refluxed with silica in dry toluene and washed free of reactants with toluene, methanol and water and then dehydrated by washing with methanol. The dried aminopropylsilica (*I*) was activated with 10 volumes of 5% aqueous glutaraldehyde for 5 h to produce an activated gel (*II*) which was retained on a filter and washed with 15 volumes of distilled water. Further derivatization by reaction with 1% aqueous solution (100 ml) of humic acid, pH 7.5, for 8 h at ambient temperature yielded SiO_2-CBHA (*III*). This material was washed with 10 volumes of 0.5 M phosphate buffer and distilled water and then treated with 0.1 M buffered ethanolamine, pH 7.5, for 3 h to block unconsumed aldehyde groups. The reaction product was washed with a large excess of distilled water and dried to yield a dark-brown product. Elemental analysis of this new stationary phase was performed in the Micro Analytical Laboratory at the University of Manchester.

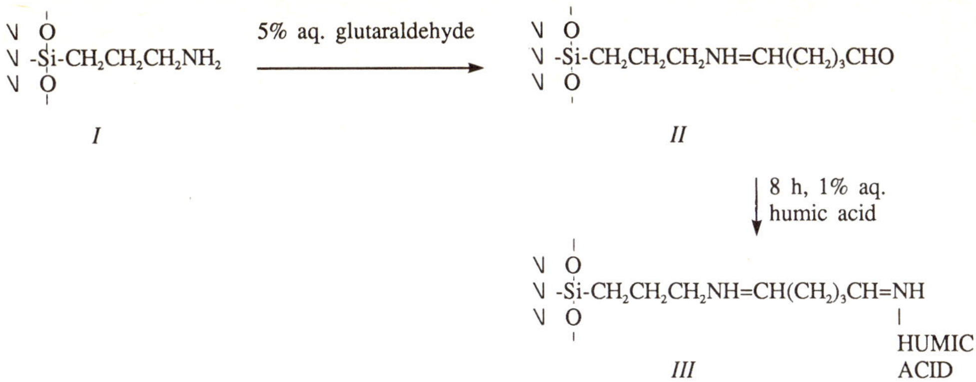

Figure 1 Synthetic route to SiO_2 - CBHA.

Chromatographic Procedures

SiO_2-CBHA was packed by Jones Chromatography, Hengoed, Wales into a 250 x 4.6 mm stainless steel column. Dissolved air was removed from HPLC grade methanol and water by helium entrainment, mixed volume/volume and pumped at a flow rate of 0.8 ml/min. A laboratory temperature of 20-23°C was used for all measurements. The mobile phase content was changed from 60% to 40% by 5% adjustments. Chromatographic retention data were determined with an LKB 2150 solvent delivery system. The test solutes were dissolved in methanol at a concentration of 100 μg/ml and the samples loaded on to the column through a Rheodyne 7125 injection valve fitted with a 20 μl loop. Chromatograms were recorded on a desk top computer using LKB Wavescan Diode-array detection. Methanol was used for the determination of the retention time (t_o) of an unretained compound. The relationship:

$$k' = (t_r - t_o)/t_o$$

was used to calculate the capacity factor, k' from the retention time (t_r) of each compound. All capacity factors reported were the mean of at least three measurements. The correlation analysis for all compounds was made by linear regression analysis of logk', or log K_{ow}, *vs* log K_{oc}. A least squares fit was made using commercially available software (Graphpad™, Instititute for Scientific Information, Philadelphia, USA). We selected from the literature, using the well established criteria of Brooke *et al* [9], log K_{oc} and log K_{ow} values for the first 10 compounds listed in Table 2.

Results and Discussion

Elemental analysis of SiO_2 - CBHA revealed a composition of 4.7% C, 0.7% N and 0.5% H. An extensive search of *Chemical Abstracts* has failed to find any report of a similar HPLC stationary phase.

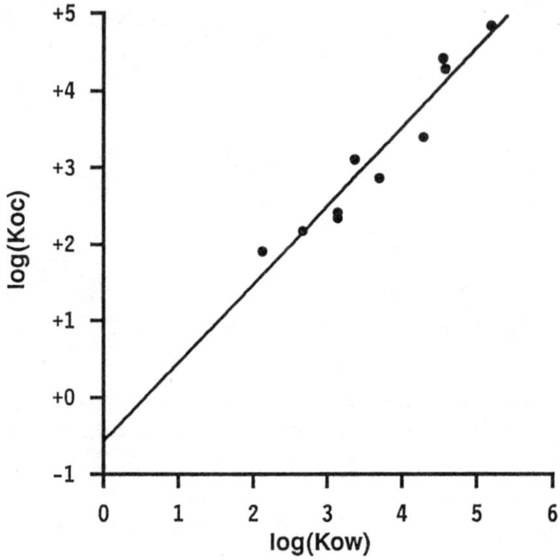

Figure 2 Relationship between soil/water partition coefficients and n-octanol/water partition coefficient.

An examination of Fig. 2 shows the correlation data between log K_{oc} and log K_{ow} values for the first 10 chemicals listed in Table 2. Analysis of the plot gave the relationship:

$$\log K_{oc} = 1.023 \log K_{ow} - 0.578; \quad r^2 = 0.922 \quad n = 10$$

At a value of 1.023, the slope parameter compares well with the observations of others [3,4,10] who reported a value of 1.0 for various substrates and solutes.

In order to eliminate selective solute-solvent interactions [4,11,12], we have used log k'_w, the capacity factor obtained by extrapolation of retention data from binary eluents to 100% water, instead of using log k', the capacity factor obtained from binary eluents. It has been shown by Snyder *et al* [13] that these two terms can be described by the linear equation:

$$\log k' = \log k'_o + S\Phi$$

where Φ is the volume fraction of organic solvent in the water-organic solvent mixture, k'_w represents the capacity factor of a solute with pure water as a mobile phase and S is a constant for a given solute-eluent combination.

The values of log k'_w, obtained by using the above equation, were calculated for the 10 standard chemicals and also for the 4 chemicals for which we were determining log K_{oc} values. These values are listed in Table 2. From an analysis of a plot of log K_{oc} against log k'_w (Fig. 3) the relationship

$$\log K_{oc} = 0.948 \log k'_w + 1.781$$

was obtained. By using the above equation log K_{oc} has been calculated for acenaphthene, acenaphthylene, fluorene and fluoranthene; the values are 3.79, 3.83, 4.15 and 4.74, respectively.

Table 2 Literature values of log K_{ow} and log K_{oc}, as well as log K_{oc} and log k'_w values for SiO_2-CBHA.

	Literature values				SiO_2-CBHA	
Chemical	Log K_{ow}	Reference	Log K_{oc}	Reference	Log K_{oc}	Log k'_w
Benzene	2.11	4	1.91	4	1.87	0.096
Toluene	2.65	9	2.18	9	2.26	0.514
Ethylbenzene	3.13	6	2.41[1]		2.52	0.782
Propylbenzene	3.69	6	2.86[1]		2.83	1.107
Butylbenzene	4.28	6	3.40	6	3.15	1.407
o-Xylene	3.13	6	2.34[1]		2.37	0.627
Naphthalene	3.36	4	3.11	4	3.15	1.452
Phenanthrene	4.57	6	4.28[1]		4.22	2.569
Anthracene	4.54	4	4.41	4	4.53	2.907
Pyrene	5.18	4	4.83	4	4.82	3.212
Acenaphthene	NA[2]		NA		3.79	2.162
Acenaphthylene	NA		NA		3.83	2.128
Fluorene	NA		NA		4.15	2.498
Fluoranthene	NA		NA		4.74	3.121

[1] Converted from literature log K_p value using the relationship $K_{oc} = K_p \times 100$ / % organic carbon

[2] NA=not available

From an examination of Table 2, it can be seen that the divergences of the two values for log K_{oc} for the 10 standard chemicals are small, from which it might be assumed that the adsorptive properties of SiO_2-CBHA are similar to those of sediment and soil. In a more detailed analysis [7] of the potential of this new HPLC stationary phase we have suggested the following order for prediction of log K_{oc} can be generated:

organic matter on soil > SiO_2-CBHA > ethylsilica > n-octanol.

In conclusion, we suggest that SiO_2-CBHA might well be a superior HPLC stationary phase for studying the interactions of a wide variety of anthropogenic organic compounds with humic substances in soils and sediments.

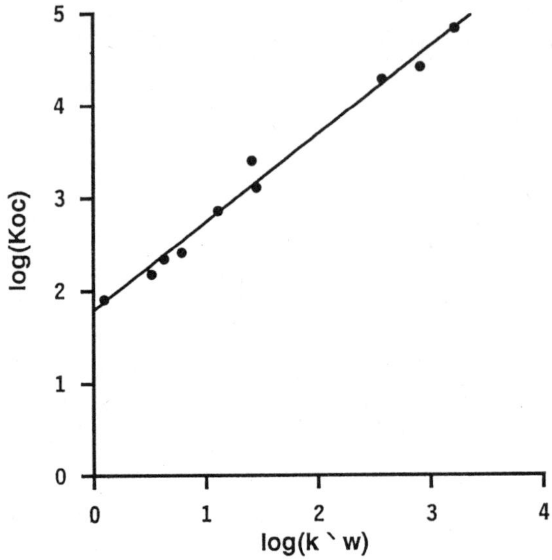

Figure 3 Relationship between soil/water partition coefficient and the theoretical capacity factor for SiO_2-CBHA.

Acknowledgements

Gy. Szabo was the grateful recipient of a training fellowship from the International Atomic Energy Agency.

References

1. Physical/Chemical Properties, Measurements and Tests. Office of Pesticide Program, Environmental Protection Agency (1978).

2. Guidelines for Testing of Chemicals, (Paris: Organization for Economic Cooperation and Development, 1981).

3. Means, J.C., S.G. Wood, J.J. Hasett and W.L. Banwort. Environ. Sci. Technol. **12**:1524 (1980).

4. Karickhoff, S.W., D.S. Brown and T.A. Scott. Water Res. **13**:241 (1979).

5. Eadsworth, C.V. Pest. Sci. **17**:341 (1986).

6. Vowles, P.D. and R.F.C. Mantoura. Chemosphere **16**:109 (1987).

7. Szabo, Gy., L.S. Prosser and R.A. Bulman. submitted to Chemosphere.

8. Bulman, R.A. and Gy. Szabo. These proceedings.

9. Brooke, D.N., A.J. Dobbs and N.A. Williams. Ecotoxicol. Environ. Sci. Safety. **11**:251 (1986).

10. Karickhoff, S.W. Chemosphere **10**:833 (1981).

11. Braumann, Th. and L.H. Grimme. J. Chromatogr. **206**:7 (1981).

12. Braumann, Th., G. Weber and L.H. Grimme. J. Chromatogr. **261**:329 (1983).

13. Snyder, L.R., J.W. Dolan and J.R. Gant. J. Chromatogr. **165**:3 (1979).

The Influence of Aquatic Humic Substances on the Octanol/Water Partition Coefficient (Kow) of Pesticides and Trace Elements at Different pH Values

Gunnhild Riise and Brit Salbu
Isotope and Electron Microscopy Laboratory, Agric. Univ. of Norway
P.O. Box 26, N-1432 Aas-NLH, Norway

Abstract

The influence of pH and aquatic humic substances on the mobility and potential bioavailability of pesticides (MCPA and dazomet) and several trace elements (e.g. Zn) have been examined. The methods used were ultrafiltration (hollow fibers) and liquid/liquid extraction (the octanol/water partition coefficient - Kow). Based on hollow fiber ultrafiltration of humic waters, about 15% of MCPA and 40-70% of ^{65}Zn were associated with high molecular weight components. The estimated relationship between log Kow and pH for MCPA and dazomet shows that the Kow value of MCPA (an ionizable organic compound) is strongly dependent on pH, whereas the neutral compound dazomet has relatively stable Kow values below pH 11. Aquatic humic substances are shown to indirectly influence the Kow value of MCPA by affecting the pH of the water samples, but there are no significant differences between the Kow values of MCPA and dazomet in the humic samples compared to the reference samples at the same pH. No significant amounts of trace elements, from different water samples (where TOC value varied from 6.54 to 230 mg C/l and pH varied from 4.43 to 3.23), were extracted into the octanol phase.

Introduction

Aquatic humic substances interact with organic [1-3] and inorganic compounds and may have a major influence on their physico-chemical properties in the environment, e.g. the mobility and bioavailability [4,5]. For low molecular weight compounds the association with humic substances will change the size distribution pattern. Based on size fractionation techniques, most trace elements in humic rich fresh waters are, to a certain extent, associated with high molecular weight organic compounds [6-8]. Biotest experiments with fish (*Salmon salar L.*) also indicate that the association with humic substances influences the biological uptake of trace metals (i.e. ^{115}Cd) [9].

The distribution of chemical compounds between an octanol and a water phase at equilibrium, the octanol/water partition coefficient (Kow), is a measure of hydrophobicity. This parameter is often used as an indicator of chemical behaviour in the environment, and has been correlated to aqueous solubility [10,11], bioconcentration in fish [12-14], and acute toxicity of waterborne chemicals [15]. Furthermore, the Kow value has been correlated to soil/sediment adsorption capacity as the sorption of neutral organic compounds is often found to be linearly related to the hydrophobicity of the *solute* and the organic carbon content of the *sorbent* [16, 17]. It has, however, also been emphasized that properties of the *aqueous* phase, i.e. the presence of macromolecules (humic acids, colloids) [18-20] and ionic surfactants [21], may influence the sorption/desorption properties of the sorbent.

The octanol/water partition coefficient of aquatic humic substances (Khow) is known to be negatively correlated to pH [22,23], which may indicate an increased bioavailability of aquatic humic substances, as well as associated compounds, with decreasing pH.

The aim of the present work was to study the influence of aquatic humic substances on the octanol/water partition coefficient (Kow) of pesticides at different pH-values. Two ^{14}C labelled pesticides (MCPA and dazomet), and radioisotopes of trace elements (e.g. $^{65}ZnCl_2$) have been added to fresh waters rich in humic substances. The influence of pH in the aqueous phase on the Kow value of the radioactive tracers has been followed. In addition, the association of pesticides and trace elements with high molecular weight components (e.g. aquatic humic substances) has been determined by using hollow fiber ultrafiltration.

Materials and Methods

Samples

Humic waters were collected from Hellerudmyra Bog and Lake Solbergvann in June 1988 and July 1989, respectively. The sampling sites are located in forest areas outside Oslo. Concentrated water samples were obtained by low temperature evaporation at low pressure; under these conditions all elements (except those that are very volatile) are concentrated at the same ratio as the organic material. Tap water (Lake Gjersjøen) and distilled water containing KCl (0.0002 M KCl) were used as reference samples.

Radioactive Tracers

Radioactive tracers used in the experiments were: 1-^{14}C labelled MCPA (4-chloro-2-methylphenoxyacetic acid) with a specific activity of 297 μCi/mg (Amersham Int. UK); ^{14}C labelled dazomet (tetrahydro-3,5-dimethyl-1,3,5-thiadiazine-2-thione) with a specific activity of 2.45 μCi/ml (BASF); $^{65}ZnCl_2$ with a specific activity of 1.25 mCi/ml (Amersham Int. UK) and 0.1 M HNO_3 solutions containing Sm, La, Br, Mo, Sb, Hg, Cr, Ce, Yb, Sc, Eu, Cd, As, Ba, Se, Sr, Cs, Co and Au irradiated for three days at a thermal neutron flux of 1.2×10^{13} n $s^{-1}cm^{-2}$ at the reactor Jeep II Kjeller, Norway.

Octanol/Water Extractions

50 ml of water samples, spiked with labelled tracers, were transferred to 100 ml glass flasks with screw caps, and 5 ml of l-octanol (PS 97% purity, Kebo Lab) was added. The flasks were shaken for 16 hours and stored for 1-2 hours before separating the two phases.

pH Adjustment/Measurement of the Water Phase

Different amounts of diluted HCl/NaOH were added to the water samples in order to obtain pH-values in the range 2-12. The water samples were stored one day prior to the pH-measurements, which were performed with an Orion model SA pH-meter connected to the Radiometer electrodes G 202 C and K 401.

Hollow Fiber Fractionation

An Amicon concentrator CH_3 equipped with H1P10-20 and H1P1-43 cartridges having molecular weight cut off levels of 10^4 Dalton and $3x10^3$ Dalton, respectively, were used for ultrafiltration. The pore size distribution of the cartridges is relativly narrow [24,25], and sorption was minimized by conditioning the system with a sample aliquot prior to the collection of ultrafiltrate [25]. Radioactive tracers were added to ultrafiltrates ($Mw<3x10^3$ Dalton), which were then stored prior to extraction. Hollow fiber fractionation ($Mw<10^4$ Dalton) of spiked solutions was also performed.

Storage/Degradation

Unfiltered and ultrafiltered water samples from Hellerudmyra Bog were spiked with labelled MCPA and stored in the dark at 4^0C for different periods of time prior to the determination of the Kow values.

Measurement

A Packard Tri-Carb 4530 liquid scintillation counter was used to measure ^{14}C. The ratio between sample and scintillation cocktail was 1:10. Packard Minaxi 500 series NaI 3" gamma counter was used to measure ^{65}Zn. Gamma spectroscopy of the multielement standards was performed with a Canberra Ge detector (20% efficiency, 1.9 keV resolution defined by means of 1332.5 keV gamma from ^{60}Co).

Results and Discussion

The two reference samples differ in pH and conductivity values (Table 1). Water samples from Hellerudmyra Bog and Lake Solbergvann have similar conductivity (3.7 and 3.8 mS m^{-1}) but different pH-values (4.27 and 6.55). About 50% of the TOC is removed from Hellerudmyra water when it is fractionated with the $3x10^3$ Dalton hollow fiber. The increase in pH is attributed to the removal of the humic substances, as this pH effect is not significant for waters with low content of organic carbon (unpubl. data).

Table 1 The pH, conductivity and total organic carbon of the aqueous phase used in the octanol/water extractions

Sample	pH	Conductivity mS/m	TOC mg C/l
0.0002 M KCl	5.47	3.0	
Tap water	7.55	16.5	3.96
Hellerudmyra Total	4.27	3.7	11.90
Hellerudmyra Mw< $3x10^3$ Dalton	4.47	2.6	6.54
Solbergvann Total	6.55	3.8	9.72

The acidic form of the phenoxyacetic acid MCPA is known to be chemically very stable [26]. However, the alkali, ammonium or amine salt, used as herbicide, decomposes in soil within 1-4 months, depending upon soil type and environmental conditions, like temperature and moisture content [27].

Table 2 Kow of MCPA added to total and filtered waters (Mw <$3x10^3$ Dalton) from Hellerudmyra bog after different time of storage (n=3).

	Storage days	Kow mean ± SD	Log Kow
Total sample	1	28.6 ± 2.7	1.45 ± 0.04
	2	31.0 ± 0.5	1.49 ± 0.006
	7	26.2 ± 0.8	1.42 ± 0.02
	29	26.0 ± 1.2	1.41 ± 0.02
	66	24.4 ± 2.0	1.39 ± 0.04
Mw <$3x10^3$ Dalton	1	10.2 ± 0.7	1.01 ± 0.03
	2	10.4 ± 0.3	1.01 ± 0.01
	7	9.5 ± 0.9	0.98 ± 0.04
	29	10.1 ± 0.4	1.01 ± 0.02
	66	8.7 ± 0.4	0.94 ± 0.02

The decrease in the Kow value of MCPA for the total and the filtered fraction after different storage time in Hellerudmyra water (Table 2), was not significant ($p < 0.05$). It may therefore be concluded that, for Hellerudmyra water, the Kow partition coefficient of MCPA is relatively stable for a storage period of 66 days.

The Kow value of MCPA is strongly dependent on the pH of the water samples (Fig.1). The differences in Kow values observed for unfiltered and filtered water samples (Tables 2 and 3) can also be attributed to this pH effect.

Table 3 pH and Kow values of MCPA for total and filtered (Mw<10^4 Dalton) water samples from Hellerudmyra Bog and Lake Solbergvann.

Sample	%MCPA	pH	Kow mean ± SD	log Kow mean ± SD
Hellerudmyra				
Total	100	4.29	26.3 ± 0.1	1.42 ± 0.01
Mw <10^4 Dalton	88 ± 4	4.37	22.1 ± 0.1	1.34 ± 0.01
Solbergvann				
Total	100	6.44	0.5 ± 0.1	-0.34 ± 0.01
Mw <10^4 Dalton	81 ± 4	6.28	0.2 ± 0.1	-0.70 ± 0.01

Phenoxy acids are weak acids with pKa values in the range 2.7-3.3 [28]. The ionization of MCPA will therefore depend on pH (Fig.1). The Kow value of hydrophobic ionizable organic compounds has earlier been demonstrated to be a non-linear function of pH, where three regions of partition behavior are observed [29]. At very low pH-values (pH<3), at which the neutral form of MCPA predominates, the Kow value of MCPA seems to be relatively independent of pH. As the ionization of MCPA increases (pH 3-6), the Kow value of MCPA seems to be a linear function of pH. In the pH-range 6-9 the Kow value is, again, relatively independent of pH. In this region the anionic form is supposed to predominate in both the aqueous and the non-aqueous phase [29]. At still higher pH levels the Kow values in the diagram are more scattered. However, the level reached seems to be rather stable and independent of pH.

The Kow value of dazomet is quite stable in the pH range 2-11 for all the water samples. This indicates that dazomet exists as neutral species below pH 11, and that the mobility of dazomet in the environment is independent of pH.

The hydrophobicity of aquatic humic substances increases with lower pH values (Figs 1 and 2). However, the humic substances are less hydrophobic than MCPA and dazomet. This should imply that MCPA and dazomet associated with humic substances would attain a more polar character and be more water soluble than "free" MCPA and dazomet. Based on hollow fiber ultrafiltration, for both waters investigated, about 85% of MCPA added is found in the ultrafiltrate (Mw<10^4 Dalton) (Table 3). This indicates that the association of MCPA with high-molecular-weight components is significant (about 15%). However, according to Figs 1 and 2 the Kow values of MCPA and dazomet in the reference samples are in agreement with the Kow values in the humic samples. This indicates that the aquatic humic substances have minor influences on the hydrophobicity of MCPA and dazomet, or,

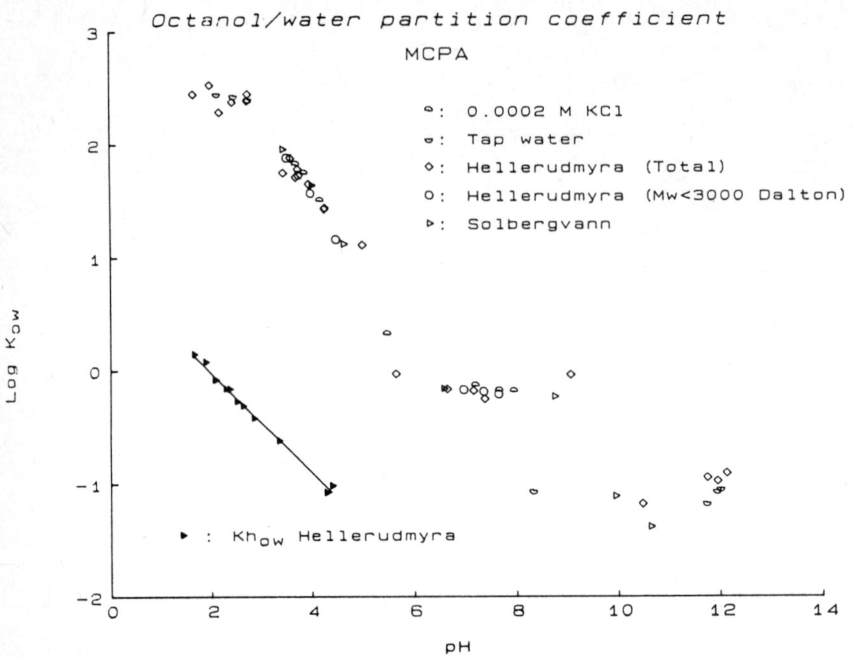

Figure 1 The logarithm of the octanol/water partition coefficient of MCPA vs pH of the aqueous phase in reference and natural water samples.

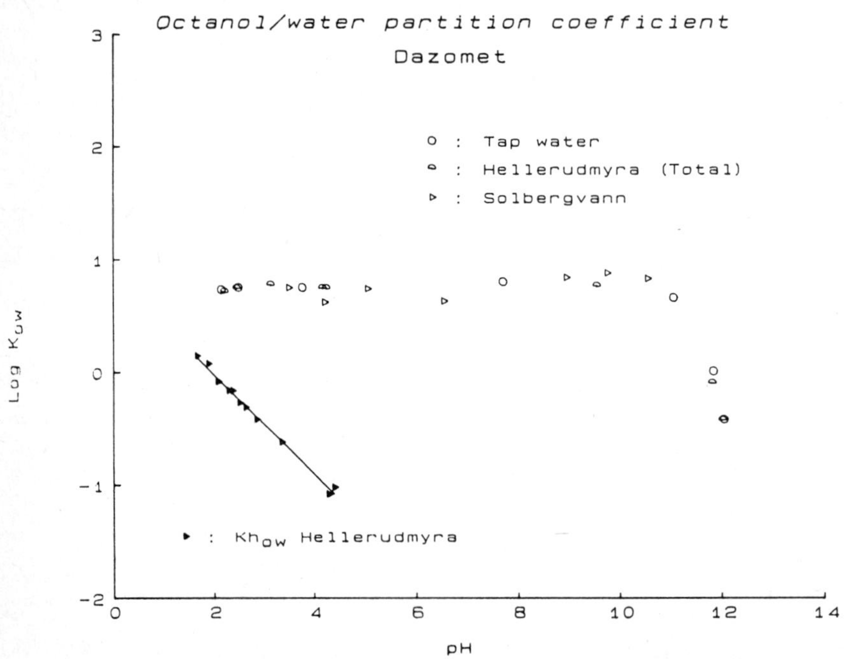

Figure 2 The logarithm of the octanol/water partition coefficient of dazomet vs pH of the aqueous phase in reference and natural water samples.

as claimed by Enfield and Bengtsson [30], that the concentration of macromolecules in the examined waters is too low to have any impact on compounds with such low Kow values as used in the present study.

Trace Elements

Standards and reference samples of aquatic humic substances, isolated according to the International Humic Substances Society standard method, contain trace amounts of several elements, e.g. Zn [31]. Thus, a transfer of humics into the octanol phase, should imply a corresponding transfer for associated trace elements. This hypothesis was tested for ^{65}Zn and multi element standards containing radioactive Sm, La, Br, Mo, Sb, Hg, Cr, Ce, Yb, Sc, Eu, Cd, As, Ba, Se, Sr, Cs, Co and Au.

Water from Hellerudmyra Bog and Lake Solbergvann was spiked with $^{65}ZnCl_2$ and stored for three days before being ultrafiltered using hollow fibers. After three days storage, a major fraction of ^{65}Zn was associated with high molecular weight components (Table 4). This effect was most pronounced for Solbergvann water which had a higher pH-value than Hellerudmyra water. Trace metals are more polar than humic substances and should attain more hydrophobic properties when associated with humic substances. According to Table 4, Zn was not extracted into the octanol phase. When the pH of Solbergvann water spiked with ^{65}Zn was adjusted to 6.2 to 3.2, no detectable amounts of ^{65}Zn could be observed in the octanol extract. With the exception of Au and, to a certain extent, Hg, the same result was obtained for the multielement standard added to ultrafiltered, unfiltered and concentrated solutions from Hellerudmyra water, for which the values for TOC varied from 6.54 to 260.3 mg C/l and pH varied from 4.43 to 3.23. This may be due to short contact time (1-2 weeks) or that trace elements associated with humic substances are not octanol extractable.

Table 4 pH and Kow values of ^{65}Zn for total and filtered ($Mw<10^4$ Dalton) water samples from Hellerudmyra Bog and Lake Solbergvann.

Sample	%^{65}Zn	pH	Kow	log Kow
Hellerudmyra				
Total	100	4.26	< 0.01	< 0.001
Mw < 10^4 Dalton	62 ± 4	4.47	< 0.01	< 0.001
Solbergvann				
Total	100	6.39	< 0.01	< 0.001
Mw < 10^4 Dalton	32 ± 4	6.02	< 0.01	~ 0.001

Conclusion

Based on ultrafiltration, using hollow fibers, the association of MCPA with high molecular weight components is significant, about 15% for both waters investigated.

MCPA is ionizable, and its octanol/water partition coefficient (Kow) is strongly dependent on pH of the water samples. This may influence the sorption/desorption properties of MCPA in soil/sediment-water systems and its potential availability for biological uptake. Due to the pH effect, aquatic humic substances are shown to indirectly influence the Kow value of MCPA. Dazomet exists as a neutral species below pH 11, and its Kow value is independent of pH in the range 2-11. Aquatic humic substances have no significant influence on the Kow value of MCPA or dazomet in the water samples used, compared to reference samples without humic material.

Based on ultrafiltration using hollow fiber, a major fraction of Zn (40-70%) is associated with a high-molecular-weight form during three days storage. Even though trace elements are to a certain extent associated with high-molecular-weight humic substances, the present results indicate that no significant amounts are extracted into the octanol phase.

Speciation studies combining results from different fractionation and extraction techniques may give increased information on physicochemical properties and bioavailability of trace components in the environment.

References

1. Hassett, J.P. and M.A. Anderson. Environ. Sci. Technol. **13:**1526 (1979).

2. Landrum, P.F., S.R. Nihart, B.J. Eadle and W.S. Gardner. Environ. Sci. Technol. **18:**187 (1984).

3. Hassett, J.P. and E. Milicic. Environ. Sci. Technol. **19:**638 (1985) .

4. Gjessing, E.T. Physical and Chemical Characteristics of Aquatic Humus. (Ann Arbor: Ann Arbor Science Publ. Inc., 1976).

5. Gjessing, E.T. Hydrobiol. **91:**144 (1981).

6. Benes, P., E.T. Gjessing and E. Steinnes. Water Res. **10:**711 (1976).

7. Lydersen, E., H.E. Björnstad, B. Salbu and A.C. Pappas. In: Lars Lander, Ed., Lecture Notes in Earth Sciences 11. pp. 85-97 (Heidelberg: Springer, 1987).

8. Lydersen, E., B. Salbu, H.E. Björnstad, J.O. Englund and J.P. Rambæk. Size Distribution Patterns for Trace Elements in Waters Draining a Small Catchment. Proc. UNESCO Int. Symp., Bolkesjö, Norway (1987).

9. John, J., E.T. Gjessing, M. Grande and B. Salbu. Sci. Total Environ. **62:**253 (1987).

10. Chiou, C.T., V.H. Freed, D.W. Schmedding and R.L. Kohnert. Environ. Sci. Technol. **11:**475 (1977).

11. Mailhot, H. and R.H. Peters. Environ. Sci. Technol. **22:**1479 (1988).

12. Neely, W.B., D.R. Branson and G.R. Blau. Environ. Sci. Technol. **13:**1113 (1974).

13. Mackay, D. Environ. Sci. Technol. **16:**274 (1982).

14. Davies, R.P. and A.J. Dobbs. Water Res. **18:**1253 (1984).

15. Hodson, P.V., D.G. Dixon and K.L.E. Kaiser. Environ. Tox. Chem. **7:**443 (1988).

16. Karickhoff, S.W., D.S. Brown and T.A. Scott. Water Res. **13** (1979).

17. Schwarzenbach, R.P. and J. Westall. Environ. Sci. Tech. **15:**1360 (1981).

18. Caron, G., I.H. Suffet and T. Belton. Chemosphere. **14**:993 (1985).

19. Voice, T.C. and W.J. Weber Jr. Environ. Sci. Technol. **19:**789 (1985).

20. Bengtsson, G., C.G. Enfield and R. Lindqvist. Sci. Tot. Environ. **67:**159 (1987).

21. Valsaraj, K.T. and L.J. Thibodeaux. Water Res. **22:**183 (1989).

22. Petersen, R.C. and A. Kullberg. Vatten **41**:236 (1985).

23. Gjessing, E.T., G. Riise, R.C. Petersen and E. Andruchow. Sci. Tot. Environ. **81/82:**683 (1988).

24. Salbu, B. Preconcentration and Fractionation Techniques in Determination of Trace Elements in Natural Waters - Their Concentration and Physico-chemical Forms. University of Oslo. (1984).

25. Salbu, B., H.E. Björnstad, N.S. Lindström, E.M. Brevik, J.P. Rambæk and P.E. Paus. Talanta. **32:**907 (1985).

26. The Agrochemicals Handbook. Royal Society of Chemistry. (Nottingham, 1983).

27. Helweg, A. Weed Res. **27:**287 (1987).

28. Matolcsky, G., M. Nadasy and V. Andriska. Pesticide Chemistry. Studies in Environmental Science 32: (Amsterdam: Elsevier, 1988).

29. Westhall, J.C. In: W. Stumm, Ed., Aquatic Surface Chemistry, pp. 3-32 (New York: Wiley-Interscience, 1987).

30. Enfield, C.G. and G. Bengtsson. Ground Water **26:**64 (1988).

31. Riise, G. and B. Salbu. Sci. Tot. Environ. **81/82:**137 (1989).

Session 4: Biological Activity

The Contradictory Biological Behavior of Humic Substances in the Aquatic Environment

Robert C. Petersen Jr.
Stream and Benthic Ecology Group, Limnology Institute
University of Lund, Box 65, S-221 00 Lund, Sweden

Abstract

Five general properties by which humic substances can affect biological processes in surface waters can be identified, and include protonation, particle formation, complexation, surfactant behavior, and metabolic interaction. All have been reported to result in either positive or negative biological effects which has led to a general confusion, even contradiction, and underestimation of the biological behavior of humic substances in aquatic systems. On closer examination the ambiguous results can be traced to differing chemical and physical conditions, difficult comparisons between field and laboratory results, failure to adequately define the humic material being studied, and the hyperbolic nature of biological responses to humic substances. In the following discussion the 5 types of biological properties are explained and examples are given. An attempt is made to show that the positive and negative biological effects that have been reported, are not contradictory but logical, when considering a biological response to a complex chemical entity such as humic substances.

Introduction

It is generally well accepted that both the chemistry and biology of humic substances (HS) are difficult to study. This is because HS are a heterogeneous and complex mixture of organic compounds, not a single chemical entity nor even a definable group of chemical compounds, and that these substances participate in a variety of chemical and biological mechanisms which make their reported behavior confusing and even contradictory. The importance of understanding the biological role of HS, and the dissolved organic compounds in general, applies not only to natural soil and aquatic systems, but to human health as well. Drinking water quality is an obvious concern but the dissolved organics that we drink as refreshments and stimulants may be an even more important concern. For example about 1.5 billion cups of coffee are drunk worldwide each day, and a recent evaluation by WHO has come to the conclusion that coffee may protect against certain kinds of cancer, but can be the cause of others as well [1]. In the following, the major chemical and biological mechanisms of HS in the aquatic environment are grouped under 5 categories including, 1) protonation, 2) particle formation, 3) complexation, 4) surfactant behavior, and 5) metabolic interaction (Table 1). For the purpose of this review no strict definition of HS will be used, nor will the published material used be limited to those studies where a strict definition of HS has been used. This is because most of the research to date has been conducted without the benefit of the clear definition of HS recently put forth by the International Humic Substances Society and now accepted by most HS researchers [2]. In addition much of the work in this area has focused on dissolved organic carbon (DOC). Some (about 50%) of the DOC pool is HS [2] and for the purposes of this review results pertaining to DOC will be included in the general discussion.

Table 1 Category headings by which humic substances can affect chemical and biological processes and their positive and negative impact.

	Mechanism	Positive	Negative
1.	Protonation	Acts as weak buffer	Lowers pH
2.	Particle formation	Contributes mass	Stabilizes DOC
3.	Complexation	More available or toxic	Less available or toxic
4.	Surfactant	Improves permeability	Decreases permeability
5.	Metabolism	As a carbon source	As a toxicant

For the most part the examples used will be drawn primarily from information on aquatic systems. However, since there is a considerably richer literature in the soil sciences, on the biological behavior of HS, it will be difficult to restrict the discussion to only aquatic examples. It is not now known whether the behavior and mechanisms of aquatic humic substances are the same as for terrestrial humic substances. Surely, there must be a great similarity in overall properties. However, it is known that chemical differences in terms of the density of important functional groups does differ between the two. Malcolm and McCarthy [3] have presented evidence that the aliphatic content of aquatic HS is considerably higher than that for terrestrial HS which tends to have a higher phenolic content. The reason for this difference is primarily due to the operational definitions of aquatic HS and terrestrial HS (Fig. 1). Aquatic HS is organic material isolated from water at pH 2 onto an exchange resin while terrestrial HS is organic material extracted from the soil matrix with 0.1 N NaOH [2]. If it is assumed that the major source of DOC into freshwaters is the terrestrial system then the HS present in surface waters has been extracted from the soil system at a pH much lower than that used to prepare terrestrial HS.

Protonation

Aquatic humic substances may be a complex mixture of organic compounds but there are some important generalities about their functional character that have important biological implications. One of the main characters is the dominance of organic acids and their carboxylic acid functional groups. Based on published values using the procedure of Leenheer and co-workers [4] the combined hydrophilic and hydrophobic organic acid fraction in a set of 5 lakes and rivers in the United States makes up from 36 to 92 % of the total DOC pool, with a concentration ranging from 2.4 to 35.1 mg/l [5]. These organic acids are able to donate hydrogen ions as a conjugated acid-base pair and have a molecular weight from 1000 to 3000 daltons [2]. Because of this high percentage of organic acids, humic substances can modify the pH of surface waters by either acting as a weak buffer and thus reducing a change in pH or by lowering the pH as an acid.

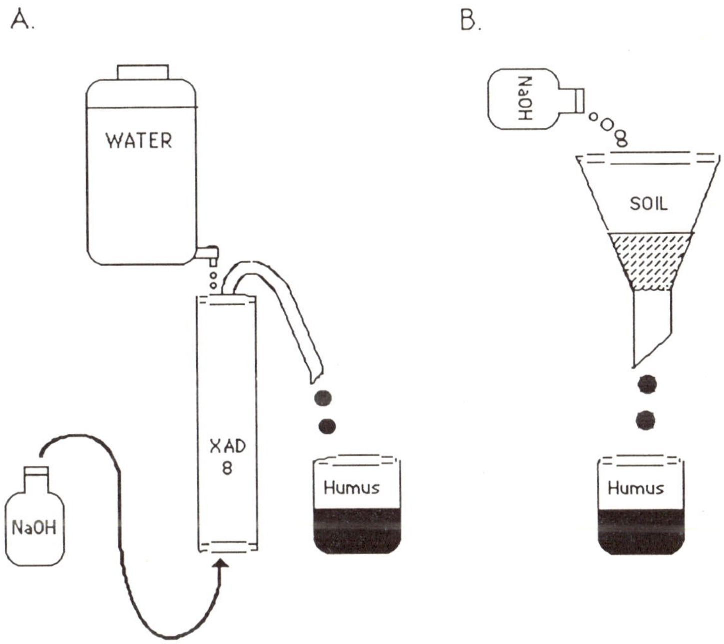

Figure 1 Comparison of the operational definition of aquatic humic substances and terrestrial humic substances. A. Aquatic HS is extracted from water at pH 2 onto an exchange resin and then released with 0.1 N NaOH. B. Terrestrial humic substances are released directly from the soil matrix with 0.1 N NaOH.

As a Weak Buffer

Humic substances can be regarded as buffers of aquatic systems which take over from the carbonate system at about pH 4.5 [6,7]. The organic acids within the HS mixture have an average pK_a (the pH at which 50% of the acids are dissociated) between 3.7 and 4.6 [8,9]. The pK_a of known simple organic acids has a much wider range, but their average is still pH 4.5 (Fig. 2). The maximum buffer intensity of any chemical compound occurs at the inflection point of the titration curve which also occurs at the pH equal to the pK_a of the material being titrated [10]. Therefore, in natural systems with a complex mixture of organic acids, maximum buffering intensity will occur at about pH 3.7-4.4.

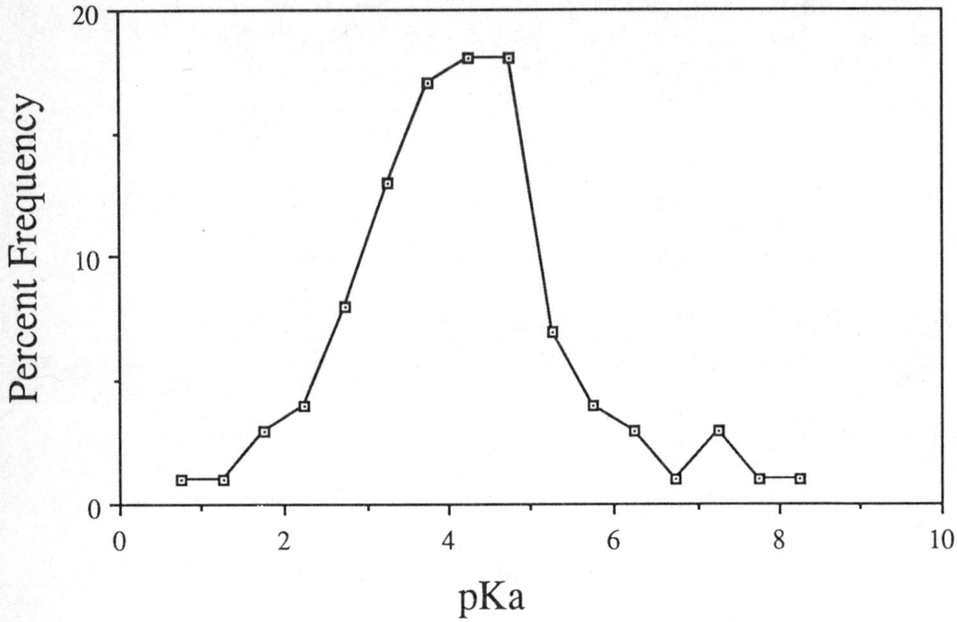

Figure 2 Percent frequency of the pK_a values of simple organic acids. (Data redrawn after Perdue [11]).

The buffering capacity of humic substances may be one explanation for the observation that humic aquatic systems seem to be less affected by acidification than similar clear water systems. In several studies throughout Scandinavia humic lakes experiencing acidification have been reported to have a richer biota than non-humic lakes subjected to acid rain [12-14]. A large part of the negative effects of acidification seems to be related to short term decreases in pH due to terrestrial runoff [15]. During these events a humic system will be more likely to ameliorate the effects of a sudden flush of low pH water than a clear water system.

As an Acid Source

The negative effect of humic substances is that their organic acids will control pH in carbonate poor waters and result in pH values of around 4.5, a value lethal to many freshwater organisms [16,17]. While it has been known that humic freshwaters generally have low pH, it has only recently been shown that in some aquatic systems in areas subjected to acidification, organic acids and not mineral acids determine pH. In a study of 33 surface waters of Nova Scotia, it was found that in peaty waters with a mean pH of 4.5 dissolved organic carbon was the best predictor of pH and sulfate, calcium or a combination of sodium, magnesium and potassium was not particularly important [18].

The fact that organic acids can lower the pH of a surface water has caused considerable disagreement over whether the lower pH of many surface waters is due to mineral acids from acid rain or due to increased organic acids caused by changes in land use [18-24]. The suggestion by Krug and Frink [19,20] that acidification of surface waters is primarily a natural process related to changing land use patterns has not been entirely supported by other studies. In Nova Scotia only those waters in peaty areas were naturally acidic due to organic acids [18]. In forested areas acidity was primarily due to anthropogenic sources. In Norway where relatively low levels of humic substances dominate aquatic systems non-marine sulfate seems to be the prime determinant of surface water pH [24]. In Finland with lakes having rather high values of DOC (mean 10 mg/l) Kortelainen and Mannio [25] concluded that TOC was a better predictor of pH than non-marine sulfate and that individual catchment characteristics was more important than atmospheric loadings.

In a study of 19 streams in southern Sweden Petersen *et al.* (unpublished data) have determined the relative contribution of organic and mineral acids to acidity and compared these findings to those from Nova Scotia. Both surface waters, those in Nova Scotia and in Sweden, had similar water chemistry with rather low pH in both regions (Table 2).

Table 2 Comparison of the range of chemical values reported for Nova Scotia [18] to 19 stream locations in southern Sweden. All values corrected for marine influence.

Chemical Parameter		Nova Scotia	This Study
n		33	19
DOC	mg/l	0.7-27	5.0-22.5
Color	mg Pt/l	7-200	10-180
H^+	μeq/l	1-98	0-66
SO_4^{2-}	μeq/l	12-203	156-482
Ca^{2+}	μeq/l	17-239	80-872
$Na^++Mg^{2+}+K^+$	μeq/l	2.4-105	59-363

However, the source of the low pH was quite different. In the Nova Scotia surface waters, DOC was the best predictor and SO_4^{2-} was not a significant predictor. In the southern Swedish surface waters DOC was not correlated with H^+ (r=0.061). Instead Ca^{2+} was the best predictor of H^+ with a simple linear correlation coefficient of -0.832. Sulfate was negatively correlated, but this was not significant (r=-0.374,

P>0.05). When sulfate and calcium were combined in a multiple correlation analysis an improvement in the correlation coefficient was observed (r=0.915). These results from Sweden tend to be intermediate between those observed in Norway and those from Finland. The observed pH of surface waters is being heavily influenced by atmospheric deposition but catchment character is also an important factor.

Particle Formation

Forms Particles

A major route of energy flux and material cycling in aquatic ecosystems is through the detritus food chain [26,27] entering as dissolved and particulate organic matter. In both marine and freshwater systems it is well documented that particulates can form from dissolved materials by both biotic and abiotic processes. Biotic mechanisms of particle formation from DOC is primarily due to microbial uptake, with the microbes themselves becoming the particulates. This is an important and indirect effect of DOC on particulates since bacteria are considered a particularly high quality food for suspension feeders [27]. The assimilation efficiency on bacteria tends to be high since bacteria have easily broken down cell walls. Petersen [28] has investigated the food selection behavior of the net spinning caddisfly and found that amorphous detrital aggregates which are composed of organic material and entrapped algae and bacteria are a preferred food item.

Several studies have suggested that the main mechanism of DOC utilization in stream systems is by biotic uptake by the microbial flora and that abiotic processes, while present, represent a lesser mechanism of DOC removal [27,29,30]. However, Petersen [31] has shown that in streams, incorporation of DOC into particles by abiotic processes such as bubble cavitation can rival rates of DOC utilization by the microbial community and suggests that abiotic particle formation is far more important to stream dynamics than previously thought.

Not surprisingly then, humic substances have also been shown to be a food resource for microorganisms [32] and in so doing incorporate humics into particles. It has been reported that the mean density of bacteria in oligotrophic brown water lakes is higher than in oligotrophic clear-water lakes and that allochthonous humic sources comprise a large part of the energy supply to bacteria [33,34]. Meyer *et al.* [35] have shown that a portion of the DOC in southeastern US blackwater rivers will support bacterial growth. Edwards [36] has suggested that the sestonic bacteria in these streams is an important food source for filter feeding invertebrates. Other studies on the same rivers have shown a high diversity and secondary production of invertebrates [37].

A second major role of particulates in aquatic systems is to provide attachment sites for both natural organics and xenobiotics. Adsorption of organics to particles have been studied and a linear relationship has been found between the dissolved organic concentration and the amount of adsorbed chemical [38-41]. This model has been frequently used in describing pesticide adsorption in soils and sediments [42,43]. The significance of this relationship in assessing the fate of hydrophobic pollutants in natural water systems is discussed by O'Connor and Connolly [43]. Di Toro [44] introduced a particle interaction model of reversible organic chemical sorption implying a negative relationship between particle concentration and the partition coefficient at high particle concentrations.

Stabilizes Particles

While there is no doubt that humic substances can and do contribute mass to particulate detritus the presence of humic substances also keeps material in suspension and prevents the genesis of particulates. Which of these mechanisms will dominate depends on the availability of calcium and possibly other cations.

Recent empirical work on the stability of particulates in hard water lakes in Switzerland has shown that calcium ions act as destabilizing agents which enhance natural coagulation and precipitation of particulates while DOC acts to stabilize particles and retard coagulation [45]. In the four lakes that were studied, an empirical stability factor, a, was defined as the probability that two particles will collide and form a particle. For a perfectly stable colloid, a=0. For a completely destabilized suspension, a=l. Particle stability varied from high (a=0.01, low probability of aggregate formation) in Lake Sempach with 4 mg C/l and 1.2 x 10^{-3} M Ca^{2+} to low (a=0.09, high probability of aggregate formation) in a similar hard water lake but with less DOC (Table 3).

Table 3 Solution chemistry, aggregate probability (a) and particle stability of 4 lakes in Switzerland [After 45].

Parameter	Zürich	Sempach	Luzern	Greifen
Ca^{2+} M	$1.2x10^{-3}$	$1.2x10^{-3}$	$0.9x10^{-3}$	$2x10^{-3}$
DOC mg/l	1	4	1	4
a[1]	0.091	0.011	0.055	0.047
Particle stability[2]	Low	High	Intermed	Intermed.

1. a refers to the probability that two particles will collide and form a larger particle.
2. Particle stability refers to the ability of a particle to remain in suspension and not form larger aggregates.

The reason for this is not fully understood but appears to be due to differences in electrostatic and steric effects. At near neutral pH a large fraction of the organic acid DOC will be dissociated and carrying a net negative charge as an organic anion. In the absence of a suitable cation the organic anions will effectively remain in solution. When cations are present, especially divalent cations such as calcium, weak bonds can form between the organic anion and the inorganic cation and a particle will be formed.

The model proposed by Weilenmann *et al.* [45] seems to work in hard water lakes and does confirm to the generally held view that humic waters are dominated by colloids which also have a low particulate load. The question then arises as to the utility of the model for running water systems and if this can be used to predict particulates or DOC if one of the two variables is available. This was tested using data collected from 20 streams in different watersheds located in southern Sweden. (Petersen *et al.*, unpublished data). In September and November all 20 streams were sampled and the DOC, POC and calcium were measured. There was a significant ($P<0.001$, F-Test) but poor ($r^2=0.50$) correlation of POC with DOC for the 40 points using a (Fig. 3). When a multiple regression analysis was performed adding in calcium the regression coefficient for calcium was not significant ($P>0.4$).

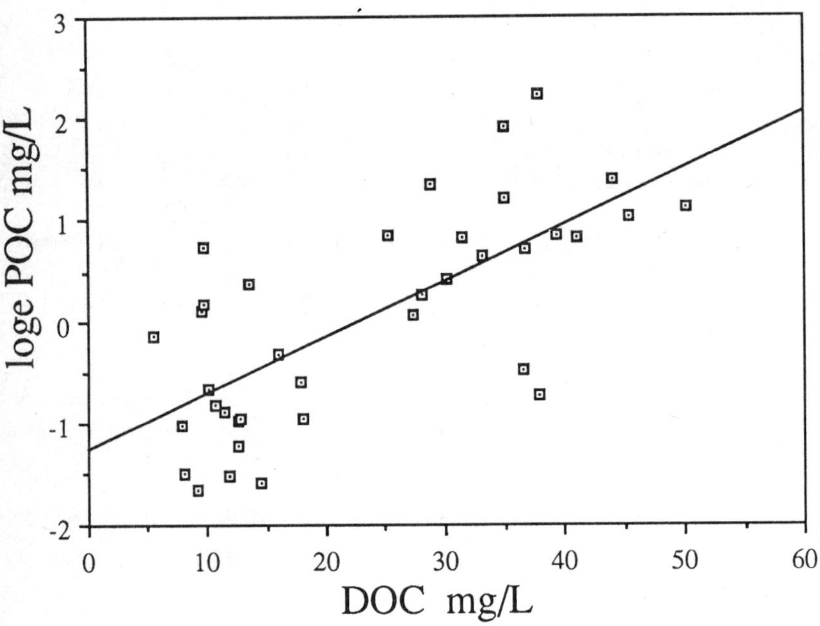

Figure 3 Relationship of total particulate organic carbon (POC) (>0.5 μm, ash free dry mass/l) to the total dissolved organic carbon (DOC) in 20 streams sampled on two occasions. The relationship was significant ($P<0.001$) but the dependence of POC on DOC was only 50% ($r^2=0.50$).

Complexation Reactions

There are several different ways that humic substances can affect biological processes indirectly through complexation reactions. These reactions either increase or decrease the availability of nutrients, the toxicity of organic xenobiotics, the toxicity of metals or activity of enzymes.

Nutrient Complexation

There is considerable evidence that humic substances complex with limiting nutrients and in most reported cases reduce their availability and retard metabolic parameters. In other reports, HS has been reported to stimulate metabolism. This seeming contradiction tends to be dependent on other chemical or physical factors coexisting in either the experiment or the system being measured.

Phosphorus.

Owing to the ecological importance of phosphorus in primary productivity, the role of HS in reducing phosphorus availability in soils (reviewed in [46]) and surface waters [2,47-49] has been suggested. This line of reasoning comes from the observation that primary productivity tends to be restricted in dystrophic systems even where phosphorus concentrations are well above those thought not to be limiting [50,51]. Stewart and Wetzel [52] have shown that alkaline phosphatase activity increases and ^{14}C assimilation decreases in the presence of humic substances. If phosphate is bound by humic substances it may only be available through enzymatic hydrolysis. Whether the hydrolysis occurs directly between a phosphate-HS bond or through an intermediary is not known.

The direct mechanism of phosphorus binding to HS is with the hydrophilic humus fraction to form inositol phosphates, the phosphoric acid esters of inositol [2]. While little is known of rates of utilization and transformation of the inositol phosphates in aquatic systems it has been demonstrated that phosphorus once bound to humic substances tends to be unavailable.

The indirect and probably more dynamic mechanism seems to involve an HS-iron-phosphate complex. Iron is easily complexed with HS [53] and it has been suggested that there is an abiotic interaction between HS, ferric iron and phosphate [49,54]. This is supported by observations by Jackson and Hecky [55] that HS depressed primary productivity of phytoplankton in a three reservoirs and a natural lake in northern Canada by making iron unavailable and that most of the phosphorus was in the form of humic-iron-phosphate complexes.

Contrary to the observation that HS retards the metabolism of phytoplankton, there are several reports that under certain conditions HS can stimulate algal growth. Prakash and Rashid [56] have reported that HS increased the yield, growth rate, and ^{14}C uptake of marine dinoflagellates. This stimulation, however, could not be linked to alteration in the availability of trace metals, N, P or vitamins alone. It was suggested that HS through a chelation mechanism could have made some nutrients more available.

Calcium

Even though calcium may be an important variable in the generation of particulates the decrease in bioavailable calcium through complexation with HS can have negative biological effects especially in acidified freshwaters.

Calcium is an important ionic regulator of freshwater organisms. MacDonald [57] has studied the effects of calcium on acid stressed fish and found increased survival in acid waters with calcium present due to a reduction of gill permeability which resulted in a decrease in the loss of sodium and chloride. In crustaceans, calcium is actively taken up from the water following moulting and production of the new exoskeleton. The rate of calcification is dependent on external calcium concentration. Wright [58] sets a lower limit of 4 mg/l for calcium, below which *Gammarus* can not produce a new exoskeleton.

In laboratory studies using *Gammarus pulex* and varying concentrations of pH, HS and calcium, Paarlberg (unpublished data) observed that the rank order of importance based on multiple linear regression for mortality was calcium > humus > pH and that mortality was highest in recently molted animals. There was a significant ($P<0.05$) interaction term between pH and humus and between humus and calcium which suggests that in these studies mortality was dependent on the combined effect of all three variables. He concluded that complexation of calcium by HS resulted in a calcium deficiency. This deficiency was made worse at the lower pH.

Organic Xenobiotic Complexation

Humic substances are known to complex with anthropogenic organics and either increase or decrease their toxicity. Gjessing and Berglind [59] have shown that humic acid decreases the analytical detectability of hexachlorobenzene (HCB) and suggested that humic acid would also decrease the biological availability of HCB. Stewart [60] has shown that humic material increases the toxicity of o-cresol, methylphenols, and compounds in an aniline methylation series from 1 to 40 fold but reduces the toxicity of benzoquinone. He suggests that modes of action may include interference with membrane permeability, mediation of one or more photochemical reactions, complexation with the organic toxicant, and changes in the cell surface.

Metal Complexation

The concentration of biologically available metals in natural waters is highly affected by the amount and composition of humic material and dissolved inorganics which in turn will control the positive or negative biological effect [61-67]. Metal-humus complexes are dissolved at low metal/humus ratios and precipitate as the chain structure of molecules grows and isolated carboxyl groups are neutralized through salt bridges. The precipitation reaction is dependent on ion strength, pH, the concentration of humic material and the metal ion. Adding cations to humic solutions results in release of bound hydrogen and metal ions by competition for binding sites, in addition to precipitation at high concentrations [68].

Humic macromolecules contain functional groups which control the reactivity with metals [69]. Metal complexation with humus is generally thought to be done via the carboxylic and phenolic hydroxyl groups [69,70]. Christman & Gjessing [65] criticised the generality of this idea and pointed out that binding at low metal saturation may take place at sites in humic material that include minor donor sites, like N and S, rather than at oxygenated sites. These authors stressed that the applicability of data obtained using high metal loading might not be significant in the many environments where metal-ion concentrations are low compared to humic concentrations.

The details of metal uptake by organisms are not known, but probably include both active transport via enzyme-substrate complexes, and passive transport including ion exchange, adsorption and diffusion [2]. The stability of the metal-humic complex is important to both active and passive transport. The outcome of competition for the metal between enzymes or adsorption sites within the organisms and chelators in the surrounding media, is dependent on the stability of the metal-chelator complex. The more stable chelators prevent uptake and the less stable ones contribute to higher concentrations of available metal ions.

The effect of humic material on the toxicity of metals to aquatic organisms can be expected to be contradictory, and actually both positive and negative effects have been found. Humus has been found to decrease the toxicity of cadmium to salmon (*Salmo salar*) and an alga (*Selenastrum capricornutum*) [71]. Other fish studies have revealed the beneficial effect of humus on the survival of fish to high concentrations of copper [Brown, 1974, in 72]. To *Daphnia* humus was found to decrease the toxicity of copper but increase the toxicity of cadmium [66]. Four size fractions of organic compounds of pond water either increased or decreased toxicity of cadmium to the cladoceran *Simocephalus serrulatus* (Giesy *et al.*, 1977, in [66]).

The discussion has so far only dealt with the hypothesis that the only toxic and biologically available metal form is the free ion. Observations such as those by Winner [66] can be explained by differences in humus binding capacity and affinity for free metal ions. It is known that HS has a higher affinity for copper than for cadmium. In addition, organic ligands can be lipophilic themselves, and able to pass biotic membranes, thus facilitating metal transport into organisms [73]. Complex binding of metals can alter the character of organic ligands from hydrophilic towards hydrophobic and increase the uptake of metals in excess to the uptake of free metal ions [74].

Aluminum

Increased levels of labile inorganic aluminum (e.g. Al^{3+} and aluminum fluoride, hydroxide and sulfate complexes) account for most of the aluminum which enters freshwater systems due to decreasing pH. Complexation reactions with HS in humic freshwaters is regarded as an important mechanism to reduce the toxic effects aluminum [75] as well as with other toxic metals [76].

In order to examine the entire response surface of the HS-aluminum interaction, Petersen *et al.* [77] used a factorial design, where instead of testing aluminum toxicity with and without HS, as is commonly done, multiple concentrations of both HS and aluminum were used. The biological response parameter was a mixed culture of two species of blackflies, *Odagmia ornata* and *Simuliun decorum,* carried out at pH 4.5. When alone both aluminum and HS were found to be toxic at pH 4.5 to blackflies as shown along the x and z axis in Fig. 4.

When together aluminum and leachate act antagonistically and cancel the toxic action of each other. The amount of aluminum bound to the 600 mg Pt/l color leachate sample was estimated by fitting the survivorship-aluminum curve at this color to a second order polynomial and determining the aluminum concentration at maximum survivorship with the second derivative. The 54 mg leachate carbon per liter was able to bind 300 μg of aluminum at pH 4.5 to form biological inactive complexes. Assuming that the base titration is a fair approximation of the available binding sites for the maple leachate, 66 microequivalents of aluminum were tied up with each milliequivalent of organic acid.

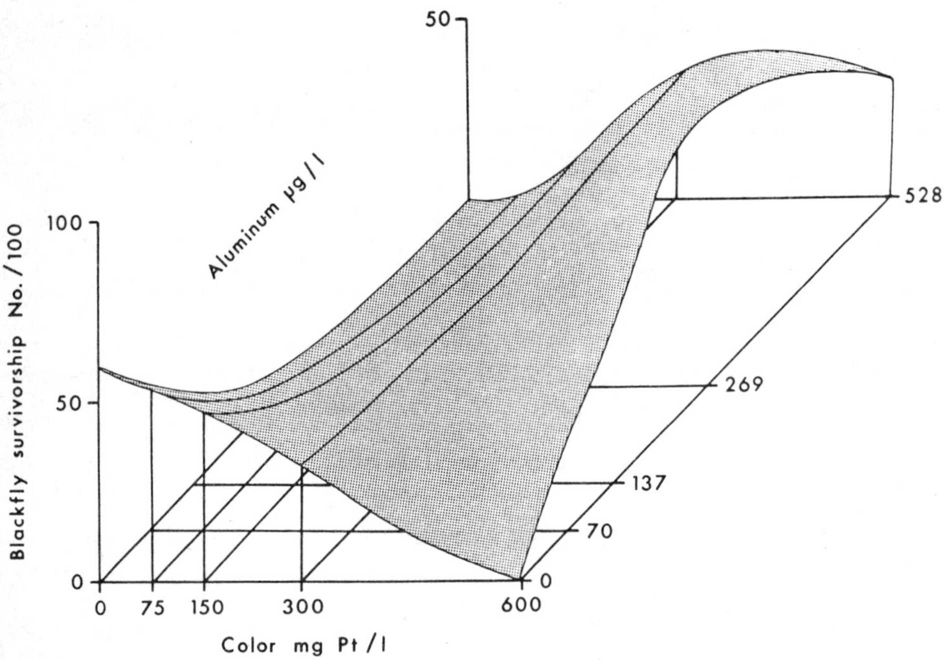

Figure 4 Survivorship of larval blackflies under conditions of increasing humic substances and aluminum at pH 4.5.

In general the alkaline earths and particularly the trivalent metals form complexes with humic substances [61]. Of particular importance is the cation exchange properties of humic complexes with calcium and aluminum which may be particularly important in acid stressed environments especially since liming of acid stressed environments has begun on a large scale. While it is clearly advantageous to limit the steady decline of surface water pH with lime additions the biological effects are not fully known. Dickson [78] has shown that liming a low colored water rich in aluminum becomes toxic to fish probably due to aluminum being released from humic substances during the liming process. Similar results can be expected if a more humic water was limed and where significant amounts of aluminum were bound to colloidal humic substances.

In order to determine in which DOC fraction aluminum was complexed, we have fractioned a humic water sample according to the IHSS procedure [3] and determined the total aluminum by ICPEMS. These results are presented in Fig. 5 and show that aluminum occurs in all fractions. Most of the aluminum (51%) occurred in the hydrophobic fraction with about an equal amount in the humic and fulvic acid fractions.

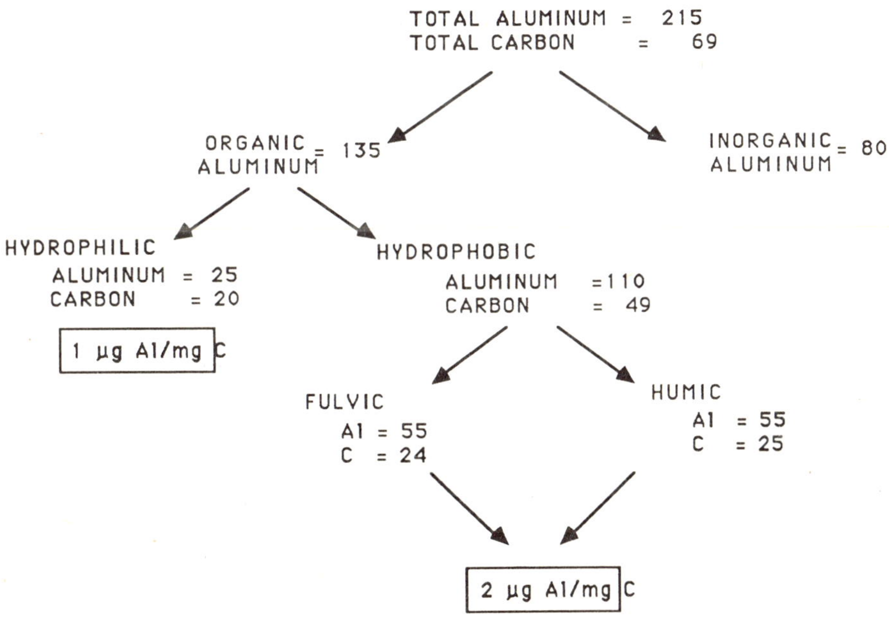

Figure 5 Partitioning of aluminum into dissolved organic fractions following the IHSS procedure For each fraction the total ICP aluminum (µg/l) and DOC (mg/l) are shown.

Humic Substances as Surfactants

Humic substances can act as surfactants which can alter membrane characteristics to either benefit or harm biological processes. Some of the effects of humus on organisms are similar to effects of synthetic surfactants, such as detergents. Visser [79] suggested that at least a part of the physiological action of humus on organisms is the result of the surfactant properties of humus.

There are several mechanisms attributable to surfactant properties that can help explain the biological role of humus in aquatic systems. In a study of different freshwater systems, rivers and groundwaters in Yugoslavia, Cosovic *et al.* [80] concluded that humic substances represent the predominant surface active substances in natural waters. Fractionation of the humus into fulvic/humic components of various molecular weight has shown that high surface activity can be associated with the low molecular weight fulvic acid fraction [81,82]. This is the same humus fraction (fulvic, low molecular weight) found to be most toxic to *Daphnia magna* in a laboratory study using Aldrich commercial humus [83]. However, a low molecular weight fraction is also the most likely to penetrate an organism and be more toxic, regardless if this was caused by other mechanisms than surface activity. The surface activity of humus increases with lowering of pH [84,85]. This suggests that any biological effects of humus caused by its surfactant characters should appear at a lower concentration and be more expressed as the water is acidified. Surfactants, like humus, are known to influence the toxicity of heavy metals and pesticides, the activity of enzymes, the metabolic rate of organisms, and are toxic [86,87].

Direct Metabolic Activity of Humic Substances

Humic substances can either be used as a carbon source thereby stimulating biological processes or act as a toxicant to damage biological processes.

Enzymes and Metabolism

Aquatic humus has been indicated as both stimulating or depressing a variety of metabolic processes depending on both a difference in the chemical involved and on the environmental conditions present.

The direct stimulatory effect of aquatic humus on bacterioplankton has frequently been reported [35,48,88,89]. Wetzel [48] suggests that the stimulation of bacterial metabolism by a humic fraction may be due to co-metabolism of a labile moiety and a more refractory compound. This mechanism can be of considerable importance in the breakdown of such refractory substances as pesticides and chlorinated hydrocarbons. In two separate reports, Eaton [90] and Bumpus *et al.* [91] have reported that a lignolytic fungus was able to breakdown polychlorinated biphenyls, compounds thought to be especially resistant to microbial degradation.

Wetzel [92] in a review of metabolic and limnological regulators has reported that there are both positive and negative interactions between algae and bacteria and between macrophytes and epiphyton. These interactions are mediated primarily by the dissolved organic carbon pool. For example it is well known that bacteria readily utilize the dissolved organic compounds released by macrophytes. However, macrophytes also have the ability to release allelochemic substances which can retard invertebrates [93]. What part of this pool should be considered HS, or proto-humic substances [46] is still open to discussion. Regardless, what we know about the chemical structure of HS [2,46,65,94-96] suggests a similarity to known biologically active chemical compounds and therefore, a strong potential for biological activity, both positive and negative.

As was briefly described above the interaction of HS and calcium tends to tie up the reactive sites on carboxylic or phenolic hydroxyl functional groups. This may have beneficial or negative effects. In dystrophic lakes where HS has been implicated in enzyme inhibition, Wetzel [92] has suggested that often observed increased productivity in humic rich waters with increased hardness may be due to cationic suppression of the ability of HS to interfere with extracellular enzymes. Schnitzer & Kahn [96] have reported that humus can promote the accumulation of reducible sugars leading to an increase in osmotic pressure in plants. Their work supports the well known fact that soil extracts stimulate growth and reproduction of many organisms held in laboratory culture.

While, some research has shown that humic acid, in small amounts, can stimulate cell metabolism [56] most work on humic substances does not clearly define which humic fraction was used in the experiments. For example most workers do not separate HS into a hydrophilic and hydrophobic fraction. Therefore, the material that is implicated as being humic and resulting in a stimulation of metabolism may really be the non-humic hydrophilic fraction. In the study done by Prakash and Rashid [56] for example, the major biological response occurred with the small molecular weight fraction. This may support the critique that HS are not involved but rather the metabolism enhancement is due to other organic fractions.

Toxicity of HS

There is ample evidence that humic substances can be toxic, have bacteriostatic properties and represent a health problem in drinking water [48]. Probably the most famous example of the bacteriostatic properties of humic material are the 690 preserved human bodies found buried in peat bogs throughout northern Europe [97]. These are iron-age people which were buried in peat deposits up to 2000 years ago. Some have remained in such perfect, undecomposed condition that an analysis of their last meal is possible. Preservation is a slow process consisting of anaerobic conditions, the elimination of decomposing organisms, and tanning by the humic acids. The completeness of the tanning process has been examined and found to be so slow that after 2,000 years it is incomplete. Approximately 1 ton of oak bark was required to complete the tanning process of the Grauballe man in Denmark [97].

There is considerable public health interest in the potential toxicity of HS, especially in drinking water and in foods and drinks [77]. Some types of tannins present in tea and medicinal herbs have been linked to the occurrence of esophageal cancer [98]. Tea is high in condensed catechin tannin, polymeric flavans, one of the two broad classes of tannins [99]. Thanks to an early warning by the British Medical Association the British have traditionally added milk to their tea to bind the tannins [100]. The Dutch do not and about 100 years ago when tea was the national drink of Holland throat diseases were common. When the Dutch switched to coffee these diseases became rare [100].

The binding properties of humic substances, especially the tannins, have been indicated as one of the major mechanisms of concern to health authorities [101]. In a study of the toxic properties of humic substances Klocking *et al.* [102] have used "coffee humic acids, polycondensates of amino acids and phenol bodies" to study the binding of humic acids to serum albumin and reported a lethal dose (LD50) of 1.1 (0.9-1.42) g/kg in intravenous injection in mice.

Humic substances contain a variety of phenolic and organic acids which depending on their pK_a will become more biologically available with a decrease in pH. Humic substances also contain tannins, polyphenols and plant defense compounds that are known to be inhibitory to biological processes. Combining these two observations it can be suggested on theoretical grounds that acidification of humic waters will mobilize some of these compounds. Recently, Moghissi [103] has pointed out that the increased bioavailability of organic compounds with reduction in pH value remains unknown and suggested that chlorinated aromatic compounds such as PCBs and dioxins may be mobilized in aquatic or terrestrial systems.

It is well known that the pH of the water modifies the toxicity of ionizable compounds to aquatic animals [104]. This is commonly referred to as the pH-partition hypothesis which states that the ionizable organic molecules only penetrate biological membranes as uncharged or undissociated molecules. The acidic humic components such as the phenolic acids have pK_as between pH 9 and 11.

Because of the pH-partition mechanism there is good agreement between the toxicity of a particular set of compounds and their octanol solubility[105-107]. Octanol is thought to best mimic the solubility of natural lipids and act as a membrane analog. As mentioned above this analogy can be extended to macroporous resins such as XAD-8 since they work by attracting the hydrophobic fraction of aquatic organics.

The response of *D. magna* to concentrations of humus at various pH was examined using the EPA recommended 21 day chronic toxicity test and a 96-h acute test [108].

In chronic tests, the instantaneous population growth rate, r, was negatively correlated to low pH and high humic color. In 96-h acute tests at pH 5, survivorship of *D. pulex* in replicates of 100 animals resulted in a curvilinear response to increasing humus (Fig. 6). The curvilinear response can be described by a second order polynomial (r^2 = 0.99, P<0.01). The first term in the polynomial is probably controlled by the complexing of humus with the inorganic metals used in the culture media. Without humus, survivorship in the 96 h acute tests was well below the control value, at pH 7 of 95% survivorship. With the increase in humus, survivorship increased until a maximum of 95% was reached at about 10 mg C/l (300 mg Pt/l). Beyond 10 mg C/l the second term in the polynomial takes over. This term is controlled by the fraction of humic material, probably phenolic in nature, which becomes biologically available at pH 5. This fraction appears to be only a small part of the total HS pool and requires rather large concentration of HS before its threshold for response can be reached. At pH 7 there was no difference at any humus concentration and all were > 90% survivorship.

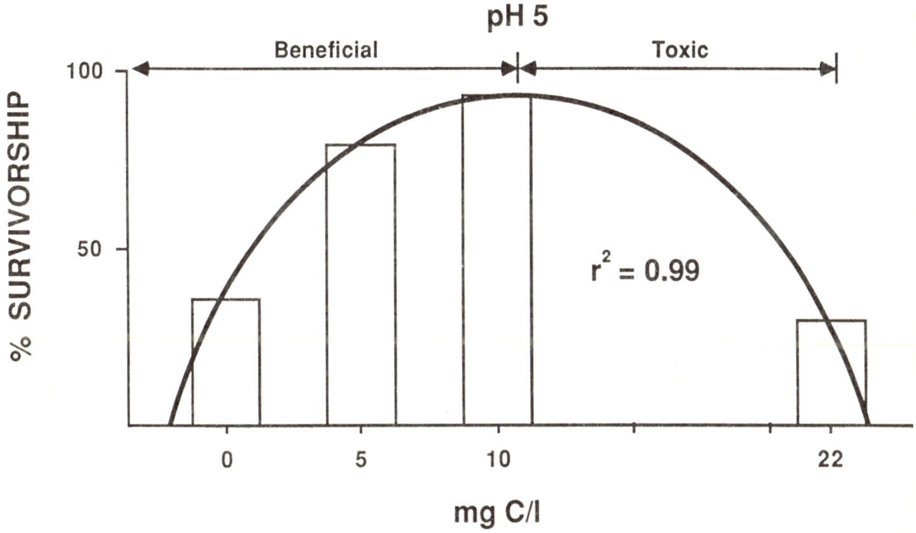

Figure 6 Curvilinear survivorship response in 96-h acute tests of D. magna (n=100) to increasing humus in a high inorganic media showing the second order effect (Redrawn after data presented in [108]).

This last example of a hyperbolic response to increasing HS best illustrates the main point of this review, that there exits both a negative and positive effect of HS on biological systems. The reason for this seeming contradiction is not contradictory at all, but the result of not one but several chemical and biological mechanisms interacting under a set of threshold conditions.

Acknowledgements

This paper is the product of the joint efforts of the Stream and Benthic Ecology Group, Limnology Institute, University of Lund and the continued unselfish and creative efforts of its members, Anders Kullberg, Anders Hargeby and Dr. Lena Petersen. Special recognition is also given to the support of and creative discussions with Dr. Egil Gjessing and Dr. Simon Visser who have influenced the ideas presented in this paper. The continued support of the Swedish Environmental Protection Board and the newly acquired support of the Swedish Council for Forestry and Agricultural Research is gratefully acknowledged.

References

1. Anon. Internat. Herald Tribune, (Paris, 9 March, 1990).

2. Thurman, E.M. Organic Geochemistry of Natural Waters, pp. 497 (Martinus Nijhoff/Dr. W. Junk, 1985).

3. Malcolm, R. L. and P. MacCarthy. Environ. Sci. Technol. **20**:904 (1986).

4. Leenheer, J.A. Environ. Sci. Technol. **15/5**:578 (1981).

5. Petersen Jr, R.C. In: E.M. Perdue and E.T. Gjessing, Eds, Organic Acids in Aquatic Ecosystems. Dahlem Konferenzen, (New York: John Wiley & Sons Ltd., 1990).

6. Wilson, D.E. Arch. Hydrobiol. **87(3)**:379 (1979).

7. Wilson, D.E. and P. Kinney. Limnol. Oceanogr. **22**:281 (1977).

8. Cronan, C.S. and G.R. Aiken. Geochim. Cosmochim. Acta **49**:1697 (1985).

9. Wright, R.F. Rain Project: Symposium on the Acidification of Waters in Kejimkujik National Park (Nova Scotia, 1988).

10. Stumm, W. and J. J. Morgan. Aquatic Chemistry (New York: Wiley Interscience, 1970).

11. Perdue, M. In: G.R. Aiken, D.M. McKnight and R.L. Wershaw, Eds, Humic Substances in Soil Sediment and Water, pp. 493-526 (New York: John Wiley & Sons. 1985).

12. Raddum, G.G. In: D. Drablos and A. Tollan, Eds, Proc. Int. Conf. Ecol. Impact Acid Precipitation, pp. 30-331 (SNSF project As-NLH, Norway, 1980).

13. Nilsson, J.P. Hydrobiologia **94**:217 (1982).

14. Raddum, G. and O. Saether. Verh. Int. Verein. Limnol. **21**:399 (1981).

15. Jacks, G., E. Olofsson and G. Werme. Ambio **5**:282 (1986).

16. Okland, J. and K.A. Okland. Experentia **42**:471 (1986).

17. Otto, C. and B. Svensson. Arch. Hydrobiol. **99**:15 (1983).

18. Gorham, E., J.K. Underwood, F.B. Martin and J.G. Ogden III. Nature **324**:451 (1986).

19. Krug, E.C. and C.R. Frink. Science **221**:520 (1983).

20. Krug, E.C. and C.R. Frink. Science **225**:1432 (1984).

21. Johnson, N.M., G.E. Likens, M.C. Feller and C.H. Driscoll. Science **225**:1424 (1984).

22. Seip, H.M. and P.J. Dillon. Science **225**:1425 (1984).

23. Wright, R.F. Science **225**:1426 (1984).

24. Brakke, D.F., A. Henriksen and S. Norton. Nature **329**:432 (1097).

25. Kortelainen P. and J. Mannio. Water Air Soil Poll. **42**:341 (1988).

26. Odum, E.P. Fundamentals of Ecology, pp. 1-546 (Philadelphia: W.B. Saunders Co. 1959).

27. Cummins, K.W., M.J. Klug, R.G. Wetzel, R.C. Petersen, K.F. Suberkropp, B.A. Manny, J.C. Wuycheck and F.O. Howard. Bioscience **22**:719 (1972).

28. Petersen, L.B-M. Proc. Int. Symp. Trichop., pp. 25-31 (5th 21-26 July, 1986).

29. Mulholland, P.J. Bioscience **22**:719 (1981).

30. Dahm, C.N. Can. J. Fish. Aquat. Sci. **38**:68 (1981).

31. Petersen, R.C. Jr. Limnol. Oceanogr. **31**:432 (1986).

32. Seepers, A.B.J. Hydrobiologia **52(1)**:39 (1977).

33. Johansson, R.-A. Hydrobiologia **101**:71 (1983).

34. Hessen, D.O. Microb. Ecol. **31**:215 (1985).

35. Meyer, J.T., R.T. Edwards and R. Risley Microb. Ecol. **13**: 284 (1987).

36. Edwards, R.T. Limnol. Oceanogr. **32(1)**:221 (1987).

37. Benke, A.C., T.C. Van Arsdall Jr. and D.M. Gillespie. Ecol. Monogr. **54**:25 (1984).

38. Lambert, S.M., P.E. Porter and H. Schieferstein. Weeds **13**:185 (1965).

39. Hague, R., D.W. Schmedding and V.H. Freed. Environ. Sci. Technol. **8**:139 (1974).

40. Karickhoff, S.W., D.S. Brown and T.A. Scott. Water Res. **13**:241 (1979).

41. Hamaker, J.W. and J.M. Thompson. In: A.I. Goring and J.W. Hamaker, Eds, Organic Chemicals in the Soil Environment, pp. 49-142 (New York: Marcel Dekker, 1972).

42. Pionke, H.B. and G. Chesters. J. Environ. Qual. **2**:29 (1973).

43. O'Connor, D.J. and J.P. Connolly. Water Res. **14**:1517 (1980).

44. DiToro, D.M. Chemosphere **14(10)**:1503 (1985).

45. Weilenmann, U., C.R. O'Melia and W. Stumm. Limnol. Oceanogr. **34**:1 (1989).

46. Stevenson, F.J. Humus Chemistry (New York: John Wiley & Sons, 1982).

47. Hutchinson, G.E. A Treatise on Limnology. Vol. I: Geography, Physics and Chemistry, pp. 1015 (New York: John Wiley, 1957).

48. Wetzel, R.G. Limnology, pp. 743 (Philadelphia: W.B. Saunders Co. 1983).

49. Jones, R.I., K. Salonen and H. De Haan. Freshwat. Biol. **19**:357 (1988).

50. Chow-Fraser, P. and H. Duthie. Archiv für Hydrobiol. **97**:109 (1983).

51. Chow-Fraser, P. and H. Duthie. Archiv für Hydrobiol. **110**:67 (1987).

52. Stewart, A.J. and R.G. Wetzel. Freshwat. Biol. **12**:369 (1982).

53. Steinberg, C. and U. Muenster. In: G.R. Aiken, D.M. McKnight, R.L. Wershaw, Eds, Humic Substances in Soil, Sediment and Water, pp. 105-145 (New York: John Wiley & Sons, 1985).

54. Devol, A.H., A. Dos Santos, B.R. Forsberg and T.M. Zaret. Hydrobiologia **109**:97 (1984).

55. Jackson T.A. and R.E. Hecky. Can. J. Fish. Aq. Sciences **37**:2300 (1980).

56. Prakash, A. and M.A. Rashid. Limnol. Oceanogr. **13**:598 (1968).

57. McDonald, D.G. Can. J. Zool. **61**:691 (1983).

58. Wright, D.A. Freshwat. Biol. **10**:571 (1980).

59. Gjessing, E.T. and L. Berglind. Vatten **38**:402 (1982).

60. Stewart, A.J. In: Synthetic Fossil Fuel Technologies, K.E. Cowser, Ed., pp. 505-521 (1984).

61. Benes, P., E.T. Gjessing and E. Steinnes. Water Res. **10**:71 (1076).

62. Mantoura, R.F.C., A. Dickson and J.P. Riley. Estaurine and Coastal Marine Science **6**:387 (1978).

63. Sholkovitz, E.R. and D. Copland. Geochim. Cosmochim. Acta **45**:181 (1981).

64. Gjessing, E.T. and A.K. Gudmundson Rogne. Vatten **38**:406 (1982).

65. Christman, R.F. and E.T. Gjessing. Aquatic and Terrestrial Humic Materials, pp. 538 (Ann Arbor: Ann Arbor Science, 1983).

66. Winner, R.W. Water Res. **19(4)**:449 (1985).

67. Winner, R.W. Aquat. Toxicol.(Amst) **5**:267 (1984).

68. Dempsey, B.A. and C.R. O'Melia. In: R.F. Christman and E.T. Gjessing, Eds, Aquatic and Terrestrial Humic Materials, pp. 239-274 (Ann Arbor:Ann Arbor Science, 1983).

69. Linnik, P.N. and B.I. Nabivanets. Hydro. J. **3**:71 (1983).

70. Alberts, J.J. and J.P. Giesey. In: R.F. Christman and E.T. Gjessing, Eds, Aquatic and Terrestrial Humic Materials, pp. 333-348 (Ann Arbor: Ann Arbor Science, 1983).

71. Gjessing, E.T. Arch. Hydrobiol. **91(2)**: 144 (1981).

72. Williams, K.A., D.W.J. Green and D. Pascoe. Arch. Hydrobiol. **102**:461 (1985).

73. Ahsanullah, M. and T.M. Florence. Marine Biology **84**:41 (1984).

74. Poldoski, J.E. Environ. Sci. Technol. **13**:701 (1979).

75. Driscoll, C.T., J.P. Baker, J.J. Bisogni and C.L. Schofield. Nature **284**:161 (1980).

76. Hutchinson, N.J. and J.B. Sparague. Environ. Toxicol. Chem. **6**:1 (1987).

77. Petersen R.C. Jr., L.B.-M. Petersen, U. Persson, A. Kullberg, A. Hargeby and A. Paarlberg. Water Qual. Bull. **11**:44 (1986).

78. Dickson, W. Vatten **39**:400 (1983).

79. Visser, S.A. Soil Biol. Biochem. **17**:457 (1985).

80. Cosovic, B., V. Vojvodic and T. Plese. Water Res. **19**:175 (1985).

81. Visser, S.A. Rev. Fr. Sci. Eau. **1**:285 (1982).

82. Visser, S.A. Plant and Soil **87**:209 (1985).

83. Pommery, J., N. Pommery, M. Imbenotte, J.-C. Lhopitault and F. Erb. Sciences de l'eau. **1**:309 (1982).

84. Chen, Y. and M. Schnitzer. Soil Sci. **125**:7 (1978).

85. Lei, W., Z. Zhu, Z. Wang and P. Chen. Ranliao Huaxue Xuebao **14**:177 (1986), (In Chinese, English abstract).

86. Helenius, A. and K. Simons. Biochim. Biophys. Acta **415**:29 (1975).

87. Swedmark, M. SNV PM 1999, pp. 115 (Solna, SNV, 1986, In Swedish, English summary).

88. de Haan, H. Limnol. Oceanogr. **22**:38 (1977).

89. de Haan, H., G. Halma, T. de Boer and J. Haverkamp. Hydrobiologia **78**:87 (1981).

90. Eaton D.C. Enzy. Microb. Techn. **7**:194 (1985).

91. Bumpus, J.A., M. Tien, D. Wright and S.D. Aust. Science **228**:1434 (1985).

92. Wetzel, R.G. Verein. Limnologie **24** (1990), in press.

93. Wium-Anderson, S.U., U. Anthoni, C. Christophersen and G. Houen. Oikos **39**:187 (1982).

94. Aiken, G.R., D.M. McKnight, R.L. Wershaw and P. MacCarthy, Eds, Humic Substances in Soil, Sediment, and Water (New York: John Wiley, 1983).

95. Gjessing, E.T. Physical and Chemical Characteristics of Aquatic Humus, pp. 1-120 (Ann Arbor: Ann Arbor Science, 1976).

96. Schnitzer, M. and S.U. Kahn. Humic Substances in the Environment (New York: Marcel Dekker, 1972).

97. Glob, P.V. Mosefolket-Jernalderens Mennesker bevaret i 2000 År, pp. 166 (Ahus, Gyldendal, 1965).

98. Morton, J.F. Econ. Bot. **32**:111 (1978).

99. Zucker, W.V. Am. Nat. **121**:335 (1985).

100. Morton, J.F. Science **204**:909 (1979).

101. Janecek, J and J. Chalupa. Arch. Hydrobiol. **65**(**4**):515 (1969).

102. Klocking, R. Arch. Exper. Vet. Med. **34**:389 (1980).

103. Moghissi, A.A. Water Qual. Bull. **11**:3 (1986).

104. Saarikoski, J. and M. Viluksela. Arch. Environ. (1981).

105. Tulp, M.Th.M. and O. Hutzinger. Chemosphere **7**:849 (1978).

106. Kenaga, E.E. Environ. Sci. Technol. **14**:553 (1978).

107. Kenaga, E.E. and C.I.A. Goring. In: J.G. Eaton, P.R. Parrish and A.C. Hendricks, Eds, Aquatic Toxicology. Am. Soc. Testing (1980).

108. Petersen, R.C. and U. Persson. Sci. Tot. Environ. **62**:387 (1987).

Effects of Humic Substances from Different Sources on Growth and Nutrient Content of Cucumber Plants

Manuel Abad[1], Fernando Fornes[2], Diego García[3], Juan Cegarra[3] and Asunción Roig[3]

[1] Universidad politécnica, Departamento de Producción Vegetal, 46020-Valencia, Spain
[2] Infertosa, Unidad de Investigación y Desarrollo, 46006-Valencia, Spain
[3] Consejo Superior de Investigaciones Científicas, CEBAS, Apartado 195, Murcia, Spain

Abstract

Humic substances prepared from different sources of organic materials were tested for their effects on nutrient uptake and growth of cucumber plants. Plants were grown in a modified Hoagland solution (iron as soluble $FeCl_3$), with the addition of 50 mg/l of carbon in the form of humic substances derived from lignite, sphagnum moss or sedge peat. Humic substances produced highly significant increases in the growth of plant tops and roots, in the stem height, in the number of flowers per plant and in the leaf size. The addition of humic substances also resulted in an increase in the contents of N, P, K, Ca, Mg and Fe in the roots and also in the N, P and Fe contents in the shoots. Variation of effects of humic substances derived from different organic materials was not significant.

Introduction

Soil organic matter can, under certain conditions, have a beneficial effect on plant growth. The favourable effects of soil humus on plant growth result either (a) from influence on the physical, chemical and microbiological properties of the soil, or (b) from possible direct effects on the physiological processes of plants [1,2].

The effects of organic substances on soil properties have been studied extensively [3, 4], but the physiological effects have not received much attention. Furthermore, most of these investigations have been limited to seed germination and shoot growth of very young seedlings [2,5], and little is known about the specific effects of humic substances on growth and on nutrient uptake of plants over long periods of time. The response to humic substances depends on the source of the organic materials, and high concentrations of such substances are usually inhibitory for plant growth [5].

The main aim of this investigation was to study nutrient uptake and growth of cucumber plants grown in nutrient solutions containing various humic substances prepared from different sources of organic materials.

Materials and Methods

Humic substances were isolated, as described below, from lignite, sphagnum moss and sedge peat. Both lignite and sphagnum moss were pretreated with 0.1 M HNO_3 and then centrifuged. The residue was subsequently extracted by shaking with 0.25 M KOH. Insoluble constituents were separated by centrifugation (7 900 g). Humic substances from sedge peat were directly extracted with 0.1 M KOH by violent shaking. After filtering through a mesh (<100 μm^2), the pH of the filtrate was adjusted to 1.5 with concentrated H_2SO_4 and then left undisturbed for 12 h. The supernatant liquid fraction was removed by centrifugation. The lignite and the sphagnum and sedge peats, respectively, have been described as LIG, P-II and P-III in another work [6]. Details on the extraction of their humic substances and some of the characteristics of these substances have also been described. All three had very high contents of humic acids in comparison to their contents of fulvic acids.

Growth trials were carried out in a heated greenhouse during the winter (November 28-February 22). Aliquots of the corresponding humic substances, to give a final organic carbon concentration of 50 mg/l, were added to a half-strength, modified Hoagland's nutrient solution (iron as soluble $FeCl_3$), and mixed thoroughly. The pH was adjusted to 6.0, and the solutions were divided into jars (900 ml/jar). The experiment was set up in 15 replicate jars with four different treatments: control (nutrient solution only) and nutrient solution with humic substances derived either from lignite, from undecomposed sphagnum peat or from strongly-decomposed sedge peat.

Seeds of cucumber (cv. Sensation F_1) were soaked in distilled water for 2 h, sown in wet vermiculite and kept in the greenhouse for 12 days. Healthy and homogeneous plants, supported by split foam plugs, were inserted into lid holes of 1 l plastic jars. Seedlings were first grown in distilled water for two days to acclimatize the system and then transferred to the treatment solutions. Solutions were replaced every two days in order to maintain oxygenation of the roots and to avoid precipitation of humic substances. Plants were grown for 8 weeks, and then 10 plants per treatment were harvested and divided into roots and shoots. Fresh and dry weights of roots and tops, stem height, number of flowers and leaves per plant and leaf size of the largest leaf were measured and recorded. Plant material was dried at 60°C for analysis of mineral nutrients. Nitrogen was determined by the micro-Kjeldahl method; Ca, Mg, K, Mn, Fe and Zn by atomic absorption spectrometry after dry combustion of the samples, and P by the molybdovanadate method. Five plants per treatment were allowed to grow for an additional 4 weeks, and then total yield and marketable yield (over 50 g/fruit) per plant were recorded.

Results and Discussion

Compared to the control, the addition of 50 mg/l of carbon, in the form of humic substances, to nutrient solutions brought about significant increases in the dry weights of the roots, tops and whole plants (Table 1), and in the height of the stems, the fresh weight of the roots, the number of flowers per plant and the leaf size (Table 2). It seemed that, under the experimental conditions used, the humic substances directly stimulated plant growth and development (short-term fertilizer effect). These results are in agreement with

Table 1 Effect of humic substances from different organic materials on the dry matter yield.

	Plant tops		Plant roots		Whole plant	
Treatment*	g/plant	% yield	g/plant	% yield	g/plant	% yield
Control	5.44	100	0.93	100	6.37	100
L I	9.26	170	1.93	208	11.19	176
S M	9.11	168	1.73	186	10.83	170
S P	8.96	165	2.28	245	11.24	176
LSD 5%	2.53	---	0.56	---	2.99	---

* LI nutrient solution with humic substances from lignite; SM, from sphagnum moss; SP, from sedge peat. LSD 5% indicates the least significant difference at the 5% probability level.

Table 2 Vegetative growth and reproductive development of cucumber plants as affected by humic substances from different organic materials.

	Stem height	Fresh weight of roots	Number/plant		Leaf length
Treatment*	(cm)	(g/plant)	Leaves	Flowers	(cm)
Control	67.1	19.05	10.6	9.4	15.3
L I	81.5	43.85	11.1	11.5	18.7
S M	78.3	41.20	12.6	11.3	17.8
S P	84.8	43.65	12.5	12.4	17.9
LSD 5%	8.4	8.91	NS	1.8	1.9

* Key symbols like in Table 1; NS = non-significant

those of Rauthan and Schnitzer [7] for cucumber in nutrient culture and with those for young seedlings on a variety of plant species (all Angiosperms) in water and nutrient-solution culture conditions [2].

The effect of humic substances on plant growth depends on the parameter measured. Humic substances enhanced the increases in fresh and dry weights of roots by 125% and 113%, respectively, whereas the dry weight of plant tops was increased by 68%, the number of flowers per plant by 25%, the height of the stems by 22%, and the leaf size by a mere 18% (Tables 1 and 2). On the other hand, the number of leaves per plant was unaffected by humic substances. Thus, it can be concluded that different parts of the plants react to humic substances to different extents. Other studies have also shown that different organs of intact plants respond differently to humic substances [7,8].

An increase in crop value may come through an increase in total yield and/or a higher proportion of marketable fruits. As these parameters were significantly enhanced by the application of humic substances prepared from sedge peat (Table 3), it may be possible

Treatment*	g/plant Total yield	Marketable yield
Control	212.6	208.5
L I	272.4	265.1
S M	282.5	271.9
S P	332.5	323.3
LSD 5%	77.1	74.1

* Key symbols like in Table 1.

Table 3 The influence of humic substances from different organic materials on total yield and marketable yield (over 50g/fruit) of cucumbers.

to increase plant yield to an extent above that based on the application of fertilizer equivalent nutrients.

Growth and development of cucumber did not show significant differences upon the addition of diverse humic substances derived from the three, above-mentioned sources of organic materials (Tables 1-3). Sedge peat did, however, cause a slight improvement in growth and development in comparison to lignite and sphagnum moss, although this effect was mostly inconsistent and not statistically significant.

The addition of humic substances to nutrient solutions yielded highly significant increases, in comparison to the controls, in the contents of N, P, K, Ca, Mg and Fe in roots and also in the N, P and Fe contents of shoots (Tables 4 and 5). Under these conditions, the content of the mentioned elements in the roots, with the exception of N and Mg, was increased by more than 55%. Also, contents of N, P and Fe in plant tops were greatly increased. Zn content in roots was significantly decreased by the administration of humic substances (Table 5). Conflicting results were obtained for the effect of humic substances on Mn content in roots, since it was significantly increased in the case of sedge peat and decreased with lignite and sphagnum moss.

Table 4 Effects of humic substances from different organic materials on the elemental contents of plant tops. Results expressed in terms of dry weight.

	Total content mg/plant					µg/plant		
Treatment*	N	P	K	Ca	Mg	Fe	Mn	Zn
Control	145.2	31.2	183.7	88.3	49.1	482.4	1,302.1	365.7
LI	211.9	51.8	267.3	87.5	47.6	1,044.2	1,010.1	364.2
SM	133.4	46.1	280.4	81.4	51.1	734.8	1,041.1	329.0
SP	282.4	30.1	253.4	96.6	58.6	715.3	1,180.6	389.3
LSD 5%	35.9	8.9	NS	NS	NS	170.1	NS	NS

* Key symbols like in Table 1. NS = non-significant.

The main effect of humic substances on the elemental nutrient content of plants is related to Fe content. Futhermore, plants grown in the absence of humic substances (controls) showed the typical interveinal chlorotic leaves which result from Fe deficiency. In plant tops, the increase in Fe content brought about by humic substances ranged between 117% (lignite) and 48% (sedge peat) (Table 4), which suggests that Fe uptake by plants depended on the nature of the carbonaceous materials from which the humic substances were isolated. Also, the higher uptake of Fe by plants grown in nutrient solution containing the humic substances from lignite could be related to the special characteristics of these substances. Lignite humic substances showed the highest contents of functional groups in comparison to the other two types of substances tested in the experiment [6].

In plant roots, Fe uptake was more than doubled by humic substances from lignite and sedge peat, but it was unaffected by sphagnum moss (Table 5). On the other hand, Fe content in roots of control plants was unusually high and probably resulted from the surface precipitation of $FeCl_3$ colloids [9]. In the same way, a portion of the Fe content in roots of plants that had received the application of the humic substances, could be superficially absorbed, owing to the precipitation of low soluble Fe-humate.

Table 5 Effects of humic substances from different organic materials on the elemental contents of plant roots. Results expressed in terms of dry weight.

	Total content mg/plant					µg/plant		
Treatment*	N	P	K	Ca	Mg	Fe	Mn	Zn
Control	66.0	6.2	59.6	3.7	3.7	1,408.7	822.6	143.2
LI	79.8	10.7	118.9	8.4	3.4	3,050.2	213.9	74.8
SM	69.3	7.0	81.2	2.7	2.9	1,173.3	397.7	27.7
SP	90.2	11.3	137.8	10.3	6.4	4,673.3	1,392.9	107.0
LSD 5%	9.2	1.8	22.8	3.7	2.4	653.7	169.7	19.4

* Key symbols like in Table 1.

The above results show that applying of humic substances to the nutrient solution in which cucumber plants are grown, significantly stimulates (a) the vegetative growth and reproductive development of plants and (b) the uptake of nutrient elements from the nutrient solution. Hence, humic substances can exert their effects indirectly and/or directly. In the former case, they can, for instance, form complexes with metal ions and thus increase their solubility and thereby their availability to plant roots. In the case of a direct effect, humic substances must be taken up by the plant, and then they can affect a number of biochemical mechanisms, such as active ion uptake or protein synthesis. Further research is needed to find out which of these explanations is most likely.

References

1. Stevenson, F.J. Humus Chemistry. Genesis, Composition, Reactions (New York: John Wiley and Sons, 1982).

2. Vaughan, D. and B.G. Ord. In: Vaughan and R. Malcolm, Eds., Soil Organic Matter and Biological Activity, pp. 1-35 (Dordrecht: Martinus Nijhoff/Dr. W. Junk Publishers, 1985).

3. Allison, F.E. Soil Organic Matter and its Role in Crop Production (New York: Elsevier Press, 1973).

4. Kononova, M.M. Soil Organic Matter. Its Nature, its Role in Soil Formation and in Soil Fertility (Oxford: Pergamon Press, 1961).

5. Vaughan, D. and R. Malcolm. In: D.Vaughan and R. Malcolm, Eds., Soil Organic Matter and Biological Activity, pp. 37-75 (Dordrecht: Martinus Nijhoff/Dr. W. Junk Publishers, 1985).

6. Cegarra, J., D. García, M. Abad, F. Fornes and A. Roig. Unpublished results.

7. Rauthan, B.S. and M. Schnitzer. Plant and Soil **63**:491 (1981).

8. Sladky, Z. Biologia Plantarum **1**:142 (1959).

9. Linehan, D.J. and M. Shepherd. Plant and Soil **52**:281 (1979).

Phytotoxic Substances in Runoff from Forested Catchment Areas

Anders Grimvall[1], Maj-Britt Bengtsson[1], Hans Borén[1] and Dan Wahlström[2]

[1]Department of Water and Environmental Studies, Linköping University
[2]Department of Biology, Linköping University, S-581 83 Linköping, Sweden

Abstract

Runoff from different catchment areas in southern Sweden was tested in a root bioassay based on solution cultures of cucumber seedlings. Water samples from agricultural catchment areas produced no signs at all or only weak signs of inhibited root growth, whereas several water samples from catchment areas dominated by mires or coniferous forests produced visible root injuries. The most severe root injuries (very short roots, discolouration, swelling of root tips and lack of root hairs) were caused by samples from a catchment area without local emissions and dominated by old stands of spruce. Fractionation by ultrafiltration showed that the phytotoxic effect of these samples could be attributed to organic matter with a nominal molecular-weight exceeding 1000 or to substances associated with organic macromolecules. Experiments aimed at concentrating phytotoxic compounds from surface water indicated that the observed growth inhibition was caused by strongly hydrophilic substances. Previous reports on phytotoxic, organic substances of natural origin have emphasized interaction between plants growing close together. The presence of phytotoxic substances in runoff indicates that there is also a large-scale dispersion of such compounds.

Introduction

It has long been known that elevated concentrations of certain metals, in particular aluminium, may inhibit root growth in higher plants [1,2]. During the past decade such effects have received increased attention, and several authors have claimed that leaching of aluminium induced by acidic deposition may be the main cause of forest decline in Central Europe [3,4]. Research on organic phytotoxins has primarily dealt with weed control [5]. However, there is now a rapidly increasing amount of research on allelopathic chemicals, i.e. substances produced by one organism that affect the growth or health of other species [6,7,8]. Several authors have emphasized that phenolic aldehydes are ubiquitous in soil and strongly phytotoxic [9,10,11]. In addition, it has been claimed that photooxidation of chlorinated solvents in the atmosphere may cause large-scale dispersion of the well-known herbicide trichloroacetic acid [12].

This study was initiated by certain, seemingly inexplicable observations of root injuries in soiless cultures. In 1986, a plant breeding company in southern Sweden (Weibullsholm, Landskrona) reported severe root injuries on seedlings of lucern grown in filter paper rolls in water. The following year, similar effects were observed, and bioassays were performed with several different types of seedlings and waters. The results indicated that the municipal tap water used at the laboratory had a phytotoxic effect. A survey of irrigation waters used in greenhouse cultures showed that several other tap waters and surface waters in southern Sweden inhibited growth of cucumber and tomato seedlings [13]. Furthermore, root injuries were observed in certain commercial cultures of cucumber and tomato on rockwool in nutrient solutions

In order to analyse the relationship between the main landuse and the presence of phytotoxic substances in runoff, cucumber seedlings were grown in paper rolls in water from different types of catchment areas. Root length, root colour, swelling of root tips and lack of root-hairs were observed. For certain water samples, solvent extraction, resin adsorption and ultrafiltration were used to isolate and test different fractions of organic compounds.

Materials and Methods

Water Samples

Surface water samples were collected at 16 different locations in southern Sweden (see Table 1). In addition, a chlorinated tap water sample from Ringsjö waterworks, which uses Lake Bolmen as raw water source, were examined. Sampling was performed from Nov. 1988 to Aug. 1989.

Table 1 Sampling sites and dominating landuse in the catchment areas.

Sample code	Sampling site	Province	Dominating landuse	Arable land (%)
A1	Nissan River	Småland	Mire	< 1
B1	Lake Fiskelösa	Bohuslän	Mire	0
C1	Herå River	Dalsland	Mire	0
D1-D2	Unnamed creek	Småland	Coniferous forest	0
E1-E8	Lake Trollsjön	Östergötland	Coniferous forest	0
F1	Kynne River	Dalsland	Coniferous forest	< 10
G1	Getbro River	Dalsland	Coniferous forest	0
H1	Lake Stora Le	Dalsland	Coniferous forest	< 10
I1-I3	Lake Bolmen	Småland	Coniferous forest	< 10
J1	Lake Vänern	Dalsland	Forest/arable land	13
K1	Lake Helgasjön	Småland	Forest/arable land	< 10
L1	Stångå River	Östergötland	Forest/arable land	10
M1	Svartå River	Östergötland	Arable land	20
N1	Skenaå River	Östergötland	Arable land	70
O1	Lidan River	Västergötland	Arable land	40
P1	Flian River	Västergötland	Arable land	~50

Milli-Q (Millipore) water was used as reference

The Cucumber Root Bio-assay

Seeds of field cucumber (Weibull Favör) were placed in a Petri dish on a piece of filter paper (Munktell 1410) moistened with the water to be tested. After 3 days in darkness at 22°C and 70% humidity, 10 seedlings of average size were selected and rolled into wet filter paper. The paper roll was put into a 300 ml beaker with 50 ml of the water sample, and the seedlings were grown for 4 days at 24°C and 70% humidity. Photoperiod (10,000 lux) was 10-12 h per day. After 2 days in the climate room, another 50 ml of the water sample was added. At the end of the bioassay, the paper roll was unrolled and the root length was measured for each seedling. Root discolouration, absence of root hairs and other visible root injuries were noted. Each water sample was tested in two of the described seedling paper rolls.

Nutrient Solution

In some experiments, plant nutrients were added to the waters to be tested. The stock solution had the following composition: KH_2PO_4 (0.20 M), $CaCl_2$ (0.15 M), $MgSO_4$ (0.10 M), $NaNO_3$ (0.32 M), KCl (0.32 M) and Fe-EDTA (0.55 g/l).

Concentration Methods

Two methods for concentrating organics in water were used: dichloromethane extraction and adsorption on a polyacrylic ester (XAD-8).

Dichloromethane extraction was performed by mixing about 10 l of sample water with 0.8 l distilled dichloromethane. After stirring with a magnet bar for 90 min, the organic phase was separated from the water phase in a separatory funnel. The organic phase was dried by freezing and then evaporated to 100 µl. Dichloromethane extraction was performed without adjusting the pH of the water sample.

Adsorption on XAD-8 was performed by passing about 10 l of sample water through a glass column with 30 ml adsorbent. The column was eluted with 25 ml ethyl acetate, which was then evaporated to 100 µl. Prior to adsorption, the water sample was acidified to pH 3.2 with concentrated HCl.

Fractionation and Filtration Methods

Molecular-weight fractionation was performed using an ultrafiltration stirred cell (Amicon, Model 8400), equipped with a YM-10 filter (Amicon, M_w cut-off 1000). Prior to the root bioassay, the pH of both the concentrate and the filtrate was adjusted to 7. The concentrate was diluted with Milli-Q water to the volume of the original water sample.

To remove different groups of organics from the waters to be tested, four water samples were filtered through activated carbon (Millipore) and two samples were treated with diethylaminoethyl cellulose in H^+ form (Cellucol DEAE Type SF).

Chemical Analyses

Total organic carbon (TOC) was measured with an Astro TOC-analyzer at Svelab, Kalmar. UV absorbance at 254 nm was measured with a Beckman DU-8 spectrometer. Adsorbable organic halogen (AOX) was analysed according to the DIN method for AOX in surface waters (DIN 38409) in a Euroglas AOX analyser. Aluminium was analysed by atomic absorption spectrometry.

Results

Root Length and Visible Root Injuries Caused by Different Types of Waters

Testing different waters with the cucumber root bioassay resulted in great variation of root growth (Fig. 1). The longest roots and the largest number of branch roots were obtained with water samples from agricultural catchment areas. Samples from catchment areas dominated by mires or coniferous forests produced, on the average, significantly shorter roots. The latter group of water samples also produced several examples of visible root injuries. Root hairs were missing, the colour of both the surface and the interior of the root was brown and swelling of root tips was pronounced.

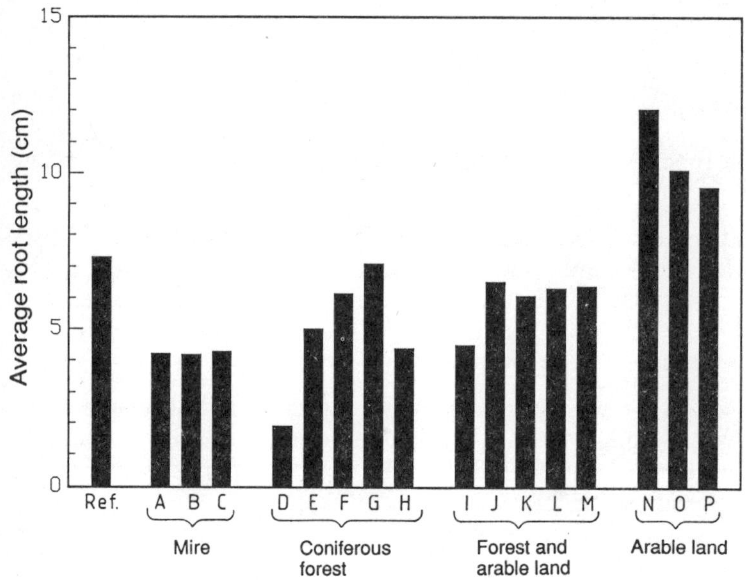

Figure 1 Average root length for surface water samples tested with the cucumber root bioassay. Ref denotes Milli-Q water and A-P different surface waters (see Materials and Methods)

Effects of pH, Aluminium and Plant Nutrients

Statistical analysis of all data from the survey of surface waters showed that the seedling root length was positively correlated to the pH of the water sample and negatively correlated to the AOX and aluminium concentrations. The strongest correlation ($r = 0.75$) was obtained for root length *vs* pH (see Fig. 2). Furthermore, a simple linear function of pH produced almost the same fit to observed root lengths as a multiple regression model with pH, AOX and aluminium. However, it is noteworthy that two of the neutral water samples caused visible root injuries. The importance of factors other

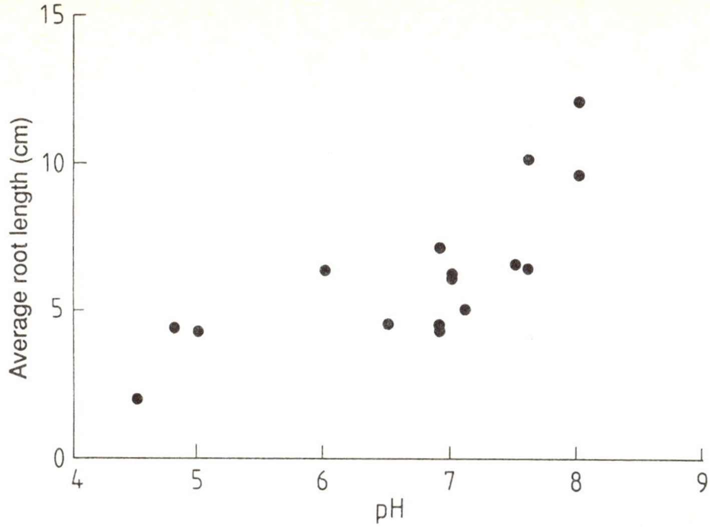

Figure 2 Average root length of cucumber seedlings *vs* pH for the sampling sites A-P.

than pH was also demonstrated by experiments in which the water samples were tested before and after pH adjustments. For example, the strong root growth inhibition caused by a sample from a spuce forest (D1) remained almost unchanged after raising the pH from 4.5 to 6.0. The average root length and the type of root injuries were nearly identical at both pH values.

Addition of aluminium chloride to Milli-Q water had a slightly negative impact on root growth (Fig. 3). However, even at the highest concentration (10 mg Al/l), there were no visible root injuries such as absence of root hairs or swelling of root tips. The Al concentration in the surface water samples varied from 0.05 to 0.5 mg/l, which strongly indicated that inorganic aluminium was not the major cause of the observed root injuries.

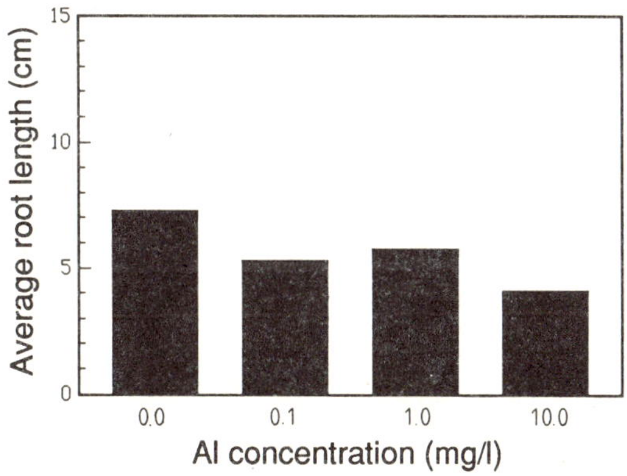

Figure 3 Average root length of cucumber seedlings grown Milli-Q water with different concentrations of aluminium.

Analysis of adsorbable organic halogens (AOX) in the tested waters showed that the samples (D1-D2) causing the most severe root injuries had an extremely high AOX concentration (180 µg/l). However, in the entire data set, there was no simple relationship between root length and AOX concentration.

In general, addition of the previously described nutrient solution markedly improved root growth (Fig. 4). Furthermore, characteristic root injuries such as swelling of root tips disappeared. This indicates that a proper composition of plant nutrients may protect the cucumber seedlings against root injuries.

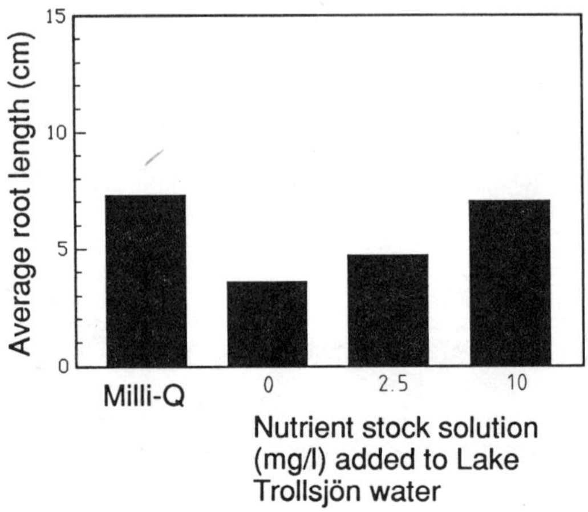

Figure 4 Average root length of cucumber seedlings grown in water from Lake Trollsjön (sample E) with different amounts of nutrient solution. Stock nutrient solution as in Materials and Methods.

Tests of Concentrates of Organic Substances and Filtered Samples

Solvent extraction by dichloromethane at pH 7 was applied to two surface water samples (E and I) and one chlorinated tap water (Ringsjö waterworks). The obtained extracts were then redissolved in Milli-Q water and tested in the root bioassay. The results gave no evidence of phytotoxicity. Even at the highest dose, which corresponded to a concentration factor of 8, when assuming 100% recovery in the extraction step, no root injuries were observed (Fig. 5). Adsorption on XAD-8 was applied to the same water samples. Two of them had no phytotoxic effects, whereas the third (sample E) caused a marked inhibition of root growth (Fig. 6).

Filtration experiments showed that activated carbon filtration had only a minor effect on the average root length. Considering that the activated carbon primarily adsorbs lipophilic organic compounds, this is consistent with the lack of phytotoxicity of dichloromethane extracts of neutral water samples. Two water samples (E and I) were treated with DEAE to remove humic substances, and, again, the results were consistent with the results obtained by testing concentrates of organic compounds. For the sample (E) producing a phytotoxic XAD-8/ethylacetate extract, DEAE treatment improved root growth. For the other sample (I), the XAD-8/ethylacetate extract was not phytotoxic and DEAE treatment had no effect on root growth.

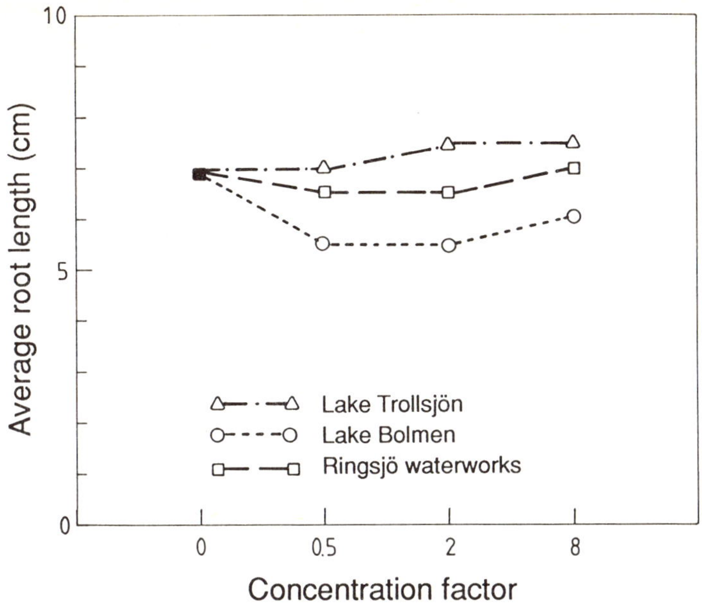

Figure 5 Average root length of cucumber seedlings. Response to different doses of dichloromethane extracts of two surface water samples (E and I) and one chlorinated tap water (Ringsjö waterworks).

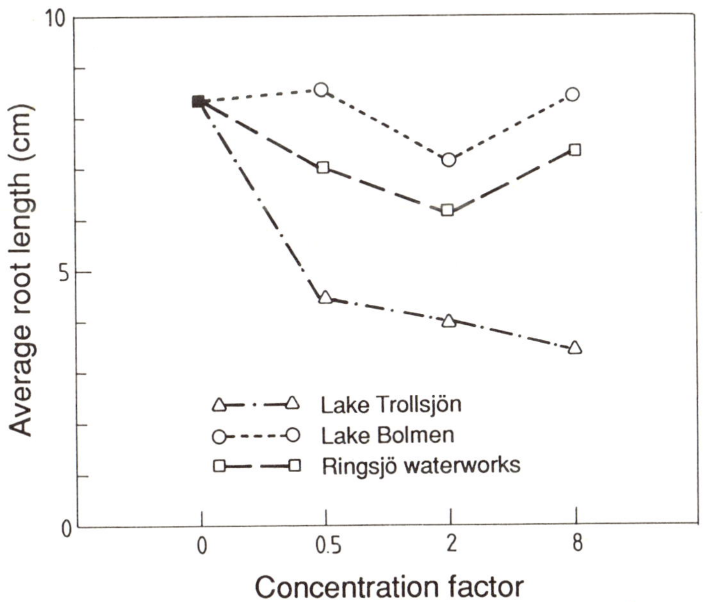

Figure 6 Average root length of cucumber seedlings. Response to different doses of XAD-8/ethylacetate extracts of two surface water samples (E and I) and one chlorinated tap water (Ringsjö waterworks).

The results of the ultrafiltration experiments are shown in Fig. 7. For the sample causing severe root injuries (D2), the phytotoxicity could be attributed to the fraction with a nominal molecular-weight exceeding 1000. For a sample causing no visible root injuries, the high- and low-molecular-weight fractions both caused the same root growth as the original water sample.

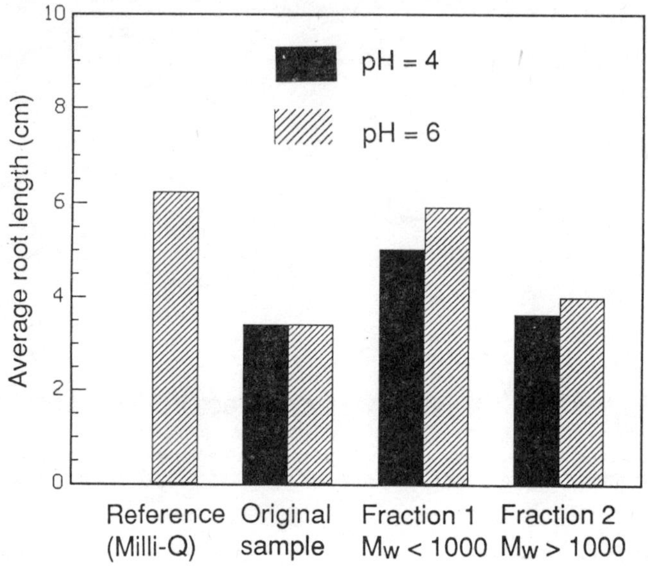

Figure 7 Average root length of cucumber seedlings grown in original water sample (Orig.) and different molecular-weight fractions ($M_W < 1000$ and $M_W > 1000$, respectively) obtained by ultrafiltration.

Discussion

Natural waters contain very complex mixtures of organic and inorganic substances, which may affect the results of root bioassays in many different ways [14]. There is a multiplicity of phytotoxic substances, and specific toxic effects can be accentuated or modified by nutritional imbalance and osmotic stress. This makes it necessary to discuss several alternative interpretations of the obtained results.

The experiments in which plant nutrients were added to some of the tested water samples showed that such additions normally had a positive impact on root growth. This indicates that the observed differences in root length between waters from agricultural and forested catchments may partly be attributed to differences in nutrient concentrations. However, the survey of surface waters also gave strong evidence of phytotoxic effects. Several samples from catchments dominated by forests and mires produced visible root injuries (discolouration and swelling of root tips) and considerably shorter roots than pure water (MilliQ-water). The results of the concentration and filtration experiments gave additional evidence of phytotoxic effects, even though the number of samples was too limited to permit any quantitative conclusions about the occurrence of phytotoxic substances in water.

The geographic distribution of surface waters causing root growth inhibition clearly showed that the observed effects did not originate from local industrial emissions or herbicide spills. Only naturally produced phytotoxins, long-range transport of air pollutants or the combined effect of these two factors may explain why water from seemingly unpolluted, forested catchment areas caused the most severe root injuries.

Chemical characterization of the substances causing injury to cucumber seedling roots is incomplete. However, it can be said that neither a low pH nor a high concentration of inorganic aluminium is sufficient to produce the observed root injuries. It is difficult to explain all the results of the concentration and filtration experiments, unless it is assumed that at least some of the samples contained hydrophilic, phytotoxic organic compounds with an apparent molecular weight exceeding 1000. Considering the special biological activity of many chloroorganic compounds [15], the negative correlation between AOX and seedling root length is worth mentioning. However, the evidence of a causal relationship between these two variables is very weak.

It has previously been shown that several low-molecular-weight organic compounds, such as phenolic acids, have phytotoxic effects [5,11]. It has also been reported that such substances can be associated with humic substances in soil. In fact, Jalal and Read [9,10] claimed that the phytotoxic substances can be so strongly associated to humic matter that they are not released from soil by aqueous leaching. The present study demonstrated that phytotoxic substances are present in runoff in concentrations high enough to cause root injuries in cucumber seedling. This shows the need for a river basin perspective on the occurrence and transport of such substances. It is not unlikely that large-scale redistributions of naturally produced phytotoxins can have thus far unknown impacts on the natural flora. It has even been suggested that phytotoxins may structure whole ecosystems [16].

Acknowledgements

The authors are grateful to Per Lundin, Weibullsholm, who introduced us to the problem area, and to AB Sydvatten and several municipalities in western Scania for financial support.

References

1. Hartwell, B. and F. Pember. Soil. Sci. **6**:259 (1918).
2. Magistad, O. Soil. Sci. **20**:181 (1925).
3. Rost-Siebert, K. Allgem. Forstzeitschr. **26/27**:686 (1983).
4. Ulrich, B. Atm. Environ. **18**:621 (1984).
5. Putnam, A.R. Weed allelopathy. In: S.O. Duke, Ed., Weed Physiology, vol I. Reproduction and Ecophysiology, pp. 131-155 (Boca Ratos, Florida: CRC Press, 1985).
6. Rice, E.L. Allelopathy. (New York: Academic Press, 1984).
7. Putnam, A.R. and C.-S. Tang. The Science of Allelopathy. (New York: Wiley-Interscience, 1986).
8. Lovett, J.V., M.Y. Ryuntyu and D.L. Liu. J. Chem. Ecol. **15**:1193 (1989).

9. Jalal, M.A.F. and D.J.Read. Plant and Soil **70**:257 (1983).

10. Jalal, M.A.F. and D.J.Read. Plant and Soil **70**:273 (1983).

11. Einhellig, F.A. Mechanisms and modes of action of allelochemicals. In: Putnam, A.R. and C.-S. Tang, Ed., The Science of Allelopathy, pp 171-188. (New York: John Wiley & Sons, 1986).

12. Frank, H. and W. Frank. Nachr. Chem. Tech. Lab. **34**:15 (1986).

13. Lundin, P., B. Wibrandt and K.O. Jönsson (In Swedish). Växtskyddsrapporter, Jordbruk, **49**:191 (1988).

14. Leather, G.R. and F.A. Einhellig. J. Chem. Ecol. **14**:1821 (1988).

15. Neidleman, S.L. and J. Geigert, J. Biohalogenation. Principles, basic roles and applications. (Chichester: Ellis Horwood Ltd, 1985)

16. McKey, D., P.G. Waterman, and J.S. Gartlan. Science **202**:61 (1978).

Interaction of Humic Acids and Humic-Acid-Like Polymers with Herpes Simplex Virus Type 1

Renate Klöcking and Björn Helbig
Institute of Medical Microbiology
Medical Academy Erfurt, DDR-5010 Erfurt, GDR

Abstract

The study was performed in order to compare the antiviral activity against herpes simplex virus type 1 (HSV-l) of synthetic humic-acid-like polymers to that of their low-molecular-weight basic compounds and naturally occurring humic acids (HA) in vitro. HA from peat water showed a moderate antiviral activity at a minimum effective concentration (MEC) of 20 µg/ml. HA-like polymers, i.e. the oxidation products of caffeic acid (KOP), hydrocaffeic acid (HYKOP), chlorogenic acid (CHOP), 3,4-dihydroxyphenylacetic acid (3,4-DHPOP), nordihydroguaretic acid (NOROP), gentisinic acid (GENOP), pyrogallol (PYROP) and gallic acid (GALOP), generally inhibit virus multiplication, although with different potency and selectivity. Of the substances tested, GENOP, KOP, 3,4-DHPOP and HYKOP with MEC values in the range of 2 to 10 µg/ml, proved to be the most potent HSV-l inhibitors. Despite its lower antiviral potency (MEC 40 µg/ml), CHOP has a remarkable selectivity due to the high concentration of this polymer that is tolerated by the host cells (>640 µg/ml). As a rule, the antiviral activity of the synthetic compounds was restricted to the polymers and was not preformed in the low-molecular-weight basic compounds. This finding speaks in favour of the formation of antivirally active structures during the oxidative polymerization of phenolic compounds and, indirectly, of corresponding structural parts in different HA-type substances.

Introduction - Historical Background

Naturally occurring humic acids (HA) and synthetic HA-like polymers derived from phenolic compounds are polyanionic substances of the humic acid type (HA-type substances). They may be characterized as negatively charged, yellow to dark brown polymers which are precipitable by heavy metals and partially degradable to low-molecular phenolic degradation products by reduction with sodium amalgam or by alkaline pressure hydrolysis.

Despite their different origins, HA and HA-like polymers have a great number of biological activities in common. For example, they exert antiviral activity against various DNA as well as RNA viruses.

The possible interaction of naturally occurring humic acids (HA) with viruses was first discussed by Schultz [1] in 1962 when he successfully used HA-containing peat-dust litter for combating an outbreak of foot-and-mouth disease in pigs. He supposed the virus inactivation to be due to the protein-denaturing effect of HA.

In the past two decades, more detailed studies with isolated HA showed that: 1) HA also affect viruses that are pathogenic for man, RNA as well as DNA viruses; 2) HA work selectively at an antiviral concentration range without impairing the viability of host cells; and 3) HA, like other polyanionic compounds, such as heparin and pentosan polysulfate, mainly interfere with an early step of virus-cell interaction [2-6]. The antiviral spectrum of HA includes herpes simplex virus type 1 (HSV-l) and type 2 (HSV-2), influenza A, coxsackie A9 and rhinovirus 1 B [5]. Vaccinia virus and adenovirus type 2 are not affected [3]. The reason for the different sensitivity of viruses to HA is still unknown. Sensitivity seems to be independent of the nucleic acid type and is not influenced by the presence or absence of viral envelope.

After discovering the antiviral properties of HA the question arose which structural regions of the HA molecule are responsible for this activity. In this context, the contribution of Hampton and Fulton [7] deserved great interest. These authors found, in 1961, that enzymatically oxidized phenols inactivate phytopathogenic viruses. In 1970, Fulton succeeded in isolating an oxidation product of chlorogenic acid from virus-infected tobacco leaves. Additionally, he was able to synthesize a protein-quinone complex by enzymatic oxidation of chlorogenic acid in the presence of albumin [8].

Modifying Fulton's procedure by omitting albumin, we started synthesizing about 30 phenolic polymers in the mid seventies; first by enzymatic and later by chemical oxidation of various o- and p-diphenolic compounds. The resulting polymers proved to be negatively charged, brown substances sharing the following properties with naturally occurring HA [9,10] (Table 1):

Table 1 Common properties of humic acids and humic-acid-like polymers (according to [10]).

Solubility in alkaline media.
Precipitation by heavy metals.
Anodic migration in an electric field.
Formation of complexes with cationic dyes.
High content of stable free radicals.
Partial degradation to low-molecular phenolic compounds by reductive cleavage with sodium amalgam or by alkaline pressure hydrolysis.

Because of the existence of more than 30 synthetic HA-like polymers produced from well-known, low-molecular compounds, it was possible, for the first time, to determine the biological activities dependent on the structure of the starting material. Furthermore, it is now possible to compare the activities of synthetic polymers to those of natural HA, as well as to those of low-molecular phenolic compounds. The antiherpetic activity served as the starting point for the present study, which attempted to answer the question whether the biological activity investigated is already present in the low molecular starting material or if it appears during oxidative polymerization.

Materials and Methods

HA were isolated from peat water by utilizing their lead(II) chelate compounds, as described previously [11]. Phenolic compounds were supplied from SERVA Feinbiochemica Heidelberg, FRG (caffeic acid, chlorogenic acid, gallic acid), FLUKA AG, Buchs SG, Switzerland (hydrocaffeic acid, gentisinic acid, nordihydroguaretic acid = 4,4'-(2,3-dimethyltetramethylene)-dicatechol), FERAK, Berlin-West (pyrogallol) and Calbiochem, San Diego, California, USA (3,4-dihydroxyphenylacetic acid).

To synthesize phenolic polymers, 10 mmoles of the phenolic compounds were dissolved in distilled water, adjusted with NaOH to pH 8.5 and oxidized with 2.5 mmoles sodium periodate. After heating a short time at 50°C, the solution was left overnight at room temperature. The next day, HA-like polymers were isolated by lead precipitation, as mentioned above for HA [11].

Monolayer cell cultures of human embryo lung fibroblasts (HELF) were used to determine the antiviral activity and cytotoxicity of the test substances. Culturing was carried out in 96-well microtest plates (MLW Polyplast Halberstadt, GDR) with 0.2 ml Eagle's minimum essential medium per well. For antiviral screening, HELF were infected with 2 $TCID_{50}$/cell HSV-l strain Kupka. Virus was allowed to adsorb for 1 h at 37°C. The infectious medium was then removed and replaced by virus-free maintaining medium. The antiviral activity of the substances was tested by means of different test designs where the cells were exposed to the test substances either 1 h before, 1 h during, or 24 h after virus adsorption. Additionally, we studied the effect of a 24-h-exposure during the entire virus replication cycle. Dependent on the experimental design, test substances were added in two-fold dilution series before, at the time of or after virus adsorption. After incubation for 24 h at 37°C, monolayers were checked microscopically for virus-induced cell alterations using a score system (0=0%, l=up to 25%, 2=up to 50%, 3=up to 75%, 4=more than 75% of the cells are altered or destroyed). The substance concentrations reducing the cytopathogenic effect (CPE) by more than 50% were within the antiviral concentration range. The substance concentration necessary for complete suppression of virus-induced CPE was defined as minimum effective concentration (MEC).

Cytotoxicity was detected using the same microtest plate containing uninfected HELF. The maximum tolerated concentration (MTC) corresponds to the highest substance concentration not inducing any morphological cell alterations.

Each experiment has been repeatedly performed. The values given in the table are the geometric means of 3 to 5 single experiments.

Results and Discussion

Table 2 illustrates the antiviral and cytotoxic activities of GENOP and KOP, respectively. If the substances were present during the whole replication cycle (0...24 h p.i.), the antiviral concentration range was between 1 and 32 µg/ml for GENOP and 0.2 and 32 µg/ml for KOP. Higher concentrations were not tolerated without morphological alterations of cells. For GENOP or KOP a concentration of 2 µg/ml was also sufficient for complete suppression of CPE if the substances were present for only 1 h during the adsorption phase. The effect could not be enhanced by prolonging the time of influence. No (KOP) or only weak (GENOP) effects were found when the substances were present before or after viral adsorption. The results confirm the high sensitivity of the early phase of virus-cell interaction, as has also been found for other polyanionic compounds [12-14]. Therefore, test substances listed in Table 3 were exclusively screened at the adsorption phase.

Table 2 Antiviral activity of GENOP and KOP dependent on the time of influence. Virus-cell system: HSV-I Kupka/HELF. Concentration range tested: 0.25-1024 µg/ml.

Test substance	Time of influence h p.i.	Antivirally active concentration µg/ml	MEC[1)] µg/ml	MTC[2)] µg/ml
GENOP	0......24	1.........32	2	32
	-1........0	32.........64	64	64
	0........1	0.5.....128	2	64
	1......24	16.........32	n.a	32
KOP	0......24	0.2.......32	2	16
	-1........0	-	-	64
	0........1	1........128	2	64
	1......24	-	-	8

[1)]Minimum effective concentration; n.a. = not achieved
[2)]Maximum tolerated concentration

Table 3 demonstrates the results for 8 of 22 tested pairs of substances (HA-like-polymers and corresponding low-molecular starting compounds) as well as results for sodium humate. The polymers more or less strongly inhibit the virus adsorption, whereas the low-molecular-weight compounds failed to do so. The only exception is 4,4'-(2,3-dimethyltetramethylene)-dicatechol (nordihydroguaretic acid), which turned brown shortly after addition to the cells. Probably, the substance reacted to the corresponding polymer and thus affected the virus-cell interaction. Both the antiviral activity and the cytotoxicity of the substances display remarkable, structure-dependent differences. GENOP, KOP, 3,4-DHPOP and HYKOP having MEC values between 2 and 10 μg/ml, proved to be the most potent antiviral substances. In spite of its lower antiviral potency CHOP shows a remarkable selectivity, since it is best tolerated by the cells (MTC > 640 μg/ml). The low-molecular phenolic compounds themselves are not antivirally active. This means that the antiviral activity is restricted to the HA-like polymers and must be initiated by fundamental alterations of the chemical structure during the oxidative polymerization.

Table 3 Screening of humic-acid-like polymers and their low-molecular-weight starting compounds for antiviral activity in vitro. Virus-cell system: Herpes simplex virus type 1/ human embryo lung fibroblasts. Concentration range tested: 0.6-640 μg/ml. Time of influence: 0-1 h post infection.

Test substance	Antiviral concentration range μg/ml	MEC[1)] μg/ml	MTC[2)] μg/ml
Caffeic acid	-	-	160
KOP	1...160[3)]	5	80[3)]
Hydrocaffeic acid	-	-	320
HYKOP	5....320	10	160
Chlorogenic acid	-	-	640
CHOP	5....640	40	>640
3,4-Dihydroxyphenyl-acetic acid	-	-	160
3,4-DHPOP	2....160	5	40
4,4'-(2,3-Dimethyltetra-methylene)-dicatechol	20....80	n.a	5
NOROP	20....80	n.a	40
Gentisinic acid	-	-	340
GENOP	1....160	2	80
Pyrogallol	-	-	2
PYROP	40....80	n.a	20
Gallic acid	-	-	640
GALOP	10...320	20	40
Sodium humate	20...320	80	80

[1)]Minimum effective concentration; n.a = not achieved
[2)]Maximum tolerated concentration
[3)]In virus-infected cells, the cytotoxicity was often shifted to higher substance concentrations.

Moreover, we would like to suggest that during these reactions new structural features are formed, which are essential for antiviral activity as well as for HA structure, and which are more or less similar in different HA-type substances. Following up this idea, a common basic structure of HA-type substances might be assumed.

The interaction of HA-type substances with viruses implies practical aspects, too. Hence it was found that adsorption to and recovery of enteroviruses from water filters are influenced by HA [15,16]. Furthermore, HA-virus interactions might be of great importance for limiting virus spread in plants, surface waters and soils and also offer new prospects for the development of antiviral drugs.

References

1. Schultz, H. Deutsche Tierärztliche Wochenschrift **69**:613 (1962).

2. Klöcking, R. and M. Sprössig. Experientia **28**:607 (1972).

3. Klöcking, R. and M. Sprössig. Zeitschrift für Allgemeine Mikrobiologie **15**:25 (1975).

4. Thiel, K.-D., R. Klöcking, H. Schweizer and M. Sprössig. Zentralblatt für Bakteriologie I. Abteilung **239**:304 (1977).

5. Klöcking, R., M. Sprössig, P. Wutzler, K.-D. Thiel and B. Helbig. Zeitschrift für Physiotherapie **33**:95 (1983).

6. Klöcking, R., K.-D. Thiel and M. Sprössig. In: Peat and Peatlands in the Natural Environment Protection Vol. 1, pp. 446-455. Proc. 5th Int. Peat Congr. (Warszawa: WCIN, 1976).

7. Hampton, R.E. and R.W. Fulton. Virology **13**:44 (1961).

8. Hampton, R.E. Phytopathology **60**:1677 (1970).

9. Helbig, B. and R. Klöcking. Zeitschrift für Physiotherapie **33**:31 (1983).

10. Hänninen, K.I., R. Klöcking and B. Helbig. Sci. Tot. Environ. **62**:201 (1987).

11. Klöcking, R., B. Helbig, and P. Drabke. Pharmazie **32**:297 (1977).

12. Vahery, A. and K. Cantell. Virology **21**:661 (1963).

13. Voss, H., H. Sensch and P. Panse. Zentralblatt für Bakteriologie I. Abteilung **192**:137 (1964).

14. Baba, M., R. Snoeck, R. Pauwels and E. De Clercq. Antimicrobial Agents and Chemotherapy **32**:1742 (1988).

15. Guttman-Bass, N. and J. Catalano-Sherman. Appl. Environ. Microbiol. **49**:1260 (1985).

16. Sobsey M.D. and A.R. Hickey. Appl. Environ. Microbiol. **49**:259 (1985).

Effects of pH and Natural Humic Substances on the Accumulation of Organic Pollutants in two Freshwater Invertebrates

Jussi Kukkonen
University of Joensuu, Department of Biology
P.O.Box 111, SF-80101 Joensuu, Finland

Abstract

The present study focused on the accumulation of benzo(a)pyrene (BaP), hexachlorocyclohexane (lindane), pentachlorophenol (PCP) and dehydroabietic acid (DHAA), from a natural humic water (DOC 18 mg/l) and a humus-free reference water, in *Daphnia magna (Cladocera)* and nymphs of the mayfly *Heptagenia fuscogrisea (Ephemeroptera)*. Effects of water pH ranging from 3.5 to 8.5 was examined. The partition coefficients (K_p) of BaP and PCP to organic material were measured by equilibrium dialysis, and in both cases increases in K_p values were noticed with decreasing pH. For neutral compounds (BaP and lindane), the bioconcentration factor (BCF) was the highest at pH 6.5 in the control water. Humic substances significantly lowered the accumulation of BaP, but had no effect on the accumulation of lindane. The lowest test pH gave the highest BCF value, and increasing pH decreased the BCF values of weak organic acids (PCP and DHAA) in the control experiments. This was because the unionized forms of these compounds accumulate better than the more hydrophilic ionized forms. The presence of dissolved organic substances lowered the accumulation of PCP in *H. fuscogrisea* between pH 4.5 and 7.5 and had no effect at pHs 3.5 and 8.5. Humic substances lowered the accumulation of DHAA in *D. magna* between pH 5.5 and 6.5 and had no effect when pH was over 7. In experiments with *H. fuscogrisea* humic substances had no effect on the accumulation of DHAA.

Introduction

Toxicity and accumulation of organic micropollutants (OMP), which are weak organic acids like pentachlorophenol (PCP) and dehydroabietic acid (DHAA), are strongly affected by pH [1-5]. In addition, pH seems to affect the toxicity and accumulation of neutral compounds, like lindane [6] or parathion [7].

It is known that pH affects the physical and chemical characteristics of natural aquatic humic substances (HS). For example the colour, molecular weight and size of humic material are dependent on pH [8-11]. Changes in pH are thought to cause aggregation of humic molecules due to hydrogen bonding and van der Waal's forces and ionization of functional groups of humic material. However, most of the explanations are based on the macromolecular structure of humic material, which is not yet completely known.

Humic substances can associate with OMP [12-15] and affect the accumulation of OMP in animals [16-18]. However, only very limited data are available about the effects of pH on this binding phenomenon and on the accumulation of OMP from humic waters in animals. The effect of pH on the binding of DDT to natural dissolved humic materials has been studied by Carter and Suffet [19], and they found that decreasing pH increases the binding capacity of the humic material. Similar results have been obtained for lindane [20] and tetracycline [21], but a commercial preparation instead of natural humic substances was used in these studies. Furthermore, the results obtained by Best *et al.* [22] and Khan [23] with methylene blue, diaquat and paraquat show the same trend. On the other hand, Guy et al. [24] and Narine and Guy [25] reported that decreasing pH decreased the binding capacity of commercial humic acid for methylene blue, diaquat and paraquat.

The main objective of this work was to demonstrate the effects of pH on the accumulation of four model compounds in *Daphnia magna* and nymphs of *Heptagenia fuscogrisea* from waters containing natural dissolved humic substances and from artificial, humus-free softwater. The model compounds were: benzo(a)pyrene (BaP), dehydroabietic acid (DHAA), hexachlorocyclohexane (lindane) and pentachlorophenol (PCP).

Materials and Methods

Animals

The *Daphnia magna* Straus population was raised in static culture of artificial soft freshwater (Ca+Mg hardness 0.5 mmol/l) at 20±1°C and pH 6.5, with a controlled day rhythm of 16/8 hrs (L/D). The population was fed every second day with a mixture of algae (*Scenedesmus obliqus* and *Monoraphidium contortum*). Young daphnids were removed from cultures every second or third day and maintained under the above conditions until they were 8-10 days old. The animals were then used in the accumulation experiments. Because of the low pH tolerance of *D. magna* especially in soft waters [26], the need for an additional test species was obvious; the mayfly *Heptagenia fuscogrisea* (*Ephemeroptera*) was found to fullfill this need. The larval form of this mayfly species can survive in a wide pH range in nature and withstands low pH in experimental conditions [27,28].

Nymphs of *H. fuscogrisea* were collected from the littoral zone of Lake Pyhäselkä (pH 6.6, temperature 9°C). The population was maintained in the laboratory in a mixture (1:1) of Lake Pyhäselkä water and artificial soft freshwater at 15°C and pH 5.5, with a day rhythm 16/8 (L/D), one and a half months before the experiments. Nymphs were fed with the same algal mixture as daphnids.

Chemicals and Experimental Waters

Stock solutions of ^{14}C-BaP (29.7 mCi/mmol; Amersham Ltd., UK), ^{14}C-lindane (62.0 mCi/mmol; Amersham Ltd., UK), ^{14}C-PCP (37.5 mCi/mmol; CEA, France) and ^{3}H-DHAA (2.3 mCi/mmol; labelled according to Kutney *et al.* [29]) were made in ethanol (Table 1).

Table 1 Characterization of the chemicals studied in this paper.

	pKa	pKow	Reference
Pentachlorophenol	4.71	5.15[a]	Cessna and Grover [31]
Dehydroabietic acid	5.7	4.8[b]	Nyrén and Back [32]
Benzo(a)pyrene	-	5.98	Miller *et al.* [34]
Hexachlorocyclohexane	-	3.7	Hawker and Connell [35]

a = Hansch and Leo [33]
b= Dr. B. Holmbom, personal communication

In the experiments on *H. fuscogrisea* the concentrations of model compounds were 1 µg/l (BaP), 2 µg/l (lindane), 10 µg/l (PCP) and 70 µg/l (DHAA). In the experiment on *D. magna* the concentrations were 1 µg/l (BaP) and 95 µg/l (DHAA). Water containing natural humic substances (DOC=18 mg/l) was sampled from Lake Louhilampi, an uncontaminated, small, forest lake in Eastern Finland (near the city of Joensuu). The humic water was filtered (Whatman GF/C glassfibre filters) immediately after sampling and stored in darkness at 4°C for 1 month before starting the experiments. Just prior to each experiment, artificial, humus-free, control water was prepared from deionized water with the following inorganic salts added: $CaCl_2$ $2H_2O$, 58.8 mg/l; $MgSO_4$ $7H_2O$, 24.7 mg/l; KCl, 1.1 mg/l and $NaHCO_3$, 13.0 mg/l (Ca+Mg hardness 0.5 mmol/l).

Determination of Partition Coefficients

Equilibrium dialysis [19] was used to determine the partition coefficient (K_p) between BaP and PCP and the DOM at pH 3.5, 6.5 and 8.5. Five ml of a filtered (0.22 µm, Nuclepore) sample was added to a dialysis bag (Spectra/Por 6, cutoff 1000 Dalton) and placed in a 200-ml glass jar containing an aqueous solution of radiolabeled BaP or PCP (180 ml). Sodium azide (0.002%) was added to inhibit microbial activity, and 2 mM phosphate buffer was added to maintain a stable pH. The jar was sealed with a teflon-lined cap and shaken in the dark at 20°C for 4 days. At least three replicate determinations were made. Equilibrium was confirmed in a parallel experiment using destilled water. Solutions inside and outside the dialysis bag were analyzed for ^{14}C-activity using scintillation cocktail (Luma Gel, Lumac) and a liquid scintillation counter (1217 Racbeta, Wallac-LKB). The outside concentration (C_o) is the freely dissolved organic pollutant, and the difference between the inside and outside concentration (C_p) is the pollutant bound to organic matter in the bag. K_p was calculated as

$$K_p = C_p/(C_o \times DOC) \tag{1}$$

where DOC is the concentration of dissolved organic carbon (kg carbon/l).

Accumulation Experiments

On the day before the accumulation experiments, the test waters were filtered (Nuclepore, 0.22 µm), and their pH was adjusted with 1.0 and 0.1 M NaOH or HCl. The pH scale used was 3.5, 4.5, 5.5, 6.5, 7.5 and 8.5. To maintain a stable pH, 2 mM phosphate buffer was added to test waters. Pilot experiments showed that this buffer concentration was sufficient to maintain a stable pH. Before an experiment each pH was checked and readjusted, if necessary.

Five nymphs of *H. fuscogrisea* were transferred to 100 ml testwater for 24 h (BaP and PCP) or 48 h (DHAA and lindane). Experiments with BaP were done in darkness, whereas the other experiments were performed using the same light conditions under which the population had been maintained. The temperature was 15±1°C.

In waterflea experiments, 50 *D. magna* were transferred to 250 ml of test solution, and the containers were placed in darkness at 20±1°C (both BaP and DHAA experiments). The experiments were run for 6 h with BaP and 12 or 24 h with DHAA. After exposure animals were netted, quickly washed with 60 ml distilled water, blotted, and weighed individually (nymphs) or in groups of 10 animals (*Daphnia*) with a microbalance (Sartorius model 4503). There were 5 replicates per pH. In each case, experiments were run at the same time with similar animals both in humic water and in control water to ensure comparable results.

Animal samples were added to 0.3 ml solubilizer (Luma Solve, Lumac), kept at 50°C for 24 h, and mixed with 10 ml scintillation liquid (Lipo Luma, Lumac). After 24 h the samples were analysed for ^{14}C or ^{3}H radioactivity by liquid scintillation counting using the automated external standard correction. The radioactivity of the water samples were analysed in triplicate by the same instrument after addition of 15 ml Luma Gel to 2 ml water. The concentrations of OMP were calculated from the specific activities of the chemicals used. The results are given as bioconcentration factor (BCF) calculated from the equation:

$$BCF = C_a/C_w \qquad (2)$$

where C_a and C_w are the concentration of the OMP in the animals (ng/g wet weight) and in the water (ng/ml), respectively. All statistical analyses (t-tests, Pearson's correlation) were done using SAS (Statistical Analysis System [30]).

Results and Discussion

Partition Coefficients

The measured K_p-values of BaP and PCP were pH dependent (Table 2). For BaP the K_p-value decreased from 18.0×10^4 at pH 3.5 to 8.4×10^4 at pH 8.5. For PCP the change in K_p as a function of pH was also distinct. At pH 3.5 the K_p was 9.8×10^3, whereas at pH 8.5 there was no measureable value at all. At pH 6.5 the K_p value was approximately 1×10^3, a more exact value cannot be given with this method, because although the difference between the concentrations of PCP on the outside and on the inside of the tube are measureable, the deviation is too high.

These K_p values indicate that pH affects the partitioning of both a neutral compound (BaP) and a weak organic acid (PCP). For BaP there was a twofold decrease in K_p as the pH was increased from 3.5 to 8.5. For a neutral OMP, this change can principally be due

Table 2 Partition coefficients (K_p) (mean ± S.D., n=3) for BaP and PCP at different pH values. The K_p values of BaP at pH 3.5 and 8.5 differ significantly ($p < 0.05$). For PCP the percentage of the nonionized compound calculated from the Henderson-Hasselbach equation is also presented for each pH value.

pH	3.5	**6.5**	**8.5**	
BaP	18.0± 5.9	14.1± 4.9	8.4± 1.2	x 10^4
PCP	9.8± 0.9	~1	~0	x 10^3
% nonionized PCP	(94%)	(1.6%)	(0.02%)	

to the effects of pH on the structure of the humic material [9,10]. At higher pHs ionized carboxyl groups of humic material may not be as effective binding sites for BaP, as it was in the case with atrazine [36]. The K_p values of PCP changed more dramatically with pH. The tenfold decrease with a pH increase from 3.5 to 6.5 is most likely due to the effects of pH both on the structure and the functional groups of the humic material, and on the form of PCP. At pH 3.5 up to 94% of PCP is in an unionized form (calculated from the Henderson-Hasselbach equation: $[HA]/C = 1/(1+10^{pH-pKa})$), at pH 6.5 only 1.6% are in an unionized form and, at pH 8.5 only minor amounts are unionized (Table 2). Because the acid form of PCP is less soluble in water than the acid form, it is not surprising that the acid form has a higher affinity for binding to humic material.

Bioavailability of Chemicals

The accumulation of PCP by *D.magna* and *H.fuscogrisea* was clearly pH dependent both in control and in natural aquatic humic substances (HS) contained water. Humic substances significantly ($P<0.05$) reduced the accumulation of PCP in *H.fuscogrisea* in the pH range from 4.5 to 7.5, but at pH 3.5 the difference was not significant, although the BCF value was lower in humic water than in control water. At pH 8.5 there were no differences between humic and control treatments (Fig. 1B). This accumulation data agrees with the K_p-values and also with the percentages of the unionized PCP at different pHs (Table 2), showing that the unionized form of PCP both accumulates more in the animals and binds more to the humic material than the ionized form. This is also in accordance with the literature stating that the unionized form of PCP is biologically much more active (toxic) [2-4]. At pH 6.5 the reduction in accumulation in humic water is much greater than would be predicted by the K_p-value. This may be because the free, unionized, PCP is totally bound to humics.

The accumulation pattern of DHAA in *H.fuscogrisea* in control water as a function of pH differs clearly from that in humic water. As observed earlier [37], the accumulation of DHAA was lower from humic water than from control water when the pH is higher than 6. At pH 5.5 or lower the opposite phenomenon was observed. However, at all pHs tested the differences between the humic water and the control water were not statistically significant (Fig. 1A). One reason for this can be the fact that the concentration of DHAA was considerably higher than the concentrations of the other compounds, because of the low specific activity of DHAA. In experiments on *Daphnia* in the pH range of 5.5 to 8.0, the results are similar to those found in the experiments on *H. fuscogrisea*. Humics

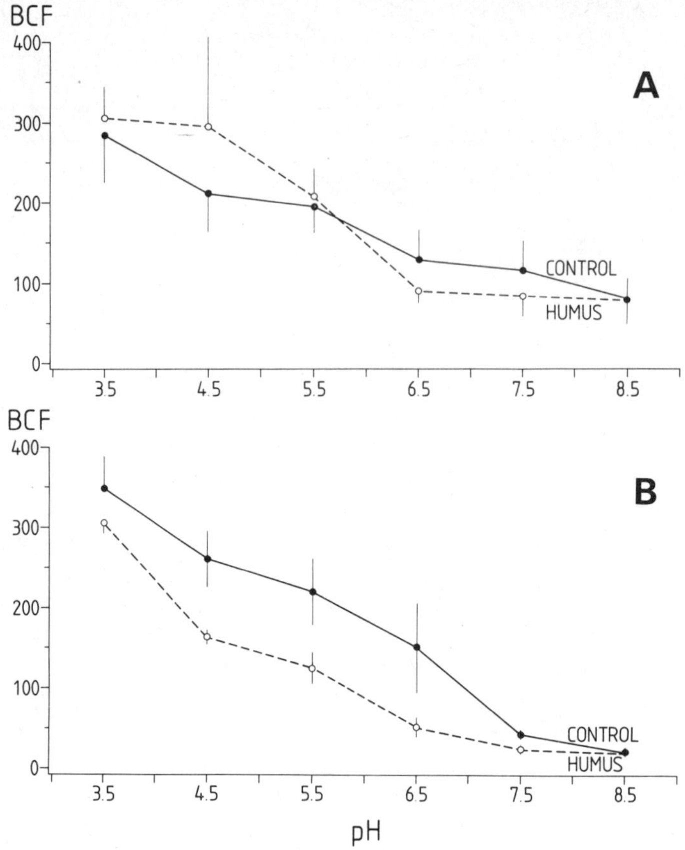

Figure 1 Accumulation of DHAA (A) and PCP (B) in *H. fuscogrisea* from the control, and humic water as a function of pH.

significantly decreased the accumulation of DHAA at pH 5.5 - 6.5 ($P<0.05$). At pH 7.0 and 8.0 there was no difference between the waters (Fig. 2). For a shorter experimental time (12 h, not a steady state), pHs 4.5 and 5.0 could also be used in *Daphnia* experiments. However, at pH 4.5 only 30% survived in humic water and 50% in control water. This result might indicate that DHAA accumulates faster from water containing humic substances than from humus-free water at low pH. On the other hand, Petersen and Persson [38] have shown that, up to a certain concentration, humus (both Aldrich humic acid and natural humic substances) had beneficial effects on the survival and reproduction of acid-stressed *Daphnia magna*, but further additions of humus incresed mortality as in the present experiment. This phenomenon was also noticed in animals that survived 12 h at pH 4.5 (Fig. 2). Also at pH 5.5, HS significantly reduced the accumulation of DHAA, but at pHs 6.5 and 7.5 there were no noticeable differences, since obtained BCF values were low due to a short experiment time, during which steady state had not yet been reached.

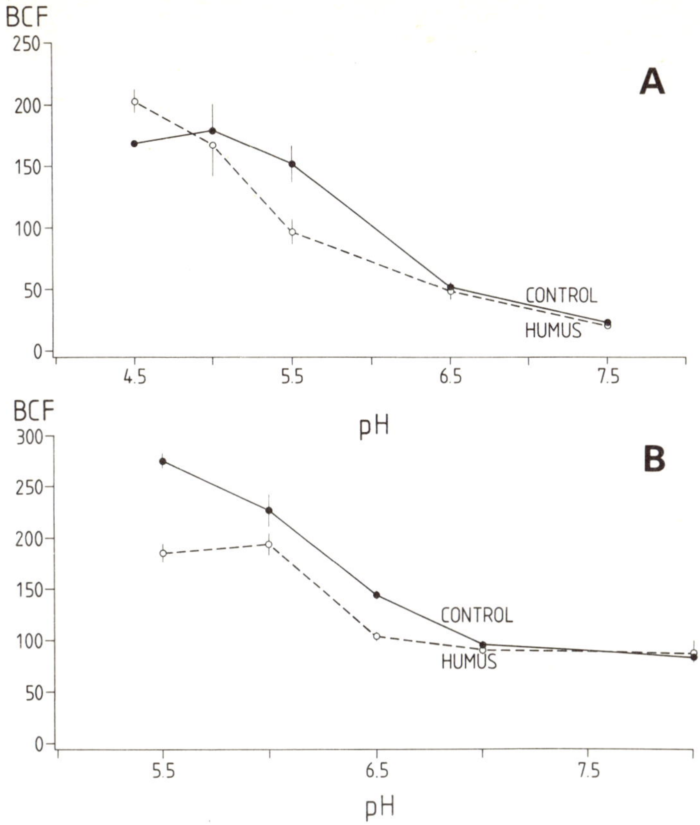

Figure 2 Accumulation of DHAA in *D. magna* from the control and humic water as a function of pH. BCF values after 12 h (A) and steady-state BCF values (24 h, B) are shown. Notice the differences between the x- and y-axes.

The accumulation of BaP was clearly reduced by HS in the experiments both on *H. fuscogrisea* and on *D. magna* (Fig. 3). In the control water, accumulation was pH dependent: the highest BCF values were obtained at pH 6.5 and the lowest at 8.5. In the humic water accumulation of BaP was slightly elevated with increasing pH, which is well in accordance with the K_p measurements (Table 2). The results from both the accumulation experiments and the K_p measurements show that the binding of BaP to dissolved organic material in the water decreases with increasing pH. The BCF value correlated with pH both in the experiments on *Daphnia* ($r=0.55$, $P<0.05$) and the experiments on *H. fuscogrisea* (0.42, $P<0.05$).

The accumulation of lindane by *H. fuscogrisea* was pH dependent in the control water in a similar way as BaP (Fig. 4). The highest BCF was observed at pH 6.5 and the lowest at pH 8.5. This pH-dependent accumulation pattern for lindane in the control water is quite similar to the pattern Fisher found for the midge *Chironomus riparius* [39]. This phenomenon was attributed to a lower permeation of lindane into the animals at pH 4 and an enhanced metabolism and excretion by the animals at pH 8 as compared to at pH 6.

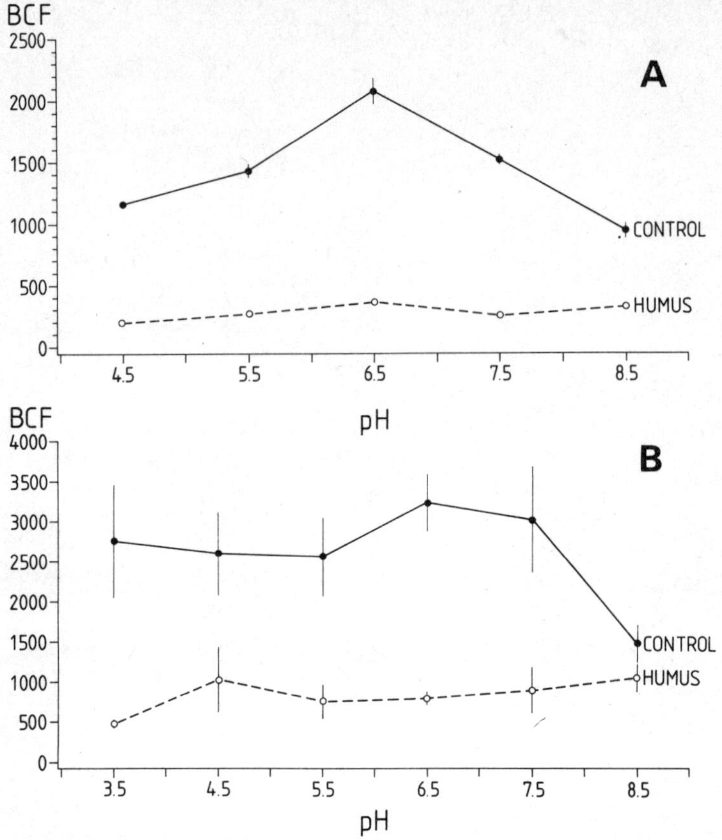

Figure 3 Accumulation of BaP in D. *magna* (A) and *H. fuscogrisea* (B) in control and humic waters as a function of pH. Notice the differences between the x- and y-axes.

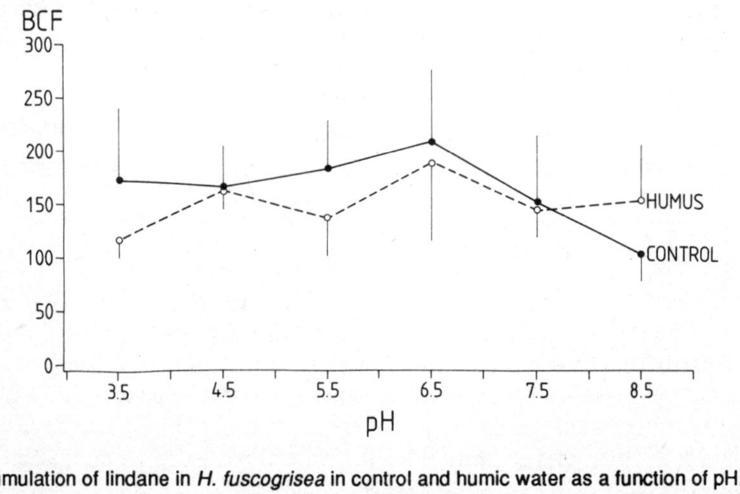

Figure 4 Accumulation of lindane in *H. fuscogrisea* in control and humic water as a function of pH.

Effects of HS on the accumulation of lindane are not clearly demonstrated by this experiment. The BCF values in the humic water are lower than in control water except at pH 8.5, but no statistically significant differences exist between waters due to high deviations (Fig. 4). However, in the humic water there is a similar trend as with BaP, i.e. accumulation of lindane increased with increasing pH. This may be due to decreased binding of lindane to humic material with increasing pH. This conclusion agrees with the results of Tramonti *et al.* [20] and Saint-Fort and Visser [40], who have found that increasing pH decreases the binding of lindane to the humic materials, although lindane does not bind very well to humics because of its rather low lipophilicity.

In conclusion, it was possible to demonstrate that the pH of the water affects the humic material by changing its ability to bind pollutants. This change in binding capacity can change the availability of pollutants to animals. On the other hand, changing pH can also affect the chemical nature (for example hydrophobicity) of pollutants, as was the case for weak organic acids, and in this way both the binding of the OMP to the humics and the ability of the OMP to accumulate is changed. Furthermore, we have to keep in mind that a change in pH affects the animals by forcing them to adapt physiologically to a new environment. These processes in the animals may affect the uptake, metabolism and excretion of xenobiotics and thus affect the net bioaccumulation of the xenobiotic. All these facts show the complex role of pH in the natural environment in controlling the fate and effects of organic pollutants.

Acknowledgments

This work has been financially supported by The Academy of Finland Research Council of the Environmental Sciences (project 06/133). The author thanks Mr. Heikki Hämäläinen for helping to collect and identify the nymphs of *H. fuscogrisea*. Dr. A. Oikari and Dr. M. Black are acknowledged for critical reading of the manuscript.

References

1. McLeay, D.J., C.C. Walden and J.R. Munro. Water Res. **13**:151 (1979).
2. Saarikoski, J. and M. Viluksela. Arch. Environ. Contam. Toxicol. **10**:747 (1981).
3. Spehar, R.L., H.P. Nelson, M.J. Swanson and J.W. Renoos. Environ. Toxicol. Chem. **4**:389 (1985).
4. Fisher, S.W. and R.W. Wadleigh. Ecotoxicol. Environ. Safety **11**:1 (1986).
5. Saarikoski, J., R. Lindström, M. Tyynelä and M. Viluksela. Ecotoxicol. Environ. Safety **11**:158 (1986).
6. Fisher, S.W. Ecotoxicol. Environ. Safety **10**:202 (1985).
7. Fisher, S.W. and T.W. Lohner. Arch. Environ. Contam. Toxicol. **16**:79 (1987).
8. Gjessing, E.T. Physical and Chemical Characteristics of Aquatic Humus. (Ann Arborii: Ann Arbor Science, 1976).
9. Chen, Y. and M. Schnitzer. Soil Sci. Soc. Am. J. **40**:866 (1976).
10. De Haan, H., G. Werlemark and T. De Boer. Plant and Soil **75**: 63 (1983).

11. Thurman, E.M. Organic Geochemistry of Natural Waters. (Dordrecht: Martinus Nijhoff, 1985).

12. Poirrier, M.A., B.R. Bordelon and J.L. Laseter. Environ. Sci. Technol. **6**:1033 (1972).

13. Hassett, J.P. and M.A. Anderson. Environ. Sci. Technol. **13**:1526 (1979).

14. Gjessing E.T. and L. Berglind. Arch. Hydrobiol. **92**:24 (1981).

15. Carlberg, G.E. and K. Martinsen. Sci. Total Environ. **25**:245 (1982).

16. Landrum, P.F., M.D. Reinhold, S.R. Nihart and B.J. Eadie. Environ. Toxicol. Chem. **4**:459 (1985).

17. Carlberg, G.E., K. Martinsen, A. Kringstad, E. Gjessing, M. Grande, T. Källqvist and J.U. Skåre. Arch. Environ. Contam. Toxicol. **15**:543 (1986).

18. Kukkonen J., A. Oikari., S. Johnsen and E. Gjessing. Sci. Tot. Environ. **79**:197 (1989).

19. Carter, C.W. and I.H. Suffet. Environ. Sci. Technol. **16**:735 (1982).

20. Tramonti, V., R.H. Zienius and D.S. Gamble. Inter. J. Environ. Anal. Chem. **24**:203 (1986).

21. Sithole, B.B. and R.D. Guy. Water Air Soil Poll. **32**:315 (1987).

22. Best, J.A., J.B. Weber and S.B. Weed. Soil Sci. **114**:444 (1972).

23. Khan, S.U. Can. J. Soil Sci. **53**:199 (1973).

24. Guy, R.D., D.R. Narine and S. deSilva. Can. J. Chem. **58**:547 (1980).

25. Narine, D.R. and R.D. Guy. Soil Sci. **133**:356 (1982).

26. Havas, M., T.C. Hutchinson and G.E. Likens. Can. J. Zool. **62**:1965 (1984).

27. Engblom, E. and P-.E. Lingdell. Bottenfaunans användbarhet som pH-indikator. Rapport SNV PM 1741. (Stockholm: Statens Naturvårdsverk, 1983) (in Swedish).

28. Engblom, E. and P-.E. Lingdell. Natl. Swedish Board Fish. Report **61**:60 (1984).

29. Kutney, J.P., M. Singh, G.M. Hewitt, P.J. Salisbury, B.R. Worth, J.A. Servizi, D.W. Martens and R.W. Gordon. Can. J. Chem. **59**:2334 (1981).

30. Statistical Analysis System. SAS User's Guide: Statistics. Version 5 edition. (Cary: 1985).

31. Cessna, A.J. and R. Grover. J. Agri. Food Chem. **26**:289 (1978).

32. Nyrén, V. and E. Back. Acta Chem. Scand. **12**:1516 (1958.)

33. Hansch, C. and A. Leo. Substituent Constants for Correlation Analysis in Chemistry and Biology. (New York: Wiley, 1979).

34. Miller, M.M., S.P. Wasik, G-.L. Huang, W-.Y. Shiu and D. Mackay. Environ. Sci. Technol. **19**:522 (1985).

35. Hawker, D.W. and D.W. Connell. Wat. Res. **22**:701 (1988).

36. Haniff, M.I., R.H. Zienius, C.H. Langford and D.S. Gamble. J. Environ. Sci. Health **B20**:215 (1985).

37. Kukkonen, J. and A. Oikari. Sci. Tot. Environ. **62**:399 (1987).

38. Petersen, R.C., Jr. and U. Persson. Sci. Tot. Environ. **62**:387 (1987).

39. Fisher, S.W. Ecotoxicol. Environ. Safety **10**:202 (1985).

40. Saint-Fort, R. and S.A. Visser. J. Environ. Sci. Health **A23**:613 (1988).

Influence of Natural Humic Acids and Synthetic Phenolic Polymers on Fibrinolysis

Hans-Peter Klöcking
Institute of Pharmacology and Toxicology
Medical Academy Erfurt, DDR-5010 Erfurt, GDR

Abstract

The influence of synthetic and natural phenolic polymers on the release of plasminogen activator was studied in an isolated, perfused, vascular preparation (pig ear). Of the tested synthetic phenolic polymers, the oxidation products of caffeic acid (KOP) and 3,4-dihydroxyphenylacetic acid (3,4-DHPOP), at a concentration of 50 µg/ml perfusate, were able to increase the plasminogen activator activity by 70%. The oxidation products of chlorogenic acid (CHOP), hydrocaffeic acid (HYKOP), pyrogallol (PYROP) and gallic acid (GALOP), at the same concentration, exerted no influence on the release of plasminogen activator. Of the naturally occurring humic acids, the influence of sodium humate was within the same order of magnitude as KOP and 3,4-DHPOP. Ammonium humate was able to increase the plasminogen activator release only at a concentration of 100 µg/ml perfusate. In rats, the t-PA activity increased after i.v. application of 10 mg/kg of KOP, Na-HS or NH_4-HS.

Introduction

So far, peat pulp has been used in the treatment of thrombophlebitis [1] without scientific substantiation. Successful local treatment of superficial thrombophlebitis requires an effective drug which is able to penetrate the skin. Acute release of tissue-type plasminogen activator (t-PA), which represents a decisive factor in fibrinolysis, from the vascular endothelium leads to the activation of the antithrombotic defence mechanism [2]. This was our starting point for studying naturally occurring humic acids and synthetically prepared phenolic polymers concerning their t-PA-releasing effect in isolated organs and in animal experiments.

Materials and Methods

Naturally occurring humic acids used in the present study were sodium humate (Na-HS, m.w. 7900) and ammonium humate (NH_4-HS, m.w. 7500), both isolated from high-moor peat (collected near Rostock, Baltic Sea coastal region of the GDR

[3]). The phenolic polymers were oxidation products from caffeic acid (KOP, m.w. 6500), hydrocaffeic acid (HYKOP, m.w. 10 000), chlorogenic acid (CHOP, m.w. 12 000), pyrogallol (PYROP, m.w. 6500), gallic acid (GALOP. m.w., 6000), and 3,4-dihydroxyphenylacetic acid (3,4-DHPOP, m.w. 8600). They were synthetized by oxidation of the basic compounds with sodium m-periodate [4]. KOP, HYKOP, PYROP, and GALOP were available from SERVA Feinbiochemica, Heidelberg, FRG.

The laboratory animals were 14 male rats (Schön: Jelei, conventional breed) with a mean body weight of 291±37 g. Anaesthesia was performed with 1.5 g ethyl urethane/kg bodyweight, intraperitoneally.

Estimation of t-PA Release in the Perfused Pig Ear

The test set-up has been previously described in detail [5]. Within 5 hours after removal, pig ears were perfused with Tyrode's solution (0.75 ml/min, 37°C, pH 7.4) via the cannulated ramus intermedius of the arteria auricularis magna, by means of a peristaltic pump. To remove blood, the vascular system was continuously perfused with Tyrode's solution for 30 min. Afterwards, the perfusate was collected at 2-min intervals. The perfusion scheme was as follows: 4 fractions with Tyrode's solution only, then, 2 fractions with compound-containing Tyrode's solution, and, finally, 6 fractions with Tyrode's solution alone.

Determination of t-PA Activity in Perfusion Solution

The t-PA content was estimated by the fibrinolytic activity (FA) on plasminogen-containing fibrin-agar plates, according to the method of Astrup and Müllertz [6]. After an incubation of 24 hours at 37°C, the lysis area was estimated in mm^2 by multiplying 2 diameters perpendicular to each other. The extent of fibrinolytic activity in mm^2 was converted into International Units (WHO Standard, from the National Institute of Biological Standards and Control, London, GB) by means of a t-PA calibration curve which was estimated on plasminogen-containing fibrin-agar plates. The t-PA release in the pig ear model was expressed as percent increase in t-PA activity of the mean value of 8 fractions, after starting the perfusion with a substance containing Tyrode's solution, compared to the mean value of the first 4 fractions perfused with Tyrode's solution alone. For significance calculation, the mean increase in activator release was related to the mean values of spontaneous release in the controls. The Student's t-test was performed to guarantee statistical reliability; $p > 0.05$ was considered insignificant.

Determination of t-PA Activity in Plasma

The t-PA activity in plasma was determined according to the method of Chmielewska and Wiman [6]. Citrated blood (1+9 v/v) was immediately mixed with acetate buffer and, starting within 2 min, centrifuged at 2000xg for 20 min. After plasma separation, 10 µl acetic acid (20%) were added to 100 µl each of already acidified plasma, in order to adjust to a pH of 4.0-4.1. 100 µl of plasma obtained in this way were diluted with 900 ml Tris buffer (0.05 mol/l, pH 8.3, 0.01% Tween 80).

To determine the t-PA activity, 200 µl of this diluted plasma were mixed with 200 µl of a solution of plasminogen-substrate (S-2251, 3.8 mM: plasminogen, 10 CU/ml: Tris buffer=1:1:3). Furthermore, 100 µl of t-PA stimulant (5 mg human fibrin(ogen) fragments/l.5 ml) were added to the mixture. After mixing, the batch was incubated at 37°C for 150 min. The reaction was stopped by adding 100 µl of acetic acid (20%) and photometrically evaluated at 405 nm. The t-PA content (IU/ml plasma) was calculated by means of a standard curve.

Determination of t-PA Inhibitor (PAI-l) Activity in Plasma

Blood (9 vol) was mixed with 0.1 M sodium citrate (1 vol) and centrifuged at 2000xg for 20 min. 25 µl t-PA solution (40 IU/ml) were added to 25 µl plasma, and the mixture was incubated at room temperature for 10 min. The batch was then diluted with 4 ml distilled water. Of this plasma solution, 200 µl were used for the determination of the remaining t-PA activity, as described above. The batch was incubated at 37°C for 70 min. Then, the reaction was stopped by the addition of 100 µl acetic acid (20%) and photometric evaluation was carried out at 405 nm. A standard curve (0-40 AU/ml) served to calculate the PAI-l activity (AU/ml plasma).

Results and Discussion

Of the naturally occurring humic acids, sodium humate, at a concentration of 50 µg/ml perfusate, caused a significant increase in t-PA release, whereas ammonium humate was effective only at a concentration of 100 µg/ml perfusate. Of synthetic phenolic polymers, the oxidation products of caffeic acid (KOP) and of 3,4-dihydroxyphenylacetic acid (3,4-DHPOP), at concentrations of about 50 µg/ml, were able to significantly increase the t-PA release within nearly the same order of potency as sodium humate. The oxidation products HYKOP, PYROP, and GALOP showed no t-PA-releasing effect. None of the tested, naturally occurring humic acids and synthetic phenolic polymers was able to reach the t-PA-releasing effect of heparin (Table 1) [7].

KOP and the naturally occurring humic acids sodium humate and ammonium humate were studied for their t-PA-releasing effect after single i.v. application each of 10 mg/kg. Of the three compounds, sodium humate proved to be the most potent releasing agent, with a maximum increase in t-PA release of 22±3 IU/ml plasma, 5 min after application, compared to the initial activity. In consequence of the increased t-PA release, the t-PA-inhibitor activity was reduced drastically, reaching a minimum 5 min after i.v. application (Fig. 1b). KOP and ammonium humate were less effective releasing agents of t-PA (Figs 1a and c).

According to the present results, the studied compounds have a different effect on t-PA release. This difference is probably due to the different structures of the humic

Table 1 Influence of humic acids and synthetically prepared phenolic polymers on plasminogen activator release in the isolated, perfused pig ear.

Compound	Concentration µg/ml perfusate	Enhancement of release[a] (%)	Significance (p)	Number of experiments (n)
Na-HS	50	65 ± 28	< 0.001	6
	100	111 ± 54	< 0.001	4
NH_4-HS	50	0	n.s.	4
	100	58 ± 19	< 0.001	4
KOP	50	70 ± 17	< 0.001	4
	100	136 ± 91	< 0.001	4
CHOP	50	30 ± 21	n.s.	4
3,4-DHPOP	50	73 ± 4	< 0.001	4
HYKOP	50	24 ± 18	n.s.	6
PYROP	50	0	n.s.	4
GALOP	50	7 ± 7	n.s.	4
Heparin	50	156 ± 48	< 0.001	3
Control	-	12 ± 10	-	33

[a] Means ± SD

acids and the synthetic phenolic polymers. Compared to Na-HS, the weaker t-PA-releasing effect of NH_4-HS is a result of the irreversible and hence nonhydrolyzable incorporation of a part of the ammonia nitrogen into the humic acid molecule [8]. In synthetic phenolic polymers, hydrogenation of the unsaturated side chains of the caffeic acid oxidation product leads to the weaker t-PA-releasing agent HYKOP. The t-PA-releasing effect is also reduced by esterification of the caffeic acid oxidation product with quinic acid (CHOP). PYROP and GALOP, the polymers of trihydroxy phenols without side chains, proved to be inactive compounds. 3,4-DHPOP is characterized by a side chain which has 1 atom of carbon less than HYKOP; hence it displays a stronger t-PA-releasing effect (for structural formulas of the parent substances of the phenolic polymers see Fig. 2).

The acute increase in t-PA release which we demonstrated in the isolated pig ear model speaks in favour of a direct action of these compounds on the vascular endothelium; their t-PA-releasing effect probably results from the displacement of a membrane-located activator. Since, on one hand, the increasing t-PA release could also be demonstrated in animal experiments, and on the other hand, t-PA plays a decisive role in the lysis of fibrin deposits in the organism [9], the t-PA-releasing effect may possibly explain the effect of peat pulp in the treatment of superficial thrombophlebitis [1]. In addition, penetration of the skin has previously been shown for KOP [10], and an antiphlogistic effect has been described for Na-HS and NH_4-HS [11]. Hence, based on these characteristics (antiphlogistic effect, t-PA-releasing effect and penetration of the skin) humic substances offer convenient possibilities for the treatment of superficial thrombophlebitis.

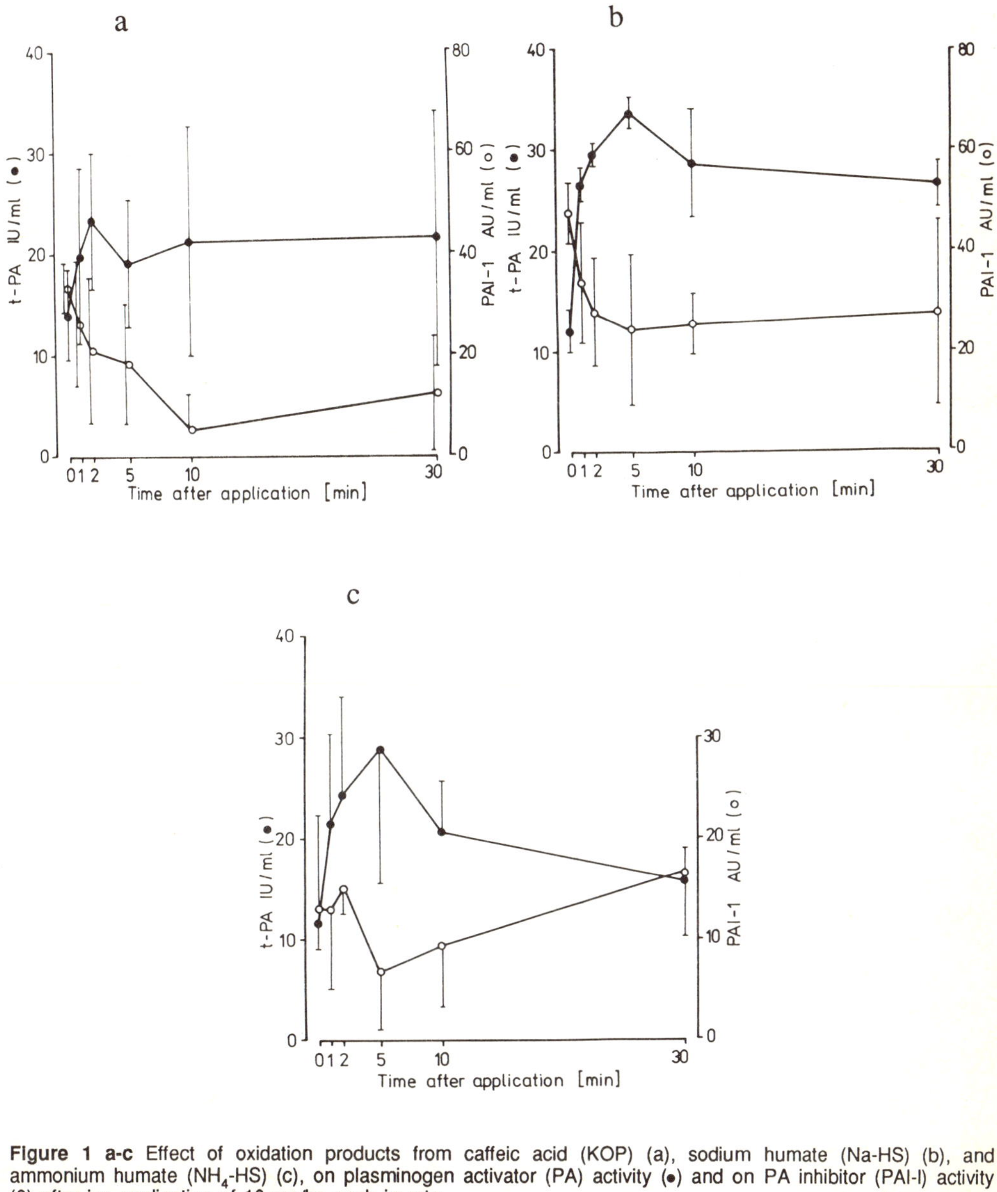

Figure 1 a-c Effect of oxidation products from caffeic acid (KOP) (a), sodium humate (Na-HS) (b), and ammonium humate (NH_4-HS) (c), on plasminogen activator (PA) activity (•) and on PA inhibitor (PAI-I) activity (0) after i.v. application of 10 mg/kg each in rats.

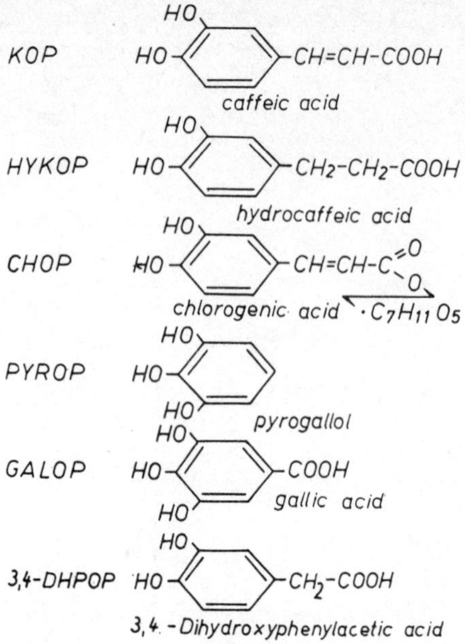

Figure 2 Structures of parent substances of phenolic polymers.

References

1. Lachmann, H. Bericht über den 8. Internationalen Kongress für universelle Moor-und Torfforschung, 5-10. Oktober 1962, Bremen, pp. 115-116 (Linz a/D:"Länderverlag" Ges.m.b.H., 1964).

2. Ronby, M. and A. Brandstrom. Enzyme **40**:130 (1988).

3. Klöcking, R., B. Helbig and P. Drabke. Pharmazie **32**:297 (1977).

4. Helbig, B. and R. Klöcking. Zeitschrift für Physiotherapie**33**:31 (1983).

5. Klöcking, H.-P., G. Bock. J. Drawert. K.-P. Hinsenbrock, B. Kaiser and U. Sedlarik. Folia Haematologica **103**:404 (1976).

6. Chmielewska, J. and B. Wiman. Clin. Chem. **32**:482 (1986).

7. Markwardt, F. and H.-P. Klöcking. Haemostasis **6**:370 (1977).

8. Wohlrab, W., B. Helbig, R. Klöcking und M. Sprössig. Pharmazie **39**:562 (1984).

9. Wiman, B. and D. Collen. Nature **272**:549 (1978).

10. Witthauer, J. Diss. Akademie der Wissenschaften der DDR, Berlin (1976).

11. Klöcking, R., R. Hofmann und D. Mücke. Arzneimittelforschung **18**:941 (1968)

Differential Inhibitory Effects of Humic Acids on Coagulation Systems of Human Blood

Helga Kübler
Institut für Bodenwissenschaften
Universität Göttingen, D-3400 Göttingen, FRG

Abstract

The present study demonstrated a strong inhibition of the intrinsic pathway of blood coagulation, when 10 µg of humic acid was added to 0.1 ml citrated plasma. An effect on the extrinsic coagulation system was only obtained after addition of more than 30 µg per 0.1 ml plasma, and was confined to a slight increase in the coagulation time. During determination of the thrombin time, humic acid, at concentrations above 40 µg per 0.1 ml plasma, acted like an antithrombin, by means of a sudden inactivation of thrombin. The humic acids were characterized by infrared spectroscopy. Humic acids were purified using a dialyzer as applied for urea clearance in nephrological therapy. The application of this method is novel.

Introduction

It is well known that one moor differs from another in its biologically active properties, and even samples from the same area can do so. For this reason, peat from the low-moor in Bad Neydharting (Upper Austria) was used in this research project [1]. Vegetation elements of low-moor grassy marshland with herbaceous plants are transformed into peat, which colours the water covering it dark brown, the so-called blackwater. This peat is available as a "moor-suspension-bath", which is obtained by suspending peat in blackwater. The suspensions are prepared by experienced personnel, thus guaranteeing fairly constant composition of the main substances. This is obviously important for research on such material. Characteristic for the Neydharting low-moor is the extraordinarily high amount of calcium present in the blackwater, 43.1 mg Ca/l, whereas, for comparision, the high-moor water from Gifhorn 4 only contains 2.5 mg Ca/l [2]. Potassium and sodium are not present.

The use of balneotherapy has been successful in healing and relieving pain, especially when treating chronic diseases. However, scientific proof of the usefulness of this type of treatment is still lacking [3-5]. This is also true for peat therapy. Therefore, this study tried to fill gaps in knowledge concerning natural medecines and furthermore to point out the importance of medicinal spas.

Humic acids have been successfully applied to prevent adhesions after abdominal and gynecological surgery. The development of adhesions may be induced by excessive fibrin formation in the area of the wound. The present study was performed to investigate whether humic acids have properties similar to antithrombins, physiological inhibitors of blood coagulation, and thus could reduce fibrin formation.

Materials and Methods

HA were extracted from the air-dried peat of the "moor-suspension-bath" with 0.1 M Na-pyrophosphate/HCl buffer in successive fractions in the pH range from 4.4 to 9.8. The pH 7.5 fraction was acidified to pH 1.2 with 6 M HCl, centrifuged, and the precipitate was used for further purification processes. Dialysis (nephrological therapy) yielded freeze-dried HA(N) with 2.3% ash. A portion of this was further precipitated with HCl and dialysed (Visking dialysis tubing), resulting in the freeze-dried retentate HA(ND), giving less than 0.5% ash. It is noteworthy that ash from native HA is always brick-red, probably due to the presence of an iron compound.

The coagulation experiments were performed on citrated human plasma. To keep conditions constant, the plasma was prepared from my own blood. It should be noted that investigations with citrated plasma must be performed within two hours of bleeding, since the activity of coagulation factors V and VIII decreases after this time. The humic acid was extracted from Neydharting peat with 0.5 M NaOH; the HA(NaOH) [6] and the model humic acids were obtained from colleges. In the case of model HA(a), the p-quinone/NaOH mixture was stirred very strongly and air was bubbled in. The obtained HA was precipitated at pH 1 with HCl, dialysed, and then washed with ethanol and ether [7]. HA(a) was easily dissolved in a buffer with a pH of 7.6. The HA(b) mixture was stirred, and then the precipitated HA(b) was dialysed, and the pH was adjusted to 7 [8]. The dried sodium-humate does not easily dissolve in a buffer with a pH of 7.

Results

Four different coagulation tests were carried out using the Behring test systems [9]. Each one could detect individual coagulation factors with great sensitivity. A standard solution of HA in sodium barbital buffer pH 7.6 was prepared and titrated with 2 M NaOH to pH 7.3. The dilution series was made up with the barbital buffer (standard buffer from Behring). Citrated plasma (0.2 ml) was pre-incubated with 0.01 ml diluted HA for 30 s at room temperature, and then 0.1 ml of this was taken for the test. The time elapsed between the triggering of the coagulation process and the formation of a fibrin clot was measured using a coagulometer (Sarstedt, Biomatic 2000). No calcium deficiency (adsorbed by HA) was observed during the test reactions. Each test was carried out three times, but on different days, and the means were calculated.

The HA were characterized by infrared (IR) spectroscopy (Fig.1). The distinct peaks at 1720 cm^{-1}, representing carbonyl groups, and at 1620 cm^{-1}, representing carboxylate anions, indicate that the HA(ND) and model HA(a) were in the acid form, whereas the spectrum of sodium-humate only has a shoulder at 1720 cm^{-1}, as do HA(NaOH) and model HA(b). In further experiments it was demonstrated that these alterations in the spectra were not associated with changes in the biological activity of the HA.

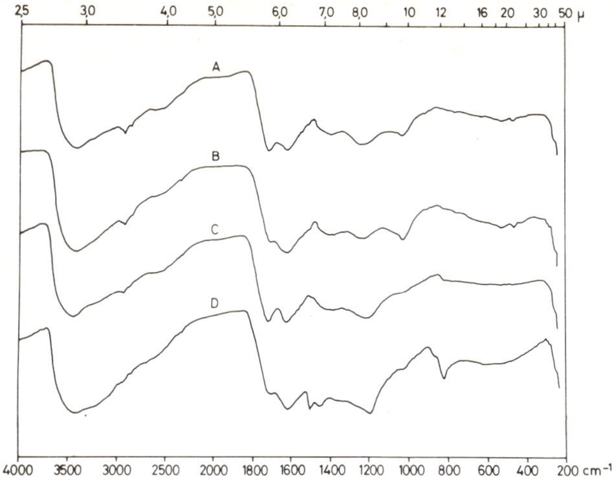

Figure 1 IR-spectra of A, native HA(ND); B, native HA(NaOH) Na-humate; C, model HA(a); D, model HA(b) Na-humate.

Determination of the Partial Thromboplastin Time (PTT)

The reagent Pathrombin enabled a screening test for coagulation disorders in the intrinsic system. Incubation of plasma with optimum amounts of phospholipids (factor P3) and the surface activator kaoline lead to activation of factor XII. The coagulation process was triggered by calcium ions.

Distinct prolongation of the PTT values was found when the plasma was pre-incubated with HA(ND) (ash 0.5%), HA(N)(ash 2.3%), and model HA(a), as shown in Fig. 2. (PTT values were prolonged to 100 s and 297 s by 6 and 10 μg HA(ND), respectively). The normal range is from 30 to 45 s. In contrast, the effects of HA(NaOH) and model HA(b) were weaker. The reduced inhibitory effect of HA(NaOH) on blood coagulation presumably indicated that alkaline hydrolysis impaired functional groups. To name all the protein-type factors which participate in the coagulation process would be redundant.

Fig. 4 shows a brief scheme (slightly altered) of the intrinsic and extrinsic system proposed by Meyer [10]. If factors XI, XII and especially VIII and IX are reduced, the PTT values are prolonged. The other factors are detected with less sensitivity.

Determination of the Prothrombin Time (PT, syn: TPZ)

The reagent Thromborel S served for the determination of coagulation disorders of the extrinsic pathway of blood coagulation according to Quick. The coagulation process was triggered by incubation of plasma with optimum amounts of thromboplastin (factor III) and calcium. It was very sensitive for the detection of factors II (prothrombin), V, VII and X. The results shown in Fig. 3 demonstrate that, compared with the intrinsic system, HA have almost no effect on the extrinsic coagulation system, within a reasonable physiological range of about 10 μg HA/0.1 ml test plasma. Here time in seconds is given intentionally, instead of percentage, as used in the clinical definition; the normal range is from 11.1 to 13.8 s. Fig. 4 illustrates the extrinsic system, in which the framed parts (interrupted lines) contain the factors added with the test reagent.

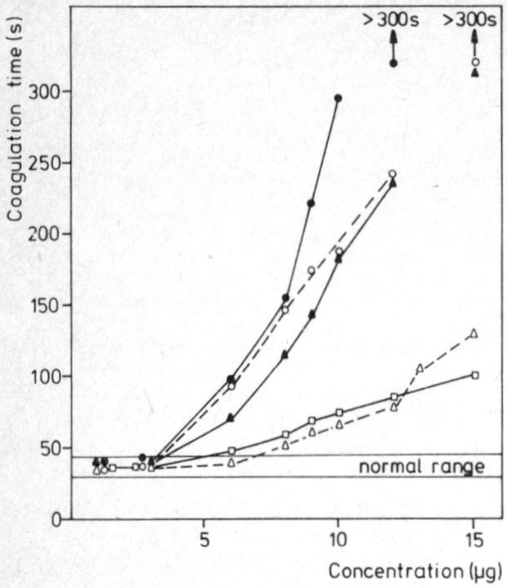

Figure 2 Determination of PTT.

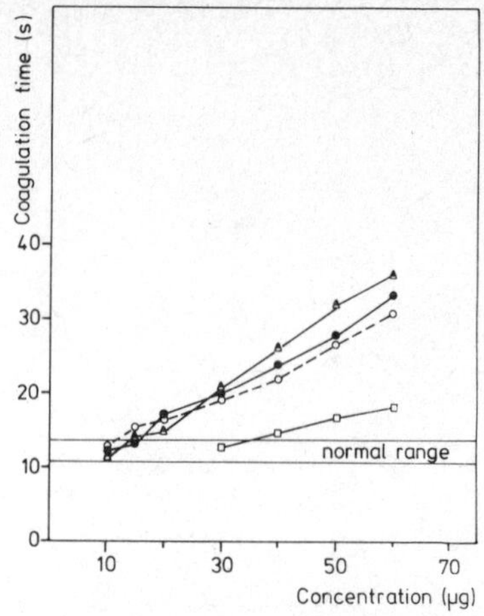

Figure 3 Determination of TPZ. Solid circles, HA(ND); open circles, HA(N); solid triangles, model HA(a); open triangles, model HA(b); open quadrangle, HA(NaOH).

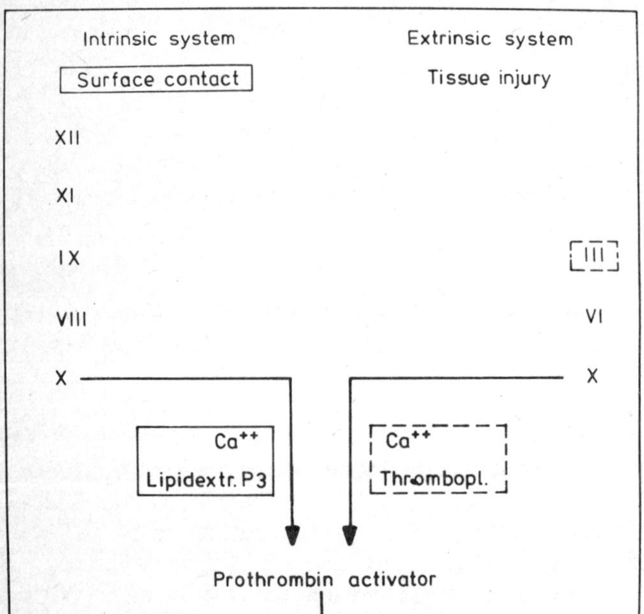

Figure 4 Schemata of the intrinsic and extrinsic coagulation systems. Areas framed with solid and dashed lines contain the factors added with the test reagent.

Fig. 5 summarizes the coagulation factors in common to both the PTT and TPZ systems. Factor I is fibrinogen, factor II is prothrombin. Thus, from both determinations, the factors I, II, V (accelerator-globulin) and X (proesterase) could be detected, and their reduction would lead to a pathological prolongation of the coagulation time in both systems. But, since the inhibitory effects of HA on the extrinsic system (TPZ) were negligible compared with those on the intrinsic system (PTT), we could conclude that HA, in concentrations of 5 or 10 μg per 0.1 ml test plasma, did not affect one of the common factors, but only one or all of the factors of the intrinsic coagulation pathway, i.e. factors VIII, IX, XI and XII.

Coagulation factors

PTT	common factors	TPZ
VIII	I	VII
IX	II	
XI	V	
XII	X	

Figure 5 Coagulation factors in common to the intrinsic system (PTT) an extrinsic system (TPZ).

Determination of the Thrombin Time (TZ)

Citrated plasma was clotted by addition of thrombin. If the plasma contained antithrombin, the added thrombin was immediately inactivated, and blood coagulation was inhibited. HA also acted like antithrombin. The sudden inactivation of thrombin occurred at concentrations of about 40 μg HA(ND) per 0.1 ml plasma (notably, out of the physiologically acceptable range), resulting in a prolongation of TZ beyond the measurement interval (300 s; Fig. 6). The standard deviations, calculated from four determinations, are indicated; the normal range is from 14 to 21 s.

Determination of Fibrinogen

Citrated plasma was clotted by the addition of relatively large amounts of thrombin. The coagulation time depended mainly upon the fibrinogen content of the plasma sample. Prior to the test, the plasma incubated with HA(ND) was diluted by a factor of ten, so that thrombin inhibitors had no effect. HA(ND) also had no effect under these test conditions (normal range 7.6 to 13 s; Fig. 7).

This test was repeated, with the diluted plasma incubated with HA(ND), in order to demonstrate the direct effect of HA(ND) on Thrombin. An increase in the HA content of only 1 μg (from 5 to 6 μg) produced a drastic prolongation of the coagulation time, from a slightly increased value to over 300 s (Fig. 7). This indicates that HA(ND) inactivated the enzyme thrombin. In contrast, even 10 μg of HA (NaOH) were without effect.

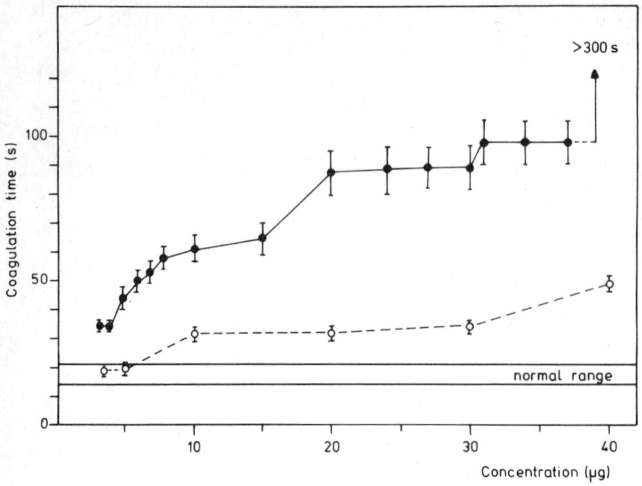

Figure 6 Determination of the Thrombin Time (TZ). Solid circles represent HA(ND), HA(N), HA (a) and HA (b); open circles, HA (NaOH).

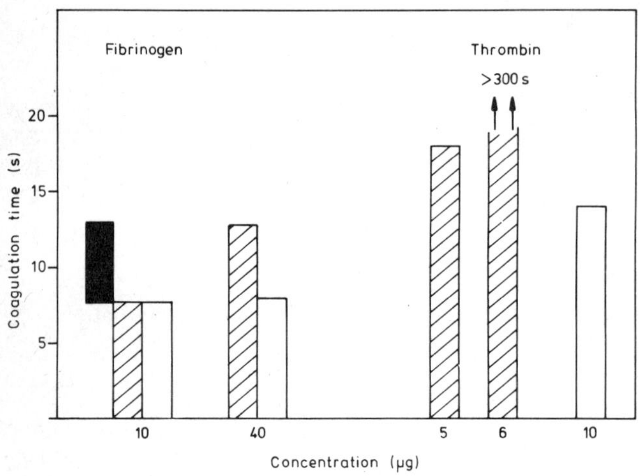

Figure 7 Determination of Fibrinogen. Inactivation of Thrombin. Solid bar, normal range; hatched bar, HA(ND); open bar, HA (NaOH).

Discussion

Guo-Fan [11] reported that HA treatment had impressive healing effects on various cornea ulcers and that HA also exhibited blood stanching properties. These results were obtained by application of thrombelastography on dogs (the formation of a fibrin clot was plotted on a graph). The HA used by these investigators had a molecular weight of less than 500 Dalton, and differs considerably from the HA used in the present experiments.

Since HA also had antiviral properties, it was interesting to learn that sulfated polysaccharides of low molecular weight showed a much lesser antiviral effect than those of a molecular weight at about 6000 Dalton [12]. The antiviral properties were mostly based on enzyme inhibition, as well as the anticoagulant effects. This may explain the discrepancy between the results of Guo-Fan and my observations, since the HA used in the present study has a molecular weight greater than 8000 Dalton.

Conclusion

The inhibitory effect of HA on the intrinsic coagulation system has been shown. The inhibitory concentrations of 60 μg and 100 μg per 1.0 ml plasma were within the same range as observed for polysaccharides with antiviral effect [12].

The similarity of the effect of HA(ND) and heparin on blood coagulation was striking. Heparin is a sulfated polysaccharide and belongs to the antithrombins. Considering the clinical experience in relation to the formation of adhesions, an analogy to a certain consumptive coagulopathy might be postulated. Here an enhanced fibrin formation was induced by activated coagulation factors of the intrinsic system. This excessive fibrin production could be interrupted with heparin, and HA would presumably serve in the same manner. There is some evidence [13] that HA forms a complex with enzymes, but it cannot automatically be assumed that this was also the case in the present study, until the HA-thrombin complex is isolated.

Postoperative adhesions were presumably induced through an enhanced production of fibrin. If coagulation factors of the intrinsic system were inhibited, which was what HA did, then the production of fibrin was reduced, and so adhesions were prevented when patients bathed in the "moor-suspension-bath". Thus the adhesions preventing and reducing effect of humic acids can be explained by the biochemical data, and the observations of clinical experience can be confirmed. The results obtained also suggest the PTT-test as an appropriate standard for determining active forms of HA and their grade of purity.

Acknowledgements

I gratefully acknowledge the gift of the test reagents by Prof. Dr. H. Rühl and Dr. M. Gabrys, Sonnenbergklinik, D-3437 Bad Sooden-Allendorf. I am also greatly indebted to Prof. Otto Stöber for providing the "moor-suspension-bath" used in these studies, Österreichisches Moorforschungs-Institut, A-4010 Linz a.D.

References

1. Stöber, O. Ewiges Moor von Neydharting (Wien-Linz-Neydharting: Stadt-Verlag, 1970).
2. Naucke, W. In: K. Göttlich, Ed., Moor- und Torfkunde, p. 181 (Stuttgart: E. Schweizerbart, 1980).
3. Baatz, H. Wissenschaftliche Schriftreihe Bad Pyrmont VI (Hameln: CW Niemeyer, 1979).
4. Lüttig, G.W. In: C. Goecke and G. Lüttig, Ed., Wirkungsmechanismen der Moortherapie, pp. 1-24 (Stuttgart: Hippokrates, 1987).
5. Quentin, K-.E. and W. Schnizer. Wissenschaftliche Reihe des Deutschen Bäderverbandes (Kassel: Hans Meister KG, 1986).
6. Ferogh, M. Preparation Nr. 16 from Neydharting peat (1987).
7. Rizk, N.J. Diss., Göttingen (1984).

8. Müller-Wegener, U. Diss., Göttingen (1976).

9. Behringwerke AG, Marburg, Ed. Nov 1986.

10. Meyer, J.G. Blutgerinnung und Fibrinolyse (Köln: Deutscher Ärzte Verlag, 1986).

11. Guo-Fan, T. In: C. Goecke and G. Lüttig, Ed., Wirkungsmechanismen der Moortherapie, pp. 89-95 (Stuttgart: Hippokrates, 1987).

12. Brede, H.D. and H. Rübsamen-Waigmann. Forum Mikrobiologie 6, X/3 (1989).

13. Ziechmann, W. Huminstoffe (Weinheim: Chemie, 1980).

Session 5:
Halogenation of Humic Substances

Mutagenic Compounds from Chlorination of Humic Substances

Bjarne Holmbom
Department of Forest Products Chemistry, Åbo Akademi, Finland

Abstract

Chlorination of natural humic substances, as well as of lignin, produces a myriad of non-chlorinated and chlorinated compounds. The identification of an important class of strongly mutagenic compounds is reviewed. The most important Ames mutagen in chlorinated drinking waters of various origin is the compound 3-chloro-4-(dichloromethyl)-5-hydroxy-2(5H)-furanone ("MX"). This compound occurs at neutral pH in the acyclic form, i.e. in the form of Z-2-chloro-3-(dichloromethyl)-4-oxobutenoic acid. Its E-isomer (E-MX) is present in chlorinated drinking waters at a similar concentration, but is less mutagenic in Ames test. Both oxidised and reduced forms of MX and E-MX are also present in chlorinated waters. The present knowledge of the chemistry and toxicology of these mutagens is examined. The formation and possible elimination of the chlorination mutagens is discussed. The need of understanding the mechanisms of formation of these mutagens from humic substances during drinking water chlorination is emphasized.

Introduction

In many industrialized countries most drinking water is produced from surface waters. Water purification is primarily intended to remove excess organic substances of natural origin present in the raw water.

Common treatment procedures are coagulation and flocculation with Al- or Fe-salts followed by sedimentation or flotation and sand filtration. Ozone is used in some water works to degrade the organic material. The water purification results in the removal of a major part, commonly 50-80%, of the organics in the raw water. However, purified water still contains a considerable residue of organics, including humic substances. In Finland, for example, surface waters used for drinking water production normally contain 10-16 mg/l TOC and the purified waters have TOC levels of 4-7 mg/l [1].

Before distribution, the drinking water has to be treated for destruction of pathogenic microorganisms. This disinfection is usually carried out by addition of chlorine, which at the prevalent pH of 7-8.5 is in the form of hypochlorous acid (HOCl) and hypochlorite ions (ClO^-). Chlorine was introduced as a disinfection agent during the first decades of this century. It is very effective for disinfection, and its use has been most important for providing safe drinking water. No doubts about the use of chlorine came up until the early

1970's. Then, in 1974, chlorination of humic substances was reported to produce haloforms, especially chloroform [2]. In extreme cases, finished drinking waters have been found to contain over 1 mg $CHCl_3$/l. Another alarm came a few years later when newly developed short-term mutagenicity tests revealed the presence of genotoxic compounds in drinking waters [3,4]. These findings lead to intensive research on chlorination disinfection by-products.

Limits for haloforms in drinking water have been set in many countries. The haloform levels have been decreased considerably through elimination of prechlorination, improved removal of organics, minimization of chlorine dosage and use of alternative disinfectants such as chlorine dioxide, monochloramine and ozone. However, there still is concern about chlorinated by-products. The fact that chlorinated waters do contain mutagenic compounds, although in extremely low concentrations, and that considerable concentrations of other chlorinated acids, mainly chloroacetic acids, are present, keeps the research on disinfection byproducts vigorous.

In this paper, the main reactions of chlorine with humic material are first shortly reviewed. The route to the identification of the key mutagens in drinking water is outlined and their chemical and toxicological properties discussed. Finally, the factors influencing the formation of mutagens in drinking water production are discussed.

Reactions of Chlorine with Humic Substances

Chlorine in water is, depending on the pH, either in the form of elementary chlorine (Cl_2), hypochlorous acid (HOCl) or hypochlorite ion (ClO^-). At pH 7 there is approximately 80% HOCl and 20% ClO^- and at 8.5 the proportions are 10% HOCl and 90% ClO^-. Elementary chlorine is not present above pH 6. These chlorine agents can react with organic material through oxidation, substitution or addition. Reactions also take place with inorganic material present in drinking water systems. Metal ions such as Fe^{2+} and Mn^{2+} are oxidised. Bromide ion is oxidised to free bromine and ammonia reacts to produce chloramines. Thus the reaction system in drinking water disinfection is extremely complex. The reaction products formed are dependent on many factors, e.g. pH, chlorine dose, reaction time, and, last but not least, the character of the organic material, which to a major part can be classified as humic material.

The reaction of chlorine with humic materials and pertinent model substances has been extensively studied, as seen, for example, in the review by Rice and Gomez-Taylor [5]. Exhaustive chlorination of aquatic humic material produces a wide variety of compounds. In fact, hundreds of compounds have been identified. Non-chlorinated and chlorinated aliphatic and aromatic acids dominate among the low-molecular substances [6-8]. The main products of fulvic acid chlorination are tri- and dichloroacetic acid and dichlorosuccinic acid. About half of the halogen-containing material is non-volatile [7]. Chloroform constitutes the major part of the volatile compounds.

Although a large number of the identified compounds in chlorinated drinking waters show some mutagenic activity, it was shown in 1985 that the compounds identified thus far in chlorinated humic water accounted for only 7-8% of the total mutagenicity in Ames test using the most sensitive strain TA 100 [9]. The active mutagens had to be identified using another approach.

Identification of Mutagenic Compounds in Chlorinated Waters

Identification of The "MX" Compound

A project with the objective to isolate and identify active Ames-mutagenic compounds formed in wood pulp chlorination was started in 1979 at the Pulp and Paper Research Institute of Canada [10]. In this work multiple step concentrations and fractionations were combined with mutagenicity monitoring of concentrates and fractions. Ames tests were made with the strain TA 100. Most of the mutagenicity was consistently found in one fraction, which indicated one single major mutagen. This "ghost" component was called "Mutagen X" or "MX". Mutagen X was progressively purified. Since MX was a true trace component a lengthy isolation scheme had to be developed. After about one year of intense work, going through nine steps of extraction, column chromatography, various modes of high pressure liquid chromatography (HPLC) and thin layer chromatography (TLC) as outlined in Fig. 1, about 200 µg of MX in fairly pure form was recovered.

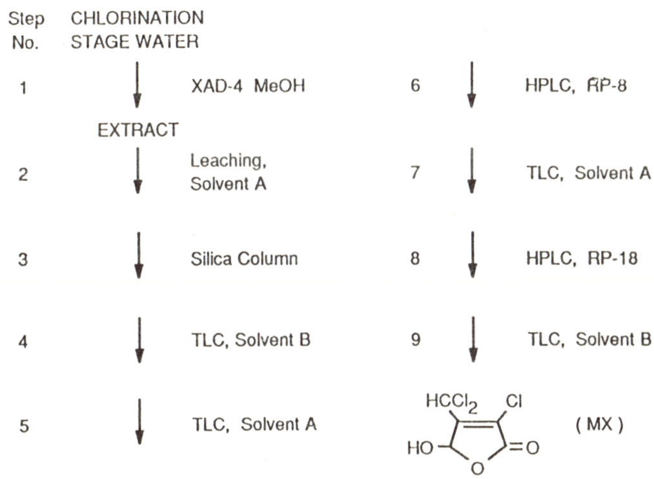

Figure 1 Scheme used for the isolation of the major mutagen (MX) in pulp chlorination waters [10,11].

On the basis of spectroscopic data (UV, IR, MS) and some chemical properties the structure 3-chloro-4-(dichloromethyl)-5-hydroxy-2(5H)-furanone was suggested (Fig. 2). This structure had not previously been described in the chemical literature. Further investigations confirmed that the compound was an important mutagen in pulp chlorination waters [11]. MX is considered one of the strongest mutagens ever tested in the Ames assay. Among a myriad of chlorinated compounds this single trace component, repre-

Figure 2 The tautomeric forms of MX.

senting only about 0.0001% of the total organics, was responsible for an astonishing 30-50% of the total mutagenicity of pulp chlorination waters.

We began research on drinking water mutagenicity at Åbo Akademi in 1981. Finnish drinking waters were found to be highly mutagenic and to show a similar response pattern on various Salmonella strains as pulp chlorination effluents [12]. Chlorination of humic-rich water in the laboratory produced similar mutagenicity. Fractionation and characterization of mutagenic extracts of drinking water and chlorinated humic waters indicated that the mutagens are predominatly nonvolatile organic acids of intermediate polarity [13].

MX was suspected to be present, but due to the lack of reference material it was not possible to develop analytical methods for analysis of MX at the expected low ng/l level concentrations. Then, in 1985, Padmapriya, Just and Lewis succeeded in the synthesis of MX [14]. The synthesized material was identical to the component isolated from bleaching waters in 1980, and thus the previously assigned structure was confirmed. The extreme mutagenicity of MX was also verified. Now, reference material was available for pursuing the identification of MX in drinking water.

An analytical method was developed based on adsorbtion on XAD 4/8 followed by elution with ethyl acetate, addition of mucobromic acid as an internal standard, methylation, extraction and final determination by GC/MS with selected ion monotoring [15] (Fig. 3). MX was found both in samples of drinking water and chlorinated humic water. Recently, studies have shown that MX is commonly occurring in chlorinated drinking waters [1,16-18]. In a study carried out in Finland, MX was detected in 23 out of 26 samples of drinking water collected from various localities [1]. MX accounted for 15-57% (average 33%) of the observed mutagenicity, although it represented an extremely small portion of the organic material, i.e. 0.5-4 ppm.

Identification of Other Mutagenic Compounds Related to MX

When extracts of chlorinated humic waters and drinking waters were fractionated, most of the mutagenicity was found in two fractions, with MX being the active compound in one of the fractions [19]. In the other mutagenic fraction a compound was present which was found to be identical to an impurity formed in the synthesis of MX. Its structure was determined to be E-(2)-chloro-3-(dichloromethyl)-4-oxobutenoic acid. The compound was the geometric E-isomer of the open form of MX and was coded E-MX (Fig. 4). E-MX was found to be mutagenic in the Ames test. However, the mutagenicity of E-MX was less than one tenth of that of MX.

E-MX is produced in chlorination and occurs in similar concentration as MX in drinking waters [1]. The contribution of E-MX to the Ames mutagenicity of chlorinated waters is at the most only a few per cent. However, E-MX can produce MX by isomerization under acidic conditions and under the influence of light [19,20].

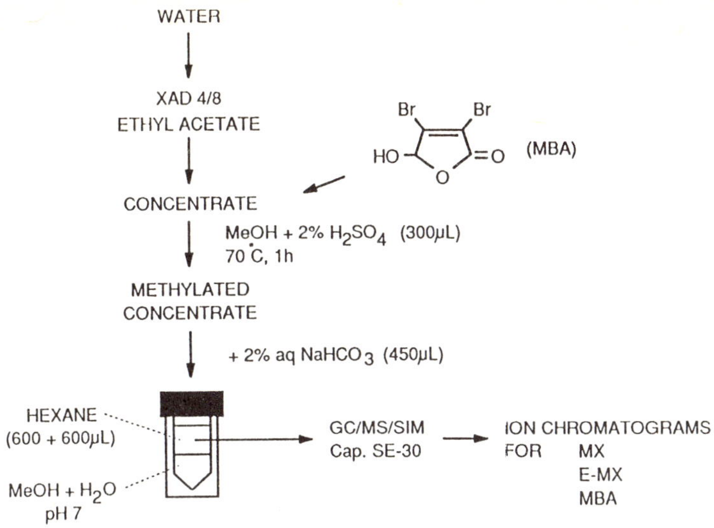

Figure 3 Analytical procedure for determination of MX (and E-MX) [15].

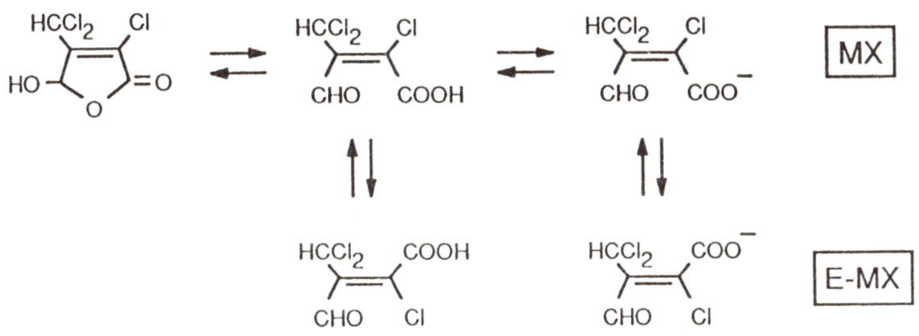

Figure 4 Transitions in the MX/E-MX system.

The aldehyde group in the open form of MX and in E-MX can be further oxidised to a carboxylic group or reduced to an alcohol group. A search for these forms of MX and E-MX has recently been conducted [21]. The oxidised forms of MX and E-MX, coded "ox MX" and "ox E-MX", as well as the reduced form of MX were synthesized from MX and E-MX. All of these compounds were found to be mutagenic, although they were weaker mutagens than E-MX and much weaker than MX (Fig. 5).

Analysis of a drinking water sample was carried out and these compounds were found to be present. The concentration of ox E-MX was 250 ng/l compared to 13 ng/l for MX and 20 ng/l for E-MX. The red MX and ox MX concentrations were 41 and 53 ng/l, respectively. The calculated contribution to the total mutagenicity of the compounds were MX: 17%, E-MX: 1.5%, red MX: 0.8%; ox MX: 1.0%, and ox E-MX: 0.4%.

In addition, it has recently been found that brominated analogues of MX may be formed by chlorination in the presence of bromide ions [22].

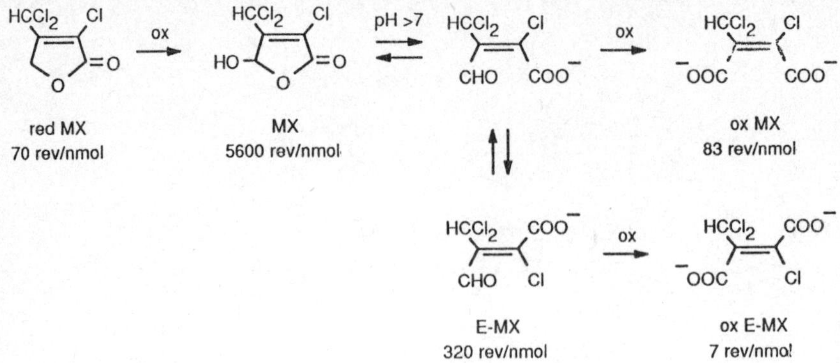

Figure 5 Structures of reduced and oxidised analogues of MX and E-MX and their mutagenicity in Ames test (TA 100, -S9).

Properties of MX and Related Compounds

Chemical Stability

MX is quite stable in water at pH 2, even at 60°C [11]. Above pH 4 ring opening and dissociation of MX appears [16,20] (Fig. 4). It can then be isomerized to E-MX, and also starts to undergo some hydrolytic degradation reactions. These degradation reactions increase with increasing pH (Fig. 6). With a half life of about 6 days at 23°C and only about 4% decrease in 6 days at 4°C [15,20], most of the MX formed during chlorination in drinking water works will also reach the consumer.

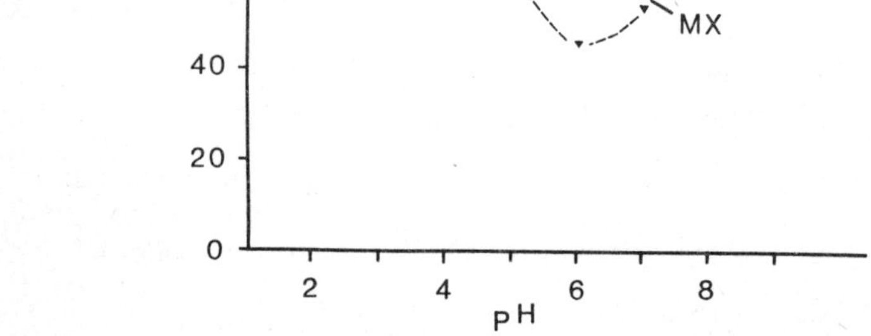

Figure 6 Isomerization of MX to E-MX and degradation of MX after storage in water for 3 days at 23°C at different pH values [20].

E-MX may convert to MX and complete conversion will gradually take place below pH 4 [20]. Thus the weakly mutagenic E-MX present in drinking water may, after ingestion, be converted in the intestine to the much more mutagenic MX. However, it seems that the conversion rate even at pH 2 and 40°C is so slow that only a minor part would be converted [23]. Boiling of drinking water for 10 minutes at pH 7 was found to degrade most of the mutagens [13].

MX reacts readily with the nucleophilic sulfite and sulfide ions in the pH range 5-9 [20]. It also reacts with glutathione [24,25] which is an important cellular nucleophile responsible for the detoxification of a variety of electrophilic xenobiotics.

The chemistry of MX and related compounds is still not well understood. For example, it is not known how these compounds react with nucleophiles. This would be essential to know since the coupling to DNA most probably is a nucleophilic reaction. Recently the DNA adduct formation of MX was investigated at US EPA [26] and a single major adduct appeared to be formed.

Toxic Properties

With a response of 6000-13000 revertants/nmole on the strain TA 100 in Ames test, MX is an extremely potent bacterial mutagen, in fact, one of the strongest ever tested in this test. MX appears to be able to induce both base-pair substitution and frameshift mutations. It has a modest acute toxicity to mice, 128 mg/kg [27]. It was also found to be a potent clastogen in mammalian cells. However, it was not genotoxic *in vivo* for mouse bone marrow.

Formation and Elimination of Chlorination Mutagens

Although the true genotoxic effects of the chlorinated mutagens are still not established, means of eliminating or minimizing the formation of these compounds need to be evaluated. During the last ten years numerous studies in many countries have shown that the mutagens in drinking water originate from reactions of chlorine with humic substances in the raw water [28 and refs. therein]. MX contributes a considerable part, in most studies ca 20-60%, of the total bacterial mutagenicity [1,15- 17]. The total mutagenicity correlates well with the concentrations of MX and E-MX [1].

The mutagenicity of chlorinated waters depends on the chlorine dose, chlorination pH, and beyond doubt also on the quality of the organic precursor material in the waters. Chlorination of humic-rich waters has revealed that the formation of mutagens is smaller at neutral pH than at acidic conditions [29,30]. The mutagenicity increases linearly with the chlorine dose at acidic pH but at pH 7 and 9 the mutagenicity goes down above a Cl_2/TOC ratio of 1:1 (Fig. 7). At higher chlorine doses MX contributes to a smaller part of the total mutagenicity.

Attempts have been made to correlate drinking water quality parameters with mutagenicity. In a recent study of Finnish drinking waters the mutagenicity was modelled as a function of the TOC and of the chlorine and ammonia doses [31]. The best fit was obtained by using, for each chlorination step (prechlorination and postchlorination), an

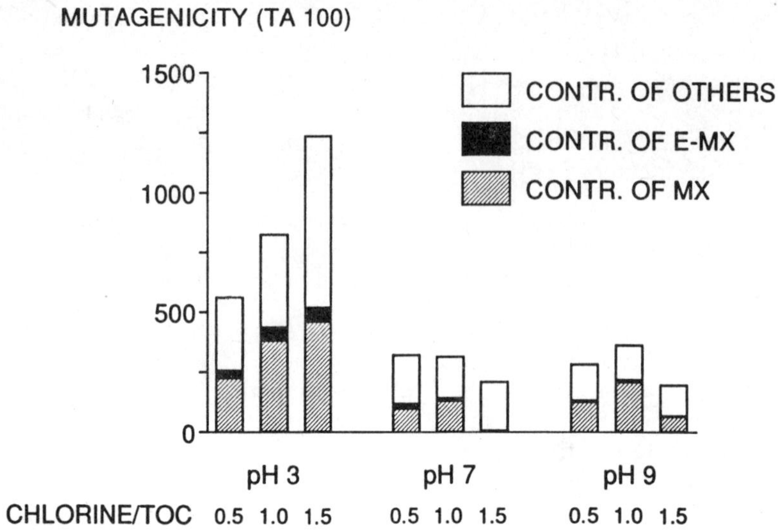

Figure 7 Mutagenicity produced by chlorination of a humic rich lake water (TOC 20 mg/l) at different pH values and with different chlorine doses [30]. The contribution of MX and E-MX to total mutagenicity was calculated from analysed concentrations using specific mutagenicity values.

exponential function of the form f = A $(1-e^{-kc})$, where f is the mutagenic activity, A and k are constants and c was defined as (TOC) x (Cl_2), both in units of mg/l.

Little is still known about the influence of the quality of the organic precursors for mutagen formation. Chlorination of humic substances from a wide range of sources has consistently been found to produce mutagenic activity by chlorination. However, because of different chlorination conditions and also different mutagenicity recovery and testing conditions, it is in most cases not possible to compare the mutagen formation potential of the different humic materials used. A Dutch river water and a humic-rich groundwater were found to produce mutagenicity on a similar level [29]. The concentration of MX and thus its contribution to the total mutagenicity was clearly higher for the humic rich water. Humic and fulvic reference materials (from the Suwannee stream) produced a similar level of mutagenic activity by chlorination [17]. As there is generally more fulvic acids than humic acids in natural waters, it was concluded that the fulvic acids may be more important than humic acids as precursors of mutagenic compounds.

It is still not clear which molecular structures in the humic substances are involved in the formation of the key mutagens, i.e. MX and related compounds. In order to gain knowledge in this area work has recently been carried out with model compounds related to humic substances. A wide variety of phenolic substances have been found to produce mutagenicity and also MX and related mutagens [17,32].

Use of alternative disinfectants may be a solution to the mutagenicity problem [28 and refs. therein]. Chlorine dioxide was found to produce less mutagenicity, and less MX, than chlorine [33]. Compared to chlorination, chloramine disinfection also produced significantly lower mutagenicity. Pretreatment with ozone slightly lowered the mutagenicity formed by chlorination.

Formed mutagens may be eliminated from drinking waters by activated carbon filtration or by chemical destruction. Granular activated carbon is effective in lowering the concentration of both mutagens and precursors for mutagens formed upon chlorination, even when the carbon is beyond its normal use for removal of organics [28 and refs. therein]. The chlorinated mutagens in drinking water can be destroyed by the addition of nucleophilic reagents, e.g. sulfite [34-36]. However, sulfite also removes the residual chlorine. Such a treatment is not feasible when a certain residual disinfectant level is considered important in the water distribution system.

Concluding Remarks

Chlorination of humic substances produces a myriad of chlorinated and nonchlorinated substances. None of the major chlorination products make a substantial contribution to the bacterial mutagenicity consistently found after chlorination of natural waters containing humic substances. The recent identification of a trace constituent, MX, as a major contributor to the overall mutagenicity in chlorinated waters has created new possibilities to achieve a fundamental understanding of the mutagenicity phenomena. We need now to explore the precursor or precursory structures to MX and the mechanism of formation. Also, there is now a single substance to concentrate on in toxicological studies.

The formation of the mutagens can be decreased both by improving the organics removal in water purification and by decreasing the chlorine doses. Alternative disinfectants such as chlorine dioxide and chloramine produce less mutagens but are also less efficient and less stable disinfectants. Mutagens can be efficiently removed by adsorbtion to activated carbon.

Achieving an understanding of the formation of the key mutagenic compounds formed by chlorination is a great challenge in the research on humic substances.

References

1. Kronberg, L. and T. Vartiainen. Mutation Res. **206**:177 (1988).

2. Rook, J.J. Water Treat. Exam. **23**:234 (1974).

3. Simmon, V.F. and R.G. Tardiff. Mutation Res. **38**:389 (1976).

4. Loper, J.C., D.R. Lang, R.S. Schoeny, B.S. Richmond, P.M. Gallagher and C.C. Smith. J. Tox. Environ. Health **4**:919 (1978).

5. Rice, R.G. and M. Gomez-Taylor. Environ. Health Persp. **69**:31 (1986).

6. Uden, P.C. and J.W. Miller. J. Am. Water Works Assoc. **75**:524 (1983).

7. Christman, R.F., D.L. Norwood, D.S. Millington, J.D. Johnson and A.A. Stevens. Environ. Sci. Technol. **17**:625 (1983).

8. E.W.B. de Leer, J.S.S. Damste and L. de Galan. In: Water Chlorination. Chemistry, Environmental Impact and Health Effects Vol. 5, p. 843 (Chelsea, MI: Lewis).

9. Meier, J.R., H.P. Ringhand, W.E. Coleman, J.W. Munch, R.P. Streicher, W.H. Kaylor and K.M. Schenck. Mutation Res. **157**:111 (1985).

10. Holmbom, B.R., R.H. Voss, R.D. Mortimer and A. Wong. Tappi **64**:172 (1981).

11. Holmbom, B.R., R.H. Voss, R.D. Mortimer and A. Wong. Environ. Sci. Technol. **18**:333 (1984).

12. Kronberg, L., B. Holmbom and L. Tikkanen. Vatten **41**:106 (1985).

13. Kronberg, L., B. Holmbom and L. Tikkanen. In: A. Björseth and G. Angeletti, Eds., Organic Micropollutants in the Aquatic Environment, p. 449 (Dordrecht: D. Deidel, 1985).

14. Padmapriya, A.A., G. Just and N.G. Lewis. Can. J. Chem. **63**:828 (1985).

15. Hemming, J., B. Holmbom, M. Reunanen and L. Kronberg. Chemosphere **15**:549 (1986).

16. Meier, J.R., R.B. Knohl, W.E. Coleman, H.P. Ringhand, J.W. Munch, W.H. Kaylor, R.P. Streicher and F.C. Kopfler. Mutation Res. **189**:363 (1987).

17. Horth. H., Aqua **38**:80 (1989).

18. Backlund, P., E. Wondergem, K. Voogd and A. de Jong. Sci. Total Environ. **84**:273 (1989).

19. Kronberg, L., B. Holmbom, M. Reunanen and L. Tikkanen. Environ. Sci. Technol. **22**:1097 (1988).

20. Holmbom, B., L. Kronberg and A. Smeds. Chemosphere **18**:2237 (1989).

21. Kronberg, L. and R.F. Christman. Sci. Tot. Environ. **81/82**:219 (1989).

22. Horth, H., M. Fielding, T. Gibson, H.A. James and H. Ross. Report BRD 2+38-M, Water Research Centre, Medmemham, U.K. 1989.

23. Kronberg, L., B. Holmbom and L. Tikkanen. In: Water Chlorination - Chemistry, Environmental Impact and Health Effects, Vol. 6, pp. 137-146 (Chelsea, MI: Lewis 1990).

24. Ishiguro, Y., R.T. Lalonde and C.W. Dence. Environ. Toxicol. Chem. **6**:935 (1987).

25. Meier, J.R., R.B. Knohl, B.A. Merrick and C.L. Smallwood. In: Water Chlorination Chemistry, Environmental Impact and Health Effects, Vol. 6, pp. 159-170 (Chelsea, MI: Lewis 1990).

26. Meier, J.R., A.B. De Angelo, F.B. Daniel, K.M. Schenck, M.F. Skelly and S.L. Huang. Paper presented at the Fifth International Conference on Environmental Mutagens, Cleveland, July 10-15, 1989.

27. Meier, J.R., W.F. Blazak and R.B. Knohl. Environ. Mutag. **10**:411 (1987).

28. Meier, J.R. Mutation Res. **196**:211 (1988).

29. Backlund, P., E. Wondergem, K. Voogd and A. de Jong. Chemosphere **18**:1903 (1989).

30. Nyman, S. M.Sc. Thesis, Åbo Akademi 1989 (in Swedish).

31. Vartiainen, T., A. Liimatainen, P. Kauranen and L. Hiisivirta. Chemosphere **17**:189 (1988).

32. Holmbom, B., L. Kronberg, P. Backlund, V-A Långvik, J. Hemming, M. Reunanen, A. Smeds and L. Tikkanen. In: Water Chlorination, Chemistry, Environmental Impact and Health Effects, Vol. 6, pp. 125-135 (Chelsea, MI: Lewis 1990).

33. Backlund, P., L. Kronberg and L. Tikkanen. Chemosphere **17**:1329 (1988).

34. Cheh, A.M., J. Skochdopole, D. Koski and L. Cole. Science **207**:90 (1980).

35. Wilcox, P. and S. Denny. In: Water Chlorination - Chemistry, Environmental Impact and Health Effects, Vol. 5, pp. 1341- (Chelsea, Lewis).

36. Croue, J-P and D.A. Reckhow. Environ. Sci. Technol. **23**:1412 (1989).

Mutagenic Activity in Disinfected Waters and Recovery of the Potent Bacterial Mutagen"MX" from Water by XAD Resin Adsorption

Peter Backlund[1], Erik Wondergem[2] and Leif Kronberg[1]

[1] Department of Organic Chemistry, Åbo Akademi, Akademigatan 1
SF-20500, Turku/Åbo, Finland
[2] National Institute of Public Health and Environmental Protection, P.O. Box 1, NL-3720, BA Bilthoven, The Netherlands

Abstract

Chlorination of humic water generated mutagenic activity in the Ames test. The formation of the potent bacterial mutagen 3-chloro-4-(dichloromethyl)-5-hydroxy-2(5H)-furanone (MX) and mutagenic activity were favoured by acidic chlorination conditions and high chlorine doses. Chlorinated humic waters from different locations differed slightly in the level of mutagenicity as well as in the proportion of activity derived from MX. Chlorination of an industrially polluted surface water with a low content of humic material generated an approximately equal level of mutagenicity (per mg of DOC) as that of chlorinated humic water, but only a minor part (26%) of the activity could be explained by the presence of MX. The mutagenicity and the amount of MX generated were substantially lower when using combined treatment methods (ClO_2+Cl_2, O_3+Cl_2) or when substituting chlorine by monochloramine or chlorine dioxide. The recovery of MX by XAD adsorption from water acidified to pH 2 was found to be quantitative.

Introduction

The formation of mutagenic activity in drinking water by chlorination is well documented [1-9]. Previous studies have indicated that the mutagens are products of reactions between chlorine and humic substances present in the raw water source [4,10,11]. An extremely potent mutagen, 3-chloro-4-(dichloromethyl)-5-hydroxy-2(5H)-furanone (MX, Fig. 1), was identified by Holmbom *et al.* in the chlorination stage effluent of pulp mills in 1981 [12]. More recently the compound has been identified in a number of chlorinated water samples [13-17]. In a study carried out in Finland [14] drinking water samples from 26 localities were analysed for MX, and the compound was found to account for 15-57% of the total mutagenicity present. The activity contribution of MX in chlorinated drinking water samples collected from three localities in the United States was estimated to 15-34% [16], and a similar activity contribution of MX was found in two drinking waters collected in Great Britain [15]. In the Netherlands, the activity contribution of MX in three chlorinated raw water samples was estimated to 20-40%, whereas, in chlorinated tap waters collected from four localities, the MX concentration was found to be below the detection limit of 1-2 ng/l [17].

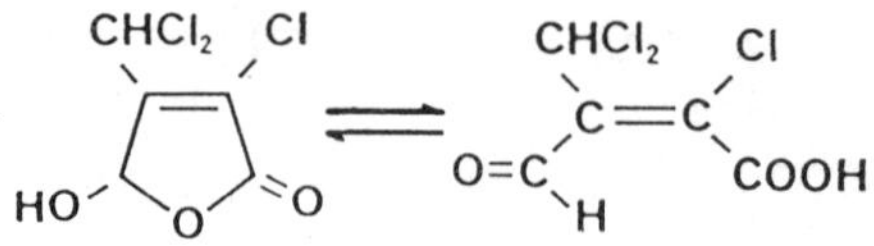

MX

Figure 1 Structure of the bacterial mutagen MX, showing its two tautomeric forms.

These studies demonstrated that MX is commonly present in chlorine-treated waters, and that the compound normally accounts for a considerable part of the observed mutagenicity.

In spite of the high mutagenicity contribution of MX in chlorinated waters the concentration level is usually low (<200 ng/l). Consequently, a concentration step is essential prior to analysis. The most frequently used concentration technique has been adsorption to XAD resins and subsequent elution with an organic solvent. Substitution of chlorine by an alternative disinfectant, or a change in the chlorination process itself (such as reaction pH, chlorine dose) are possible ways of decreasing or preventing the formation of mutagenic compounds.

In the present paper we review results from our studies of the effects of chlorination on mutagenic activity and on MX concentrations in water, of the effects of different water disinfectants and disinfection conditions on the formation of mutagenicity and MX, and of the recovery of MX from water using XAD resin adsorption.

Mutagenicity in Chlorinated Waters from Different Sources

Mutagenicity according to the Ames assay (especially with strain TA100 without metabolic activation) is known to be generated upon chlorination of raw and purified surface waters, wood pulp, and aqueous solutions of isolated humic and fulvic acids and certain amino acids and phenolic model compounds.

The similar characteristics observed in the mutagenicities generated in these waters [6, 18-21] indicate that they may contain common structural moieties. Amino acids and phenolic compounds represent structural components of humic material and, furthermore, structural similarities are known to exist between the lignin portion of wood pulp and aquatic humic material [20].

In previous studies a substantially higher mutagenicity has been reported after chlorination of waters with a high content of humic material than after chlorination of waters containing mainly man-made organic pollutants [2,6]. However, a direct comparison of results obtained in different studies has been hampered by the lack of a standard concentration method. Even a correct comparison between studies in which XAD resin adsorption has been used as concentration technique is difficult because of differences in experimental conditions, e.g. flow rate, extraction pH and resin type.

To be able to make a direct comparison of the level of mutagenicity and of the amount of MX produced during chlorination of waters from different sources we collected samples from Lake Savojärvi in Finland, and the Meuse River and a ground water source (St Jansklooster) in the Netherlands. The waters from Lake Savojärvi and St Jansklooster are rich in humic material, but relatively unaffected by industrial or municipal effluents (DOC: 20 mg/l and 6.5 mg/l, respectively). The water from the Meuse River contains only a low concentration of humic material but is heavily polluted by industrial waste and effluents from sewage treatment plants (DOC: 4.2 mg/l).

The samples (2-10 l, depending on the source) were treated identically with respect to the chlorination conditions and the concentration/testing technique. Chlorination was carried out at pH 7 (phosphate buffer) for 65 hours resulting in a residual chlorine concentration of <0.15 mg/l. A constant Cl_2/DOC ratio of 1/1 was used in each experiment. The results were evaluated on a DOC weight basis (mutagenicity/mg DOC and amount of MX/mg DOC, respectively). After chlorination the samples were acidified to pH 2 (4 M HCl) and passed over a column of XAD 4/8 (v/v:1/1) using a flow rate of 0.5 bed volumes per minute (one bed volume = 40 ml). Excess water was expelled by purging the column with nitrogen and the adsorbed organics were eluted with 5 bed volumes (200 ml) of ethyl acetate. The eluate was divided into two portions, one to be used for the mutagenicity testing (*Salmonella* typhimurium TA100 -S9) and one for the gas chromatographic/mass spectrometric (GC/MS) determination of MX. Just prior to the mutagenicity tests the ethyl acetate was evaporated to dryness and the residue was dissolved in DMSO. The procedures used in the mutagenicity assay and in the chemical analysis of MX have been described elsewhere [13,17].

All the waters studied exhibited mutagenicity after chlorination (Fig. 2). The mutagenicity (expressed as TA100 net rev./mg DOC) and the contribution by MX varied slightly between St Jansklooster water (HW1) and Lake Savojärvi water (HW2). HW2 exhibited a higher mutagenicity, and in this sample MX accounted for 50% of the activity. In HW1 the activity contribution of MX was 40%. This indicates that the amount of molecular structures generating mutagenic compounds during reaction with chlorine is higher in the dissolved humic material of HW2 than in that of HW1.

The mutagenic activities presented for HW1 and HW2 in Fig. 2 correspond to levels of 2400 rev./l and 10400 rev./l, respectively, and the MX mutagenicity contributions correspond to MX concentrations of 40 and 200 ng/l, respectively. These activities are comparable to those reported in earlier studies on chlorinated humic waters using the same concentration/testing technique [6,22] and demonstrate the high potential of aqueous humic material to generate mutagenic compounds.

The activity (per mg of dissolved carbon) observed in chlorinated the Meuse River water (RW) was approximately equal to that of chlorinated humic water from St Jansklooster (HW1). The number of 1500 rev./l is 6 to 15 times higher than that previously found for this water by other researchers [2,21]. This discrepancy is most probably due to differencies in the work up procedures. The MX mutagenicity contribution in the chlorinated Meuse River water was 26% of the total activity, which corresponds to a concentration of 15 ng MX/l.

Thus, in spite of its low humic content, river Meuse water seems to contain important precursors generating mutagenic compounds, readily extracted at acidic conditions. On the other hand, no mutagenicity could be extracted at neutral pH from chlorinated purified river water (results not shown), indicating that the major part of the still unexplained activity is caused by acidic compounds.

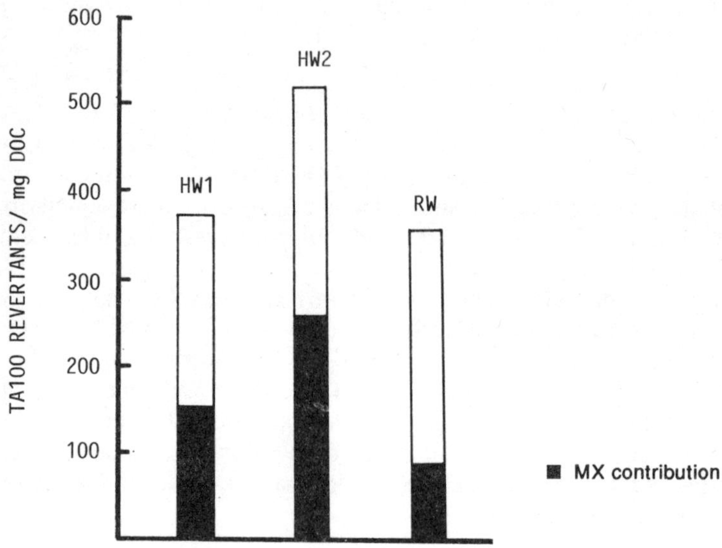

Figure 2 Mutagenicity and calculated MX mutagenicity contribution in chlorinated waters.

HW1 = Chlorinated humic rich ground water from St Jansklooster, The Netherlands

HW2 = Chlorinated humic rich water from lake Savojärvi, Finland

RW = Chlorinated water from the Meuse River, The Netherlands

Mutagenicity Produced by Alternative Disinfectants

The observation that mutagenic activity is produced during chlorine disinfection of water focused our attention on alternative treatment methods. Would it be possible to reduce or prevent mutagenicity from being formed simply by changing the disinfection technique? Lake Savojärvi water was treated with various disinfectants and combinations of disinfectants. The mutagenicities and the MX concentrations generated were determined and comparisons were made between the methods of disinfection. The methods for preparation of the disinfectants used, the procedures for disinfection and mutagenicity testing, and the methods for analysing MX have been described earlier [13,22,23].

Monochloramine is a less effective and less reactive disinfectant than free chlorine. Consequently, in order to guarantee a safe biological quality of the drinking water during distribution, monochloramine is normally added as a complement to a more efficient disinfectant (such as ozone). A correct and clear-cut comparison of the effects of chlorine and chloramine on mutagenicity is difficult to obtain. On one hand, owing to the relatively low reactivity of chloramine with dissolved organic matter, the initial dose should be low enough to avoid an unreasonably high residual concentration in the tap. On the other hand, owing to its lower disinfection efficiency (as compared to chlorine), the initial chloramine dose should be high enough to give a degree of disinfection comparable to that obtained by chlorination. The situation is further complicated by the fact that MX is unstable at the normal chlorination pH of 6-8 [24], and, for this reason, the contact time should be kept as short as possible and of equal length in the two experiments. A direct comparison of the effects of free chlorine and chlorine dioxide is easier to obtain because the reaction rates as well as the disinfection efficiences of these agents are approximately equal.

In our study free chlorine was dosed at a Cl_2/DOC (w/w) ratio of 0.5 and 1.0, respectively. Chlorine dioxide was dosed at a ratio of 1.0 and monochloramine at a ratio of 0.5 (expressed as Cl_2). In the combined Cl_2+ClO_2 treatment the disinfectants were mixed just prior to application at a total disinfectant/DOC ratio of 1.0, and in the combined O_3+Cl_2 treatment ozonation was carried out 2 h prior to chlorination. In a previous study [25] we demonstrated that preozonation has no significant influence on the chlorine consumption in humic water, and for this reason the chlorine dose used in the combined O_3+Cl_2 treatment was equal to that used in the chlorine treatment without a preceding ozonation step.

All the disinfectants studied generated mutagenicity and some amount of MX (Fig. 3). Chlorination gave rise to by far the highest activity and the highest MX concentration. When chlorination was preceded by an ozonation step, the mutagenicity, as well as the MX concentration, were slightly decreased. No mutagenicity was generated when the water was treated with ozone alone (results not shown), which is in agreement with results obtained by other groups [21,26]. Cognet and co-workers [27], on the other hand, observed a complex relationship between ozone dose and mutagenicity. They found that the activity of ozonated water was reduced by using high and low ozone concentrations, but was enhanced using an intermediate ozone concentration. Ozonated water from another sampling station occasionally exhibited an increased activity.

Disinfection with a combination of chlorine and chlorine dioxide generated less mutagenicity and a lower MX concentration than did disinfection with chlorine alone. The higher the proportion of chlorine dioxide applied, the lower the mutagenicity and the amount of MX generated. The activity observed after treatment with chlorine dioxide alone was only 15% of that generated during chlorination. A decrease as compared to chlorination-derived mutagenicity was also achieved when monochloramine was used as a substitute for chlorine. Chloramination produced less than 50% of the mutagenicity and less than 25% of the MX concentration found in chlorinated waters.

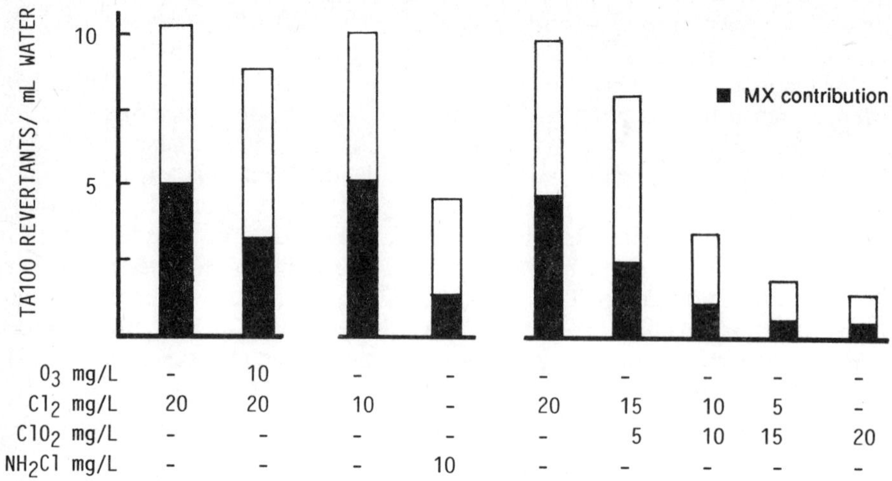

Figure 3 Mutagenicity of humic-rich water treated with Cl_2, NH_2Cl, ClO_2 and combinations of O_3 and Cl_2, and of ClO_2 and Cl_2.

Influence of Chlorine Dose and Chlorination pH on Mutagenicity

To evaluate the influence of different chlorination conditions on mutagenic activity and MX formation, the humic-rich ground water from St Jansklooster was treated at four different pH values (pH 2, 4, 7 and 9), using three different chlorine doses at each pH (Cl_2/DOC (w/w) ratio: 0.5, 1.0 and 2.0).

Because of the complex, pH-dependent transition reactions of MX in water [24] the choice of reaction time is of great importance when studying the formation of MX and mutagenicity at different chlorination conditions. During a long contact time the formed MX will be subjected to different degrees of isomerisation and degradation reactions at different pHs. Furthermore, the combination of high chlorine doses and short contact times may create artifacts caused by reactions between residual chlorine and organic material, as the water is acidified prior to extraction. We wanted to avoid adding any reducing agent to remove excess chlorine in the water because such agents have been found to deactivate MX and other electrophilic mutagens [24,28]. Therefore, we carried out our experiments at two different contact times in order to achieve a chlorine residual of <0.15 mg/l before sample acidification/concentration.

The formation of mutagenicity and of MX was normally favoured by chlorination at acidic conditions using high chlorine doses (Fig. 4). However, when excess chlorine was used (Cl_2/DOC: 2.0) at neutral or alkaline reaction conditions, no MX could be detected. This is probably explained by the poor stability of MX at neutral and alkaline conditions, as reported previously by Meier *et al.* [16] and Holmbom *et al.* [24]. The contribution of MX to the total TA100 mutagenicity in the examined waters chlorinated at Cl_2/DOC ratios of 0.5 and 1.0 was approximately 30-40% at all pH values, which is similar to previously reported results obtained after chlorination at neutral conditions [14,22,23].

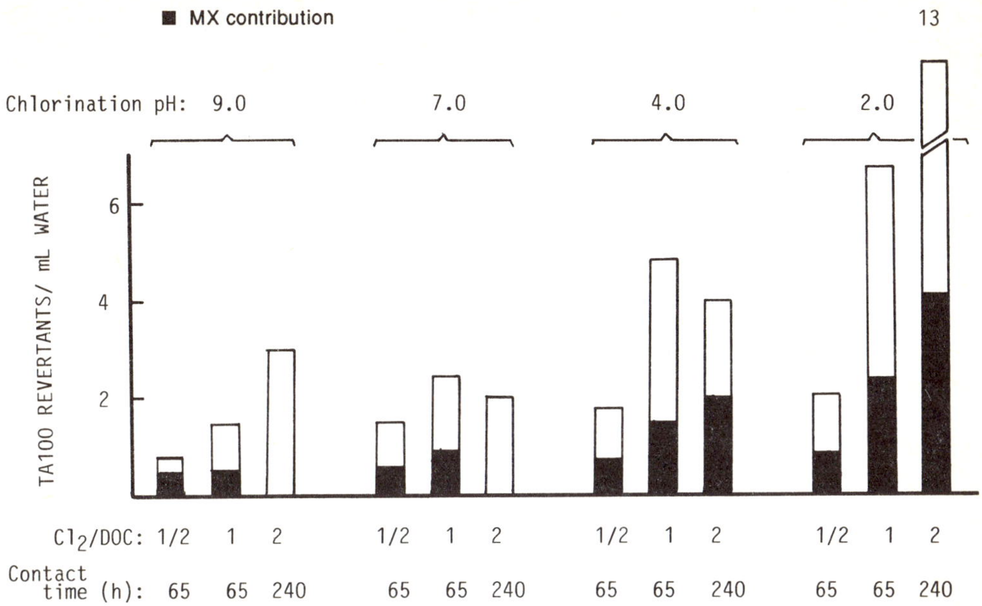

Figure 4 Mutagenicity of humic water chlorinated at various pH:s and chlorine doses.

Recovery of MX from Water by XAD Resin Adsorption

Several studies have demonstrated that the mutagenic activity of water extracts obtained at pH 2 is higher than of those obtained at neutral pH [2,6,9,18,19]. As far as the recovery of MX is concerned, a pH-dependent extraction efficiency is expected, as the compound will be in its ionized, open form at neutral pH.

To evaluate the recovery of MX on a mixed column of Amberlite XAD 4 and XAD 8 (v/v: 1/1), MX was spiked at a concentration of 1 μg/l into 5 l each of three different waters: clean water , chlorinated surface water (river Meuse) and chlorinated, humic-rich ground water (St Jansklooster). The clean water was prepared by double distillation of demineralized water in the presence of 5 ml 0.1 M potassium permanganate and 1 ml 4 M sulphuric acid.

The river and the ground waters were chlorinated before spiking in order to simulate normal conditions. Chlorination was performed at pH 7 for 65 h in the dark using a chlorine to DOC ratio of 1/1, resulting in a residual chlorine concentration of <0.15 mg/l.

One liter of the spiked and unspiked water samples was passed over the XAD column using the same procedure as described above. The mutagenic activity derived by the amount of MX recovered from the chlorinated spiked waters was calculated by subtracting the activity of the blanks (unspiked chlorinated waters) from that of the spiked waters. The unspiked clean water was non-mutagenic. The recovery of MX was thereafter determined by comparing the mutagenic activity in the extracts of spiked water (after subtracting the activity of the blank) with the activity of a solution of DMSO to which MX had been added in the same theoretical concentration/test-volume ratio (="100%

recovery"). The theoretical MX concentrations tested per plate were 20, 40, 60 and 100 ng.

The XAD 4/8 procedure turned out to be extremely efficient for the extraction of MX. The recovery was approximately 100% from all waters studied (Fig. 5) which is higher than previously reported by other groups using XAD adsorption [14,16]. The recovery obtained by conventional diethyl ether extraction at pH 2 was approximately equal to that obtained by the XAD procedure, but the reproducibility was poor (results not shown).

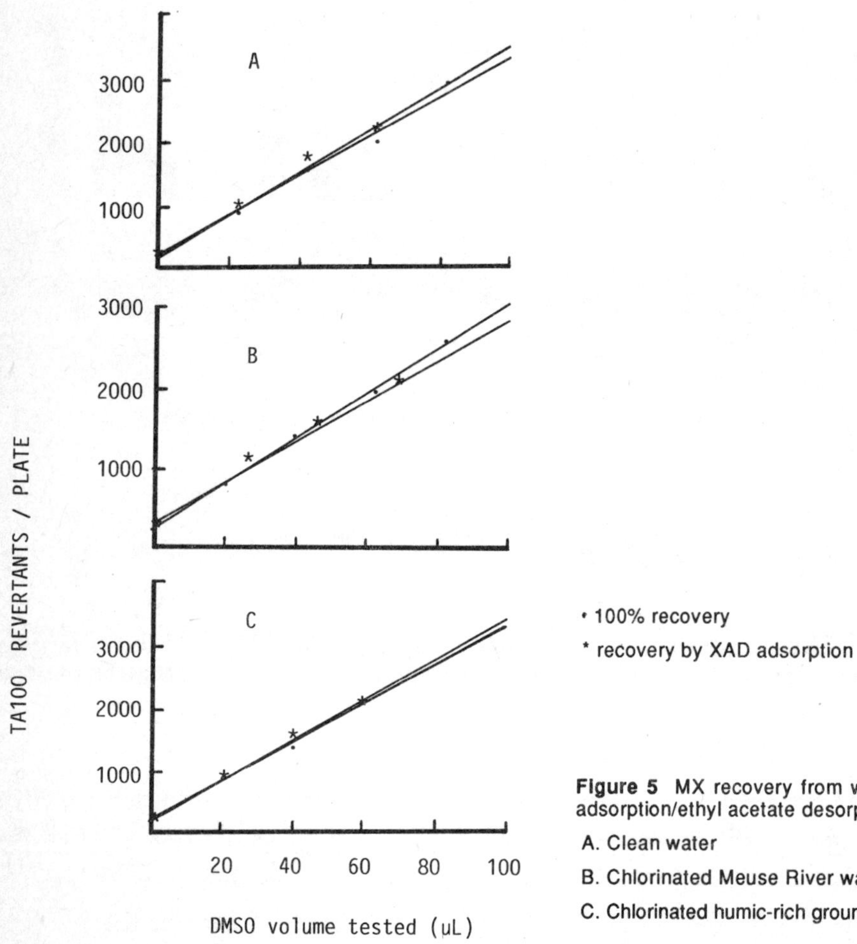

Figure 5 MX recovery from water by XAD adsorption/ethyl acetate desorption.

A. Clean water

B. Chlorinated Meuse River water

C. Chlorinated humic-rich ground water

Conclusion

Mutagenic activity and the potent bacterial mutagen 3-chloro-4-(dichloromethyl)-5-hydroxy-2(5H)-furanone (MX) are produced upon chlorine treatment of humic-rich waters from various locations. The observed mutagenicity (expressed as TA100 revertants per mg DOC) as well as the calculated mutagenicity contribution of MX vary slightly between the samples, indicating that humic substances of different origin may contain different amounts of precursors or different structural precursor moieties. The mutagenic activity generated during chlorination of the industrially polluted water from Meuse River is approximately similar to that in chlorinated humic water from St Jansklooster (The Netherlands). The MX mutagenicity contribution, however, is considerably lower than that in the chlorinated humic water, demonstrating that important mutagen precursors are present in both waters, but that the structural differences in the DOC gives rise to different mutagenic byproducts during chlorination.

The formation of mutagenic activity is favoured by acidic reaction conditions and high chlorine doses. Using chlorine doses which may be regarded as common in drinking water preparation, the formation of MX follows much the same trend as that of the mutagenicity. However, using a surplus of chlorine at alkaline conditions and a long contact time, the amount of MX generated will be below the detection limit.

Substitution of chlorine by chlorine dioxide or monochloramine results in a decrease in mutagenicity as well as in the concentration of MX. A decrease is also achieved if chlorination is preceded by an ozonation step or used in combination with chlorine dioxide. The higher the proportion of chlorine dioxide in the combined Cl_2+ ClO_2 treatment, the lower the mutagenicity and the MX concentration generated.

MX is quantitatively extracted from water at pH 2 by the XAD 4/8 resin adsorption/ethyl acetate desorption technique.

References

1. Loper, J. C. Mutat. Res. **76**: 241 (1980).

2. Van der Gaag, M. A., A. Nordsij and J. P. Oranje. In: M. Sorsa and A. Vainio, Eds., Mutagens in Our Environment, pp. 277-286 (Alan R. Liss: New York, 1982).

3. Kool, H. J., C. F. van Kreijl, E. de Greef and H. J. van Kranen. Environ. Health Perspect. **46**:207 (1982).

4. Meier, J. R., R. D. Lingg and R. J. Bull. Mutat. Res. **118**:25 (1983).

5. Kringstad, K. P., P. O. Ljungquist, F. de Sousa and L. M. Strömberg. Environ.Sci. Technol. **17**:553 (1983).

6. Kronberg, L., B. Holmbom and L. Tikkanen. Vatten **41**:106 (1985).

7. Marouka, S. Wat. Supply **4**:103.

8. Vartiainen, T and A. Liimatainen. Mutat. Res. **167**:29 (1986).

9. Ringhand, P. H., J. R. Meier, F. C. Kopfler, K. M. Schenck, and D. E. Mitchell. Environ. Sci. Technol. **21**:382 (1987).

10. Bull, R. J., M. Robinson, J. R. Meier and J. Strober. Environ. Health Persp. **46**:215 (1982).

11. Coleman, W. E., J. W. Munch, W. H. Kaylor, R. P. Streicher, H. P. Ringhand and J. R. Meier. Environ. Sci. Technol. **18**:674 (1984).

12. Holmbom, B., R. H. Voss, R. D. Mortimer and A. Wong. Tappi **53**:172 (1981).

13. Hemming, J., B. Holmbom, M. Reunanen and L. Kronberg. Chemosphere **15**:549 (1986).

14. Kronberg, L. and T. Vartiainen. Mutat. Res. **206**:177 (1988).

15. Horth, H., M. Fielding, H. James, M. Thomas, T. Gibson and P. Wilcox. Paper presented at the sixth conference on Water Chlorination, Oak Ride, TN, May 3-8, 1987. In: R.L. Jolley *et al.*, Eds., Water Chlorination: Chemistry, Environmental Impact and Health Effects, Vol. 6. In press.

16. Meier, J. R., R. B. Knohl, W. E. Coleman, H. P. Ringhand, J. W. Munch, W. H. Kaylor, R. P. Streicher and F. C. Kopfler. Mutat. Res.**189**:363 (1987).

17. Backlund, P. E., Wondergem, K. Voogd and A. de Jong. Sci. Total Environ. **84**:273 (1989).

18. Monarca, S., J. K. Hongslo, A. Kringstad and G. E. Carlberg. Water Res. **19**:1209 (1985).

19. Wigilius, B., H. Borén, G. E. Carlberg, A. Grimvall and M. Möller. Sci. Total Environ. **47**:265 (1985).

20. Ertel, J. R., J. I. Hedges and E. M. Perdue. Science **223**:485 (1984).

21. Kool, H. J. and C. F. van Kreijl. Water res. **18**:1011 (1984).

22. Backlund, P., L. Kronberg, G. Pensar and L. Tikkanen. Sci. Total Environ. **47**:257 (1985).

23. Backlund, P., L. Kronberg, and L. Tikkanen. Chemosphere **17**:1329 (1988).

24. Holmbom, B., L. Kronberg and A. Smeds. Chemosphere. In press 18:2237 (1989).

25. Backlund, P. In: Hallikas, J, Ed., VTT Symposium 65, Technical Research Centre of Finland, Espoo, Finland, pp. 194-203 (1986).

26. Lykens, B.W. J. Am. Water Assoc., **78**:66 (1986).

27. Cognet, L., Y. Courtois and J. Mallevialle. Environ. Health Persp. **69**:165 (1986).

28. Cheh, A. M., J. Skochdopole, P. Koski and L. Cole. Science **207**:90 (1980).

Formation of 3-Chloro-4-(Dichloromethyl)-5-Hydroxy-2(5H)-Furanone (MX) and Mutagenic Activity by Chlorination of Phenolic Compounds

Vivi-Ann Långvik[1], Osmo Hormi[1], Leif Kronberg[1],
Leena Tikkanen[3] and Bjarne Holmbom[2]

[1] Dept. of Organic Chemistry, Åbo Akademi, SF-20500 Turku/Åbo 50, Finland.
[2] Laboratory of Forest Products Chemistry, Åbo Akademi, SF-20500 Turku/Åbo 50, Finland.
[3] Food Research Laboratory, Technical Research Centre of Finland, SF-02150 Espoo/Esbo, Finland.

Abstract

Homovanillic acid, vanillin, protocatechualdehyde, 3,4-dihydroxyphenylalanine (L-DOPA) and caffeic acid were chlorinated at pH 2 in aqueous solutions with various chlorine doses. The occurance of Ames-mutagenicity was tested, and the formation of 3-chloro-4-(dichloromethyl)-5-hydroxy-2(5H)-furanone (MX) and its isomer (E)-2-ch'oro-3-(dichloromethyl)-4-oxo-butenoic acid (EMX), was determined by combined gas chromatography/mass spectrometry in the selected ion monitoring (SIM) mode. Chlorination of the studied phenols resulted in Ames-mutagenicity and the formation of the mutagens MX and EMX. The molar ratios of MX to EMX varied from about 30-50 for caffeic acid and L-DOPA and from about 0.5 to 2 for homovanillic acid, vanillin and protocatechualdehyde. The results of our work show that MX and EMX in chlorinated drinking water may originate from phenolic subunits of the humic macromolecules.

Introduction

The occurrence of Ames-mutagenicity in drinking water disinfected with chlorine is well documented [1,2]. The mutagenic compounds are derived from the reactions between chlorine and humic substances in the raw water [3]. The strong mutagen 3-chloro-4-(dichloromethyl)-5-hydroxy-2(5H)-furanone (MX) and its isomer (E)-2-chloro-3-(dichloromethyl)-4-oxo-butenoic acid EMX (Fig. 1), have been identified in chlorinated drinking waters [4-7]. MX is a major mutagen, although a minor constituent, of these waters, occurring in concentrations of 5-67 ng/l [2].

MX and EMX are consistently formed by chlorination of natural humic waters and of isolated humic and fulvic acid fractions [4-7]. In addition, we have found [8,9], as have Horth *et al.* [10], that MX and EMX are formed upon chlorination of various phenolic compounds. Thus, MX and EMX might be products of reactions of chlorine and phenolic subunits in humic macromolecules.

The objective of the present investigation was to further study the potential of MX and EMX formation upon chlorination of those phenolic compounds that formed significant amounts of MX and EMX in the preliminary studies [8,9]. Mutagenicity of the chlorinated compounds was studied by the Ames-test, using tester strain TA 100 without metabolic activation [11].

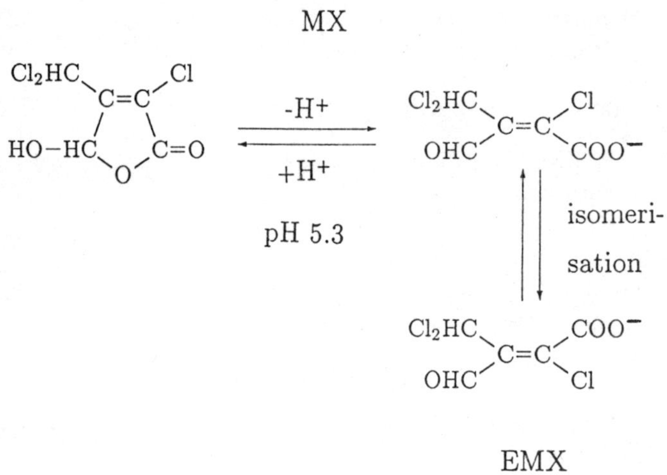

Figure 1 MX and its isomer EMX.

Materials and Methods

The chlorinated phenols were homovanillic acid (HVA), caffeic acid (CA), 3,4-dihydroxyphenylalanine (L-DOPA), vanillin (VAN) and protocatechualdehyde (PCA) (Fig. 2). The chlorine used was freshly prepared by bubbling chlorine gas (generated by drop-wise addition of concentrated HCl to $KMnO_4$, with subsequent cleaning and drying) into a solution of 1.12% NaOH, until a pH of 7-8 was obtained. Immediately before use, the chlorine concentration was determined by iodometric titration. The chlorination was performed in stoppered glass bottles at ambient temperature in the dark, until the chlorine residual was 0.1 mg/l or less (contact time 2-3 days). The chlorine dose applied was determined by chlorination of each phenol at a chlorine to TOC ratio of 10. Following 24 h contact time, the chlorine residual was determined by iodometric titration, and moles chlorine consumed per mole of phenol was calculated. Following chlorination, the aqueous solutions (50 mg of each phenol in 500 ml) were extracted with newly distilled diethyl ether (150 +100 +100 ml). The solvent was evaporated, and the residue was dissolved in ethyl acetate (25 mg/ml). The extracts were stored in a freezer (-20°C). The sample extracts were methylated by acidic methanol and then analyzed by gas chromatography/mass spectrometry (GC/MS) in the selected ion monitoring (SIM) mode. Mucobromic acid (MBA) was used as internal standard. Further details of the analysis have been published previously [2,4].

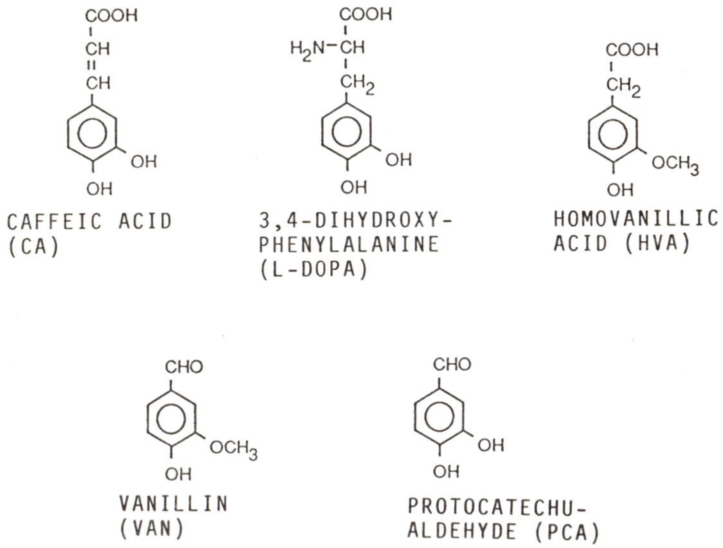

Figure 2 Structures of the phenolic compounds studied.

Mutagenicity of the extracts was determined with the Ames-test, using tester strain TA 100 without metabolic activation [11]. Just before the test, the solvent was changed to dimethyl sulfoxide. The tests were performed in duplicates at three dose levels. The tests included positive (2.5 μg Na-azide per plate) and negative controls. MX and EMX contribution to mutagenicity was calculated from the lowest dose level, as there were some toxicity problems with the higher doses. The contribution of MX and EMX to the total mutagenicity was calculated from a value of 5600 net revertants per nmole of pure MX, and EMX was calculated from a value one tenth of the MX value [2].

Results

For 50 mg/l solutions (approx. 0.25 mM; pH 2) of homovanillic acid (HVA), vanillin (VAN), protocatechualdehyde (PCA), L-DOPA and caffeic acid (CA), the chlorine consumption was very rapid during the first hour (Fig. 3). The highest consumption on a molar basis was observed for homovanillic acid (HVA).

The phenols studied formed MX and EMX upon chlorination (Table 1).

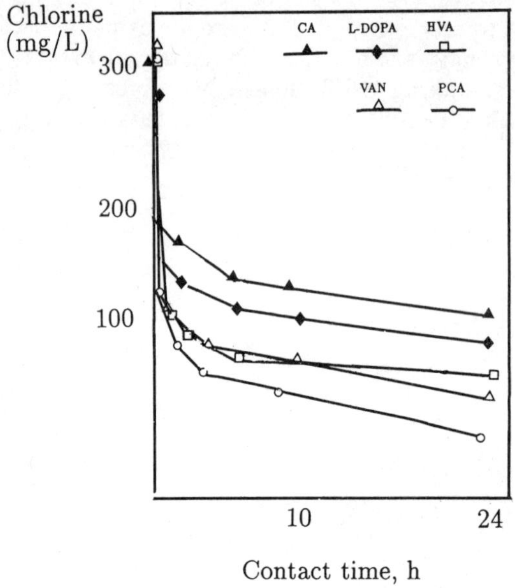

Figure 3 The chlorine consumption of 50 mg/l solutions of pure compounds. CA=caffeic acid; HVA=homovanillic acid; L-DOPA=3,4-dihydroxyphenylalanine; VAN=vanillin; PCA=protocatechualdehyde.

Table 1 Formation of MX and EMX from chlorinated phenolic compounds. Percentages in brackets express the sum of contributions of MX and EMX to the total mutagenity.

Compound	Molar ratios				Mutagenicity
	CL_2/model compd.	mmol MX/ mol model compd.	mmol EMX/ mol model compd.	MX/EMX	rev/nmol model compd.
Caffeic acid (CA)	8.9	1.11	0.04	27	7.0 (89%)
L-DOPA	9.4	1.34	0.04	31	7.7 (98%)
Homo-vanilic acid (HVA)	11.6	0.84	0.35	2.4	
Vanilin (VAN)	10.6	0.08	0.07	1.2	0.5
Proto-catechu-aldehyde (PCA)	10.3	0.12	0.33	0.37	1.1 (78%)

CA, L-DOPA and HVA had the highest potential for the formation of MX, whereas the highest amounts of EMX were obtained from PCA. On a molar basis, MX and EMXwere produced at trace levels. At most, the yields were 1.34 mmol/mol of MX (L-DOPA) and 0.33 mmol/mol of EMX (PCA). The molar ratio of MX to EMX ranged from 31 for L-DOPA to 0.4 for PCA.

The extracts of chlorinated CA and L-DOPA generated 7.0 and 7.7 rev/nmol, respectively. This is at least 7 times more mutagenic activity than the amount generated by the extracts of PCA and VAN.

The calculations of the activity contribution of MX and EMX showed that together the compounds account for 78% or more of the observed mutagenicity.

Discussion

Several research groups have studied the reaction products formed by chlorination of phenols as well as of various amino acids [10, 12-14]. Most of these studies have focused on the major reaction products formed and not on major mutagens. Rapson *et al.* [15] chlorinated a great number of phenolic model compounds and showed that many of them form mutagenic compounds detected in the Ames-test.

In the present study we have shown that the two mutagens MX and EMX, which seem to be ubiquitously present in chlorine disinfected drinking water, are formed as by-products of aqueous chlorination of various phenols under acid conditions. We chose acid conditions as we were interested in studying the potential of these phenols to form MX and EMX. Although the acid chlorination conditions should favor the formation and stability of MX, we found that the compounds were formed in trace amounts only. On a molar basis the yield of MX was at most 0.14% and of EMX 0.4%. Because of this, and because of the enormous amount of other by-products formed, we had to carry out the analyses with the highly selective method of GC/MS in the SIM mode. In comparison to the amount of MX produced upon acid (pH 3) chlorination of water with a high content of humic material (TOC=20 mg/l), CA, L-DOPA and HVA produced 15-20 times more MX on a TOC basis [16].

The molar ratios of MX to EMX (Table 1) show the relative formation of MX contra EMX. CA and L-DOPA form mainly MX, whereas PCA forms mainly EMX. Thus it seems that some phenolic subunits of the humic macromolecule account for MX formation, whereas others account for EMX formation.

Our results can be compared to those of Horth *et al.* [9] for chlorination of the amino acid tyrosine. Tyrosine is an amino acid with the same structure as L-DOPA, except for the lack of the second hydroxy group in the three position of the aromatic ring. Horth *et al.* found that 0.02 mmole of MX and 0.17 mmole of EMX were produced from one mole of tyrosine. We found that much higher amounts of MX (1.34 mmol/mol) and lower amounts of EMX (0.04 mmol/mol) were formed by the chlorination of L-DOPA. This discrepancy may be explained by differences in chlorination pH and by the structural differences of the two compounds. Horth and coworkers performed the chlorination at pH 6.2, a pH at which MX slowly isomerizes to EMX (and partly undergoes degradation) [17]. The second hydroxy group in the aromatic ring of L-DOPA may render the compound more susceptible to chlorine attack and thereby create a greater potential for MX production.

The highest mutagenicity was seen for CA and L-DOPA. These phenols also have the greatest potential for MX formation. The calculated mutagenicity contribution of MX and EMX show that MX, in particular, accounts for the major portion of the observed mutagenicity, and other mutagens are thus of minor importance (Table 1). However, the mutagenicity reported is based on the activity observed at the lowest dose and not on the slope of a linear dose response curve, which make the values only approximative.

Conclusions

In conclusion, our work shows that MX and EMX are formed in trace amounts when phenolic compounds are chlorinated. This finding supports the assumption that phenolic subunits of the humic macromolecule are responsible for the presence of MX and EMX in chlorinated drinking water. The observation that the molar ratio of MX/EMX clearly varies for the different phenols, indicates that the compounds are mainly produced from differnt phenolic subunits. Most of the mutagenicity generated by chlorination of the phenols is derived from MX.

Acknowledgement

Maa ja vesi tekniikan tuki r.y. is thanked for financial support.

References

1. Meier, J.R. Mutat. Res. **196:**211 (1988).

2. Kronberg, L. and T. Vartiainen. Mutat. Res. **206**:177 (1988).

3. Meier, J.R., R.D. Lingg and R.J. Bull. Mutat. Res. **118:**25 (1983).

4. Hemming, J., B. Holmbom, M. Reunanen and L. Kronberg. Chemosphere **15:**549 (1986).

5. Meier, J.R., W.F. Blazak and R.B. Knohl. Environ. Molecular Mutagenesis **10:**411 (1987).

6. Horth, H. Aqua **38:**80 (1989).

7. Backlund, P., E. Wondergem, K. Voogd and A. de Jong. Sci. Tot. Environ., in press.

8. Långvik, V. Suomen Akatemian Julkaisuja **5**:130 (1989).

9. Holmbom, B., L. Kronberg, P. Backlund, V-A. Långvik, J. Hemming, M. Reunanen and L. Tikkanen. Presented at the 6th Conference on Water Chlorination. Environmental Impact on and Health Effects. May 3-8 1987, Oak Ridge.

10. Horth H., M. Fielding, H.A. James, M.J. Thomas, T. Gibson and P. Wilcox. Presented at the 6th Conference on Water Chlorination. Environmental Impact and Health Effects. May 3-8 1987, Oak Ridge.

11. Ames, B.N., J. McCann and E. Yamasaki. Mutat Res. **31**:347 (1975).

12. Norwood, D.L., J.D. Johnson, R.F. Christman, J.R. Hass and M.J. Bobenrieth. Environ. Sci. Technol. **14**:187 (1980).

13. Boyce, S.D. and J.F. Hornig. Environ. Sci. Technol. **17**:202 (1983).

14. Trehy, M.L., R.A. Yost and C.J. Miles. Environ. Sci. Technol. **20**:1117 (1986).

15. Rapson, W.H., M.A. Nazar and V.V. Butsky. Environ. Contam. Toxicol. **24**:590 (1980).

16. Kronberg, L. (personal communication).

17. Holmbom, B., A. Smeds and L. Kronberg. Chemosphere, in press.

Iodinated Humic Acids

Jesper V. Christiansen and Lars Carlsen*
Chemistry Department, Risø National Laboratory,
DK-4000 Roskilde, Denmark

Abstract

Humic acids are iodinated by elemental iodine and, if the iodine is present as iodide, by peroxidase-mediated reactions. It is demonstrated that iodination of humic acids leads to a product with a uniform distribution of iodine. It could not be unambiguously verified whether the enzymatically mediated iodination is a direct reaction between a peroxidase-iodine complex and the humic acid molecule or a two-step reaction in which the enzyme creates elemental iodine, which consecutively reacts with the humic acid. Based on a simple model of a reaction between sites in the humic acids available for iodination and the electrophilic iodinating species, it was concluded that the reaction should be described as an equilibrium with a logarithmic equilibrium constant of approximately 4. The number of sites available for iodination was, in the humic acids studied, determined to be approximately $4x10^{-4}$ per gram humic acid. The different parameters influencing the enzymatically controlled iodination of humic acids are discussed.

Introduction

During the past two decades, the increasing concern about the possible transport of long-lived radionuclides with groundwater has drawn attention to the presence of organoiodine compounds in the terrestrial environment [1]. Strong evidence of microbial formation of organic iodine compounds has been presented [2-4], whereas direct reactions between inorganic iodide and soil organic matter have been considered less plausible.

In a recent study [5] we demonstrated the enzymatically controlled iodination of phenol, using iodide/hydrogen-peroxide/lactoperoxidase as the iodinating reagent. Owing to the polyphenolic nature of humic acids, incorporation of iodine into humic substances by peroxidase-mediated reactions appears reasonable. The present paper summarizes our studies on iodination of commercially available humic acids, that use lactoperoxidase/hydrogen peroxide as the enzymatic catalyst.

---

* Corresponding author

Materials and Methods

Chemicals

The chemicals and their sources were as follows: humic acid (Aldrich); H_2O_2 (J.T. Baker Chemicals); NaI (Merck p.a.); ^{131}I as NaI; lactoperoxidase (EC.1.11.1.7.) (Sigma L-2005), acetate buffer pH 5 and NaOH (Merck p.a.); $NaHSO_3$ (J.T. Baker Chemicals); and HCl (FERAC zur analyse).

Verification of the Existence of Iodinated Humic Acids

In a final volume of 5 ml 0.05 M acetatebuffer pH 5, the concentration of iodide (spiked with $^{131}I^-$) was $5x10^{-5}$ M and of humic acid 0.1 g/l. The reaction between iodide and humic acid was initiated by addition of lactoparoxidase (appr. 5 units) and hydrogen peroxide corresponding to a final concentration of $2x10^{-4}$ M. The reaction was allowed to proceed for 30 min. The reaction mixture was analyzed directly by liquid chromatography (Column: 250x4.6 mm Ultrahydrogel™. Eluent: 0.1 M sodium acetate, pH adjusted to 9.6 with sodium hydroxide. Flow rate: 0.6 ml/min.).

Incorporation of Iodine in Humic Acids

The amount of iodine incorporated in humic acids was determined as a function of the initial concentrations of iodide, enzyme, humic acids and hydrogen peroxide, respectively. The effect of variation in reaction time was studied. In general the following procedure was used. In a final volume of 5 ml (acetate buffer pH 5), iodide (spiked with $^{131}I^-$) and humic acids were mixed. The reaction was initiated by addition of the enzyme and hydrogen peroxide. The exact concentrations of the chemicals are listed in the text of the figures corresponding to the individual experiments. After the allowed reaction time (typically 10 min.) 1 ml of $NaHSO_3$ (0.1 M) was added to terminate the reaction. The reaction mixture was transferred to a clean vessel in order to eliminate iodine adsorbed to the internal surface of the original reaction vessel, and 1 ml of HCl (12 M) was added. The reaction mixture was left undisturbed for 5 min. and subsequently centrifuged (3000 rpm) for 25 min. Four ml of the supernatant was subjected to gamma-counting. Two ml of NaOH (2 M) was added to the vessel containing the remaining supernatant and the precipitated humic acid to dissolve the humic material; the resulting solution was subsequently subjected to gamma-counting. The distribution of iodine in humic acid and in the solution was then calculated based on the mutual counting rates.

The inhibiting effect of humic acid on the enzyme function was investigated allowing contact between lactoperoxidase and humic acid before initiating the iodination reaction by adding hydrogen peroxide.

The equilibrium was investigated allowing the reaction to proceed for 20 hours while applying varying initial concentrations of iodide in the range from $2x10^{-5}$ to $2x10^{-4}$ M. The distribution between bound and free iodine was carried out as described above.

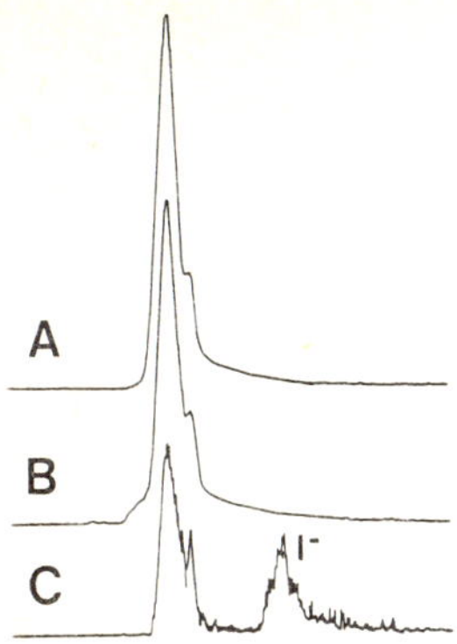

Figure 1 Chromatographic trace of A: humic acid prior to iodination (UV detection), B: humic acid after iodination (UV detection), C: humic acid after iodination (^{131}I detection). Initial concentrations: humic acid, 0.1 g/l; iodide, 30 µM; hydrogen peroxide, 200 µM; enzyme, 10 µg/ml. Reaction time: 20 min.

Iodination of Humic Acids

In the presence of lactoperoxidase (LP) and hydrogen peroxide, humic acids (HA) dissolved in an acetate buffer (pH 5) reacted readily with iodide. Analysis of the HA by liquid chromatography (Fig. 1) demonstrated that iodine was incorporated into all molecular weight fractions of the HA. Hence, the reaction products can be described as iodinated humic acids.

The fact that the iodine incorporation took place in all molecular weight fractions of the HA indicated that humic acids are composed of certain "basic units", the number of which determines the molecular weight of the single humic acid molecule. Model experiments with phenols indicated that phenolic sites would be available for iodination [5]. However, all types of phenolic units are not readily available for iodination. This was demonstrated in a series of model experiments with polysubstituted phenolic structures believed to reflect at least some of the phenolic moieties in HA [6,7].

It is important to note that HA may be iodinated by elemental iodine. By means of liquid chromatography (not shown) it was demonstrated that this reaction also leads to the same molecular weight distribution of the iodinated humic acids, as does the above described enzymatically controlled iodination. However, under the conditions normally prevailing in the environment, iodine will predominantly exist as anionic iodide [1]. Furthermore, in the concentration range investigated ($[I^-] \leq 2.5 \times 10^{-4}$ M), hydrogen peroxide ($[H_2O_2] \leq 2 \times 10^{-4}$ M) was not able to convert iodide into elemental iodine in the course of the reaction times studied. This shows that the incorporation of iodine into HA cannot be attributed to a simple chemical oxidation. The exact mechanism of the enzymatically controlled iodination is not known. The formation of iodinated HA can be a result of a reaction directly involving a lactoperoxidase-iodine complex or a reaction of elemental iodine generated during primary enzyme activity. Possibly, a composite reaction should be considered, in which the formation of elemental iodine operates initially, whereas the direct engagement of an enzyme-iodine complex prevails for low iodide concentrations, e.g. towards the end of the reaction.

In the following the influence of the different parameters on the iodination reaction shall be discussed.

Influence of Humic Acid Concentration

In the presence of excess iodide, the amount of incorporated iodine was expected to increase linearly with the humic acid concentration. However, the observed incorporation increased less than theoretically predicted. More precisely, the amount of incorporated iodine increased only from 8.6×10^{-6} to 3.5×10^{-5} M, whereas the initial HA concentration increased from 0.05 to 0.4 g/l. This tendency was even more pronounced when the initial humic acid concentration was increased to 1.0 g/l. The corresponding iodine incorporation was then found to be 3.4×10^{-5} M. All experiments were carried out with initial concentrations of iodide, hydrogen peroxide and lactoperoxidase equal to 1×10^{-4} M, 2×10^{-4} M and 10 μg/ml, respectively.

A reasonable explanation of the apparently decreasing incorporation of iodine with increasing humic acid concentration seems to be a deactivation of the enzyme function due to complexation between the humic acids and the enzyme [8]. In order to obtain support for this hypothesis, we carried out a series of experiments in which the humic acids were in contact with the enzyme for 1 h prior to the addition of iodide and hydrogen peroxide. Compared to the above experiments, a significant decrease in enzyme activity was observed in all cases. For initial concentrations of humic acids equal to 0.1, 0.2 and 0.4 g/l, the decrease in enzyme activity was determined to 7, 11 and 21%, respectively. We concluded that the concentration of "active enzyme" plays an important role.

Influence of Enzyme and Hydrogen Peroxide Concentrations

The incorporation of iodine into HA increased with increasing enzyme concentrations. For given initial concentrations of iodide, hydrogen peroxide and HA (3×10^{-5} M, 2×10^{-4} M and 0.1 g/l, respectively), the amount of incorporated iodine was found to be 1.6, 3.5, 7.1 and 8.2×10^{-5} M/g HA for enzyme concentrations equal to 2, 4, 10 and 20 μg/ml, respectively. However, further experiments showed that the hydrogen peroxide concentration also plays a dominant role in determining the enzyme activity. Increasing the initial hydrogen peroxide concentration from 5×10^{-5} to 8×10^{-4} M caused a 60% decrease in iodine incorporation. A priori, one would have expected an increased enzyme activity when increasing the hydrogen peroxide concentration, due to an increased rate of formation for the oxidized form of the peroxidase (LP_{ox}). However, increasing the hydrogen peroxide concentration above a certain limit apparently caused a deactivation, or possibly destruction, of the enzyme function. The results are summarized in Figs 2 and 3.

Influence of Iodide Concentration

As expected, the amount of iodine incorporated into the HA varied with the initial iodide concentration. In Figs 4 and 5 the amount of iodine consumed by the HA is related to the initial iodide concentration. The curves in Fig. 4 show a pronounced tendency

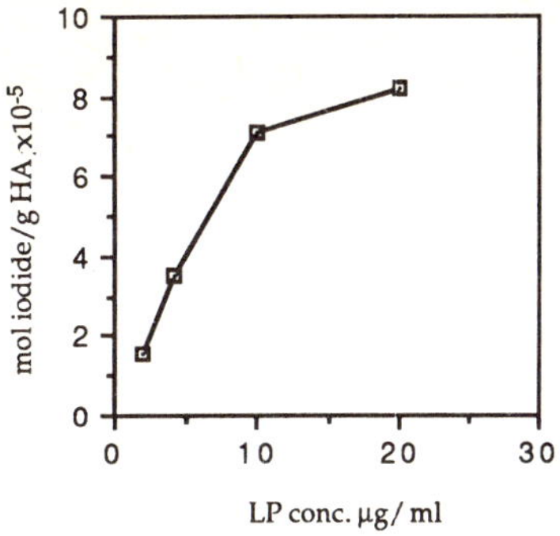

Figure 2 Iodine incorporation as a function of enzyme concentration (Initial concentrations: iodide, 30 µM; hydrogen peroxides, 200 µM; humic acid, 0.1 g/l).

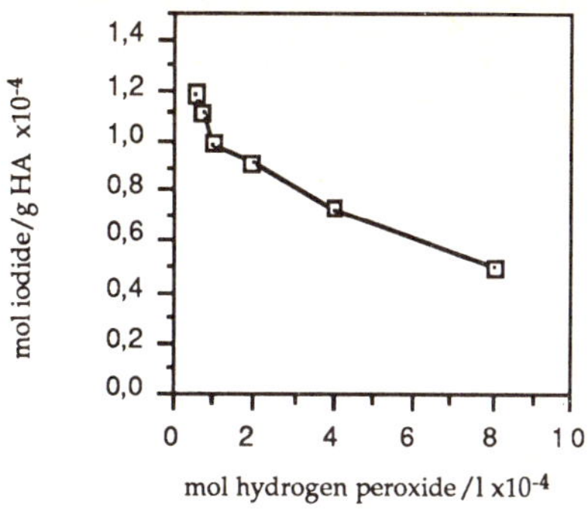

Figure 3 Iodine incorporation as a function of hydrogen peroxide concentration (Initial concentrations: iodide, 50 µM; enzyme, 10 µg/ml; humic acid, 0.1 g/l).

toward levelling-off phenomena. This effect is further elucidated by Fig. 5, which displays the percentage of the initial iodide being incorporated in the humic acids. The curves depicted in Fig. 5 correspond to the first derivative of the corresponding curves displayed in Fig. 4. Optimal iodide consumption apparently takes place for initial iodide concentrations around 2.5×10^{-5} M. The mechanistic aspects of these findings are outlined below.

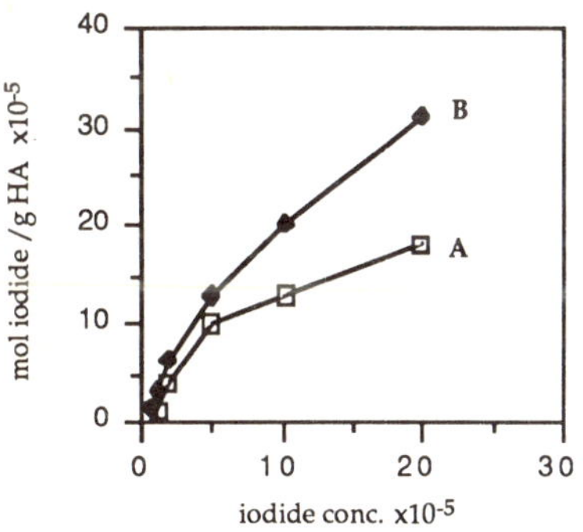

Figure 4 Iodine incorporation as a function of initial iodide concentration (Initial concentrations: hydrogen peroxide, 200 µM; enzyme, 10 µg/ml; humic acid, 0.1 g/l). Reaction time: A, 10 min.; B, 4 h.

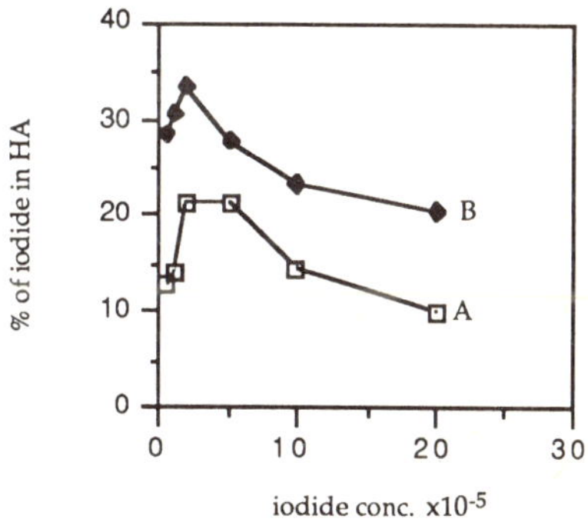

Figure 5 Percentage of iodine incorporated as a function of initial iodide concentration (Initial concentrations: hydrogen peroxide, 200 µM; enzyme, 10 µg/ml; humic acid, 0.1 g/l). Reaction time: A, 10 min.; B, 4h.

Influence of Reaction Time

As already noted from the above figures (Figs 4 and 5), the incorporation of iodine into humic acids increased with the contact time. However, in order to obtain further insights in the kinetics of the humic acid iodination reaction, we studied the reaction over a period of time up to 24 hours. In Fig. 6 the time profiles are visualized for initial iodide

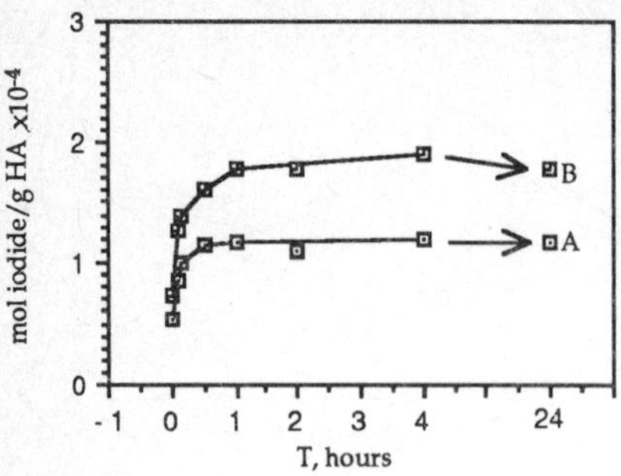

Figure 6 Incorporation of iodine into humic acid as a function of reaction time for initial iodide concentrations equal to 50 μM (A) and 100 μM (B).

concentrations equal to $5x10^{-5}$ and 10^{-4} M. It is immediately seen that the maximum incorporation of iodine is reached rather rapidly, and at approx. 4 h the final level of iodine consumption has virtually been reached.

Mechanism of Humic Acid Iodination

It is interesting to note that a prolonged reaction time did not lead to more than approximately 35-40% consumption of the initially applied iodide (cf. Fig. 5). This is in contrast to previous studies on the iodination of phenol, in which a virtually complete iodide consumption was obtained under reaction conditions comparable to those of the present study. The levelling-off phenomena observed for increasing initial iodide concentrations (Fig. 4) and the time profiles in Fig. 6 strongly suggest that the iodination of HA should be considered as an equilibrium reaction.

Owing to the above described findings, we have to consider a wide range of reactions which jointly may be responsible for the iodination of humic acids. These reactions comprise: oxidation of LP to LP_{ox} by hydrogen peroxide; reaction of LP_{ox} with I^-, leading to the iodinating enzyme iodine complex LP-I; and reaction of LP-I with iodide and/or humic acids to form elemental iodine and/or iodinated humic acids, respectively. As previously stated it has not been possible to elucidate to what extent the two latter reactions operate.

In an attempt to explain the above results, we propose a simple equilibrium model, taking into account only the concentration of sites available for iodination, [S], the concentration of the iodinating species, ["I"], and the concentration of iodinated sites, [SI].

$$S + \text{"I"} \rightleftharpoons SI$$

The initial concentration of the iodinating species are proposed to be equal to the initial iodide concentration, i.e. $[\text{"I"}]_0 = [I^-]_0$.

The proposed equilibrium system gives rise to the following expression, where β is the equilibrium constant.

$$\frac{[SI]}{[S]["I"]} = ß$$

This can be rewritten into

$$\frac{[I^-]_o}{1 + Q} = q\frac{[HA]}{Q} - \frac{1}{ß}$$

where Q is equal to the ratio between the concentrations of incorporated iodine and free iodide in solution, and q is the number of sites available for iodination per gram HA.

Obviously, a plot of $[I^-]_o/(1 + Q)$ vs. [HA]/Q should give a straight line, where the slope would be the number of sites available for iodination per gram humic acid, q. The equilibrium constant may be derived from the intercept $-1/\beta$.

The applicability of this rather simple model for the enzymatically controlled iodination of humic acids seems verfied by the plot displayed in Fig. 7. A linear relationship between $[I^-]_o/(1 + Q)$ and [HA]/Q was obtained for initial iodide concentrations $[I^-]_o$ ranging from $2x10^{-5}$ to $2x10^{-4}$ M. Based on a least-square procedure, the number of sites available for iodination was found to be $q = 4.28x10^{-4} \pm 0.22x10^{-4}$ sites per gram HA. The equilibrium constant was estimated to $\beta = 1.32x10^4 \pm 0.15\xi 10^4$.

A priori, the number of sites available for iodination may seem rather low. However, looking at proposed models for humic acid structures, it appears obvious that only few "free" aromatic hydrogens, i.e. potentially available sites, are present [9]. Furthermore, taking into account that a significant number of the potentially available sites must, in fact, be considered as non-available, probably due to an enzyme-inhibiting effect [7], the obtained number appears reasonable.

Finally, it can be mentioned that a logarithmic equilibrium constant in the range of 4 is in full agreement with an observed rather slow release of iodide from isolated iodinated humic acids [10].

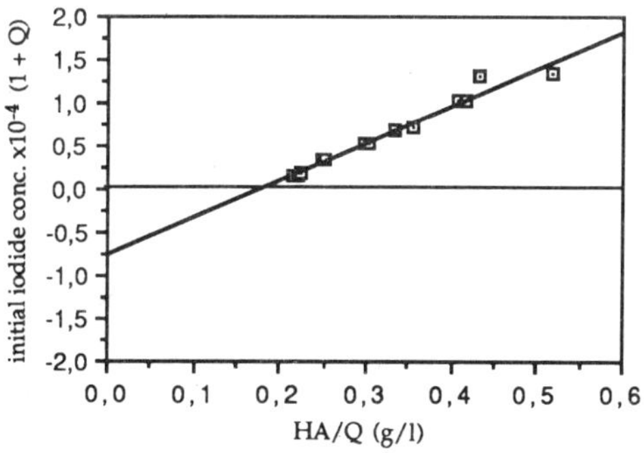

Figure 7 Plot of Initial iodide concentration/(1 + Q) as a function of HA/Q.

Conclusion

It has been demonstrated that iodinated humic acids can be obtained either by a direct reaction between humic acids and elemental iodine or with iodide in the presence of lactoperoxidase and hydrogen peroxide. The reaction, which is regarded as an electrophilic aromatic substitution, can, at least for a certain range of concentrations of reactants, satisfactorily be described as a simple equilibrium between humic acids and an electropositive iodinating species.

References

1. Christiansen, J.V. and L. Carlsen. Iodine in the environment revisited. An evaluation of the chemical and physico chemical processes possibly controlling the migration behaviour of iodine in the terrestrial environment, Risø-M-2791 (Roskilde: Risø National Laboratory, 1989).
2. Behrens, H. In: Environmental Migration of Long-Lived Radionuclides, pp.27-40 (Vienna: IAEA, 1982).
3. Behrens, H. In: R.A. Bulman and J.R. Cooper, Eds, Speciation of Fission and Activation products in the Environment (London: Elsevier, 1985).
4. Behrens, H. In: The Effects of Natural Organic Compounds and of Microorganisms on Radionuclide Transport (Paris: OECD/NEA, 1986).
5. Christiansen, J.V., A. Feldthus, and L. Carlsen. Risø-M-2850 (Risø National Laboratory, 1990).
6. Cooksey, R.C., E. Gaitan, R.H. Lindsay, J.B. Hill and K. Kelly. Org. Geochem. **8**: 77 (1985)
7. Christiansen, J.V. Ph.D. Thesis, Risø-M-2851 (Risø National Laboratory, 1990).
8. Serban, A. and A. Nissenbaum. Soil Biol. Biochem. **18**:41 (1986).
9. Grauer, R. Zur Koordinationschemie der Huminstoffe, PSI-Bericht Nr. 24 (Würenlingen: Paul Scherrer Institut, 1989), and references therein.
10. Christiansen, J.V. and L. Carlsen. Radiochim. Acta, in press.

Soil Peroxidase-Mediated Chlorination of Fulvic Acid

Gunilla Asplund[1], Hans Borén[1], Uno Carlsson[2] and Anders Grimvall[1]

[1] Department of Water and Environmental Studies
[2] Department of Chemistry
Linköping University, S-58183 Linköping, Sweden

Abstract

Humic matter has recently been shown to contain considerable quantities of naturally produced organohalogens. The present study investigated the possibility of a non-specific, enzymatically mediated halogenation of organic matter in soil. The results showed that, in the presence of chloride and hydrogen peroxide, the enzyme chloroperoxidase (CPO) from the fungus *Caldariomyces fumago* catalyzes chlorination of fulvic acid. At pH 2.5 - 6.0, the chlorine to fulvic acid ratio in the tested sample was elevated from 12 mg/g to approximately 40-50 mg/g. It was also shown that this reaction can take place at chloride and hydrogen peroxide concentrations found in the environment. An extract from spruce forest soil was shown to have a measurable chlorinating capacity. The activity of an extract of 0.5 kg soil corresponded to approximately 0.3 enzyme units, measured as CPO activity. Enzymatically mediated halogenation of humic substances may be one of the mechanisms explaining the widespread occurrence of adsorbable organic halogens (AOX) in soil and water.

Introduction

Natural production of organohalogens is generally assumed to be small, and group parameters such as adsorbable organic halogens (AOX) have been widely used as indicators of industrial pollution [1,2]. However, during the past few years, investigations of background levels of organohalogens in air, soil and water, have shown that the natural production is considerable. Harper [3] showed that chloromethane in the atmosphere is mainly of natural origin. More recently, Asplund and co-workers [4] showed that humic matter contains considerable pools of naturally produced organohalogens. Fulvic acids isolated from old (1300 - 5200 yrs) groundwaters contained measurable amounts of AOX. Furthermore, mass balance calculations for a raised bog showed that anthropogenic sources only accounted for a small fraction of the total pool of AOX in the bog.

A large number of organisms are known to produce halometabolites [5-7]. However, attempts to identify specific organohalogens in humic matter have thus far had limited success [8]. Furthermore, molecular-weight fractionations indicate that the halogens are almost uniformly distributed in humic matter [9]. This inspired the search for a non-specific, enzymatic halogenation mechanism in the environment.

Haloperoxidases have been thoroughly investigated, most extensively through studies on chloroperoxidase (EC 1.11.1.10 or CPO) from the fungus *Caldariomyces fumago* [10,11]. These studies show that haloperoxidase-mediated halogenation differs from most other enzymatic reactions by being nonspecific [7].

This study focused on two issues:

- Can CPO catalyze chlorination of fulvic acids?
- Can chloroperoxidases be extracted from soil?

Materials and Methods

CPO, *i.e.* chloroperoxidase (chloride hydrogen peroxide oxidoreductase: EC 1.11.1.10) from the fungus *Caldariomyces fumago*, was purchased from Sigma, in solution with 4945 enzyme units/ml. One enzyme unit is defined as the amount of CPO that catalyzes the conversion of 1.0 μmole of monochlorodimedone (MCD) to dichlorodimedone (DCD) per min under standard assay conditions [12]. DEAE-cellulose (Cellucol, Type SR, Biochemical Corporation) was used both for isolation of fulvic acid and purification of soil extract.

A Euroglas AOX analyzer Model 84/85 was used for the determination of organically bound halogens (AOX) in water and fulvic acid (FA). Analysis of MCD and its reaction products was performed with a Hewlett-Packard 5880 gas chromatograph, equipped with flame ionization detection (FID) and a fused silica column DB-1 (J&W) 0.25 μm (60 m x 0.32 mm i.d.). A Shimatzu QP 2000 GC/MS system was used to identify reaction products of DCD in water. Soil enzyme extracts were concentrated using an Amicon ultrafiltration stirred cell, Model 8400, equipped with a YM-10 filter, MW cut-off 10,000 (Amicon).

Sampling and Analytical Procedures

Chloroperoxidase (CPO) Mediated Chlorination of Fulvic Acid

FA was isolated from surface water using an anion-exchange resin, DEAE-cellulose (Cellucol, Type SF, Biochemical Corporation) [13]. Sampling took place in a small creek in a spruce forest 4 km N Värnamo in SW Sweden.

CPO (50 enzyme units) was added to FA (0.5 mg) dissolved in 20 ml potassium phosphate (0.1 M) with potassium chloride (0.1 M). Enzymatically mediated chlorination was then studied at pH 2.5, 3.0, 4.0, 5.0, 6.0 at 20°C. The chlorination reaction was started by adding hydrogen peroxide (0.1 ml, 40 mM). After 10 s, a 5 ml sample was withdrawn, and the reaction in this sample was quenched by adding a sodium thiosulphate solution (45 ml, 0.7 mM). Additional hydrogen peroxide (0.1 ml, 40 mM) was added to the enzyme-FA solution every ten minutes for two hours. After 2 h another 5 ml sample was withdrawn, and thiosulphate was added as above. Control experiments were performed with solutions that did not contain CPO but were otherwise treated as above.

The pH of the thiosulphate-FA samples was adjusted to 2.8, and the amount of organohalogens formed was determined by AOX measurements. The AOX-analysis was performed according to the DIN method [14] for AOX in surface water, with the exception that extra nitrate was added because of the high chloride concentration in the samples. The thiosulphate-FA solution (50 ml) was mixed with activated carbon (50 mg), HNO_3 (5 ml, 0.02 M) and $NaNO_3$ (0.8 g) in an 100 ml Erlenmeyer flask and shaken for 1 h. The mixture was then filtered through a 0.45 μm polycarbonate filter. Remainders of inorganic halide were removed by washing the filter with an acidic nitrate solution (0.02 M HNO_3, 0.2 M $NaNO_3$). The filter was incinerated at 1000°C and the halides thus formed were determined by microcoulometric titration with silver ions.

Chlorination of Fulvic Acid at Low Hydrogen Peroxide and Chloride Concentrations

CPO (100 enzyme units) was added to a mixture of FA (1 mg) and potassium chloride (0.1 mM) in a 50 ml sodium phosphate buffer (0.1 M, pH 3.5). The chlorination reaction was started by addition of hydrogen peroxide (0.1 ml, 40 mM). Aliquots of 5 ml were taken after 3, 7, 11, 16, 29, 60 and 120 s, and the reaction was immediately quenched by mixing the aliquots with a sodium thiosulphate solution (45 ml, 0.7 mM). The amount of AOX formed was determined as described above.

Extraction, Concentration and Detection of Soil Chloroperoxidase

Soil (about 10 kg) was collected in June in a spruce forest 4 km N Värnamo in SW Sweden. After removal of the litter layer in a 0.5 x 0.5 m square, the 5 cm thick organic layer was collected with a spade. To stimulate microbial activity in the soil, tap water was added to the soil, whereafter it was kept in a plastic bag at room temperature for one week. The soil was then sifted through an 8 mm sieve. Loss-on-ignition (600°C), pH_{KCl} (2.0 M KCl) and pH_{H2O} were determined.

Extraction of soil chloroperoxidase was carried out according to a standard peroxidase extraction procedure [15]. However, in order to optimize the recovery of chloroperoxidase, the soil extraction was performed at pH 3.6 instead of at pH 6.0; the isoelectric points of isoenzyme 1 and 2 in CPO are attained at pH 3.5 and 3.8, respectively [16]. The extraction solution was a 0.2 M sodium phosphate buffer (pH 3.6) with 1.0 M KCl added. Soil (5 kg) and buffer solution (5 l) were mixed in a plastic bucket and gently stirred for 15 minutes at 3°C. The soil-buffer mixture was then filtered through a Whatman filter no. 3. The pH of the extract (about 4 l) was adjusted to 4, and 100 mg DEAE cellulose was added. The suspension was stirred for 30 min and filtered through a Whatman filter no. 3. The filtrate was dialyzed (MW cut-off 12,000) over night against distilled water and was finally concentrated in an ultrafiltration cell to 300 ml. The whole filtration procedure was performed on an ice bath.

Detection of Chloroperoxidase Activity in Soil Enzyme Extracts

In the presence of chloride and hydrogen peroxide, CPO catalyzes the conversion of MCD to DCD. However, our studies showed that DCD is not stable in water, and the formation of degradation products is pH dependent. At pH 3.5 mainly one product is formed, which was tentatively identified by GC-MS analysis as 6,6-dichloro-3,3-dimethyl-5-oxo-hexanoic acid (DDHA) (Fig.1).

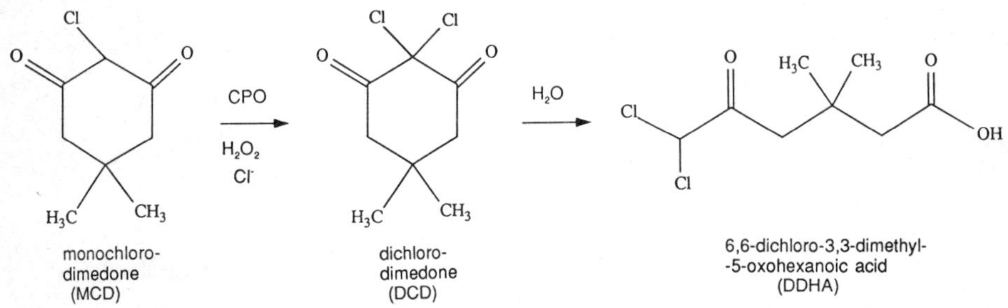

Figure 1 In the presence of chloride and hydrogen peroxide, chloroperoxidase catalyzes the conversion of monochlorodimedone (MCD) to dichlorodimedon (DCD). In water, DCD is then converted to a compound tentatively identified as 6,6-dichloro-3,3-dimethyl-5-oxo-hexanoic acid (DDHA).

CPO activity in soil extracts was estimated by measuring the amount of DDHA formed. A calibration curve of CPO activity was determined by chlorinating MCD with different doses of CPO. Enzyme (0.04 - 0.40 enzyme units) was added to 80 ml of a sodium phosphate buffer (0.1 M, pH 3.5) with KCl (0.1 M) and MCD (1 mg/ml). Hydrogen peroxide (0.1 ml, 40 mM) was added every 10 min for 2 h, and then the solution was left over night. Extraction of the solution was performed with 2 ml of dichloromethane, whereafter 10 µl of an internal standard, 1-chlorodecane (10 µg/µl), was added to 1 ml of the extract. Formation of DDHA was measured by gas chromatography.

The soil enzyme extract (30 ml) to be tested with respect to CPO activity was added to 50 ml sodium phosphate buffer (0.1 M, pH 3.5) with KCl (0.1 M) and MCD (1 mg/ml). The reaction was started by addition of hydrogen peroxide (0.2 ml, 20 mM). Equal additions were made every 10 min for 2 h, whereafter the solution was left over night.

Control experiments were performed both in the absence and in the presence of soil extract that had been heated to 70°C for 20 min. Determination of CPO activity was repeated at pH values from 2.5 to 5.0 (0.1 M sodium phosphate buffer) in an attempt to determine possible pH dependence.

Results

The experiments with CPO (chloroperoxidase) showed that, in the presence of chloride and hydrogen peroxide, this enzyme is able to chlorinate FA. At pH 2.5-6.0 the AOX-FA ratio increased from 12 mg/g to approximately 40-50 mg/g within 2 h. Further exposure of the FA to CPO, chloride and hydrogen peroxide did not increase the AOX content, indicating that the upper limit of enzymatic incorporation of chloride into the FA had been reached (Fig. 2). The AOX-FA ratios measured after 10 s indicated that the enzymatically mediated AOX incorporation into FA had a maximum reaction velocity at pH 3, which is in accordance with the pH optimum reported by Shaw and Hager [17].

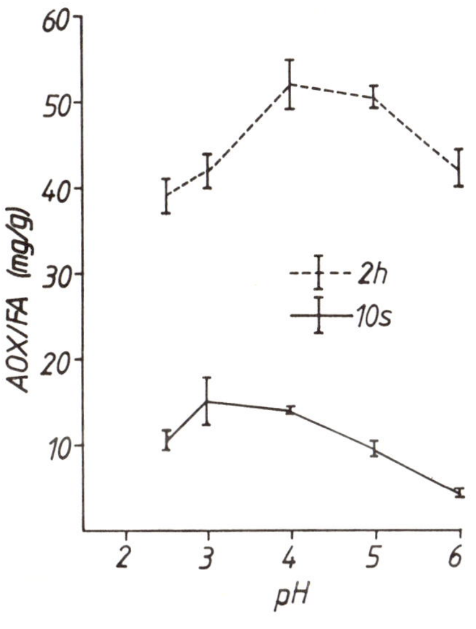

Figure 2 Net formation of AOX by CPO-mediated chlorination of fulvic acid (0.5 mg dissolved in 20 ml phosphate buffer) at different pH values. Chloride concentration 0.1 M. Hydrogen peroxide (0.1 ml, 40 mM) added every 10 min for 2 h. The curves show observed mean values ± s.d. (n=3).

The velocity of the enzymatically mediated chlorination of FA varied strongly with the potassium chloride and hydrogen peroxide concentrations. The curve in Fig. 3 shows that measurable amounts of AOX were formed, even at hydrogen peroxide and potassium chloride concentrations as low as 0.1 mM.

The soil used for extraction of enzymes had pH_{H2O} 3.8 and pH_{KCl} 2.9. Loss-on-ignition was approximately 75%. Chloroperoxidase activity is usually measured spectrophotometrically as loss of absorbance at 278 nm during transformation of MCD to DCD. Preliminary experiments showed that the chloroperoxidase activity of the soil extract was too low to be determined by this method. However, gas chromatographic analysis of one of the degradation products of DCD proved to be

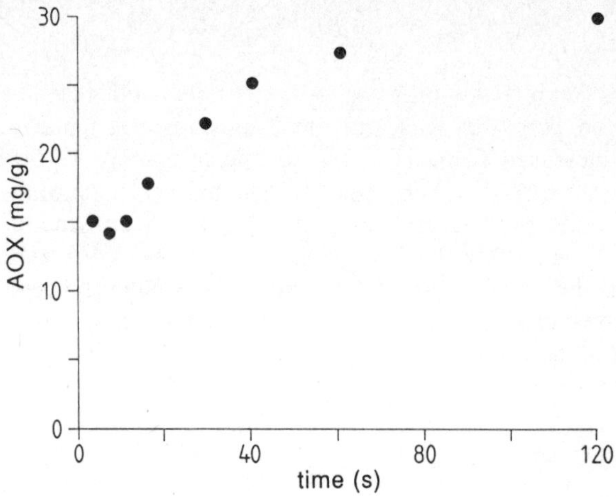

Figure 3 Net formation of AOX by CPO-mediated chlorination of fulvic acid in 0.1 mM potassium chloride and 0.1 mM hydrogen peroxide at pH 3.5 (0.1 M phosphate buffer).

sufficiently sensitive. Using this method, the activity of an enzyme extract from 0.5 kg soil was estimated to 0.3 enzyme units (Fig. 4).

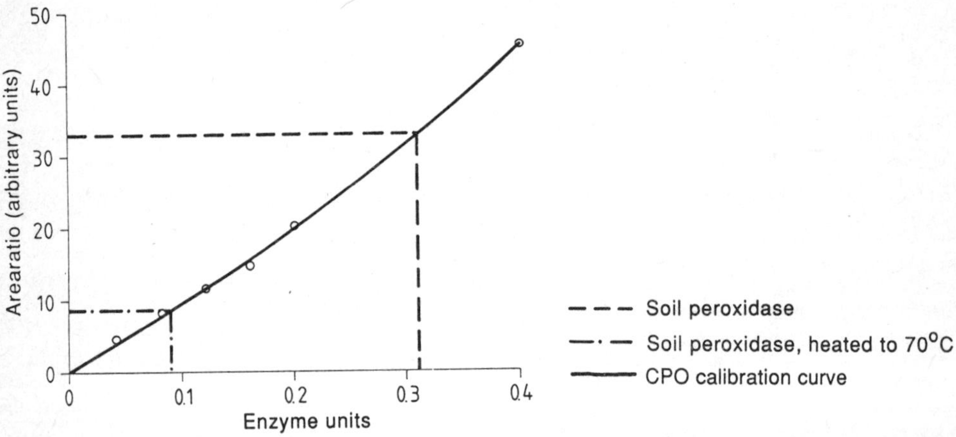

Figure 4 Estimation of CPO activity in 0.5 kg acidic spruce forest soil. The activity of soil extract and CPO was measured by gas chromatography as amount of DDHA formed. The experiment was performed in 0.1 M phosphate buffer at pH 3.5 with 0.1 M KCl.

Measurements of CPO activity of the soil extract at different pH values showed that the amount of DDHA formed peaked at pH 3.5 (Table 1). However, the velocity of the transformation of DCD to DDHA in water is also pH dependent. Therefore, activity measurements based on gas chromatographic analysis of DDHA are not suitable for the determination of the pH dependence of CPO activity.

Table 1 Amount of DDHA formed at pH values 3.0, 3.5, 4.0, 4.5 and 5.0 from a) MCD used as substrate for the soil extract and b) DCD dissolved in water. Amount of DDHA was measured by GC as the area ratio between DDHA and an internal standard (1-chlorodecane).

	pH 3.0	pH 3.5	pH 4.0	pH 4.5	pH 5.0
a) soil extract + MCD	0.14	0.33	0.16	0.11	0.03
b) DCD in H_2O	1.5	1.0	0.57	0.31	0.044

Discussion

It has previously been shown that several different organic compounds, such as ethylene, anisole and MCD, can be chlorinated by CPO [7]. The present study shows that chlorination of one of the major constituents of organic matter in the environment can be enzymatically catalyzed in the presence of potassium chloride and hydrogen peroxide. The maximum AOX-FA ratio obtained by enzymatically mediated chlorination was 40-50 mg/g which corresponds approximately to 1 chlorine atom in 30 carbon atoms.

Shaw and Hager [17] showed that the maximum velocity of CPO-mediated synthesis of chlorolevulinic acid was reached at 2 mM chloride and 1 mM hydrogen peroxide, i.e. at hydrogen peroxide concentrations much higher than those found in the environment. However, the experiments in this study showed that measurable amounts of chlorine were incorporated in FA at concentrations as low as 0.1 mM chloride and 0.1 mM hydrogen peroxide. Chloride concentration in that range is not uncommon in surface waters and precipitation in Sweden. Hydrogen peroxide concentrations up to 0.1 mM have been observed in precipitation in Long Island, NY [18]. This implies that chloroperoxidase-mediated chlorination of FA can take place at chloride and hydrogen peroxide concentrations that may occur in the environment.

The present experiments showed that soil extract may have a measurable chlorinating capacity. This strongly indicates that enzymatically mediated chlorination of fulvic acids occurs in soil. Nothing is known, however, about the efficiency of the extraction procedure or the enzyme activity needed to give a substantial contribution of AOX to the environment. Therefore, the importance of enzymatic halogention reactions in soil could not be quantified.

The sampling site was selected because of the high AOX-TOC ratio in the runoff from the area: 17 mg/g as compared to 2-4 mg/g for most surface waters in Sweden. Surveys of AOX in soil and water in Sweden have shown that samples from acidified areas in southwest Sweden have the highest AOX-TOC ratios [4,19]. Since CPO-mediated chlorination is favoured by a low pH, this is the geographic pattern that would be expected, if enzymatically mediated chlorination was the major source of AOX. One may therefore hypothesize that the occurrence of naturally produced organohalogens in humic matter is partly due to soil acidification induced by human activity.

Acknowledgements

We express our appreciation to Dr Jan Landin for offering valuable comments on the manuscript and to Lisbeth Samuelsson for making the drawings.

References

1. von Keller, M. Deutsche Gewasserkundliche Mitteilungen **31**:38 (1987).

2. Hoffmann, H.-J., G. Bühler-Neiens and D. Laschka, Vom Wasser **71**:125 (1988).

3. Harper, D.B. Nature **315**:55 (1985).

4. Asplund, G., A. Grimvall and C. Pettersson. Sci. Tot. Environ. **81/82**:239 (1989).

5. Siuda, J.F. and J.F. de Bernardis. Lloydia **36**:107 (1973).

6. Fenical, W. 1981. Natural Halogenated Organics. In: Duursma, E.K. and Dawson, R. Eds., Marine Organic Chemistry. Elsevier Oceanography Series 31, pp. 375-393 (Amsterdam: Elsevier 1981).

7. Neidleman, S.L. and J. Geigert. Ann. Proc. Phytochem. Soc. Eur. **26**:267 (1985).

8. de Lijser, H.J.P, C. Erkelens, A. Knol, W. Pool and E.W. de Leer. Natural organochlorine in humic soils. GC and GC/MS studies with soil pyrolysates. Lecture Notes in Earth Sciences, this volume.

9. Wigilius, B., B. Allard, H. Borén, and A. Grimvall. Chemosphere **17**:1985 (1988).

10. Hewson, W.D. and L.P. Hager. Peroxidases, catalases and chloroperoxidase. In: D. Dolphin, Ed., The Porfyrins, Vol VII (New York: Acad. Press 1979).

11. Neidleman, S.L. and J. Geigert. Biohalogenation. Principles, Basic Roles and Applications. (Chichester: Ellis Horwood Ltd. 1986).

12. Sigma Chemical Company. P.O. BOX 14508, St. Louis, MO 63178 U.S.A (1989).

13. Pettersson, C., I. Arsenie, J. Ephraim, H. Borén and B. Allard. Sci. Tot. Environ. **81/82**:287 1989).

14. DIN 38409, Teil 14, Summarische Wirkungs- und Stoffkenngrössen (Gruppe H). (Berlin: Beuth-Verlag, 1985)

15. Peterson, N.V. and K. Kuryrulak. Soviet Soil Science **14**:22 (1982) translated from Pochvovdeniye **5**:60 (1982).

16. Sae, S.W. Biological Halogenation - Chloroperoxidase. PhD Thesis, Kansas State University, USA (1969).

17. Shaw, P.D. and L.P. Hager. J. Biol. Biochem **236**:1626 (1961).

18. Lee, Y.-N., J. Shen and P.J. Klotz. Water Air Soil Poll. **30**:143 (1986).

19. Enell, M., L. Kaj and L. Wennberg. Long-distance distribution of halogenated organic compounds (AOX). In: H. Laikari, Ed., River Basin Managemant (Pergamon Press PLC 1989).

Natural Organochlorine in Humic Soils. GC and GC/MS Studies of Soil Pyrolysates

H.J. Peter de Lijser, Corrie Erkelens, Adri Knol, Wim Pool and Ed W.B. de Leer*
Delft University of Technology, Department of Analytical Chemistry, De Vries van Heystplantsoen 2, 2628 RZ Delft, The Netherlands

Abstract

Organic Halogen (OX) matter of natural origin in soils forms an integral part of the macromolecular humic materials and, therefore, it cannot be removed by simple washing and extracting processes. In this study pyrolysis at 400°C in a nitrogen atmosphere was used to break down the humic macromolecules to gas chromatographable compounds, which were collected in a cold trap system. According to coulorimetric analysis, 10-60% of the original soil OX was removed by the pyrolysis process but only a small part of the OX removed (2-40%) was collected in the cold trap system. The bulk of the OX was found in a precipitate that was formed between the furnace and the cold trap system. Analysis of the pyrolysates with GC/ECD indicated the presence of organohalogen compounds. GC/MS-NCI/SIM gave evidence for the presence of at least 20 different organochlorine compounds. Identification of individual organochlorine compounds with GC/MS and GC/MS-SIM was not successful, possibly due to the relatively low concentrations of these compounds and the overlap with other major non-chlorinated components. GC/MS-CNL indicated the presence of dichloropropane or chlorobenzene, and chlorophenol as of several unidentified chloroalkanes and/or chloroalkenes.

Introduction

The amount of naturally produced Organic Halogen (OX) compounds present in the environment is a subject of great interest. More than ten years ago, Siuda and DeBernardis [1] and Strunz [2] showed that there is a large variety of OX compounds both in the terrestrial and in the aquatic environment. It has also been shown that OX compounds can be naturally produced in very large quantities. The macroalgae have been implicated as an important source of volatile organobromine compounds released into the atmosphere (approximately 10^4 tons/year) [3]. Other natural sources, such as forest fires and decomposition of seaweeds, release up to 100 times more chloromethane than that factured by the chemical industry [4]. Recent studies in Sweden by Wigilius *et al.* [5] and Asplund *et al.* [6] showed that surface water, groundwater and soil far from industrial activities contain large quantities of OX. In addition, these studies showed that the observed back-

* Corresponding author. Present address: MT-NTO, Department of Analytical Chemistry, P.O. Box 217, 2600 AE Delft, The Netherlands

background concentrations could be attributed to halogens incorporated into humic substances. However, the mechanism of halogen incorporation is still a matter of discussion.

Furthermore, it is still unknown or uncertain whether humic materials contain organically bound chlorine or bromine. It may be suggested that after the production of OX compounds by a haloperoxidase-catalyzed reaction [7], another peroxidase-catalyzed reaction takes place, thus incorporating the OX compounds into humic substances [8,9]. A second possibility is the haloperoxidase-catalyzed substitution or addition of halide ions to humic constituents, as shown by Asplund *et al.* [10].

The almost uniform distribution of halogens in humic matter [5] indicates that the OX is an integral part of the macromolecular humic material. If this is the case, the OX cannot be removed by simple washing or extraction processes. Identification of individual OX compounds or structures is preferably performed by degrading the macromolecules into smaller (gas chromatographable) units.

We investigated pyrolysis of OX-containing soils of Dutch and Swedish origins as a method for this process. The pyrolysis products were analyzed with GC and GC/MS techniques, including Negative Chemical Ionization Mass Spectrometry with Selected Ion Monitoring (NCI/SIM) and Constant Neutral Loss (CNL) scans.

Materials and Methods

Ten different soils from Holland and Sweden were heated in a tube oven in a nitrogen atmosphere (flowrate = 1.5-2.0 ml/min) at 400°C (see Fig. 1). The pyrolysis products were collected during 45 min in a cold trap system consisting of two pentane filled tubes (30 ml). The first tube was kept at ambient temperature, while the second one was cooled to approximately -20°C. The combined pentane solutions were concentrated to 2 ml and analyzed by coulometry and several GC and GC/MS techniques.

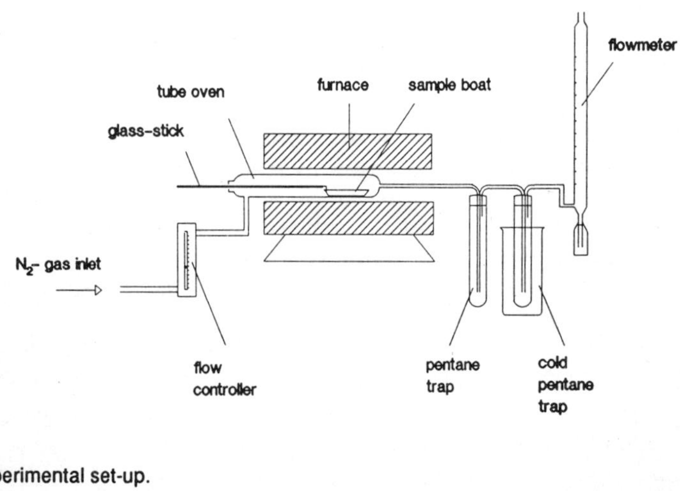

Figure 1 Experimental set-up.

Coulometry

An Euroglas micro-coulometer was used for the determination of OX in the soils and pyrolysates.

Gas Chromatography

GC/ECD (Electron Capture Detection) was carried out on a Packard Model 439 gas chromatograph equipped with a CP-SIL-5-CB column (25 m; i.d. 0.20 mm; film thickness 0.33 µm). The oven temperature was programmed from 50°C (3 min isothermal) with 8°C/min to 300°C (10 min final time).

GC/NPD (Nitrogen Phosphorous Detection) was carried out on a Chrompack Model 438A gas chromatograph equipped with a CP-SIL-19-CB column (50 m; i.d. 0.32 mm; film thickness 1.2 µm). The oven temperature was programmed from 50°C (5 min isothermal) with 8°C/min to 250°C (20 min final time).

GC/FPD (Flame Photometric Detection) was carried out on a Carlo Erba HRGC 5300 gas chromatograph equipped with a CP-SIL-5-CB column (25 m; i.d. 0.32 mm; film thickness 0.13 µm). The oven temperature was programmed from 50°C (3 min isothermal) with 8°C/min to 300°C (10 min final time).

GC/HECD (Hall Electrolytic Conductivity Detection) was carried out on a Hewlett Packard Model 5890A gas chromatograph equipped with a CP-SIL-8-CB column (25 m; i.d. 0.22 mm; film thickness 0.12 µm) and a Tracor Model 1000 HECD. The oven temperature was programmed from 50°C (3 min isothermal) with 8°C/min to 300°C (10 min final time).

In all cases 1 µl samples were injected in a Grob capillary injector (split mode, split ratio 1:20, injection temperature 280°C). Helium was used as the carrier gas, except in the case of GC/ECD where nitrogen was used.

Gas Chromatography/Mass Spectrometry

GC/MS (full scan) and GC/MS-SIM (Selected Ion Monitoring) were carried out on a HP 5890A gas chromatograph and a HP 5790B mass selective detector with a HP chemstation. The GC was equipped with a CP-SIL-5-CB column (25 m; i.d. 0.20 mm; film thickness 0.33 µm). The oven temperature was programmed from 50°C (3 min isothermal) with 8°C/min to 300°C (10 min final time).

GC/MS-NCI/SIM (Negative Chemical Ionization with Selected Ion Monitoring) was carried out on a Finnigan 3500 GC/MS combination at the National Institute of Public Health and Environmental Protection, Bilthoven, The Netherlands. The GC was equipped with a CP-SIL-5-CB column (25 m; i.d. 0.20 mm; film thickness 0.12 µm). The oven temperature was programmed from 50°C (3 min isothermal) with 8°C/min to 300°C (10 min final time).

GC/MS-CNL (Constant Neutral Loss scan) was carried out with a HP 5890A gas chromatograph coupled to a VG Analytical Model 70-250SE double focussing mass spectrometer. The GC was equipped with a CP-SIL-5-CB column (25 m; i.d. 0.32 mm;

film thickness 0.10 μm). The oven temperature was programmed from 40°C (2 min isothermal) with 4°C/min to 310°C (5.5 min final time).

Results and Discussion

To study the occurrence of OX in soils, samples were collected in several protected areas. Two samples (Sorghvliet and Overbosch) were collected in sandy areas (dunes) near the Northsea. The three other sampling sites consisted of peat (De Haeck), marsh forest (Ackerdijkse Plassen) and unmanured hayland (Blauwgrasland) areas. All but two samples were taken at 10-20 cm depth (Table 1). Swedish soils were a gift from G. Asplund (Linköping University, Sweden).

Table 1 OX concentrations (in μg Cl/g dry weight) of soils before and after pyrolysis and in the pentane extract.

Soil Sample	OX before pyrolysis	OX after pyrolysis	OX in pentane extract
Sorghvliet	30	20	3
Overbosch	30	20	1
Ackerdijkse Plassen	80	55	3
De Haeck	105	95	4
Blauwgrasland	245	190	4
Lyby Forest (S)	415	225	12
Lyby Bog (S)	140	55	9
Komosse (250-270 cm depth) (S)	775	410	7
Komosse (310-330 cm depth) (S)	300	240	6

The soils marked (S) originate from Sweden.

Coulometry

Before and after the pyrolysis experiments, the OX concentrations of the soils were measured with coulometry. Results are given in Table 1. The results indicate that only a small part of the original soil OX was collected in the pentane trap system, thereby making the identification of individual chlorine-containing compounds difficult. Part of the missing OX can be explained by the production of volatile compounds, such as HCl and CH_3Cl [11,12], which cannot be retained in the pentane traps. Furthermore, during the pyrolysis of soils high in OX, a tarry precipitate was formed between the oven and the first pentane trap. Coulometric analysis of the precipitates showed that they contained up to 90% of the missing OX. The tarry products could not be analyzed by GC.

All pentane concentrates were investigated with GC/FID and GC/ECD, producing similar chromatograms that differed in relative peak heights but not in composition. The highest signals were found for the Lyby Forest pyrolysate, which was chosen for further studies with specialized GC and GC/MS techniques. Only the presence of chlorinated compounds was investigated, since bromine is a minor halide in soil organic matter.

Gas Chromatography

GC/ECD (Fig. 2) indicated the presence of a large number of compounds with electron capting groups; these are potential halogenated compounds. However, these peaks could also be caused by compounds with, for example, nitro, sulphide or 1,2-diketo groups.

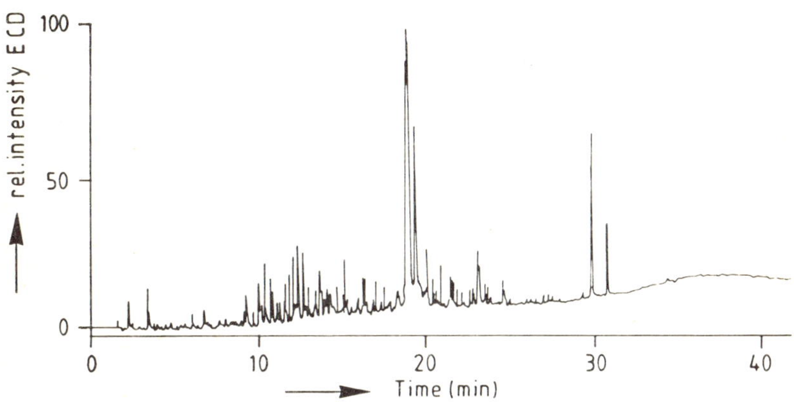

Figure 2 GC/ECD chromatogram of the Lyby Forest pyrolysate.

To find out if the pyrolysate contained nitro- and/or sulfur-containing compounds, GC/NPD and GC/FPD were used. Both chromatograms gave only minor peaks in the beginning of the chromatograms, thus ruling out nitrogen- or sulphur-based functional groups.

A more selective but less sensitive method, GC/HECD, was used to confirm the GC/ECD results. Although a slight positive signal was obtained, the sensitivity of the method was not good enough to confirm the presence of halogenated compounds.

Gas Chromatography/Mass Spectrometry

To give definitive evidence of the presence of chlorinated compounds in the pyrolysate, GC/MS-NCI/SIM was used. Literature studies showed that in favourable cases NCI with SIM can lead to sub-picogram detection levels [13]. We selectively detected the $^{35}Cl^-$ and the $^{37}Cl^-$ ions, which are produced by fragmentation reactions of organochlorine compounds. The two chromatograms thus obtained showed the presence of at least 20 different organic chlorine compounds (Fig. 3).The intensity ratio of 3:1 between the two chromatograms is in excellent agreement with the natural abundance of ^{35}Cl and ^{37}Cl.

For routine analyses GC/ECD can be used instead of the more complicated and more expensive GC/MS-NCI/SIM method. However, GC/ECD gives several artefact peaks, such as for high concentrations of substituted phenols. In addition GC/MS-NCI/SIM has

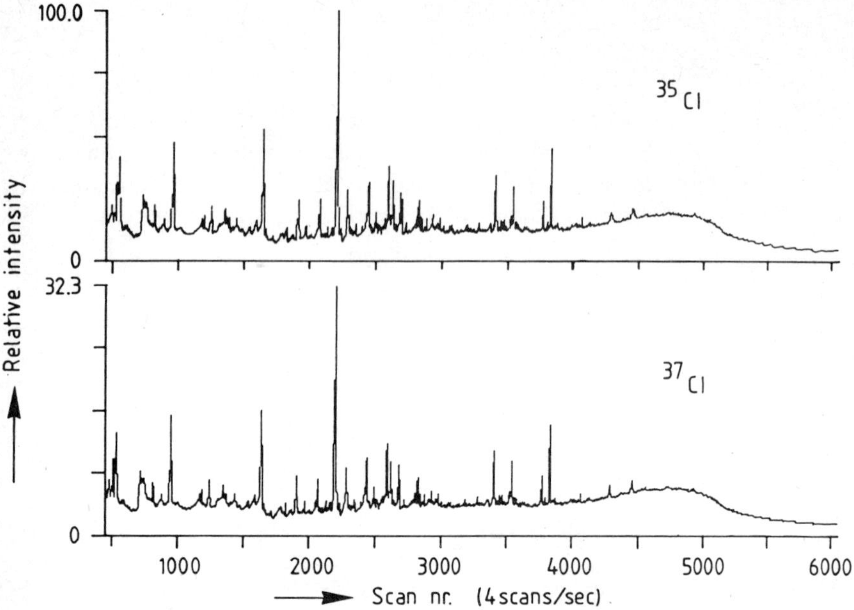

Figure 3 GC/MS-NCI/SIM chromatogram of the Lyby Forest pyrolysate.

proven to be a more sensitive method, which is able to differentiate between chlorine- and other halogen-containing compounds.

GC/MS (full scan; Fig. 4) was used for the identification of individual chlorine-containing compounds. Although structures could be assigned to more than 100 individual pyrolysis products no chlorinated compounds could be identified, probably due to their relatively low concentration and the overlap with other, more predominant compounds.

Van Loon *et al.* [14] showed that pyrolysis-GC/MS of chlorinated lignin isolated from pulp mill effluents gave several alkylated chlorophenols. This indicates that if halogenation of humic materials takes place by a peroxidase-mediated chlorine substitution of phenolic structures, it can be expected that chlorinated compounds as shown in Table 2 will be formed during the pyrolysis of these materials. GC/MS-SIM was used to detect these potential pyrolysis products. The intensity of the basepeak ($M^{+\cdot}$) and the ^{37}Cl-isotope peak ($M^{+\cdot}$+2) of the compound were recorded. A correct abundance ratio of 3 : 1 and a correct retention time were used as the criteria for a correct identification. Since none of the chromatograms matched the criteria, it must be concluded that the compounds shown in Table 2 are present below the detection limit.

The last method used was GC/MS-CNL. The CNL scan is a so-called linked-scan in which only fragment ions, i.e. ions that have lost a neutral fragment in the first field free region of the mass spectrometer, are detected. The electric sector and the magnetic sector of the high resolution mass spectrometer are scanned in a specific mode ($(B/E)x(1-E/E_0)$ = constant). In this linked-scan mode, the mass spectrometer loses its double focussing character. Hexachlorobenzene (HCB) was used to test the mass spectrometer in the CNL

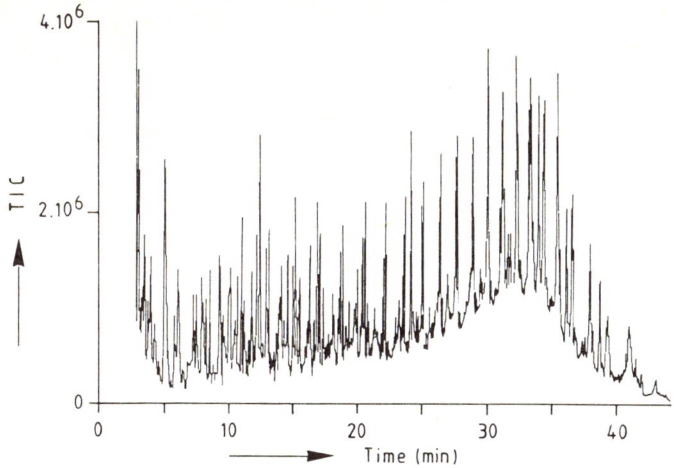

Figure 4 GC/MS chromatogram (full scan) of the Lyby Forest pyrolysate.

Tabel 2 Possible chlorine-containing products and their specific m/z- values after pyrolysis of humic materials.

Compound	m/z	Compound	m/z
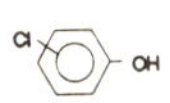	128 130	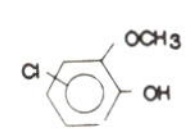	158 160
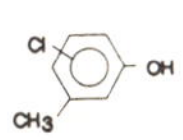	142 144	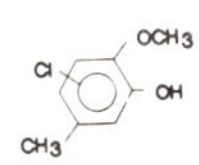	172 174

mode by measuring the output of the detector when the neutral mass loss was varied from 30 to 40 amu. The resolution in the CNL mode was rather low (Fig. 5), which means that during a 35 (Cl) GC/MS-CNL run ions resulting from neutral mass losses of 32 - 38 will also contribute positively to the total signal. The detection limit for HCB proved to be about 1 ppm when helium was used as a collision gas in the first field free region. Therefore, the GC/MS-CNL chromatograms of the soil pyrolysates contained many artefact peaks. We tried to improve the selectivity by doing seperate CNL-scans for neutral mass losses 35 and 37. Both chromatograms (Fig. 6) were compared, and only the mass spectra of those peaks with (almost) identical scan numbers and a correct intensity ratio were inspected. This procedure gave 17 compounds which showed chlorine-containing fragments at m/z 92, 104, 106, 112, 116, 126, 128, 132, 146, 161, 167 and 402. Chemical structures for these compounds can be given only very tentatively, since alternative interpretations are possible, and due to the potential of interferences by other compounds. Dichloropropane or chlorobenzene (Fig. 7a), several chlorinated alkanes and/or chlorinated alkenes (Fig. 7b) and chlorophenol were tentatively identified. For a

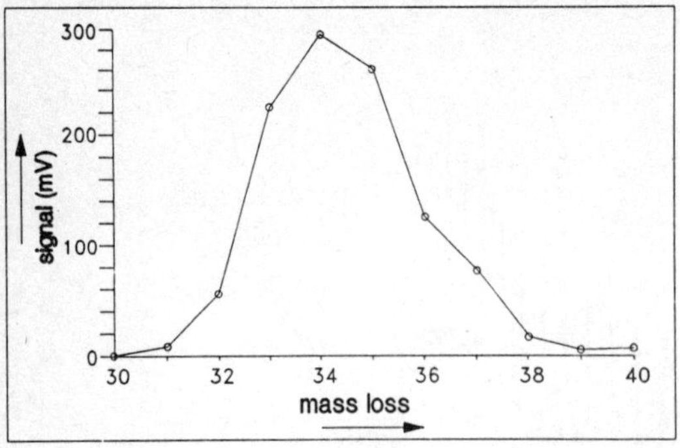

Figure 5 Resolution of the mass spectrometer in the linked-scan mode.

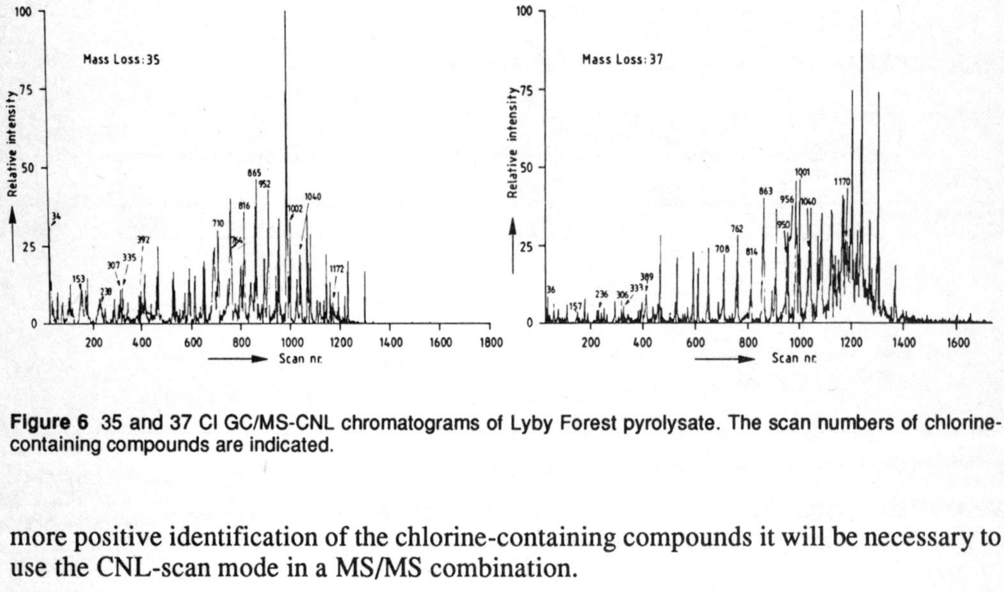

Figure 6 35 and 37 Cl GC/MS-CNL chromatograms of Lyby Forest pyrolysate. The scan numbers of chlorine-containing compounds are indicated.

more positive identification of the chlorine-containing compounds it will be necessary to use the CNL-scan mode in a MS/MS combination.

Conclusions

Pyrolysis of soils containing high amounts of AOX at 400°C in a nitrogen atmosphere results in a partial volatilization of the OX. However, most of the volatilized OX is not amenable for GC analysis.

GC/ECD and GC/MS-NCI/SIM gave definitive evidence for the presence of a large number of chlorinated compounds in the pyrolysate. However, their structural assignment with GC/MS was hampered by the great chromatographic overlap with other components. GC/MS-CNL was used as a more selective structural assignment method, which indicated that both chlorinated aromatic and aliphatic compounds are present. The

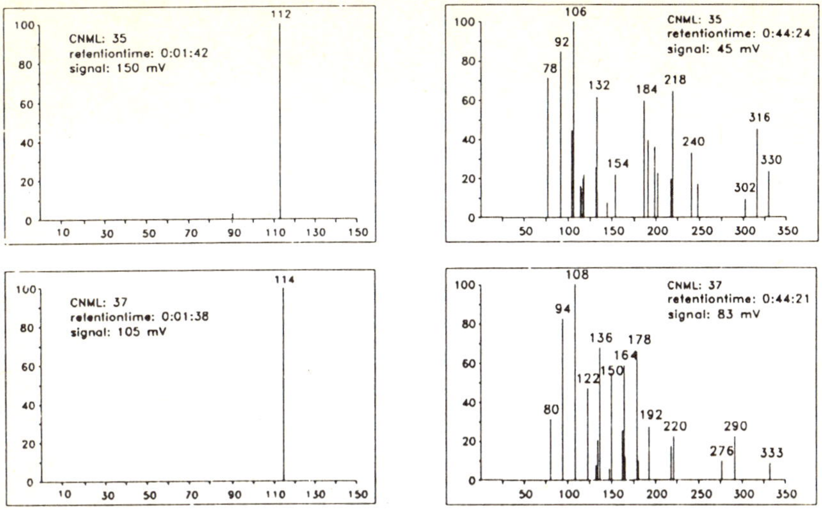

Figure 7 CNL-mass spectra of chlorine-containing compounds:**a**. Chlorobenzene or dichloropropane (M = 112); **b**. Chloroalkane or chloroalkene (fragments m/z 78/80, 92/94, 106/108).

high AOX concentrations in the soils probably result from a large number of different OX structures incorporated in the humic materials.

If the haloperoxidases are responsible for the high OX concentrations in humic soils, it is to be expected that a large number of different OX structures are produced, because the haloperoxidases can catalyze reactions with a large variety of substrates, such as alkanes, alkenes, aromatic compounds, nitrogen-containing compounds, sulfur-containing compounds, diketones etc. [7]. For more detailed studies it will be necessary to increase the yield of gas chromatographable products and to use more sensitive analytical methods such as GC/MS-NCI (full scan) and GC/MS/MS.

More insight in the structures of the OX could be obtained by separating the macromolecular humic substances into specific fractions, followed by pyrolysis. Pyrolysis at higher temperatures increased the amount of HCl formed and is therefore not suitable.

References

1 Siuda, J.F. and J.F. DeBernardis. Lloydia **36**: 107 (1973).

2 Strunz, G.M. In: A.I. Laskin and H.A. Lechevalier, Eds, CRC Handbook of Microbiology, Vol. III: Microbial Products, pp 415-443 (CRC Press Inc., 1973).

3 Gschwend, P.M., J.K. MacFarlane and K.A. Newman. Science **227**: 1033 (1985).

4 Leisinger, T. Experientia **39**: 1183 (1983).

5 Wigilius,B., B. Allard, H.Borén and A. Grimvall. Chemosphere **17**: 1985 (1988).

6 Asplund, G., A. Grimvall and C. Pettersson. Sci. Tot. Env. **81/82**:239 (1989).

7 Neidleman, S.L. and J. Geigert. Biohalogenation: Principles, Basic Roles and Applications (Ellis Horwood Limited Publishers Chichester, 1986).

8 Simmons, K.E., R.D. Minard and J-M. Bollag. Environ. Sci. Technol. **21**: 999 (1987).

9 Simmons, K.E., R.D. Minard and J-M. Bollag. Environ. Sci. Technol. **23**: 115 (1989).

10 Asplund, G., H. Borén, U. Carlsson and A. Grimvall. In: Humic Substances in the Aquatic and Terrestrial Environment (Linköping, 1990).

11 Saiz-Jimenez, C. Origin and Chemical Nature of Soil Organic Matter (Delft University Press, 1988).

12 Meent, D. van de. Particulate Organic Matter in the River Rhine Delta: Its Role in Trace Metal Scavenging (Delft University Press, 1982).

13 Jennings, K.R. Phil. Trans. R. Soc. Lond. **A293**: 125 (1979).

14 Loon, W.M., G.M. van, A. Pouwels, P. Veenendaal and J.J. Boon. Ultrafiltration and Pyrolysis Gas Chromatography Mass Spectrometry of Chlorolignins in Pulp Mill Effluent. Int. J. of Anal. Envir. Chem. Accepted for publication.

Organic Halogens in Danish Groundwaters

Christian Grøn
Institute for Applied Geology, Groundwater Research Centre,
Technical University of Denmark, DK-2800
Lyngby & Water Quality Institute, Agern Allé 11, DK-2870 Hørsholm, Denmark

Abstract

The occurrence and properties of halogenated organic compounds determined as adsorbable organic halogens (AOX) in Danish groundwaters were investigated. Preceded by a thorough evaluation of analytical procedures, a nation-wide survey of 145 wells was conducted. AOX was found in all wells but one, at levels from 1 µg Cl/l to 80 µg Cl/l. The occurrence of AOX was strongly correlated to the amount of organic carbon in the samples. Resampling of 6 wells and further characterization of these samples revealed, that the AOX consisted of organically bound chlorine, bromine and iodine of molecular weight distribution and polarity/pK_a properties resembling those of the bulk mass of dissolved organic carbon. Except for a few locations, where inorganic iodide may have biased the AOX determinations, there was no indication of analytical artefacts. The ubiquitous occurrence of AOX in Danish groundwaters is not likely to be caused by man-made pollutants, but is attributed to naturally occurring halogenated, humic compounds.

Introduction

Until recently, halogenated organic compounds found in the environment were almost exclusively assumed to be of anthropogenic origin. However, the research within natural products chemistry has demonstrated, that halogenated metabolites are produced by a diversity of organisms from bacteria to man [1]. The predominating species capable of producing halogenated organics are marine [2, 3], and the occurrence and potential impacts of marine, biogenic halogenated organics have lately received considerable attention [4, 5, 6]. Other studies have emphasized the occurrence of adsorbable organic halogens (AOX) of natural origin in surface water, groundwater and soil [7, 8].

It has long been recognized, that humic and fulvic acid preparations contain small amounts of chlorine [9]. The occurrence of organic chlorine and bromine in lake sediments has also been demonstrated [10, 11, 12]. Inorganic iodine occurring in oil and water from gas fields has been attributed to iodine associated with humic organic matter [13]. Chlorophenols have been found in sediments from unpolluted lakes [14] and in lake sediments deposited in the 13th century [12].

In many groundwater monitoring programmes and investigations of pollution incidents, the presence of organic halogens in water has been measured by group parameters.

The most commonly employed analytical methods differ primarily by the preconcentration step employed: adsorption of organics onto activated carbon (AOX), extraction into a non-polar solvent (EOX) or stripping of volatile organics (VOX). Normally, the concentration step is followed by combustion and coulometric detection of the hydrogen halides formed [6]. Whereas AOX covers most halogenated (X=Cl,Br,I) organics, the EOX expresses the content of non-polar haloorganics such as PCB's, PCA's, chlorophenols and many chloropesticides. VOX reflects the content of volatile haloorganics, such as chlorinated solvents and trihalomethanes.

AOX has frequently been found in seemingly unpolluted groundwaters [15-18], but has also been absent in some areas and in some investigations [17, 19]. Whether the AOX found in unpolluted groundwaters was due to the presence of halogenated organics of natural origin, whether it was a result of non-point source pollution, or whether it was merely a result of analytical artefacts has been subject to some controversy.

EOX has been detected occasionally in groundwaters and has mainly been attributed to pollution [20], though the occurrence of naturally originating EOX in some recently deposited marine strata has been suggested [21].

VOX has generally been found only in aquifers, where pollution, mostly with chlorinated solvents, has been verified [21]. In several groundwater monitoring programmes, chlorinated solvents have been detected in 5-15% of the sampled wells [22, 23].

This study was primarily aimed at verifying the presence and elucidating the nature of the AOX found in unpolluted groundwaters.

Experimental

Groundwater samples were taken from drinking water taps prior to all treatments or from monitoring wells. Only materials compatible with analysis of organic trace elements (teflon, glass and metals) were used. Samples were taken in 5 l glass flasks cleaned by heating to 450°C over night. Samples were stored at 1-5°C for no more than three days before organic chemical analysis. Analysis of labile chemical parameters, such as ammonia, was performed at local laboratories, generally on the day of sampling.

Chemicals used were of analytical grade, unless otherwise stated, and reagent grade water was obtained from a Milli-Q installation (Millipore Corp.). XAD-8 resin was purchased from Sigma (Rohm & Hass), and organic solvents were HPLC grade from Rathburn or LiChrosolv from Merck.

The determinations of AOX, EOX and VOX were performed employing methods developed or adapted to the analysis of unpolluted groundwater samples [24, 25]. Organic carbon was determined by UV-light/acidic persulphate oxidation at ambient temperature followed by IR detection utilizing a Dohrmann DC-80 TOC Analyzer. Humic samples were analysed by thermic oxidation, reduction of CO_2 to methane and flame ionization detection on a Dohrmann DC-52A TOC Analyzer, as the recovery of humics proved to be insufficient in the UV/persulphate method.

General groundwater chemical analyses were performed employing standard procedures [26, 27] on local laboratories. A flow injection method (Tecator equipment) was used for the determination of iodide [28], and bromide was analysed for by ion chromatography on a Dionex 4000i ion chromatograph [29].

The apparent molecular weight distribution of organic solutes in water samples was characterized by ultrafiltration of samples first filtered through a 0.45 μm membrane filter (cellulose acetate/nitrate, Millipore). Using an Amicon 2000C ultrafiltration unit, the ultrafiltration was performed as a cascade process [30, 31] with membranes of cut-off 10 000 D (Amicon YM10) and 500 D (Amicon YC05), respectively. In each step, ultrafiltration was terminated, when the volume had decreased to 10% of the original sample volume. Washing of the retenate was performed by the addition of potassium nitrate solution (0.5% KNO_3 aq., 10% of the original sample volume) followed by ultrafiltration to 10% of the original sample volume. This washing procedure was repeated three times.

The organics of the water samples were partitioned according to polarity and pK_a of acidic or basic functional groups by column chromatography on XAD-8 resin [32]. A 25 ml resin bed was packed in a Pharmacia SR10 column, and the column flow was established by means of a peristaltic pump (Ismatec) equipped with Tygon pumping tubes (Technicon, SMA). A sample volume of 1000 ml was employed. The partitioning procedure was carried out as described previosly [32], yielding fractions containing hydrophobic bases, hydrophobic acids and hydrophilic compounds. In order to assess the organic halogens of the hydrophobic, neutral fraction, the resin was Soxhlet extracted (thimbles (Macherey-Nagel) cleaned in advance by Soxhlet extraction with methanol) over night with methanol (50 ml). Subsequently, 50 ml pentane was added, and the organic phase was washed three times in a separatory funnel with 100 ml of a sodium sulphate solution (Na_2SO_4, cleaned by heating to 450°C over night, aq., 20 g/l). The extract was dried with sodium sulphate (sic., cleaned as described) and concentrated to 0.5 ml after the addition of 0.05 ml hexadecane. Quantification was done with automatic injection in an Euroglas combustion/coulometer unit applying standard conditions [25].

Chlorinated C_1- and C_2-compounds were determined by extration of the water samples with pentane followed by gas chromatography with electron capture detection.

Humic acid was prepared from soil of the B-horizon from Skarrild, a moor area in Western Jylland, by the IHSS procedure [33] and by a modified procedure employing nitric acid and potassium nitrate as reagents, whenever the IHSS procedure prescribes the use of hydrochloric acid and potassium chloride.

Neutron activation analyses were performed at the Isotope Laboratory, Risø, Denmark, on carbon from the adsorption process of the AOX procedure. Concomitant combustion/coulometric detection was applied to carbon from parallel experiments. The inorganic halide removal washing procedure of the AOX method was here extended with a reagent grade water washing step in order to remove nitrate from the carbon prior to irradiation.

Tritium analyses were performed by the Danish Isotope Center.

Results

In a nation-wide survey of organohalogens in groundwater, VOX was detected (>0.5 µg Cl/l) and verified by reanalysis in 10 of 145 wells. In seven of these wells, chlorinated solvents or trihalomethanes were detected, and a probable source of pollution was tentatively identified. In the remaining three wells, the initially observed VOX concentrations were close to the limit of detection, and no source of pollution could be identified.

EOX (>0.5 µg Cl/l) was detected in three of the 145 wells of the survey. In one case, this was attributed to pollution with trichloroethene (80-90 µg/l). In another case, the observed concentration was low (1.5 µg Cl/l) in the original sample, as well as after resampling (0.3 µg Cl/l, duplicate analyses). No source of groundwater pollution could be identified. In the third case, a humic water sample gave an observed EOX concentration of 12 µg Cl/l. However, this result could not be verified by resampling. Analytical artefacts might explain the original observation.

AOX (>1 µg Cl/l) was detected in all but one of the 145 wells. In wells not affected by known point sources, the concentrations were in the range 1-40 µg Cl/l. The geographical distribution of the AOX concentrations is illustrated in Fig. 1.

In Table 1, the wells of the groundwater survey have been classified according to AOX concentration and geology of the aquifer. All results obtained from wells suspected of contamination (VOX found and verified) and from wells of geology of even the slightest ambiguity are omitted from this classification, which is consequently covering 122 wells only.

Figure 1 The distribution of AOX in Danish aquifers. Results from a nation-wide survey of 145 wells.

Table 1 Frequency distribution of 122 Danish wells classified according to geology of the aquifers. Data from a nation-wide survey of 145 wells. Wells of ambiguous geology or suspected of being contaminated (confirmed detection of VOX) omitted in the table.

Geology of aquifer	Number of wells	Percentage of wells with AOX concentration (μg Cl/l)		
		< 15	15-30	>30
Postglacial, marine sand	8	63	25	13
Glacial sand and gravel	60	93	7	0
Miocene sand	18	78	0	22
Limestone from Selandien and Senon	36	97	3	0

The general groundwater chemistry of all wells and of wells with elevated AOX concentrations (>15 μg Cl/l) is illustrated by the Piper diagrams [34] in Fig. 2. Observed values of AOX and of general chemical parameters (pH, Na^+, K^+, Ca^{++}, Mg^{++}, NH_4+, total Fe and Mn, F^-, Cl^-, NO_3-, HCO_3-, SO_4--, total phosphorous, organic carbon) were analysed statistically. AOX was correlated primarily to the concentration of organic carbon (correlation coefficient, r=0.4885), and slightly to pH (r=-0.2797) and nitrate (r=0.2805). Comparing AOX in low chloride (< 100 mg Cl^-/l) and high chloride samples (≥ 100 mg Cl^-/l), no significant difference was observed.

In order to further elucidate the presence and origin of AOX in Danish groundwaters, 6 of the wells in the national survey were selected for resampling and further charac-

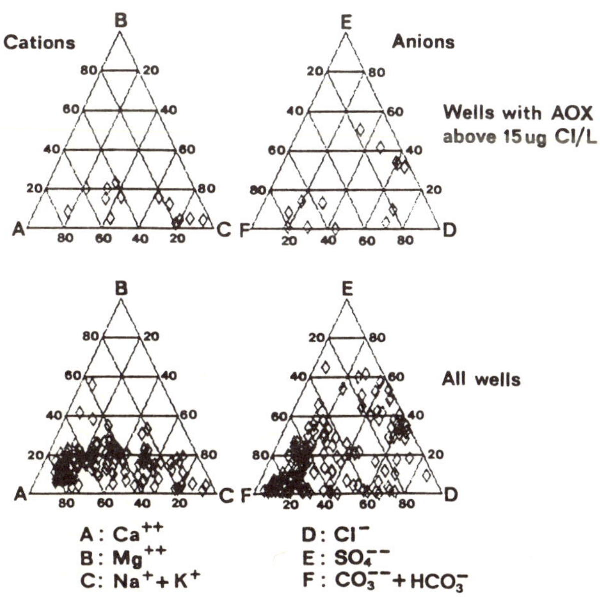

Figure 2 Piper diagrams of general groundwater chemistry of all wells and of wells with elevated AOX(>15 μg Cl/l).

Table 2 Data on geology, vulnerability and hydrochemistry for the 6 wells of the special study

Site	Fjand	Grindsted	Bramming	Læsø	Kalundborg	Skagen
DGU code	72.606	114.852	131.826	12.65	196.277	1.193
Geology	Deep, alternating limnic and marine Miocene deposits	Shallow, limnic Miocene sand deposits	Deep sand aquifer deposited by the interglacial Holstein Sea	Shallow marine sand from the postglacial period deposited by the Litorina Sea	Deep glacial sand aquifer with risk of sea water intrusion	Shallow marine sand aquifer deposited by the postglacial Litorina Sea; sea water intrusion possible
Percolation time (yrs)	37	0.8	41	1.6	49	1.3
Tritium (t.u.)	3±1	63±5	1±1	29±3	1±1	25±3
AOX (µg Cl/l)	82	6.5	32	19	13	32
NVOC (mg C/l)	33	0.83	32	3.4	5.6	11
Chloride (mg/l)	26	17	20	24	368	300
Bromide (mg/l)	0.11	0.047	<2	<2	2	1.2
Iodide (mg/l)	0.03	<0.02	0.03	<0.02	0.12	0.45
Hydrochemical type	Na^+ HCO_3^-	mixed	$Na^+HCO_3^-$	mixed	mixed	Na^+ Cl^-
Redox conditions	reduced with methane	oxic with nitrate	reduced with methane	reduced with sulphate reduction	reduced with sulphate reduction	mixed with ammonia, sulphate and methane

terization in a special study. Selected data on geology, vulnerability to pollution and hydrochemistry are given in Table 2.

AOX results obtained with two different detection methods: combustion/coulometric titration (AOX) and neutron activation analyses (NAA-AOX) are presented in Table 3. The fractions of AOX, that could be attributed to Cl, Br and I, respectively, are presented in Fig. 3.

The six samples were reanalysed with respect to AOX after addition of 5 g $NaNO_3$ per L sample and after addition of different amounts of inorganic chloride. The results of this study are shown in Table 4.

The apparent molecular weight distributions of AOX and non-volatile organic carbon (NVOC) in three different wells are presented in Fig. 4. In the same figure, the results of fractionations with respect to polarity/pK_a are also presented. It was assumed in calculating the molecular weight distribution on each fraction, that the AOX lost in each

Table 3 AOX in the 6 wells described in Table 2. AOX determined by combustion/coulometric titration (AOX) and by neutron activation analyses (NAA-AOX).

Site	AOX µmol/l	NAA-AOX µmol/l
Fjand	2.7	2.4
Grindsted	0.15	0.15
Bramming	1.1	0.95
Læsø	0.46	0.49
Kalundborg	0.82	1.4
Skagen	2.8	3.6
Control [a]	0.61	0.62

[a] 2,4,6-trichlorophenol

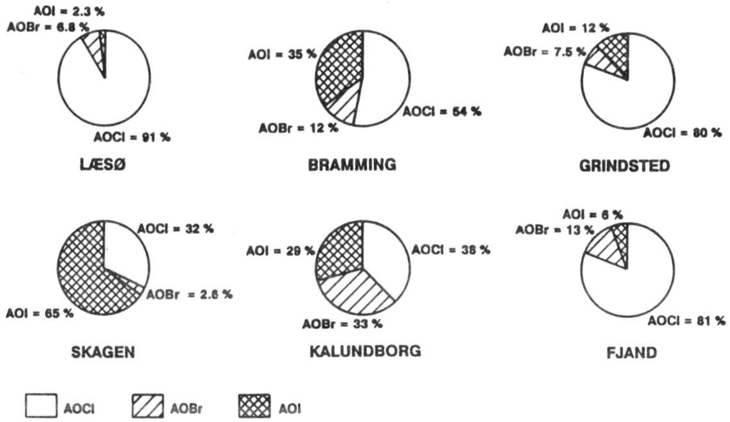

Figure 3 Fractions of AOCl, AOBr and AOI in the six wells listed in Table 2. Detection by neutron activation analysis (NAA).

Table 4 Observed AOX concentrations (µg Cl/l) before and after the addition of sodium nitrate and sodium chloride. Samples from the 6 wells listed in Table 2.

Sample treatment	Fjand	Grindsted	Bramming	Læsø	Kalundborg	Skagen
Original sample	82	6.5	32	19	13	32
5 g $NaNO_3$/l added	89	6.3	33	18	12	34
50 mg Cl^-/l added	86	-	-	19	-	-
250 mg Cl^-/l added	87	6.9	31	20	14	-
500 mg Cl^-/l added	-	-	-	-	14	31

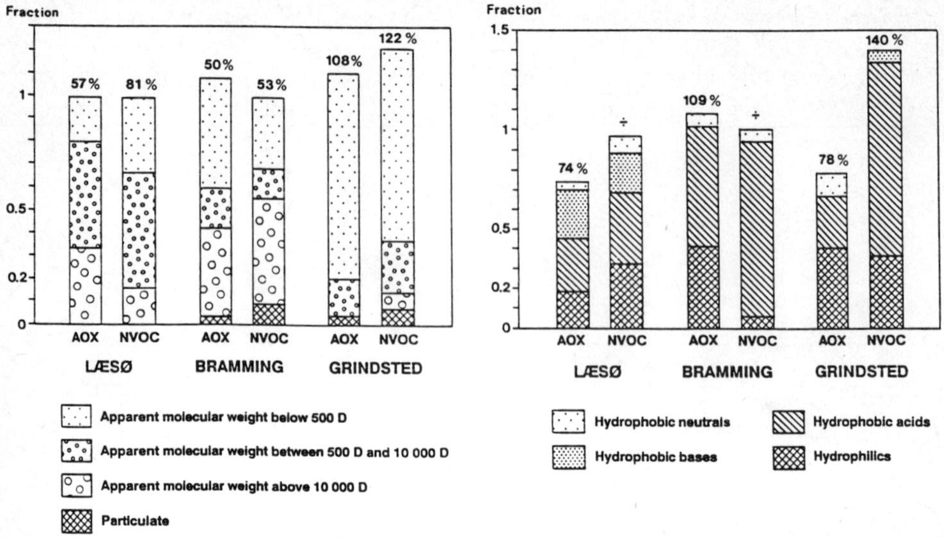

Figure 4 Relative distribution of AOX and NVOC on different molecular weight fractions and polarity/pKa-fractions. Data for the three wells are given in Table 2.

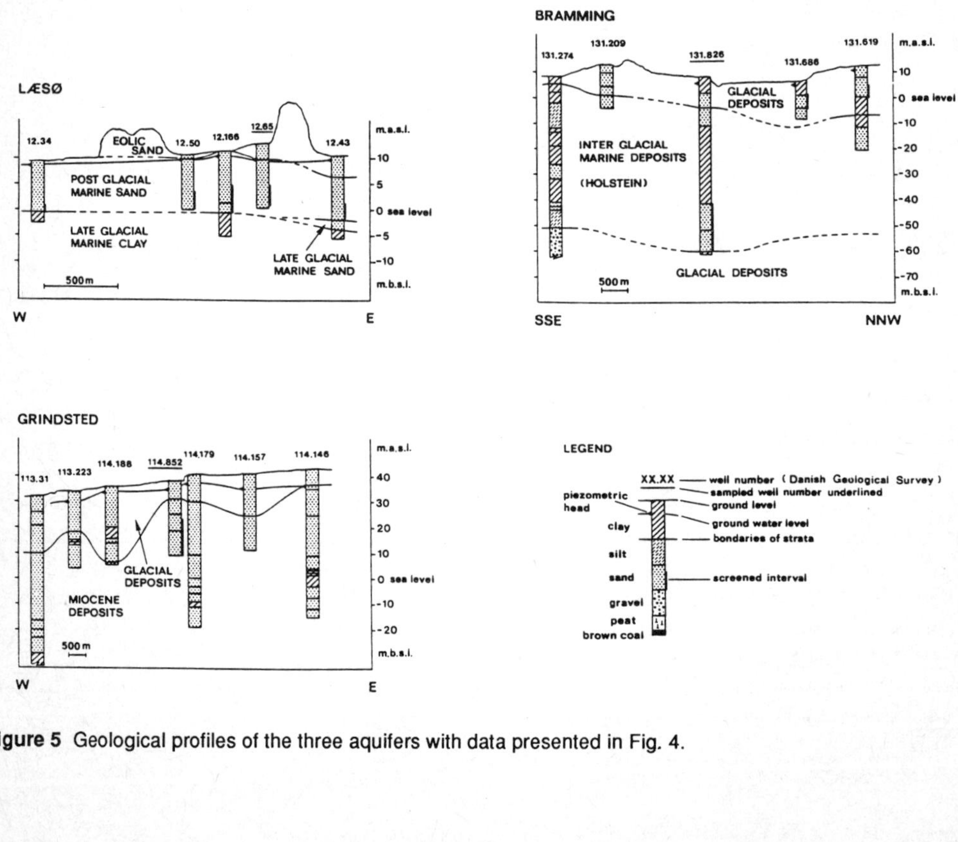

Figure 5 Geological profiles of the three aquifers with data presented in Fig. 4.

ultrafiltration step was due to precipitation of high molecular weight species in the retenate. The overall recoveries of the full fractionation procedures are given at the top of each column in Fig. 4, in order to enable an evaluation of the performance of the fractionation procedures. The geological profiles of the three studied wells are shown in Fig. 5.

The elementary compositions of soil humic acids prepared by the IHSS method and by the modified method are given in Table 5.

Table 5 Elementary composition of soil humic acids

	C %	H %	N %	S %	Cl %	Br %	I %
Humic acid, IHSS procedure	53.09	4.15	2.06	0.42	0.16	0.08	0.009
Humic acid, special procedure	54.33	4.15	2.15	0.47	0.15	0.08	0.01

Discussion

AOX was found in all aquifers (Fig. 1), whereas EOX and VOX occurred only occasionally. The concentration of AOX varied with the geology of the aquifers (Table 1). The general background level was 1-15 µg Cl/l. Elevated values (>15 µg Cl/l) were primarily found in postglacial, marine aquifers, and, occasionally, in shallow glacial and Miocene aquifers on the outwash plains of Western Jylland. Humic groundwaters from deep, confined, interglacial and Miocene reservoirs exhibited the highest concentrations of AOX (30-80 µg Cl/l). Limestone aquifers had low AOX concentrations (< 10 µg Cl/l) and a low content of organic carbon.

Statistical analysis of the data from the national survey showed, that the AOX content was primarily a function of the organic carbon content. Chloride concentrations were of less importance. In the special study, there was no indication of an interference from inorganic chloride, neither directly, nor via formation of complexes with humics (Table 4). Furthermore, the Piper diagrams of Fig. 2 did not support the hypothesis, that elevated AOX concentrations were related to sea water intrusion.

The inorganic bromide concentrations measured in the special study (Table 2) were all two orders of magnitude below the highest concentration acceptable to the AOX method [25]. Inorganic iodide interference seems to be a more serious problem. The maximum tolerable concentration of iodide is 0.06 mg/l [25], and this was exceeded in two samples of the special study (Kalundborg and Skagen). Furthermore, the fraction of AOI was high for these two samples (Fig. 3), and AOX obtained by combustion/coulometric detection (recovers 30% of adsorbed iodine [25]) was substantially lower (Table 3) than NAA-AOX (recovers all iodine adsorbed). Iodide contents of Danish groundwaters are generally low (< 0.05 mg/l) [35, 36], but elevated iodide (and both chloride and bromide) concentrations are to be expected in aquifers with sea water intrusion and near salt domes [37]. Consequently, AOX measured in groundwater samples can in part be an analytical artefact caused by inorganic halides only in such aquifers, where elevated iodide concentrations occur.

Experiments comparing AOX results obtained with the element specific NAA detection technique and with combustion/coulometric detection demonstrated (Table 3), that there was no significant contribution from elements other than the halogens (e.g. sulphur or nitrogen compounds) to the AOX measured in the special study.

The molecular weight distributions of AOX and NVOC differed between the samples, but AOX and NVOC corresponded closely within each sample (Fig. 4). The distribution of AOX and NVOC on polarity/pK_a-fractions differed from sample to sample, but the correspondence within each sample was reasonable. The trend in the distributions suggests, that AOX was associated with the bulk mass of organic matter, i.e. the humic compounds. The fact, that the prepared soil humic acids were indeed halogenated (Table 5), supports this suggestion. High and medium molecular weight, acidic compounds dominated the AOX and the NVOC of the Bramming sample (old groundwater, see below), and low to medium molecular weight compounds of more mixed functionalities dominated the AOX and the NVOC of the Læsø and the Grindsted samples (younger groundwaters).

From the geological profiles, from the percolation times calculated and from the tritium values obtained it is seen (Table 2 and Fig. 5), that the groundwaters of Læsø and Grindsted were quite young (formed later than 1960), whereas the groundwater of Bramming was older than 1960 and pressumably much older. The high AOX concentration of the Bramming sample can not be caused by point source pollution, as no potential source could been identified, and it cannot be due to non-point source pollution, as the intensive use of chlorinated solvents and pesticides did not occur until after 1960.

Conclusions

In a nation-wide survey of Danish groundwaters, halogenated organic compounds measured as AOX occurred ubiquituosly at levels from 1 to 80 µg Cl/l. The concentration differed with the geology of the aquifer, but depended only to a limited extent on the general groundwater chemistry. The highest AOX values were found in aquifers of marine, postglacial origin and in humic groundwaters.

At some locations, the AOX measured could, in part, be due to interference from inorganic iodide. There was no indication, that any other inorganic halide contributed to the measured AOX. Nitrogen and sulphur compounds did not bias the measured AOX values. Consequently, the measured AOX concentrations of Danish groundwaters do reflect an actual content of halogenated organic compounds and are not the results of analytical artefacts.

The AOX found is not likely to be caused by halogenated pollutants, but can rather be attributed to halogenated humic compounds formed by natural processes. Hypothetically, either halogenated organics produced by marine organisms at the time of sedimentation of the aquifer material, or haloorganics formed during the humification process could be suggested as the origin of the AOX found.

Acknowledgements

This work has been supported by grants from the Technical University of Denmark. I also wish to thank M. Sc. B. Raben-Lange of the Chemistry Department, the Royal Veterinary and Agricultural University of Denmark, for preparing humic acids, Ph. D. H. Holst and M. Sc. C. Bilbo of the Institute of Mathematical Statistics and Operational Research, Groundwater Research Centre, Technical University of Denmark, for performing the statistical analyses and M. Sc. S. Genders of the Danish Isotope Center for doing the tritium analyses. Furthermore, I am indebted to the Danish counties and the National Agency of Environmental Protection/the Danish Geological Survey for assistance in collecting the groundwater samples and for doing general chemical analyses on some samples from the monitoring programme.

References

1. Siuda, J.F. and J.F. DeBernardis. Lloydia Journal of Natural Products **36**:107 (1973).
2. Faulkner, D.J. Natural Products Reports **1**:251 (1984).
3. Faulkner, D.J. Natural Products Reports **1**:551 (1984).
4. Fogelquist, E. Low Molecular Weight Chlorinated and Brominated Hydrocarbons in Sea Water, Ph. D. Thesis (University of Göteborg, Sweden, 1984).
5. Lawrence M.M., S.A. Macko, W.H. Mook and S. Murray. Organic Geochemistry **3**:37 (1981)
6. Grøn, C. Vatten **44**:205 (1988).
7. Wigilius, B., B. Allard, H. Borén and A. Grimvall. Chemosphere **17**:1985 (1988).
8. Asplund,G., A. Grimvall and C. Pettersson. Sci. Tot. Environ. **81/82**:239 (1989).
9. Huffman, E.W.D. and H.A. Stuber. In: G.R. Aiken, D.M. McKnight, R.L. Wershaw and P. MacCarthy, Eds, Humic Substances in Soil, Sediment, and Water, pp. 433-455 (Wiley, 1985).
10. Brenner, S., R. Ikan, N. A. Agron and A. Nissenbaum. Soil Sci. **125**:226 (1978).
11. Hutchinson, G.E. and U.M. Cowgill. Arch. Hydrobiol. **72**:145 (1973).
12. Paasivirta, J., J. Knuutinen, P. Maatela, R. Paukku, J. Soikkeli and J. Särkkä. Chemosphere **17**:137 (1988).
13. Kudelskiy, A.V. International Geological Reviews **20**:362 (1978).
14. Salkinoja-Salonen, M.S., R. Valo, J. Apajalahti, L. Silakoski and T. Jaakkola. In: M. J. Klug and C.A. Reddy, Eds, Current Perspectives in Microbiological Ecology, pp. 668-676 (American Society for Microbiology, 1984).
15. Williams, D.T., J.A. Coburn and J. J. Bansci. Environment International **10**:39 (1984).
16. Sorrell, R.K., D.L. Boyerand and H.J. Brass. In: R.L. Jolley, R.J. Bull, W.P. Davis, S. Katz, M.H. Roberts and V.A. Jacobs, Eds, Water Chlorination, Volume 5, pp. 1123-1133 (Lewis, 1984).
17. Kool, H.J., C.F. van Kreijl and H.van Oers. Toxicol. Environ. Chem. **7**:111 (1984).
18. Kerndorff, H., V. Brill, R. Schleyer, P. Friesel and G. Milde. WaBoLu Hefte **5** (1985).
19. Kenrick, M.A.,P. L. Clark, K.M. Baxter, M. Fleet, H.A. James, T.M. Gibson and M. B. Turner. Trace Organics in British Aquifers.-A Baseline Survey (Water Research Centre, Technical Report 223, 1985).

20. Veenendaal, G., C.G.E.M. van Beek and L.M. Puijker. Organische Stoffen in Grondwater (KIWA Mededeling nr. 97, 1986).

21. Duijvenbooden, W. van and J. Taat. Landelijk Meetnet Grondwaterkwaliteit: eindrapport van de inrichtingsfase (RIVM nr. 840382001, 1985).

22. Hanson, H.F. In: R.G. Rice, Ed., Safe Drinking Water, pp. 161-166 (Lewis, 1985).

23. Montiel, A., S. Rauzy, D. Tricard and Y. Penverne. Water Supply **3**:157 (1985).

24. Grøn, C. International Journal of Environmental Analytical Chemistry (1990), in press.

25. Grøn, C. Chemosphere (1990), in press.

26. Danish Council of Standardization. Methods of Water Analysis, in Danish (1973-1988).

27. American Public Health Association, American Water Works Association and Water Pollution Control Federation. Standard methods for the Examination of Water and Wastewater (1985).

28. Gundersen, V. Automatization of Quantitative Iodine Analyses of Biological Tissues by Application of Flow Injection Analysis, in Danish (Danish Academy of Engineering, Chemistry Department, 1985).

29. Dionex. Application Sheets.

30. Buffle, J., P. Deladoey and W. Haerdi. Anal. Chim. Acta **101**:339 (1978).

31. Aiken, G. R. Environ. Sci. Technol. **18**:978 (1984).

32. U. S. Geological Survey. Methods for the Determination of Organic Substances in Water and Fluvial Sediments. Fractionation of Dissolved Organic Carbon (U. S. Government Printing Office, Washington D. C. 1987).

33. Proceedings 1st Int. Humic Subs. Soc. Conf., Estes Park, Colorado, USA (IHSS Press, 1983).

34. Piper, A.M. In: American Geophysical Union Twenty-Fifth Annual Meeting (Washington D. C. 1944).

35. Grøn, C. Organic Halogens in Danish Ground Waters, Ph. D. Thesis (Technical University of Denmark, 1989).

36. Laursen, G. Iodine in the County of Århus, report in Danish (1989).

37. Dinesen, B. Saline Waters from Deep Danish Aquifers, in Danish (Reitzel, Copenhagen , 1961).

Appendix

List of Participants

Alberts, James	Univ. of Georgia, Marine Institute, Sapelo Island, GA 31327, USA
Allard Bert	Dept. of Water and Environmental Studies, University of Linköping, S-58183 Linköping, Sweden
Andersson, Tord	Naturgeogr. Avd., Umeå Universitet, S-901 87 Umeå, Sweden
Arsenie, Irina	Dept. of Water and Environmental Studies, University of Linköping, S-58183 Linköping, Sweden
Asplund, Gunilla	Dept. of Water and Environmental Studies, University of Linköping, S-58183 Linköping, Sweden
Backlund, Peter	Dept. of Organic Chemistry, Åbo Akademi, Akademig. 1, SF-20500 Åbo, Finland
Becher, Georg	Dept. of Toxicology, National Institute of Public Health, Geitmyrveien 75, 0462 Oslo 4, Norway
Bengtsson, Göran	Dept. of Ecology, University of Lund, Helgonav. 5, S-22362 Lund, Sweden
Berggren, Dan	Växtekologiska avd, Lunds Universitet, Östra Vallgatan 14, S-22361 Lund, Sweden
Blaser, Peter	Swiss Federal Inst. of Forestry Research, CH-8903 Birmensdorf, Switzerland
Borén, Hans	Dept. of Water and Environmental Studies, University of Linköping, S-58183 Linköping, Sweden
Bringmark, Lage	Naturvårdsverket, Box 7050, S-750 07 Uppsala, Sweden
Burba, Peter	Institut für Spektrochemie und angewandte Spektroskopie, Bunsen-Kirchhoff-Strasse 11, D-4600 Dortmund 1, F.R.G.
Caceci, Marco	CEA-CEN-FAR, DRDD/SESD Bat 52, B.P. 6 F-92265, Fontenay-aux-Roses, France
Carlsen, Lars	Chemistry Department, Risø National Laboratory, DK4000 Roskilde, Denmark
Cegarra, Juan	Consejo Superior de Investigaciones Cientificas, C E B A S, Avenida De La Fama No 1, 30080 Murcia, Spain
Christman, Russel F.	The School of Public Health, Dept. of Sciences and Engineering, The University of North Carolina at Chapel Hill, CB# 7400, Rosenau Hall, Chapel Hill, N.C. 27599-7400, USA

Christiansen, Jesper	Chemistry Department, Risø National Laboratory, DK-4000 Roskilde, Denmark
Ciavatta, Claudio	Inst. of Agricultural Chemistry, University of Bologna, Via S. Giacomo 7, 40126 Bologna, Italy
Clarholm, Marianne	Dept. of Ecology and Environmental Sciences, Swedish University of Agric. Sciences, S-75007 Uppsala, Sweden
Dahlman, Olof	STFI, Box 5604, S-11486 Stockholm, Sweden
Danielsson, Lars-Göran	Dept. of Analytical Chemistry, Royal Inst. of Technol., S-100 44 Stockholm, Sweden
Derome, John	Dept. of Soil Science, Finnish Forest Research Institute, P.O. Box 18, SF-01301 Vantaa, Finland
Duarte, Armando	Dept. of Chemistry, University of Aveiro, 3800 Aveiro, Portugal
Ephraim, James	Dept. of Water and Environmental Studies, University of Linköping, S-58183 Linköping, Sweden
Falck, W. Eberhard	British Geological Survey, Keyworth, Nottingham NG 12 5GG, U.K.
Forsberg, Curt	Dept. of Limnology, Box 557, S-75122 Uppsala, Sweden
Forsgren, Gunilla	Naturgeografiska Inst, Umeå Universitet, S-90187 Umeå, Sweden
Frimmel, Fritz, H.	Bereich Wasserchemie, Engler-Bunte-Institut der Universität, Richard-Willstätter-Allee 5, 7500 Karlsruhe, BRD
Gjessing, Egil	NIVA, P.O. Box 23, N-0313 Blindern, Oslo 3, Norway
Gobran, George	The Swedish University of Agric. Sci., Dept of Ecology and Environ. Res., Box 7072, S-75007 Uppsala, Sweden.
Gomez, Paloma	CIEMAT, Division Técnicas Geologicas, Avda.Complutense, 22, 28040 Madrid, Spain
Gonzalez-Vila, Francisco, J	Instituto De Recursos Naturales, Y Agrobiologia, C.S.I.C.,Apartado 1052 E.P., 41080-SEVILLA, Spain
Grimvall, Anders	Dept. of Water and Environmental Studies, University of Linköping, S-58183 Linköping, Sweden
Grøn, Christian	Inst. of Applied Geology, The Technical University of Denmark, Bygning 204, DK-2800 Lyngby, Denmark
Haglund, Åke	Norrköpings Kommun, Gatukontoret, Lindöv. 65, S-60181 Norrköping, Sweden
Heikkinen, Kaisa	Oulun Vesi- Ja Ympäristöpiiri, PL 124, SF-90101 Oulu, Finland
Hernebring, Claes	VIAK AB, S-58101 Linköping, Sweden

Hodin, Fredrik	Dept. of Water and Environmental Studies, University of Linköping, S-58183 Linköping, Sweden
Holmbom, Bjarne	Lab of Forest Products Chemistry, Åbo Akademi, SF-20 500 Åbo, Finland
Hongve, Dag	National Inst of Public Health, Geitsmyrsveien 75, N-0462 Oslo 4, Norway
Huang, P.M.	Dept. of Soil Science, Univ. of Saskatchewan, Saskatoon SK, Canada, S7N OWO
Håkansson, Karsten	Dept. of Water and Environmental Studies, University of Linköping, S-58183 Linköping, Sweden
Ingman, Folke	Dept. of Analytical Chemistry, Royal Inst.of Technology, S-100 44 Stockholm, Sweden
Jansson, Mats	Avd för Naturgeografi, Umeå Universitet, S-901 87 Umeå, Sweden
Johansson, Maj-Britt	Swedish University of Agricultural Sciences, Dept. of Forest Soils, P.O. Box 7001, S-750 07 Uppsala, Sweden
Jonsson, Susanne	Dept. of Water and Environmental Studies, University of Linköping, S-58183 Linköping, Sweden
Järvinen, Ari	Lab of Sanitary and Env. Eng., Helsinki University of Technology, Rakentajanaukio 4, SF-02150 Espoo, Finland
Karlsson, Fred	SKB, Box 5864, S-10248 Stockholm, Sweden
Karlsson, Susanne	Dept. of Water and Environmental Studies, University of Linköping, S-58183 Linköping, Sweden
Klöcking, Renate	Institut für Virologie, Medizinische Akademie, Nordhauser Str. 74, DDR-5010 Erfurt
Kockum, Karin	VBB, Geijersg. 8, S-21618 Malmö, Sweden
Kortelainen, Pirkko	National Board of Waters and the Environment, Water and Environment Research Institute, P.O. Box 250, SF-00101 Helsinki, Finland
Kronberg, Leif	Inst för Organisk Kemi, Åbo Akademi, Akademig. 1, SF-20500 Åbo, Finland
Krosshavn, Marit	Department of Chemistry, University of Trondheim, AVH, N-7055 Dragvoll, Norway
Kukkonen, Jussi	University of Joensuu, Department of Biology, P.O. Box 111, SF-80101 Joensuu, Finland
Kübler, Helga	Inst. für Bodenwissenschaften, Abt. Biochemie im System Boden, Von-Siebold Str. 2, D-3400 Göttingen, BRD

Ledin, Anna	Dept. of Water and Environmental Studies, University of Linköping, S-58183 Linköping, Sweden
Lee, Ying-Hua	Swedish Environmental Research Institute, P.O. Box 47086, S-402 58 Göteborg, Sweden
de Leer, Ed	Delft University of Technology, Dept. of Analytical Chemistry, De Vries van Heystplantsoen 2, 2628 RZ Delft, The Netherlands
Lexmond, Ruud	TU Delft, De Vries van Heystplantsoen 2, 2628 RZ Delft, The Netherlands
de Lijser, Peter	TU Delft, De Vries van Heystplantsoen 2, 2628 RZ Delft, The Netherlands
Lundin, Lars	Dept. of Forest Soils, SLU, Box 7001 S-750 07 Uppsala, Sweden
Lundström, Ulla	Dept. of Forest Site Research, The Sw. Univ. of Agricultural Sciences, S-90183 Umeå, Sweden
Långvik, Vivi-Ann	Dept of Org Chemistry, Åbo Akadmi, SF-20500 Åbo 50, Finland
Lövgren, Lars	Dept. of Inorg Chemistry, University of Umeå, S-901 87 Umeå, Sweden
Maes, Andre	Laboratorium voor Colloidale Scheikunde, Kardinal Mercierlaan 92, B-3030 Heverlee, Belgium
Malcolm, Ronald, L.	US Geological Survey, Box 25046, M.S. 408, Federal Center, Denver, Colorado 80 225, USA
McCarthy, John	Environmental Sciences Division, Oak Ridge National Laboratory, P.O. Box 2008, Oak Ridge TN 37831-6036 USA
Mopper, Ken	Univ. Miami, 4600 Rickenbacker Cswy, Miami FL 33149, USA
Moulin, Valérie	Commissariat à l'Energie Atomique, CEA-DRDD/SESD/ SCPCS, B.P. 6, 92265 Fontenay-aux-Roses, CEDEX, France
Mäkinen, Irma	National Board of Waters and the Environment, Kyläsaarenkatu 10, SF-00550 Helsinki, Finland
Mörck, Roland	STFI, Box 5604, S-11486 Stockholm, Sweden
Niemeyer, Juergen	Abt. Chemie, Institut für Boden-wissenschaften, 3400 Goettingen, BRD
Nilsson, Nils	Dept. of Inorganic Chemistry, University of Umeå, S-901 87 Umeå, Sweden
Nilsson, Åke	Naturgeogr. Avd., Umeå Universitet, S-901 87 Umeå, Sweden
Nordén, Maria	Dept. of Water and Environmental Studies, University of Linköping, S-58183 Linköping, Sweden

Nørnberg, Per	Institute of Geology, University of Aarhus, Ny Munkegade, DK-8000 Aarhus c, Denmark
Palma, Achille	Metapontum Agrobios, S.S. Jonica 106, Km. 448,2, 75010 Metaponto (MT), Italy
Paxéus, Nicklas	GRYAAB, Karl IX:s väg, S-41722 Göteborg, Sweden
Pellinen, Jukka	STFI, Box 5604, S-114 86 Stockholm, Sweden
Pempkowiak, Janusz	Inst. of Oceanology, Polish Academy of Sciences, P.O. Box 68, Sopot, Poland
Petersen, Robert	Dept of Ecology/Limnology, University of Lund, Box 117, S22100 Lund, Sweden
Pettersson, Catharina	Dept. of Water and Environmental Studies, University of Linköping, S-58183 Linköping, Sweden
Peuravuori, Juhani	Dept of Chemistry, University of Turku, SF-20500 Turku, Finland
Pihlaja, Kalevi	Dept of Chemistry, University of Turku, SF-20500 Turku, Finland
Raben-Lange,Birgitte	Royal Veterinary and Agricultural University, Chemistry Department, Thorvaldsenvej 40, DK-1871 Fredriksberg C, Denmark
Reimann, Anders	STFI, Box 5604, S-11486 Stockholm, Sweden.
Riise, Gunnhild	Isotope and Electron Microscopy Lab, Agric. Univ of Norway, P.O. Box 26, N-1432 Aas, NLH, Norway
Rydén, Bengt-Erik	Dept. Phys. Geogr., Univ. of Uppsala, P.O. Box 554, S-75122 Uppsala, Sweden
Saez-Granero, Francisco	CIEMAT, Division de Quimica, Avda. Complutense. 22, 28040 Madrid, Spain
Sàiz-Jimenez, Cesàreo	Instituto De Recursos Naturales Y Agrobiologia, Apartado De Cooreos 1052 E.P., 41080-Sevilla, Spain
Saleh, Farida	Environmental Chemistry Program, University of North Texas, P.O. Box 13078, Denton, Texas 76203, USA
Santos, Eduarda	Dept. of Chemistry, University of Aveiro, 3800 Aveiro, Portugal
Senesi, Nicola	Istituto di Chimica, Agraria-Universita di Bari, Via Amendola, 165/A, 70126-Bari, Italy
Sequi, Paolo	Institute of Agricultural Chemistry, University of Bologna, Viale Berti Pichat, 10, 40127 Bologna, Italy

Skoog, Annelie	Dept of Analytical and Marine Chemistry, University of Göteborg and Chalmers University of Technology, S-412 96 Göteborg, Sweden
Skyllberg, Ulf	Institute of Forest Site Research, Swedish University of Agricultural Sciences, S-901 83 Umeå, Sweden.
Smed-Hildmann, Raija	Institut für Wasser-, Boden- und Lufthygiene Des BGA, Paul-Ehrlich-Strasse 29, D-6070 Langen, BRD
Suortti, Anna-Mari	National Board of Water and the Environment, Kyläsaarenkatu 10, SF-00580 Helsinki, Finland
Szabo, Gyula	"FJC" National Research Institute for RadioBiology, H-1775 Budapest, P.O. Box 101, Hungary
Sävenhed, Roger	Dept. of Water and Environmental Studies, University of Linköping, S-58183 Linköping, Sweden
Tits, Jan	K U Leuven, Laboratory of Colloid Chemistry, Kard. Mercierlaan 92, B-3030 Heverlee, Belgium
Tjeder,Anders	Norrköpings Kommun, Gatukontoret, Lindöv. 65, S-601 81 Norrköping, Sweden
Vittori Antisari,Livia	Institute of Agricultural Chemistry, University of Bologna, Via S. Giacomo 7, 40126 Bologna, Italy
Vuorinen, Ulla	Technical Research Centre of Finland, Reactorlaboratory, Otakaari 3 A, SF-02150 Espoo, Finland
Wallberg, Maria	Dept. of Water and Environmental Studies, University of Linköping, S-58183 Linköping, Sweden
Wedborg, Margareta	Dept of Analytical and Marine Chemistry, University of Göteborg and Chalmers University of Technology, S-412 96 Göteborg, Sweden
Xu, Hao	Dept. of Water and Environmental Studies, University of Linköping, S-58183 Linköping, Sweden